Engineering Emergence

A Modeling and Simulation Approach

Engineering Emergence

A Modeling and Simulation Approach

Edited by
Larry B. Rainey
Mo Jamshidi

CRC Press
Taylor & Francis Group
Boca Raton London New York

CRC Press is an imprint of the
Taylor & Francis Group, an **informa** business

CRC Press
Taylor & Francis Group
6000 Broken Sound Parkway NW, Suite 300
Boca Raton, FL 33487-2742

First issued in paperback 2020

ISBN 13: 978-0-367-65611-9 (pbk)
ISBN 13: 978-1-138-04616-0 (hbk)

Library of Congress Cataloging-in-Publication Data

Names: Rainey, Larry B., author. | Jamshidi, Mohammad, author.
Title: Engineering emergence : a modeling and simulation approach / Larry B. Rainey and Mo Jamshidi.
Description: Boca Raton : Taylor & Francis, CRC title, part of the Taylor & Francis imprint, a member of the Taylor & Francis Group, the academic division of T&F Informa, plc, 2018. | Includes bibliographical references.
Identifiers: LCCN 2018010556| ISBN 9781138046160 (hardback : acid-free paper) | ISBN 9781138046412 (ebook)
Subjects: LCSH: Systems engineering. | Engineering models. | System design.
Classification: LCC TA168 .R336 2018 | DDC 620/.0042--dc23
LC record available at https://lccn.loc.gov/2018010556

Visit the Taylor & Francis Web site at
http://www.taylorandfrancis.com

and the CRC Press Web site at
http://www.crcpress.com

This text is dedicated to ***Mr. Frank Dies*** *who had retired from SPARTA Corporation but had been the major contributor to a proposed research effort to investigate emergent behavior in the Missile Defense Agency's Ballistic Missile Defense System, a major instantiation of a system of systems engineering application. This effort would have been a significant contribution to achieving the research objectives addressed in Chapter 22 of this text.*

Contents

Foreword

The world is getting more interconnected every day as more systems are networked together, forming various systems of systems. Whether these individual systems come together by design or by happenstance, the result is the emergence of unique capabilities and functions that cannot be achieved by single systems alone. Emergence is defined in "Engineering Emergence: A Modeling and Simulation Approach" as the potential outcome/result (negative or positive) of the interaction of dissimilar and nonconnected systems. It is not a matter of if emergence will occur, but rather how, when, and to what extent. As such this phenomenon is classified as stochastic in nature and is a phenomenon that can occur in different domains.

Emergence, in and of itself, is neither positive nor negative, but is the registered effect upon the system of systems as noted by the perspective, whether positive or negative, that it is compared against. The resulting system of systems exhibits new and often unpredictable properties and behaviors that need to be understood to employ the system of systems effectively and efficiently. The goal of studying emergent behavior is three-fold: (1) define methods to detect, assess, measure, and predict all emergent behavior present within a system of systems; (2) identify the unintentional (negative) consequences of the emergent behavior, and develop avenues that will either eliminate or mitigate the unwanted behavior where possible; and (3) identify the intentional (positive) consequences of emergent behavior and harness it to its fullest potential.

Heretofore, system of systems emergence is a phenomenon that has been not widely studied or understood. In "Engineering Emergence: A Modeling and Simulation Approach," Larry Rainey and Mo Jamshidi assembled many of the thought leaders in the system of systems domain to establish a baseline of knowledge, theoretical perspectives of emergence, and model-based applications of emergence.

This book is unique in that it lays the foundation for the rigorous development of theoretical underpinnings and provides model-based approaches for system of systems emergence. The study of system of systems is frequently criticized for the lack of a theoretical basis, or the "physics of the domain," as referred to by Larry Rainey in Chapter 1. The absence of a theoretical underpinning inhibits the development of technical solutions befitting the complexity of the emergence problem.

The study of system of systems emergence requires a consistent baseline, including a common definition, understanding of causes and characteristics, and an ontology, to serve as the point of departure. Section II comprises the most comprehensive collection of thought pieces published to date, and establishes the theoretical underpinnings of system of systems emergence. This section includes 12 of the book's 22 chapters, and introduces the notion, characteristics, and sources of emergence. Conceptually, emergence has been considered both subjectively and objectively, but not consistently. For emergence to be studied and harnessed, a solid understanding of the concept is essential.

Chapters 3, 4, 6, and 8 establish the conceptual bedrock for the understanding of engineering emergence. These chapters discuss the characteristics, sources of emergence, and the prevalence of combinatorial complexity in system of systems. Combinatorial complexity is found in systems in which the complexity lies in finding the best solution (i.e., the best configuration of elements or components) from an extremely large number of possible system combinations within a system of systems. Understanding combinatorial complexity is essential to the engineering and management of positive and negative emergence.

Another concept that is critical to the understanding of emergence is ontology. An ontology is a collection of standardized, defined terms or concepts and the rules and relationships among the terms and concepts. A solid understanding of the rules and relationships identified by the ontology is as important to an understanding of emergence as combinatorial complexity.

In addition to the lack of a theoretical underpinning, system of systems theory and applications are often inhibited by the limitations of application of single systems analysis. Systems engineering methods and applications are often inappropriately used to solve system of systems problems, because the applications often lack the ability to address the complexity associated with system of systems emergence problems.

The study of system of systems emergence requires an in-depth application of Model-Based Systems Engineering (MBSE) to understand, predict, and manage the dynamic nature of emergence and its inherent complexity. Model-Based Systems Engineering is the formalized application of modeling (static and dynamic) to support system design and analysis throughout all phases of the system lifecycle, through the collection of modeling languages, structures, model-based processes, and presentation frameworks used to support the discipline of systems engineering in a model-based or model-driven context. This definition differs from MBSE definitions previously posed by the systems engineering community, in that it includes not only the familiar static representations common to systems engineering, but also the dynamic nature of emergence characterized by various mathematical methods that can be executed alone or via modeling and simulation.

In the spirit of the MBSE definition, this book treats system of systems emergence from a static modeling method as well as a wide variety of dynamic modeling methods. Chapter 11 explores how to characterize emergence from a static perspective by introducing five viewpoints, allowing for emergence to be considered from a holistic perspective. This chapter is a "must read" for those readers who are interested in systems architecting.

While static architectures help to understand the complexity, emergence is best described by behavior. Chapter 18 presents a simulation approach on how to detect, classify, predict, and control emergent behaviors in system of systems known as Monterey Phoenix. This approach is typically performed in conjunction with system architectures; therefore a bridge between the viewpoints described in Chapter 11 may lend itself to additional insights.

Section III provides several different theoretical dynamic modeling perspectives and practical methods and applications that have been proven in the systems domain and other domains; hence it provides an excellent starting point for considering the complexity and emergence in system of systems. A wide variety of modeling and simulation techniques (e.g., agent-based modeling, MATLAB/Simulink, Monterey Phoenix Behavioral Model) are introduced to detect, classify, predict, control, and visualize behavior in system of systems. These discussions are the heart of this book, because optimizing the positive consequences and mitigating the negative consequences of emergence can only be accomplished through robust mathematical modeling due to the nature, scope, and complexity of system of systems. However, while the discussion of mathematical methods is comprehensive in this book, it is not meant to be an exhaustive discussion, as other methodologies, both familiar and new, should be explored to further explore emergence.

Finally, Chapter 22 highlights the lessons learned by the various thought leaders who have authored the chapters. This is a very nice compendium to concepts and methods discussed throughout the chapters in this book. In addition to the lessons learned, it is a proposed way-ahead, outlining eight areas that are ripe for establishing a research agenda. These issues are the result of the Engineering Emergence Workshop, attended by researchers and practitioners from across academia, government, federally funded research and development centers, and industry.

Often the challenge of approaching a book so rich in content as "Engineering Emergence: A Modeling and Simulation Approach" is where to begin. I recommend reading Chapters 1 and 2 first to establish the foundation for this effort in the literature of engineering emergence. Chapter 1 provides an excellent guide to the remainder of the book and thus should be consulted frequently. If establishing an engineering emergence research agenda is desired, read Section 22.3, "Proposed Way-Ahead" next to understand the identified research problems. Chapter 22 provides the lessons learned for each chapter and therefore should be reviewed after reading each chapter. If a sound theoretical baseline of system of systems emergence is desired, read Chapters 3, 4, 6, and 8 next. The remaining chapters should be read based on your research questions and the technical approach that is desired.

As the world becomes more interconnected through system of systems, emergence takes on a new importance. Emergence must be properly detected, classified, and predicted to understand its implications, and controlled to effectively capitalize on the positive effects or mitigate the negative effects within a system of systems.

"Engineering Emergence: A Modeling and Simulation Approach" is the first book on this topic, and is designed to serve as the stimulus for system of systems emergence basic and applied research efforts. The hope is that this book is only the beginning, and more scholarly papers and books will be published on the findings and lessons learned from basic and applied research in the near future. The importance of emergence cannot be overstated. It is certainly well timed for consideration by academics, researchers, and practitioners.

Electronic copies of the appendices will be hosted on the book's companion website. Visit the book's CRC Press website for further details: https://www.crcpress.com/Engineering-Emergence-A-Modeling-and-Simulation-Approach/Rainey-Jamshidi/p/book/9781138046160.

Warren K. Vaneman
Systems Engineering Department
Naval Postgraduate School
Monterey, California

MATLAB® is a registered trademark of The MathWorks, Inc. For product information, please contact:

The MathWorks, Inc.
3 Apple Hill Drive
Natick, MA 01760-2098 USA
Tel: (508) 647-7000
Fax: (508) 647-7001
E-mail: info@mathworks.com
Web: http://www.mathworks.com

Editors

Larry B. Rainey, PhD is the Vice President for Engineering at Integrity Systems & Solutions of Colorado, LLC, a consulting firm that specializes in modeling and simulation within missile defense and space operations domains. Dr. Rainey has also worked for the Missile Defense Agency and other U.S. Department of Defense organizations that address system of systems challenges. He has also been an assistant professor of systems engineering at the Colorado Technical University in Colorado Springs, Colorado and a visiting assistant professor of astronautical engineering at the Air Force Institute of Technology at Wright-Patterson Air Force Base, Ohio. He is the executive editor of four other major texts on the application of modeling and simulation and the author of many technical articles addressing the applications of systems theory and cybernetics to real-world problems.

Mo Jamshidi is a Life Fellow-Institute of Electrical and Electronic Engineers, Fellow-American Society of Mechanical Engineers, A. Fellow-American Institute of Aeronautical and Astronautics, Fellow-American Association for Advancement of Science, Fellow-The World Academy of Science, and Fellow-New York Academy of Science. He received BS (Cum Laude) in electrical engineering, Oregon State University, Corvallis, OR, USA in 1967, and the MS and PhD degrees in electrical engineering from the University of Illinois at Urbana-Champaign, IL, USA in June 1969 and February 1971, respectively. He holds honorary doctorate degrees from the University of Waterloo, Canada, 2004, Technical University of Crete, Greece in 2004, and Odlar Yourdu University, Baku, Azerbaijan in 1999. Currently, he is the Lutcher Brown Endowed Distinguished Chaired Professor at the University of Texas, San Antonio, TX, USA. He was an advisor to NASA for 10 years (including 1st MARS Mission, landed on July 4, 1997 and 7 years with NASA HQR), 9 years with US Air Force Research Laboratory, 8 years with USDOE and 1 year with EC/EU (Brussels). He has over 780 technical publications including 71 books (11 textbooks), research volumes, and edited volumes in English and five foreign languages. He is the Founding Editor or co-founding editor or Editor-in-Chief of five journals including *IEEE Control Systems Magazine* and the *IEEE Systems Journal*. He has graduated or advising 70 PhD and 75 MS students, so far, during his 50-year engineering career. In October 2005, he received the IEEE SMC society's highest honor—Norbert Weiner Research Achievement Award. In 2014, he received the IEEE-USA Career Award in Systems Engineering. He is the recipient of IEEE Centennial Medal 1984, World Automation Congress (WAC) Medal of Honor, 2014, among many other awards and honors. He is a member of the University of the Texas System Chancellor's Council since 2011. He is currently involved in research on system of systems engineering with emphasis on cloud computing, robotics, unmanned aerial vehicles, biological and sustainable energy systems. He has over 9600 citations on Google Scholar.

Contributors

Mikhail Auguston
Naval Postgraduate School
Monterey, California

Leo Carlos-Sandberg
University College London
London, United Kingdom

Christopher D. Clack
University College London
London, United Kingdom

John M. Colombi
Air Force Institute of Technology
Dayton, Ohio

Saikou Y. Diallo
Virginia Modeling, Analysis and Simulation Center
Norfolk, Virginia

Umut Durak
German Aerospace Center (DLR)
Braunschweig, Germany

Timothy L.J. Ferris
Cranfield University
Shrivenham, United Kingdom

Kristin Giammarco
Naval Postgraduate School
Monterey, California

Thomas Holland
Naval Surface Warfare Center
Norfolk, Virginia

Mo Jamshidi
University of Texas
San Antonio, Texas

Moath Jarrah
Jordan University of Science and Technology
Irbid, Jordan

John J. Johnson IV
Systems Thinking & Solutions
Broadlands, Virginia

Polinpapilinho F. Katina
Old Dominion University
Norfolk, Virginia

Charles B. Keating
Old Dominion University
Norfolk, Virginia

Gary O. Langford
Portland State University
Portland, Oregon

Christopher J. Lynch
Virginia Modeling, Analysis and Simulation Center
Norfolk, Virginia

Tuncer Ören
University of Ottawa
Ottawa, Ontario, Canada

Jose J. Padilla
Old Dominion University
Norfolk, Virginia

Thorsten Pawletta
Univesity of Wismar
Wismar, Germany

Mikel D. Petty
University of Alabama in Huntsville
Alabama

John J. Quartuccio
Naval Postgraduate School
Monterey, California

Larry B. Rainey
Integrity Systems and Solutions of Colorado, LLC
Colorado Springs, Colorado

Donna H. Rhodes
Massachusetts Institute of Technology
Cambridge, Massachusetts

Josef Schaff
Naval Air Systems Command
Patuxent River, Maryland

Carol C. Woody
Software Engineering Institute
Pittsburgh, Pennsylvania

Bernard P. Zeigler
RTSync
Tucson, Arizona

Section I

Introduction and Overview

1

Introduction and Overview for Engineering Emergence: A Modeling and Simulation Approach

Larry B. Rainey

CONTENTS

1.1 Rationale for This Text

In general, the term "emergence" is associated with the field of complexity. Yaneer Bar-Yam (2011) of the New England Complex Systems Institute has stated:

> *Emergence* refers to the existence or formation of collective behaviors—what parts of a system do together that they would not do alone.
>
> In describing collective behaviors, emergence refers to how collective properties arise from the properties of parts, how behavior at a larger scale arises from the detailed structure, behavior and relationships at a finer scale. For example, cells that make up a muscle display the emergent property of working together to produce the muscle's overall structure and movement. A water molecule has emergent properties that arise out of the properties of oxygen and hydrogen atoms. Many water molecules together form river flows and ocean waves. Trees, other plants and animals form a forest.
>
> When we think about emergence we are, in our mind's eye, moving among views at different scales. We see the trees and the forest at the same time, in order to see how the trees and the forest are related to each other. We might consider particularly those details of the trees that are important in giving rise to the behavior of the forest.
>
> In conventional views the observer considers either the trees or the forest. Those who consider the trees consider the details to be essential and do not see the patterns that arise when considering trees in the context of the forest. Those who consider the forest do not see the details. When one can shift back and forth between seeing the trees and the forest one also sees which aspects of the trees are relevant to the description of the forest. Understanding this relationship in general is the study of emergence.
>
> Emergence can also describe a system's function—what the system does by virtue of its relationship to its environment that it would not do by itself.

> In describing function, emergence suggests that there are properties that we associate with a system that are actually properties of the relationship between a system and its environment.
>
> Consider a key. A description of a key's structure is not enough to show us that it can open a door. To know whether the key can open a door, we need descriptions of both the structure of the key and the structure of the lock. However, we can tell someone that the function of the key is to unlock the door without providing a detailed description of either.
>
> One of the problems in thinking about complex systems is that we often assign properties to a system that are actually properties of a relationship between the system and its environment. We do this for simplicity, because when the environment does not change, we need only describe the system, and not the environment, in order to describe the relationship. The relationship is often implicit in how we describe the system.
>
> The concept of emergence as referring to function in an environment is related to the concept of emergence as the rise of collective behaviors, because any system can be viewed along with the parts of its environment as together forming a larger system. The collective behaviors due to the relationships of the larger system's parts reflect the relationships of the original system and its environment.

There are no texts available called "Engineering Emergence: Principles and Applications." The reason for this is that the physics of the domain space of emergence has not been defined, specified, quantified, and documented. Therefore, the equations of motion for emergent behavior cannot be written. As such, this text takes the perspective of modeling and simulation by which the domain space or operational environment can be examined.

To start with we define emergence as the potential outcome/result (negative or positive) of the interaction of dissimilar and nonconnected systems. It's not a matter of *if* emergence will occur, but rather *how and when*. As such, emergence, in and of itself, is not negative or positive per se. Rather, it is the registered effect upon the system of systems and noted by the observer that classifies it as positive or negative. Therefore, it is observer dependent as to its classification, based upon the observer's understanding of imputed knowledge. As such, this phenomenon is classified as stochastic in nature. I posit that this phenomenon can occur in different domains where there is the intersection of dissimilar and non-connected systems. Also, it is postulated that this phenomenon can occur in different environments/disciplines (e.g., engineering, physics, computer science, etc). As such, three basic questions arise for examination: (1) How does one detect the existence or presence of emergence? (2) If the emergence is deemed to be positive, how does one capitalize on this effect? and (3) If the emergence is deemed to be negative, how does one control this effect?

1.2 Purpose

This text deals with emergence found in man-made systems, in general, and in system of systems engineering applications, specifically. For the former application, the above definition applies. For the latter definition, Maier (1998) has defined emergent behavior as follows:

> The system of systems performs and carries out purposes that do not reside in any component system. These behaviors are emergent properties of the entire system of

systems and not the behavior of any component system. The principal purposes supporting engineering of these systems are fulfilled by these emergent behaviors.

The reason for or cause of emergent behavior in system of systems has also been addressed by Maier (1998). He states: "Emergent properties can appear only if information exchange is sufficient." This, of course, refers to his "five principal characteristics of a system of systems: (1) Operational Independence of the Elements, (2) Managerial Independence of the Elements, (3) Evolutionary Development, (4) Geographic Distribution, (5) Emergent Behavior." Emphasis is on the first four of these five principal characteristics. This author suggests that "information exchange" refers to that software that facilitates (i.e., communicates) a common or single primary mission being performed among all the operational independent elements. It is this "mission software" that gives the system of systems in question a single or primary purpose. This issue is addressed further in Chapter 12.

As such, emergent behavior, in its full context, is a controls problem. However, it is beyond the scope of modern/classical control theory, digital control theory, and estimation theory. Therefore, we want to represent the operational domain space mathematically and, in turn, interrogate/explore the domain space via modeling and simulation to, ideally, facilitate the understanding of the functions of detection, classification, prediction, and control of the phenomenon. The definitions of these functions are provided below:

Detection: The initial discovery of the presence/existence of emergence.

Classification: Four types of emergence have been identified to date:

1. Simple: According to Page (2009), "*simple emergence* is generated by the combination of element properties and relationships and occurs in non-complex or 'ordered' systems."
2. Weak: Page (2009) describes "*weak emergence* as expected emergence which is desired (or at least allowed for) in the system *structure*. However, since weak emergence is a product of a complex system, the actual level of emergence cannot be predicted just from knowledge of the characteristics of the individual system *components*."
3. Strong: "The term *'strong emergence'* is used to describe unexpected emergence; that is, emergence not observed until the system is simulated or tested or, more alarmingly, until the system encounters in operation a situation that was not anticipated during design and development. Strong emergence may be evident in failures or shutdowns." Guide to the Systems Engineering Body of Knowledge.
4. Spooky: "An emergent property that is inconsistent with the known properties of the system's components. The property is not reproduced in any model of the system, even one with complexity equal to that of the system itself, even one that appears to be precisely simulating the system itself in all details." Rainey and Tolk (2015).

Prediction: Based upon initial detection of the existence of emergence, then being able to postulate the potential future states of emergence.

Control: Specification of a modeling and simulation technique, possibly in concert with another analytical technique, used either to manage negative cases of emergence or to capitalize on positive cases of emergence.

1.3 Organization of the Text

There are four sections to this text. Section I, Introduction and Overview, provides a context or foundation for the rest of the text. Section II, Theoretical Perspectives, addresses those chapters that are based solely on solid academic principles. Section III, Theoretical Perspectives with Practical Applications, goes the next step to provide definitive applications that are based upon the academic principles provided earlier. Section IV, Summary, provides the reader with those major tenets that can be derived from this text and what steps can yet be taken.

1.3.1 Introduction and Overview

The main objective of Section I is found in Chapter 2, "System of Systems Engineering—An Overview." This chapter is the foundation to understanding the context for the subject of this text. In this chapter, Mo Jamshidi provides an introduction to the subject. Given that this field is relatively new, he starts with definitions of the term "system of systems engineering" to illustrate the various perspectives on the subject, then identifies the various types of system of systems that currently exist to provide a taxonomy for discussion purposes and show the diversity of thought on the subject. The author then turns his attention to identifying the various problems that exist in this relatively new discipline. The impact of system of systems engineering is the next subject addressed. The chapter concludes with a summary while presenting the challenges to consider for the future in the new field of system of systems engineering.

1.3.2 Theoretical Perspectives

Section II is the major portion of this book. It contains 12 chapters that are solely theoretical in purpose.

In Chapter 3, "DEVS-Based Modeling and Simulation Framework for Emergence in System of Systems," Bernard Zeigler introduces the reader to the subject of emergence as it has been recently treated in both its subjective and objective aspects. Then he shows how Discrete Event Systems (DEVS) provides the components and coupling for models of complex systems. Next, he addresses how DEVS supports dynamic structure for genuine adaptation and evolution. In his fourth section, he explains how the experimental frame supports emergence behavior observation. The author then addresses how DEVS Markov models support prediction of emergence. In the sixth section, he explains how DEVS enables fundamental emergence modeling. The next section addresses conditions for positive emergence and gives examples of DEVS emergence modeling. The chapter ends with Zeigler's conclusions.

Chapter 4, "Sources for Emergence and Development of System of Systems," is written by John Colombi. His point of departure is an introduction that makes the case for his subject through several aerospace examples. Next, he addresses the Bak-Sneppen evolutionary model that captures concepts of fitness as applied in a MATLAB simulation. Interoperability is the subject next considered, followed by a presentation of system of systems evolution with interoperability. The chapter closes with discussion and conclusions.

"Leveraging Deterministic Chaos to Mitigate Combinatorial Explosions" is the title of Chapter 5 by Joe Schaff. Following the introduction, the author provides an overview of complex systems by addressing emergent patterns in these systems. Then he addresses three general methods to manage complexity. This is followed by an overview of complexity science where some system of systems instantiations are provided that have emergent behavior. Next, he considers the subject of emergent order from the point of view of both complexity and chaos. Following is a discussion of some complex problems. The author then draws the attention of the reader to the change from chaos to iterative order. Recursive generating functions, also known as Iterated Function Systems, is then discussed. The author closes with a presentation of self-organizing systems, also called non-predetermined parametric random iterated function systems, as well as his conclusions and next steps.

In Chapter 6, "Phenomenological and Ontological Models for Predicting Emergence," Gary Langford first examines emergence through the lens of ontology and phenomenology. Then he expounds on the subject of emergence to let the reader understand its breadth and then its depth. This is followed by a brief discussion of paths of ontological investigation and a research approach. The author then turns his attention to a presentation of foundational mereological structure. Next, he addresses an interpretive integrative framework, after which a phenomenological perspective is presented. The chapter closes with a section that addresses the Gladstone bag, which is equipped for ontological exploration and investigation.

Chapter 7, "System of Systems Process Model," is also written by Gary Langford. After giving the reader an overview of the content to be covered, in the first major section he addresses the topics of objects, functions, behaviors, processes, cognition, mechanisms, and modeling as being major considerations for his proposed process model for system of systems. The essential foundation for system of systems and systems is the next major subject for consideration. The chapter ends with the author's conclusions.

Chapter 8, "An Ontology of Emergence," by John Johnson and Jose Padilla, begins with an introduction, followed by a literature review of their subject. They then explain what is involved in building an ontology of emergence, and report the detailed findings associated with studying the ontology of emergence. They consider the implications for modeling and simulation and provide an example of modeling the New York Stock Exchange Flash Crash based on the ontology of emergence, then finish the chapter with their conclusions.

Mikel Petty is the author of Chapter 9, "Modeling and Validation Challenges for Complex Systems." After an introduction, the author provides a thorough presentation of complex systems to provide a solid foundation for the subject of this chapter. He then presents a discussion of the three defining characteristics of complex systems that have an effect on modeling and validation: (1) sensitivity to initial conditions, (2) the subject of this text, namely, emergent behavior, and (3) composition of components. The chapter ends with a summary.

Chapter 10, "Foundations for the Modeling and Simulation of Emergent Behavior Systems," is authored by Thomas Holland. The author first defines emergent behavior systems, then addresses the importance of modeling emergence. He then poses the question, for discussion, of whether emergence can be measured. Next, three examples of emergent systems are considered. The chapter ends with the author's conclusions.

In Chapter 11, "Characterizing Emergent Behavior in System of Systems," after an introduction, Donna Rhodes addresses the subject of the nature of complexity. She then turns her attention to viewpoints for understanding emergent behavior, stating that there are five means to do so: structural, operational, contextual, temporal, and perceptual. The last subject covered by her chapter is that of re-architecting that occurs as a result of both positive and negative emergence in a system of systems.

Carol Woody, the author of Chapter 12, "Engineered to Be Secure," begins the chapter by discussing the five principal characteristics of a system of systems, as defined by Mark Maier (1998), and addresses the relationship of each characteristic to software engineering. In addition, a statement is made as to what software engineering actions are required to preclude emergent behavior. In the second section, the author addresses the transitioning security perspective from secure isolated systems to security of highly networked software intensive systems of systems. Next, she addresses how security of system of systems requires consideration of software and its multi-layer composition. In the fourth section, Woody discusses how system of systems development and sustainment relies on a global multi-organizational software supply chain. Next, she establishes the mandate that systems must be engineered for system of systems security. In the sixth section, the author provides an electrical grid example by modeling mission impact of security risks to engineer system of systems security, then ends the chapter with a summary of topics covered.

Chapter 13, also by Carol Woody, is entitled "Cyber Insecurity Is Growing." As in her previous chapter, Woody addresses the relationship between her subject and each of Maier's five principal characteristics of a system of systems. She states that emergent behavior of the system of systems arises from efforts by acquirers and providers to reduce technology costs and increase flexibility by shifting functionality and information flow to software. In the second section, she addresses the transition from standalone systems to highly networked software-intensive systems of systems. The growth of cyber security risk is the subject of the next section, followed by a discussion of how security controls can add more software and more vulnerabilities. In the fifth section, Woody addresses the subject of modeling the system of systems ecosystem that supports cyber insecurity. She then concludes the chapter with a summary.

Timothy Ferris, in Chapter 14, "The Challenge of Performing Research Which Will Contribute Helpful Engineering Knowledge Concerning Emergence," starts off by contrasting engineering with science and mathematics. He then turns the discussion specifically to emergence by providing various views on the nature of emergence found in the systems engineering community, after which he addresses the pros and cons with respect to framing research into the field in the larger context of systems engineering. He then provides a framework to assist with the discussion of how to organize research objectives. This is followed by a discussion of methods to investigate emergent effects. Next, he addresses the implications for research in terms of what is considered publishable. Finally, he considers what constitutes funded research.

1.3.3 Theoretical Perspectives with Practical Applications

Section III consists of those chapters that are deemed to address theoretical perspectives that have accompanying practical applications. The seven chapters in this section discuss two types of applications—those that are quantitative in nature and those that are qualitative in nature. There are five chapters that are of the former type and two that are of the latter type. The quantitative type is that which can address detection, classification, prediction, and control or some combination of these functions as described in Section 1.2 above. The qualitative type is that which can characterize the presence of emergence via example or scenario description. Table 1.1 below is a summary of only the quantitative type of application.

The first chapter of this section, Chapter 15, "Examination of Emergent Behavior in the Ballistic Missile Defense System: A Modeling and Simulation Approach," is by Saikou Diallo and Christopher Lynch. Following their introduction, the authors address

TABLE 1.1

Use of Modeling and Simulation to Evaluate Emergence

Chapter Number	Number of Examples Used	Type of Modeling & Simulation or Methodology Used	Functions of M&S for Emergence Described
16	1	MATLAB/Simulink	Control
17	1	Agent-based	Detection, Classification
18	1	Monterey Phoenix Behavior Model (Objective: Automated Event Trace Generation)	Detection, Classification, Prediction, Control—all conducted post execution by human operator
19	1	Monterey Phoenix Behavior Model (Objective: Automated Event Trace Generation)	Detection, Classification, Prediction, Control—all conducted post execution by human operator
20	1	Agent-based	Detection, Classification, Control, Visualization

the development of their agent-based and discrete event model of the Missile Defense Agency's Ballistic Missile Defense System (BMDS). They then turn their attention to what is required to configure their simulation. Next, they discuss the exploration of emergence in the BMDS using their model, and complete the chapter with their conclusions.

Chapter 16, by Umut Durak, Thorsten Pawletta, and Tuncer Ören, is entitled "Simulating Variable System Structures for Engineering Emergence." After providing a background and motivation for the chapter, these authors address Organic Computing and Simulation. Then the authors turn their attention to the details of actually simulating emergence, ending the chapter with their conclusions.

Chapter 17, "Emergence as a Macroscopic Feature in Man-Made Systems," by Moath Jarrah starts off with an introduction of what is to be addressed in his chapter. He then turns to the development of a multi-agent simulation framework for man-made systems and provides a labor market example. A case study examining congestion on the Internet is provided, and the chapter ends with his conclusions.

In Chapter 18, "Monterey Phoenix—Behavior Modeling Approach for the Early Verification and Validation of System of Systems Emergent Behaviors," by Kristin Giammarco and Mikhail Auguston, the authors begin with an introduction that provides the reader with a roadmap of the topics to be addressed in the chapter. The second section, Motivation, makes a clear argument to the reader concerning the rationale for their approach to the chapter. A discussion of Behavior Modeling Concepts for Engineering Emergence is found the next section. The authors then provide an introduction to Monterey Phoenix as a foundation for their chapter. This is followed by three examples of emergent behaviors in Monterey Phoenix models. The major thrust, for application purposes, of this chapter is then addressed in a section entitled Detection, Classification, Prediction, and Control of Emergent Behavior. The chapter closes with a conclusion and way-ahead.

John Quartuccio and Kristin Giammarco are the authors of Chapter 19, "A Model-Based Approach to Investigate Emergent Behaviors in Systems of Systems." In the introduction section, the authors address the language of behavior models and provide some further definitions applicable from systems engineering that apply to their chapter. In the second section, Background, they address enabling technologies that implement emergent behavior within a system of systems, potential sources of unfavorable emergent behavior, and an analysis of emergent behavior within system of systems. In Methodology, the third section, they discuss using the Monterey Phoenix-Firebird Analyzer and the associated six steps to implement their methodology. The fourth section consists of an example problem. The fifth section is their discussion of results, which is followed by their concluding remarks.

Chapter 20 is entitled "InterDyne: A Simulation Method for Exploring Emergent Behavior Deriving from Interaction Dynamics" and is written by Christopher Clack and L. Carlos-Sandberg. These authors first provide an introduction to the chapter in which they address emergent behavior deriving from interaction dynamics and the subject of feedback loops. The next section addresses their research focus, followed by a section that addresses their method of analysis. The fourth section addresses their research method, which is to model discrete time at the finest level of detail possible. Then they turn their attention to the development of their InterDyne Simulator, and in the next section, discuss a specific case study to illustrate the use of their simulator. The authors end the chapter with a summary.

Chapter 21 is the last chapter of this section, written by Charles Keating and Polinpapilinho Katina and entitled "Emergence in the Context of System of Systems." In the introduction, the authors provide a background and describe the general problem domain for the chapter. They build upon this by addressing the problem domain associated with system of systems. The details of system of systems are then specifically addressed. Building upon this discussion, the authors turn their attention to emergence, providing an introduction and addressing the central tenets associated with this topic. Next, the authors unpack the subject of emergence in the context of its implications for system of systems. They then address applications of emergence in the context of system of systems, and end the chapter with conclusions and implications.

1.3.4 Lessons Learned and the Proposed Way-Ahead

Section IV, the last part of this text, consists of only one chapter. The purpose of Chapter 22, "Lessons Learned and the Proposed Way-Ahead," by Larry Rainey and Charles Keating, is twofold: first, to capture lessons learned and the associated rationale from each of Chapters 3 through 21; and second, to address a proposed research plan for the future of engineering emergence. The first purpose constitutes the foundation stones upon which this text was developed from the perspective of each of the authors. The second purpose becomes the superstructure for this text to suggest what needs to be built upon the foundation to standout for serious consideration for future development of emergent behavior in the context of system of systems engineering applications.

References

"Concepts: Emergence." Yaneer Bar-Yam of New England Complex Systems Institute, 2011. Web. 24 March 2017. http://necsi.edu/guide/concepts/emergence.html.

Guide to the Systems Engineering Body of Knowledge: Emergence. http://sebokwiki.org/wiki/Emergence.

Maier, M. (1998). Architecting Principles for Systems-of-Systems. Systems Engineering 1(4), 267–284.

Page, S.E. (2009). *Understanding Complexity*. The Great Courses. Chantilly, VA, USA: The Teaching Company.

Rainey, L. and Tolk, A. (Eds.). (2015). *Modeling and Simulation Support for System of Systems Engineering Applications*. Hoboken, NJ: Wiley and Sons.

2

System of Systems Engineering—An Overview

Mo Jamshidi

CONTENTS

2.1 Introduction

This chapter introduces the concept of system of systems (SoS) and the challenges ahead to extend systems engineering (SE) to system of systems engineering. The birth of a new engineering field may be on the horizon—system of systems engineering (SoSE). A SoS is a collection of individual, possibly heterogeneous, but functional and operational systems integrated together to enhance the overall efficiency and robustness, lower the cost of operation, and increase the reliability of the overall complex (SoS) system. Having said that, the field has a large vacuum from basic definition, to theory, to management and implementation. Many key issues like architecture, modeling, simulation, identification, emergence, standards, net-centricity, control, etc. are all begging for attention. In this review chapter, we will go through all these issues briefly and bring the challenges to the attention of interested readers.

This growing interest in SoS as a new generation of complex systems has opened a great many new challenges for systems engineers. Performance optimization, robustness, and reliability among an emerging group of heterogeneous systems in order to realize a common goal has become the focus of various applications including military, security, aerospace, space, manufacturing, service, environment, disaster management, transportation, just to name a few [1–5]. There is an increasing interest in achieving synergy between these independent systems to achieve the desired overall system performance [6]. In the literature, researchers have addressed the issue of coordination and interoperability in a SoS [7,8]. SoS technology is believed to more effectively implement and analyze large, complex, independent, and *heterogeneous* systems working (or made to work) cooperatively [7]. The main thrust behind the desire to view the systems as a SoS is to obtain higher capabilities and performance than would be possible with a traditional system view. The SoS concept presents a high-level viewpoint and explains the interactions between each of the independent systems. However, the SoS concept is still at its developing stages [1,2].

The next section presents some definitions out of many possible definitions of SoS. However, a practical definition may be that a system of systems is a "super system" comprised of other elements which themselves are independent complex operational systems and interact among themselves to achieve a common goal [9]. Each element of a SoS achieves well-substantiated goals even if it is detached from the rest of the SoS. For example, a Boeing 747 airplane, as an element of a SoS, is not a SoS, but an airport is a SoS; or a rover on Mars is not a SoS, but an airport or a robotic colony (or a robotic swarm) exploring the Red Planet, or any other place, is a SoS. As will be illustrated shortly, associated with SoS are numerous problems and open-ended issues which need a great deal of fundamental advances in theory and verifications. It is hoped that this chapter will be a first effort towards bridging the gaps between an *idea* and a *practice*.

2.2 Definitions of System of Systems

Based on the literature survey of system of systems, there are numerous definitions whose detailed discussion is beyond the space allotted to this topic [1,2,10]. Here we enumerate only 7 of many potential definitions:

Definition 1: Systems of systems exist when there is a presence of a majority of the following 5 characteristics: operational and managerial independence, geographic distribution, emergent behavior, and evolutionary development [10].

Definition 2: Systems of systems are large-scale concurrent and distributed systems that are comprised of complex systems [1,2,10].

Definition 3: Enterprise Systems of Systems Engineering is focused on coupling traditional systems engineering activities with enterprise activities of strategic planning and investment analysis [10].

Definition 4: System of Systems Integration is a method to pursue development, integration, interoperability, and optimization of systems to enhance performance in future battlefield scenarios [1,2,11–13].

Definition 5: SoSE involves the integration of systems into systems of systems that ultimately contribute to evolution of the social infrastructure [14].

Definition 6: In relation to joint war-fighting, system of systems is concerned with interoperability and synergism of Command, Control, Computers, Communications, and Information (C4I) and Intelligence, Surveillance, and Reconnaissance (ISR) Systems [15].

Definition 7: SoS: The sum of the whole is greater than the sum of the individual parts [11]—the parts are integrated (i.e., have interfaces); the parts may or may not be members of a common domain (such as a product line, for example, surface ship radars).

Detailed literature survey and discussions of these definitions are given in [1,2,10,11,16]. Various definitions of SoS have their own merits, depending on their application. The favorite definition of this author and the volume's editor is: *Systems of systems are large-scale integrated systems which are heterogeneous and independently operable on their own, but are networked together for a common goal.* The goal, as mentioned before, may be cost, performance, robustness, etc.

2.3 Types of System of Systems

As was mentioned before, Maier [17] has proposed 5 characteristics of SoS, as shown in Figure 2.1. Also, there are 4 types of SoS, as shown by Dahmann [18]. These are described as follows:

Directed. In a Directed SoS, there is a defined design paradigm in which there is a clear leader, similar to a military mission where a commander is in charge of the entire mission.

Acknowledged. In an Acknowledged SoS, there is a pre-contractual agreement between systems to share information and on a need-basis to share data. An example of this could be America's future smart grid.

Collaborative. In a Collaborative SoS, the constituents fully agree to cooperate and exchange information and data to achieve the overall objective of the SoS.

Virtual. In a Virtual SoS, there is no clear coordinator nor any agreement to cooperate and/or exchange data across the enterprise. An example of this SoS is the Internet.

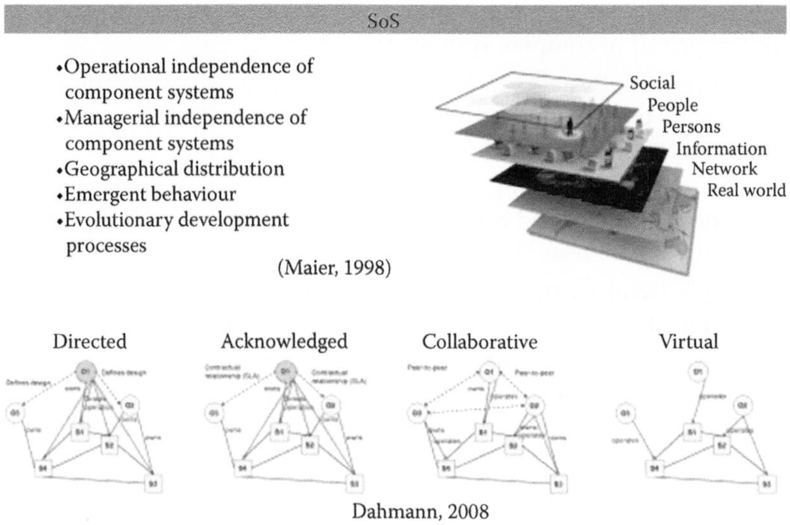

FIGURE 2.1
Types of system of systems [18].

2.4 Challenging Problems in System of Systems Engineering

In the realm of open problems in SoS, there is an unsolved problem just about anywhere one looks, and immense attention is needed from many engineers and scientists. No engineering field is more urgently needed in tackling SoS problems than SE—system engineering. On top of the list of engineering issues in SoS is the "engineering of SoS," leading to a new field of SoSE [19]. How does one extend SE concepts like analysis, control, estimation, design, modeling, controllability, observability, stability, filtering, simulation, etc. so they can be applied to SoS? Among numerous open questions is how one can model and simulate such systems by Mittal et al. [14]. More readings about these issues can be found here [1,2,14–16,19]. Figure 2.2 shows 8 aspects of SoSE that have been briefly brought up for the reader's attention. These are areas that system engineering, as a 6-decade-old engineering discipline, needs to be considered redefined or needs to be revisited.

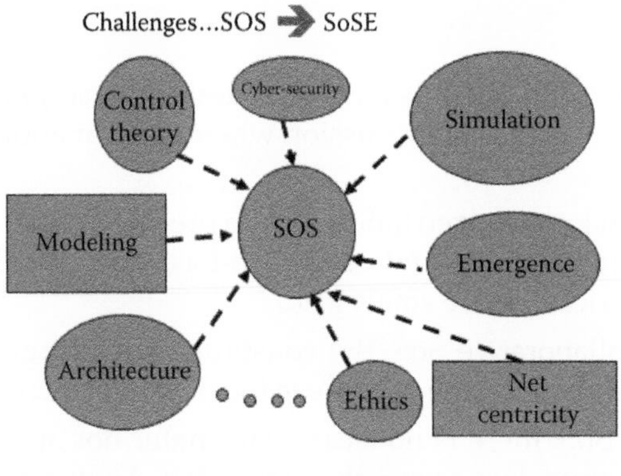

FIGURE 2.2
Areas of systems engineering that need to be revisited to move to system of systems engineering.

2.4.1 Theoretical Problems

The Internet came into realization about 1994 in the United States and elsewhere, whereby for the first time, connected people of the world got connected together. System of systems (or cyber-physical systems) is considered a generalization of the interconnectivity of the Internet. While system of systems has received a great deal of attention in the past 15 years, SoS engineering still remains an "opportunity" for the field of systems engineering (SE) to expand its domains from a single system to SoS. Figure 2.2 shows a potential set of areas that systems engineering needs to revisit to extend many aspects of SE to be able to accommodate SoS for the 21st century.

In this section we discuss some urgent problems facing SoS and SoSE. The major issue here is that a merger between SoS and engineering needs to be made. In other words, systems engineering (SE) needs to undergo a number of innovative changes to accommodate and encompass SoS.

2.4.1.1 Open Systems Approach to System of Systems Engineering

Azani [20] discusses an open systems approach to SoSE. The author notes that SoS exists within a continuum that contains ad-hoc, short-lived, and, relatively speaking, simple SoS on one end of the continuum, and long lasting, continually evolving, and complex SoS on the other end. Military operations and less sophisticated biotic systems (e.g., bacteria and ant colonies) are examples of ad-hoc, simple, and short-lived SoS, while galactic and more sophisticated biotic systems (e.g., ecosystem, human colonies) are examples of SoS at the opposite end of the SoS continuum. The engineering approaches utilized by galactic SoS are at best unknown and perhaps forever inconceivable. However, biotic SoS seem to follow, relatively speaking, less complicated engineering and development strategies, allowing them to continually learn and adapt, grow and evolve, resolve emerging conflicts, and have more predictable behavior. Based on what the author already knows about biotic SoS, it is apparent that these systems employ robust reconfigurable architectures, enabling them to effectively capitalize on open systems development principles and strategies such as modular design, standardized interfaces, emergence, natural selection, conservation, synergism, symbiosis, homeostasis, and self-organization. Azani [20] provides further elaboration on open systems development strategies and principles utilized by biotic SoS, discusses their implications for engineering of man-made SoS, and introduces an integrated SoS development methodology for engineering and development of adaptable, sustainable, and interoperable SoS based on open systems principles and strategies.

2.4.1.2 Engineering of SoS

Emerging needs for a comprehensive look at the applications of classical systems engineering issue in SoSE will be discussed in detail in this volume. The thrust of the discussion will concern the reality that the technological, human, and organizational issues are each far different when considering a system of systems or federation of systems and that these needs are very significant when considering system of systems engineering and management.

As we have noted, today there is much interest in the engineering of systems that are comprised of other component systems, and where each of the component systems serves organizational and human purposes. These systems have several principal characteristics that make the system family designation appropriate: operational independence of the individual systems; managerial independence of the systems; often large geographic and temporal distribution of the individual systems; emergent behavior, in which the system family performs functions and carries out purposes that do not reside uniquely in any of the constituent systems but which evolve over time in an adaptive manner and where these behaviors arise as a consequence of the formation of the entire system family and are not the behavior of any constituent system. The principal purposes supporting engineering of these individual systems and the composite system family are fulfilled by these emergent behaviors. Thus, a system of systems is never fully formed or complete. Development of these systems is evolutionary and adaptive over time, and structures, functions, and purposes are added, removed, and modified as experience of the community with the individual systems and the composite system grows and evolves. The systems engineering and management of these systems families poses special challenges. This is especially the case with respect to the federated systems management principles that must be utilized to deal successfully with the multiple contractors and interests involved in these efforts. Please refer to the paper by Sage and Biemer [21] and DeLaurentis et al. [22] for the creation of a SoS Consortium (the ICSOS) of concerned individuals and organizations by the author of this chapter.

2.4.1.3 Standards of SoS

System of systems literature, definitions, and perspectives are marked by great variability in the engineering community. Viewed as an extension of systems engineering to a means of describing and managing social networks and organizations, the variations of perspectives leads to difficulty in advancing and understanding the discipline. Standards have been used to facilitate a common understanding and approach to align disparities of perspectives to drive a uniform agreement to definitions and approaches. Having the ICSOS—International Consortium on System of Systems [23]—represent to the IEEE and INCOSE for support of technical committees to derive standards for system of systems will help unify and advance the discipline for engineering, healthcare, banking, space exploration, and all other disciplines that require interoperability among disparate systems [23–25].

In recent years, INCOSE (International Council of Systems Engineering) has formed a working group to advance and promote the application of Systems Engineering to Systems of Systems (SoSE) [24,25]. Standardization of SoSE is among a key area of the INCOSE Working Group. Some of the initial results are reflected in Body of Knowledge and Curriculum to Advance Systems Engineering (BKCASE) [24,25]. BKCASE describes SoS engineering as *"an opportunity for the systems engineering community to define the complex systems of the 21st Century. While systems engineering is a fairly established field, SoSE represents a challenge for the present systems engineers at the global level. In general, SoSE requires considerations beyond those usually associated with engineering to include socio-technical and sometimes socio-economic phenomena."* Many authors, including this one, has contributed to the BKCASE [24].

2.4.1.4 System of Systems Architecting

Dagli and Kilicay-Ergin [26] provide a framework for SoS Architectures. As the world is moving towards a networked society, the authors assert, the business and government applications require integrated systems that exhibit intelligent behavior. The dynamically changing environmental and operational conditions create a need for system architectures that will be effective for the duration of the mission but evolve to new system architectures as the mission changes. This challenging new demand has led to a new operational style: Instead of designing or subcontracting systems from scratch, business or government gets the best systems the industry develops and focuses on becoming the lead system integrator to provide a system of systems (SoS). SoS is a set of interdependent systems that are related or connected to provide a common mission. In the SoS environment, architectural constraints imposed by existing systems have a major effect on the system capabilities, requirements, and behavior. This fact is important, as it complicates the systems architecting activities. Hence, architecture becomes a dominating but confusing concept in capability development. There is a need to push system architecting research to meet the challenges imposed by new demands of the SoS environment. This chapter focuses on system of systems architecting in terms of creating meta-architectures from collections of different systems. Several examples are provided to clarify the system of systems architecting concept. Since the technology base, organizational needs, and human needs are changing, the system of system architecting becomes an evolutionary process. Components and functions are added, removed, and modified as owners of the SoS experience and use the system. Finally, the authors discuss the possible use of artificial life tools for the design and architecting of SoS. Artificial life tools such as swarm intelligence, evolutionary computation, and multi-agent systems have been successfully used for the analysis of complex adaptive systems. The potential use of these tools for SoS analysis and architecting are discussed by the authors, using several domain application specific examples. Figure 2.3 shows meta-architecture generation for financial markets [26].

2.4.1.5 SoS Simulation

Sahin et al. [27] have presented a SoS architecture based on Extensible Markup Language (XML) in order to wrap data coming from different systems in a common way. The XML can be used to describe each component of the SoS and their data in a unifying way. If XML-based data architecture is used in a SoS, the only requirement is for the SoS components to understand/parse the XML file received from the components of the SoS. In XML, data can be represented in addition to the properties of the data such as source name, data

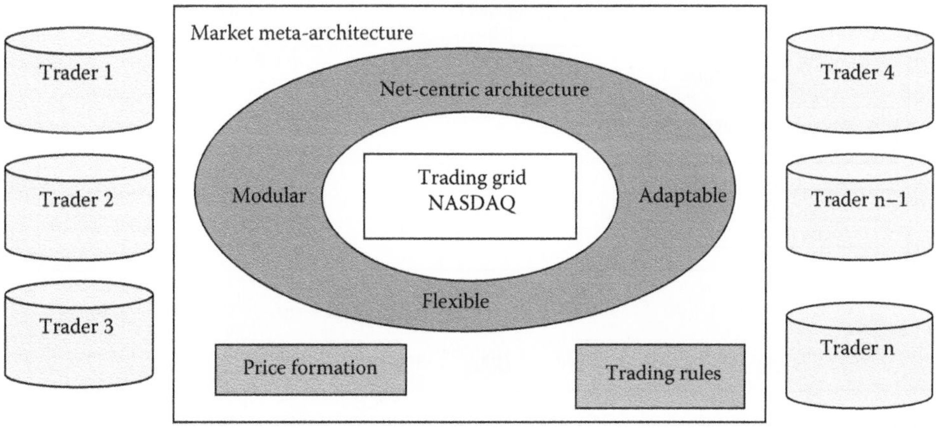

FIGURE 2.3
Meta-architecture generation for financial markets [26].

type, importance of the data, and so on. Thus, it does not only represent data but also gives useful information which can be used in the SoS to take better actions and to understand the situation better. The XML language has a hierarchical structure whereby an environment can be described with a standard and without a huge overhead. Each entity can be defined by the user in the XML in terms of its visualization and functionality. As a case study in this effort, see Mittal et al. [14,28] where the DEVS (discrete event systems) is presented as a platform for modeling and simulation of SoS. Also presented is architecture for a master-scout rover combination representing a SoS where first a sensor detects a fire in a field. The fire is detected by the master rover, which commands the scout rover to verify the existence of the fire. It is important to note that such an architecture and simulation does not need any mathematical model for members of the systems. Figure 2.4 shows a DEVS-XML simulation framework for a system of robots seeking to warn of a disaster waiting to occur [27]. For further insight on DEVS-based modeling of system of systems engineering, refer to Chapter 3 by Zeigler.

2.4.1.6 SoS Integration

Integration is probably the key viability of any SoS. Integration of SoS implies that each system can communicate and interact (control) with the SoS regardless of its hardware, software characteristics, or nature. This means that each system needs to have the ability to communicate with the SoS or a part of the SoS without compatibility issues such as

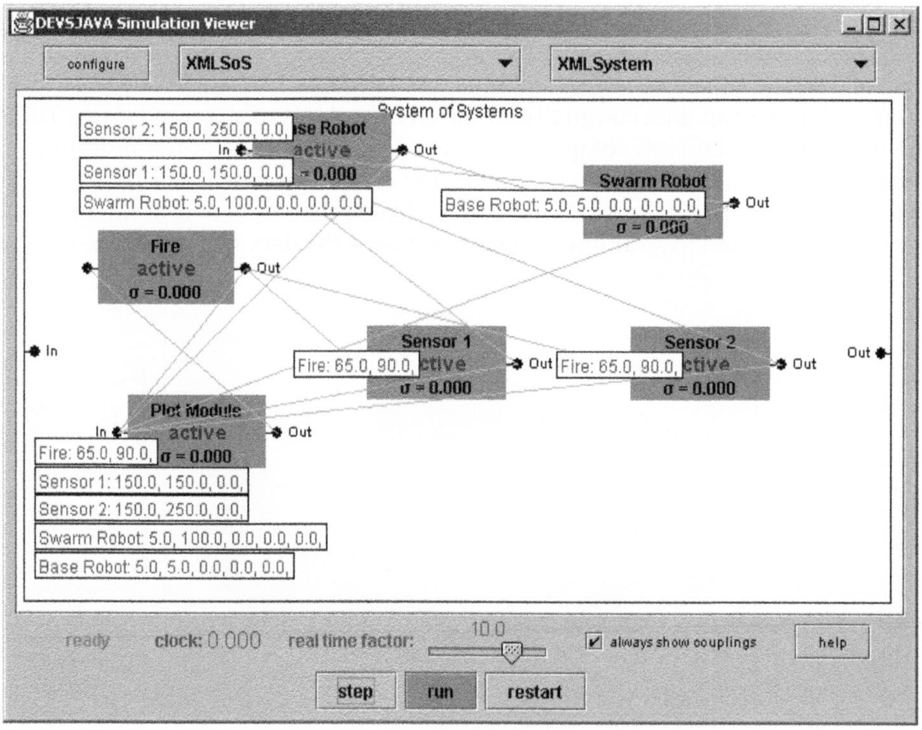

FIGURE 2.4
A DEVS-XML simulation framework for a system of robots finding start of a fire [27].

operating systems, communication hardware, and so on. For this purpose, a SoS needs a common language that its systems can speak. Without having a common language, the systems of any SoS cannot be fully functional and the SoS cannot be adaptive in the sense that new components cannot be integrated to it without major effort. Integration also implies the control aspects of the SoS because systems need to understand each other in order to take commands or signals from other SoS systems. See Cloutier et al. [29] for more on network-centric architecture of SoS.

2.4.1.7 Emergence in SoS

Emergence is the key theme of this volume, with many topics on emergence being presented. Emergent behavior of a SoS resembles the slowdown of traffic going through a tunnel, even in the absence of any lights, obstacles, or accidents. A tunnel, automobiles, and the highway, as systems of a SoS, have an emergent behavior or property in slowing down [1,2]. Fisher [30] has noted that a SoS can not achieve its goals dependent on its emergent behaviors. The author explores "interdependencies among systems, emergence, and interoperation" and develops maxim-like findings such as these: (1) Because they cannot control one another, autonomous entities can achieve goals that are not local to themselves only by increasing their influence through cooperative interactions with others. (2) Emergent composition is often poorly understood and sometimes misunderstood because it has few analogies in traditional systems engineering. (3) Even in the absence of accidents, tight coupling can ensure that a system of systems is unable to satisfy its objectives. (4) If it is to remain scalable and affordable no matter how large it may become, a system's cost per constituent must grow less linearly with its size. (5) Delay is a critical aspect of systems of systems. Keating [31] provides a detailed perspective into the emergence property of SoS, which is also found elsewhere in this volume.

2.4.1.8 SoS Management: The Governance of Paradox

Sauser and Boardman [32] present a SoS approach to the management problem. They note that the study of SoS has moved many people to support their understanding of these systems through the groundbreaking science of networks. The understanding of networks and how to manage them may give one the fingerprint which is independent of the specific systems that exemplify this complexity. The authors point out that it does not matter whether one is studying the synchronized flashing of fireflies, space stations, the structure of the human brain, the Internet, the flocking of birds, a future combat system, or the behavior of red harvester ants. The same emergent principles apply: large is really small; weak is really strong; significance is really obscure; little means a lot; simple is really complex; and, complexity hides simplicity. The conceptual foundation of complexity is paradox, which leads us to a paradigm shift in the SE (systems engineering) body of knowledge.

Paradox exists for a reason, and there are reasons for systems engineers to appreciate paradox even though they may be unable to resolve paradoxes as they would a problem specification into a system solution. Hitherto paradoxes have confronted current logic only to yield at a later date to more refined thinking. The existence of paradox is always the inspirational source for seeking new wisdom, attempting new thought patterns, and ultimately building systems for the "flat world." It is our ability to govern, not control, these paradoxes that will bring new knowledge to our understanding of how to manage the emerging complex systems called system of systems.

Sauser and Boardman [32] establish a foundation in what has been learned about how one practices project management, establish some key concepts and challenges that make the management of SoS different from our fundamental practices, present an intellectual model for how they classify and manage a SoS, appraise this model with recognized SoS, and conclude with grand challenges for how they may move their understanding of SoS management beyond the foundation.

2.4.1.9 Control of System of Systems

As was shown in Figure 2.2, all aspects of systems engineering need to be revisited by the systems engineering community. Analysis and design of system of systems control is one of these aspects. In this section, control of SoS is considered. The main challenge in designing a controller for SoS is the difficulty or impossibility of developing a comprehensive SoS model, either analytically, through simulation, or with machine learning data-based modeling. By and large, SoS control remains an open problem and is, of course, different for each application domain. Should a mathematical model be available, some control paradigms are available, which was the focus of work here [33]. Moreover, real-time control—which is required in almost all application domains—of interdependent systems poses an especially difficult problem. Nevertheless, several potential control paradigms are briefly considered by Nanayakkara et al. [34]. In this section, three of the most promising control approaches—consensus-based, cooperative, and networked—are briefly discussed.

2.4.1.9.1 Consensus-Based Control

Ren and Beard [35] have proposed a cooperative control paradigm based on "consensus" among constituents of a SoS. The primary motivation of this control paradigm, like all other paradigms, is extracting greater benefits to all systems of a SoS. Here, the constituents agree to and follow a common goal. Sub-goals of systems in SoS may have differences, but through communication among them, they can come into a "consensus" and follow it in their set formation. Joordens and Jamshidi [36] have applied this control approach for some 50 submerged vehicles in an underwater swarm control.

2.4.1.9.2 Cooperative Control

Cooperative control of a SoS assumes that it can be characterized by a set of interconnected systems or agents with a common goal. Classical techniques of control design, optimization, and estimation could be used to create parallel architectures for, as an example, coordinating underwater gliders [36]. However, many issues dealing with real-time cooperative control have not been addressed, even in non-SoS structures. A critical issue concerns controlling a SoS in the presence of communication delays to and among the SoS systems. Ren and Cao [37] have presented some categorization for cooperative control. These include approaches like leader-follower, behavioral, virtual structure/leader, etc. Application areas of cooperative control, just as in any viable control paradigm of SoS, are autonomous or semi-autonomous vehicles, satellites, space crafts, automated highways, Earth observation, air traffic, border/port security, environment (oil spills, rural areas, forest fires, wildlife, etc.).

2.4.1.9.3 Networked Control

Control systems where a real-time communication network is in the feedback path is called *Network Control System* (NCS) [37]. Figure 2.5 shows an adaptive network control system of systems. The AD HOC network can be implemented through a number of

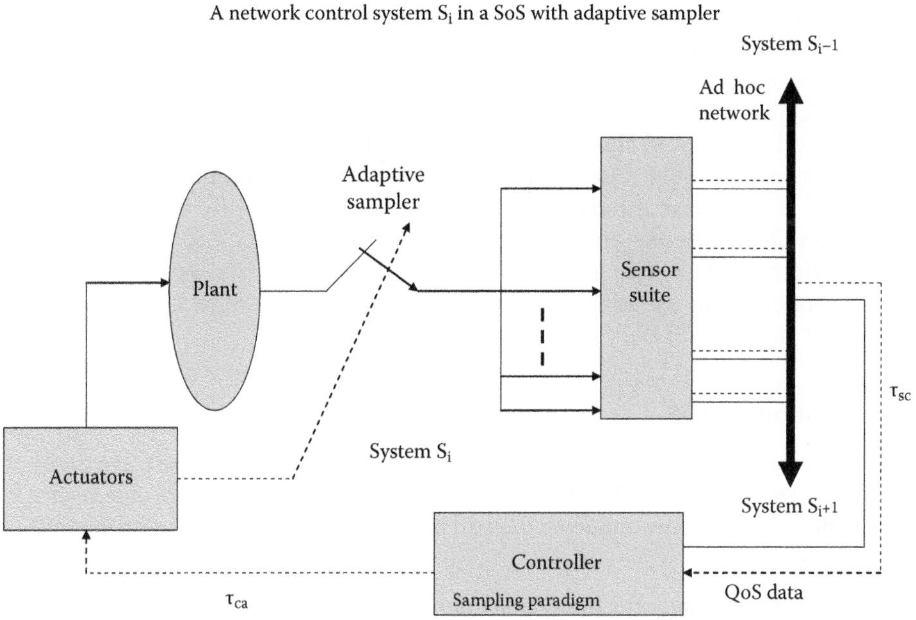

FIGURE 2.5
A network or distributed control system for System S_i of a SoS.

alternatives like Ethernet, FireWire, etc. However, one often has a time varying channel dependent on a fixed capacity for the total amount of information that can be communicated at any one-time instance to the collection of autonomous UVs controllers. One of the main challenges in a NCS is the loss of or delays in transmission and receipt of data from sensors to controllers τ_{sc} and from controllers to actuators τ_{ca}. The challenge in SoS networked control is to develop a SoS distributed control system which can tolerate lost packets, partially decoded packets, delays, and fairness issues—i.e., add robustness to the control paradigm. Here fairness means that certain systems can maximize their total wireless system capacity if we transmit data to their controller more frequently. Values of parameters like sensor-controller and controller-actuator delays will be keys to how much the fairness issue must be practiced. These communication infractions can be compensated (i) by adjusting control power and controlling distances between systems (power control); (ii) by trading off modulation, coding, and antenna diversity versus throughput (adaptive modulation coding); (iii) where the (non-wireless) intra-feedback (on-board hardware) loop of the autonomous control within S_i is lower latency than the inter-wireless distributed control loop between S_i and S_j or the inter-wireless SoS controller and the S_i controller.

For an alternative approach to check on the communication infractions and guarantee a high level of QoS (quality of service) and QoP (quality of control performance), the design of WNCS (wireless networked control system) needs to fully take all aspects of the attributes of the ad hoc network into account. When the sensor measurements reach the controller (see Figure 2.5), the following tasks need to take place: (1) compute the control action u(kT), and (2) adjust control action depending on QoS parameters, where T is the sampling period [34]. Here, at each sampling period, the distributed control will generate two components, i.e., $u(kT) = u_l(kT) + u_c(kT)$, where the first component

$u_l(kT)$ is the local controller using classical or modern control techniques such as PID (Proportional + Integral + Derivative) or LQG (Linear Quadratic Gaussian), etc., and $u_c(kT)$ is the correction component of the controller determined to compensate for ad hoc network QoS parameters. The latter correction control component will depend on sampling period T(k+1) and the sampling policy of the WNCSOS (wireless networked control system of systems). One way to make sure the correction component $u_c(kT)$ is determined is to devise a performance index to minimize the effects of ad hoc net parameters on QoS is to use an LQG problem and use a policy to determine the next sampling period $T(k+1) = T(k) + \Delta T(k)$ [38]. A combination of a local controller, a correction component, and an adaptive sampling period will enhance both stability and robustness for the WNCSOS.

2.4.2 Applications of SoS

Aside from the many theoretical and essential difficulties with SoS, there are many implementation challenges facing SoS. Here some of these implementation problems are briefly discussed and references are made to the full coverage of some of these problems.

2.4.2.1 Systems Engineering for the Department of Defense System of Systems

Dahmann [18] has addressed the national defense aspects of SoS. Military operations are the synchronized efforts of people and systems toward a common objective. In this way, from an operational perspective, defense is essentially a "systems of systems" (SoS) enterprise. However, despite the fact that today almost every military system is operated as part of a system of systems, most of these systems were designed and developed without the benefit of systems engineering at the SoS level factoring the role the system will play in the broader system of systems context. With changes in operations and technology, the need for systems that work effectively together is increasingly visible. Dahmann [18] outlines the changing situation in the Defense Department and the challenges it poses for systems engineering. Further insights into defense applications of system of systems can be found in the 2008 systems engineering guide for system of systems [39].

2.4.2.2 E-Enabling and SoS Aircraft Design via SoSE

A case of aeronautical application of SoS worth noting is that of E-enabling in aircraft design as a system of a SoS at Boeing Commercial Aircraft Division [40]. The project focused on developing a strategy and technical architecture to facilitate making the airplane (Boeing 787, see Figure 2.6) network-aware and capable of leveraging computing and network advances in industry. The project grew to include many ground based architectural components at the airlines and at the Boeing factory, as well as other key locations such as the airports, suppliers, and terrestrial Internet Service Providers (ISPs).

Wilber [39] points out that the E-enabled project took on the task of defining a system of systems engineering solution to the problem of interoperation and communication with the existing, numerous, and diverse elements that make up the airlines' operational systems (flight operations and maintenance operations). The objective has been to find ways of leveraging network-centric operations to reduce production, operations, and maintenance costs for both Boeing and the airline customers.

(a)

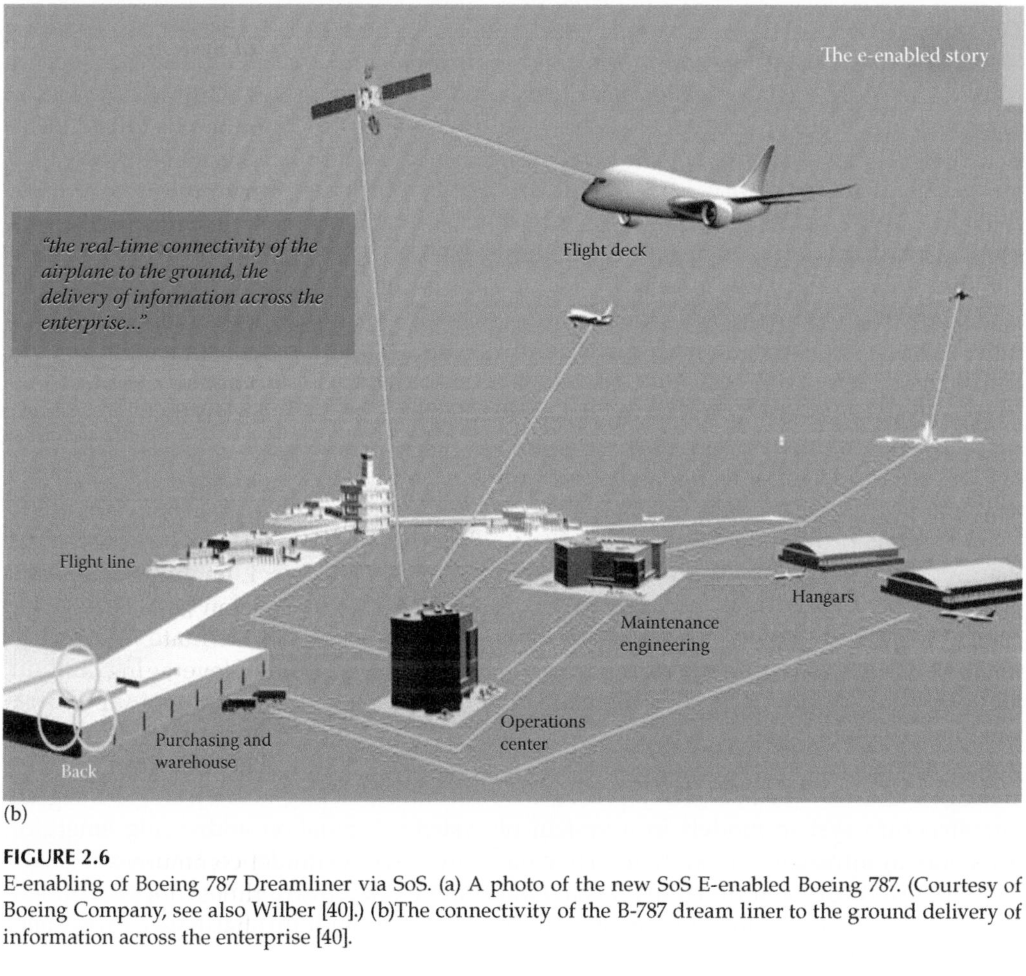

(b)

FIGURE 2.6
E-enabling of Boeing 787 Dreamliner via SoS. (a) A photo of the new SoS E-enabled Boeing 787. (Courtesy of Boeing Company, see also Wilber [40].) (b)The connectivity of the B-787 dream liner to the ground delivery of information across the enterprise [40].

> One of the key products of this effort is the "e-Enabled Architecture". The e-Enabling Architecture is defined at multiple levels of abstraction. There is a single top-level or "Reference Architecture" that is necessarily abstract and multiple "Implementation Architectures". The implementation architectures map directly to airplane and airline implementations and provide a family of physical solutions that all exhibit common attributes and are designed to work together and allow re-use of systems components. The implementation architectures allow for effective forward and retrofit installations addressing a wide range of market needs for narrow and widebody aircraft.
>
> The 787 "Open Data Network" is a key element of one implementation of this architecture. It enabled on-board and off-board elements to be networked in a fashion that is efficient, flexible and secure. The fullest implementations are best depicted in Boeing's GoldCare Architecture and design.

Wilber [40] presents architecture at the reference level and how it has been mapped into the 787 airplane implementation. *GoldCare* environment is described and is used as an example of the full potential of the current E-enabling (see Figure 2.4).

2.4.2.3 A Systems of Systems Perspective on Infrastructures

Thissen and Herder [41] touch upon a very important application in the service industry (see also Tien [42]). Infrastructure systems (or infrasystems) providing services such as energy, transport, communications, and clean and safe water are vital to the functioning of modern society. Key societal challenges with respect to our present and future infrastructure systems relate to, among other things, safety and reliability, affordability, and transitions to sustainability. Infrasystem complexity precludes simple answers to these challenges. While each of the infrasystems can be seen as a complex system of systems in itself, increasing interdependency among these systems (both technologically and institutionally) adds a layer of complexity.

One approach to increased understanding of complex infrasystems that has received little attention in the engineering community thus far is to focus on the commonalities of the different sectors, and to develop generic theories and approaches such that lessons from one sector could easily be applied to other sectors. The system of systems paradigm offers interesting perspectives in this respect. The authors present, as an initial step in this direction, a fairly simple three-level model distinguishing the physical/technological systems, the organization and management systems, and the systems and organizations providing infrastructure related products and services. The authors use the model as a conceptual structure to identify a number of key commonalities and differences between the transport, energy, drinking water, and ICT sectors. Using two energy related examples, the authors further illustrate some of the system of systems related complexities of analysis and design at a more operational level. The authors finally discuss a number of key research and engineering challenges related to infrastructure systems, with a focus on the potential contributions of systems of systems perspectives.

Grogan and de Weck [43] have presented a formal modeling framework to integrate infrastructure system models in a system of systems simulation addressing emerging behaviors in infrastructure systems. They have proposed a simulation framework based on graphical theory and functional behavioral framework capturing the temporal dimension of infrastructure. One of their case studies has to do with infrastructure investment systems with emphasis on urban areas.

2.4.2.4 Sensor Networks

The main purpose of sensor networks is to utilize the distributed sensing capability provided by tiny, low-powered and low-cost devices. Multiple sensing devices can be used cooperatively and collaboratively to capture events or monitor space more effectively than a single sensing device [44,45]. The realm of applications for sensor networks is quite diverse, and include military, aerospace, industrial, commercial, environmental, and health monitoring applications, to name a few. Applications include: traffic monitoring of vehicles, cross-border infiltration detection and assessment, military reconnaissance and surveillance, target tracking, habitat monitoring, and structure monitoring, etc.

The communication capability of these small devices, often with heterogeneous attributes, makes them good candidates for system of systems. Numerous issues exist with sensor networks such as data integrity, data fusion and compression, power consumption, multi-decision making, and fault tolerance, which all make these SoS very challenging just like other SoS. It is thus necessary to devise a fault-tolerant mechanism with a low computation overhead to validate the integrity of the data obtained from the sensors ("systems"). Moreover, a robust diagnostics and decision making process should aid in monitoring and control of critical parameters to efficiently manage the operational behavior of a deployed sensor network. Specifically, Sridhar et al. [44] have focused on innovative approaches to deal with multi-variable multi-space problem domain as well as other issues, in wireless sensor networks within the framework of a SoS. Figure 2.7 shows that the components in the SoS, which are themselves systems, are sufficiently *complex* as shown in Exhibit 1.

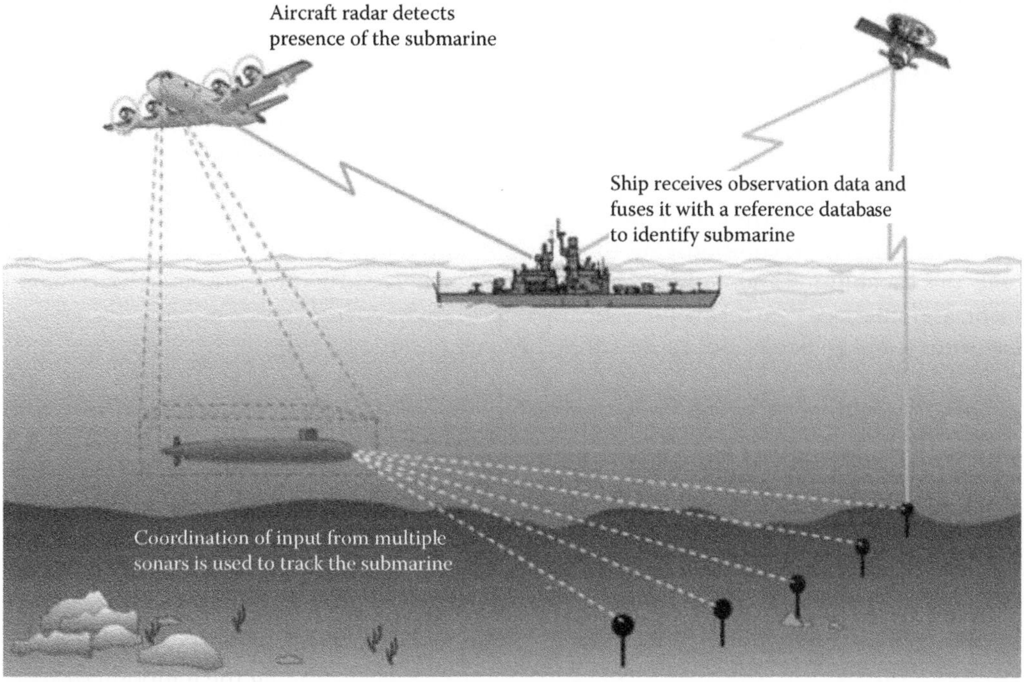

FIGURE 2.7
A classical system of systems application. (Image courtesy of [44].)

2.4.2.5 A System of Systems View of Services

Tien [42] covers a very important application of SoS in today's global village—the *service industry*. The services sector employs a large and growing proportion of workers in the industrialized nations, and it is increasingly dependent on information technology. While the interdependences, similarities, and complementarities of manufacturing and services are significant, there are considerable differences between goods and services, including the shift in focus from mass production to mass customization (whereby a service is produced and delivered in response to a customer's stated or imputed needs). In general, a service system can be considered to be a combination or recombination of three essential components: people (characterized by behaviors, attitudes, values, etc.), processes (characterized by collaboration, customization, etc.), and products (characterized by software, hardware, infrastructures, etc.). Furthermore, inasmuch as a service system is an integrated system, it is, in essence, a system of systems whose objectives are to enhance its efficiency (leading to greater interdependency), effectiveness (leading to greater usefulness), and adaptiveness (leading to greater responsiveness). The integrative methods include a component's design, interface, and interdependency; a decision's strategic, tactical, and operational orientation; and an organization's data, modeling, and cybernetic consideration. A number of insights are also provided, including an alternative system of systems view of services; the increasing complexity of systems (especially service systems), with all the attendant lifecycle design, human interface, and system integration issues; the increasing need for real-time, adaptive decision making within such systems of systems; and the fact that modern systems are also becoming increasingly more human-centered, if not human-focused—thus, products and services are becoming more complex and more personalized or customized.

2.4.2.6 System of Systems Engineering in Space Exploration

Jolly and Muirhead [46] cover SoSE topics that are largely unique for space exploration with the intent to provide the reader a discussion of the key issues, the major challenges of the 21st century in moving from systems engineering to SoSE, potential applications in the future, and the current state of the art. Specific emphasis is placed on how software and electronics are revolutionizing the way space missions are being designed, including both the capabilities and vulnerabilities introduced. The role of margins, risk management, and interface control are all critically important in current space mission design and execution—but in SoSE applications they become paramount. Similarly, SoSE space missions will have extremely large, complex, and intertwined command and control and data distribution ground networks, most of which will involve extensive parallel processing to produce tera- to petabytes of products per day and distribute them worldwide. Figure 2.8 indicates NASA's space constellation project as a SoS.

2.4.2.7 Communication and Navigation in Space SoS

Bahsin and Hayden [47] have taken on the challenges in communication and navigation for space SoS. They indicate that communication and navigation networks provide critical services in operation, system management, information transfer, and situation awareness to the space system of systems. In addition, space systems of systems are requiring system interoperability, enhanced reliability, common interfaces, dynamic operations, and autonomy in system management. New approaches to communications and navigation

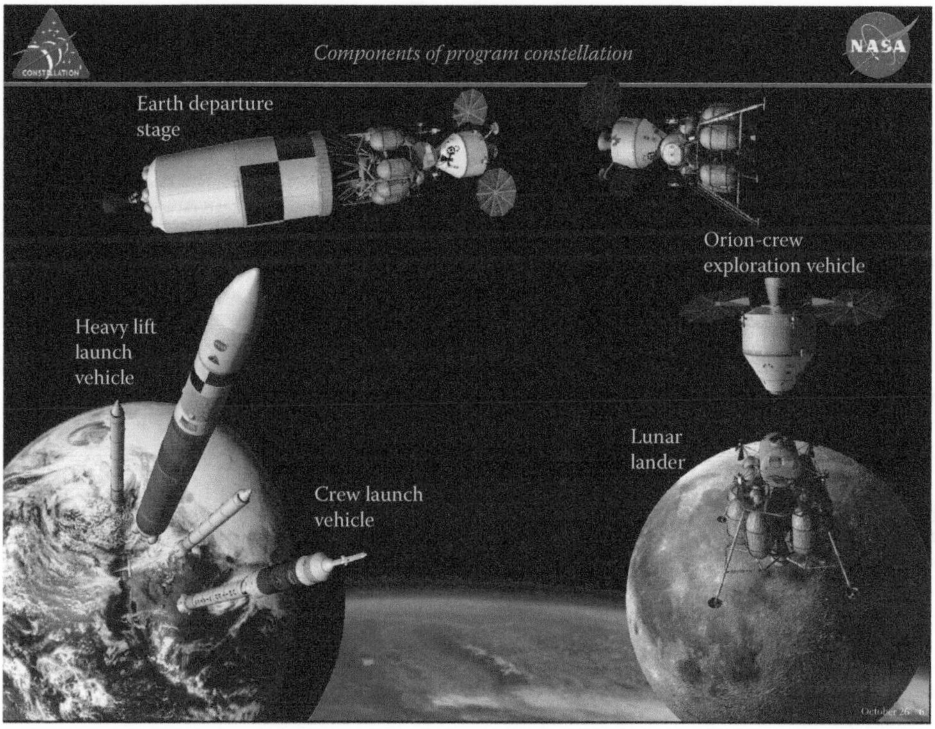

FIGURE 2.8
NASA's space constellation project as a system of systems. (Photo courtesy of NASA.)

networks are required to enable the interoperability needed to satisfy the complex goals and dynamic operations and activities of the space system of systems. Historically space systems had direct links to Earth ground communication systems, or they required a space communication satellite infrastructure to achieve higher coverage around the Earth. It is becoming increasingly apparent that many systems of systems may include communication networks that are also systems of systems. These communication and navigation networks must be as nearly ubiquitous as possible and accessible on the demand of the user, much like the cell phone link is available at any time to an Earth user in range of a cell tower. The new demands on communication and navigation networks will be met by space Internet technologies. It is important to bring Internet technologies, Internet Protocols (IP), routers, servers, software, and interfaces to space networks to enable as much autonomous operation of those networks as possible. These technologies provide extensive savings in reduced cost of operations. The more these networks can be made to run themselves, the less humans will have to schedule and control them. The Internet technologies also bring with them a very large repertoire of hardware and software solutions to communication and networking problems that would be very expensive to replicate under a different paradigm. Higher bandwidths are needed to support the expected voice, video, and data transfer traffic for the coordination of activities at each stage of an exploration mission.

Existing communications, navigation, and networking have grown in an independent fashion with experts in each field solving the problem just for that field. Radio engineers designed the payloads for today's "bent pipe" communication satellites. The Global Positioning System (GPS) satellite design for providing precise Earth location determination is an extrapolation of the LOng RAnge Navigation (LORAN) technique of the 1950s where precise time is correlated to precise position on the Earth. Other space navigation techniques use artifacts in the RF communication path (Doppler shift of the RF and transponder-reflected ranging signals in the RF) and time transfer techniques to determine the location and velocity of a spacecraft within the solar system. Networking in space today is point-to-point among ground terminals and spacecraft, requiring most communication paths to/from space to be scheduled such that communications is available only on an operational plan and is not easily adapted to handle multidirectional communications under dynamic conditions.

Bahsin and Hayden [47] begin with a brief history of the communications, navigation, and networks of the 1960s and 1970s in use by the first system of systems, the NASA Apollo missions; it is followed by short discussions of the communication and navigation networks and architectures that the DoD and NASA employed from the 1980s onwards. Next is a synopsis of the emerging space system of systems that will require complex communication and navigation networks to meet their needs. Architecture approaches and processes being developed for communication and navigation networks in emerging space system of systems are also described. Several examples are given of the products generated in using the architecture development process for space exploration systems. The architecture addresses the capabilities to enable voice, video, and data interoperability needed among the explorers during exploration, while in habitat, and with Earth operations. Advanced technologies are then described that will allow space system of systems to operate autonomously or semi-autonomously. The authors [47] end with a summary of the challenges and issues raised in implementing these new concepts.

2.4.2.8 Electric Power Systems Grids as SoS

Korba and Hiskins [48] provide an overview of the systems of systems that are fundamental to the operation and control of electrical power systems. Perspectives are drawn from industry and academia, and reflect theoretical and practical challenges that are facing power systems in an era of energy markets and increasing utilization of renewable energy resources (see also [49]). Power systems cover extensive geographical regions and are composed of many diverse components. Accordingly, power systems are large-scale, complex, dynamical systems that must operate reliably to supply electrical energy to customers. Stable operation is achieved through extensive monitoring systems and a hierarchy of controls that together seek to ensure total generation matches consumption, and that voltages remain at acceptable levels. Safety margins play an important role in ensuring reliability but tend to incur economic penalties. Significant effort is therefore being devoted to the development of demanding control and supervision strategies that enable reduction of these safety margins, with consequent improvements in transfer limits and profitability. Recent academic and industrial research in this field is also addressed by Korba and Hiskins [48]. Figure 2.9 shows a wide-area monitoring of electrical power systems using synchronized PMU (Phasor Measurement Units) [49].

Smart grids and micrograms are typical examples of SoS, also called cyber-physical systems, which consist of the following constituents: PV, Load, Satellite, Sky Imager, smart meter, battery, converters, and data flow across the entire system (see Figure 2.10). Unlike classical grids, in smart grids the power flows in both directions, as does information and data, which flows everywhere. In such case, the utility companies collect data among their consumers of power and try to make generation and distribution of power much more efficient and give them energy conservation guidelines.

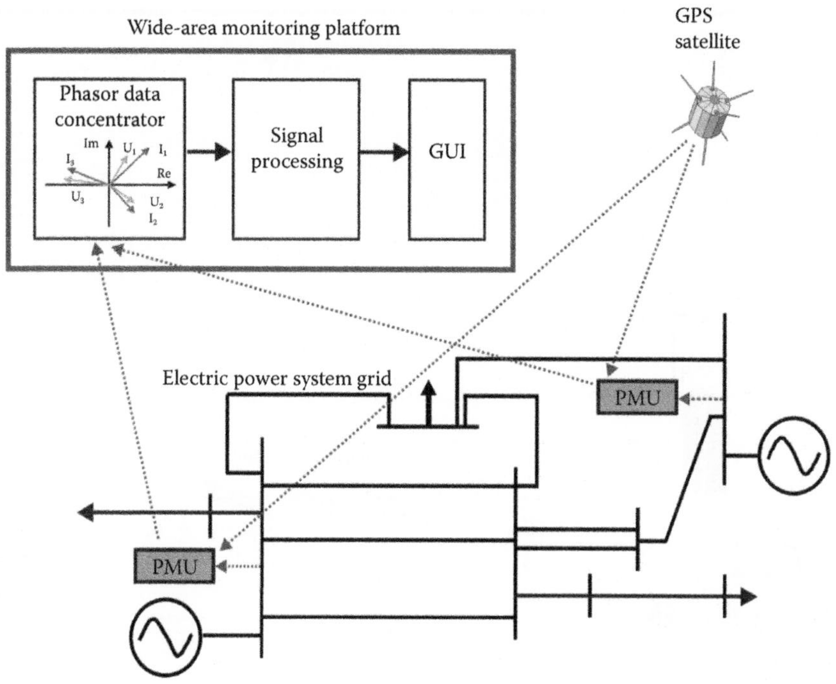

FIGURE 2.9
A wide-area monitoring of electrical power systems using synchronized phasor measurement units (PMU) [49].

Sherick and Yinger [49] have presented an evolutionary vision of the smart grid of the future, as planned in the state of California. They present and review a set of disruptive technologies which have impacted generation, transmission, and distribution of electricity. We are already witnessing monitoring, control, prediction, and optimization in the smart grid. In monitoring, the utility companies can have "Real-time situational awareness," "Power quality awareness," and "Distribution load flow analysis" [49]. In control, utilities will be able to perform "Auto circuit reconfiguration," "Distributed energy resources (DER) dispatch" and "Microgrid management." In prediction, the companies can have capabilities for "Short-term DER forecasting," [50], "Long-term DER forecasting," and "Contingency Analysis." Finally, in optimization, they can perform "voltage optimization," "Power flow optimization," and "Adaptable protection" [49].

2.4.2.9 SoS Approach for Traditional Energy Systems

Duffy et al. [51] have provided the SoS approach to sustainable supply of energy. They note that over one-half of the petroleum consumed in the United States is imported, and that percentage is expected to rise to 60 percent by 2025. America's transportation system of systems relies almost exclusively on refined petroleum products, accounting for over two-thirds of the oil used. Each day, over 8 million barrels of oil are required to fuel over 225 million vehicles that constitute the U.S. light-duty transportation fleet. The gap between U.S. oil production and transportation oil needs is projected to grow, and the increase in the number of light-duty vehicles will account for most of that growth. On a global scale, petroleum supplies will be in increasingly higher demand as highly-populated

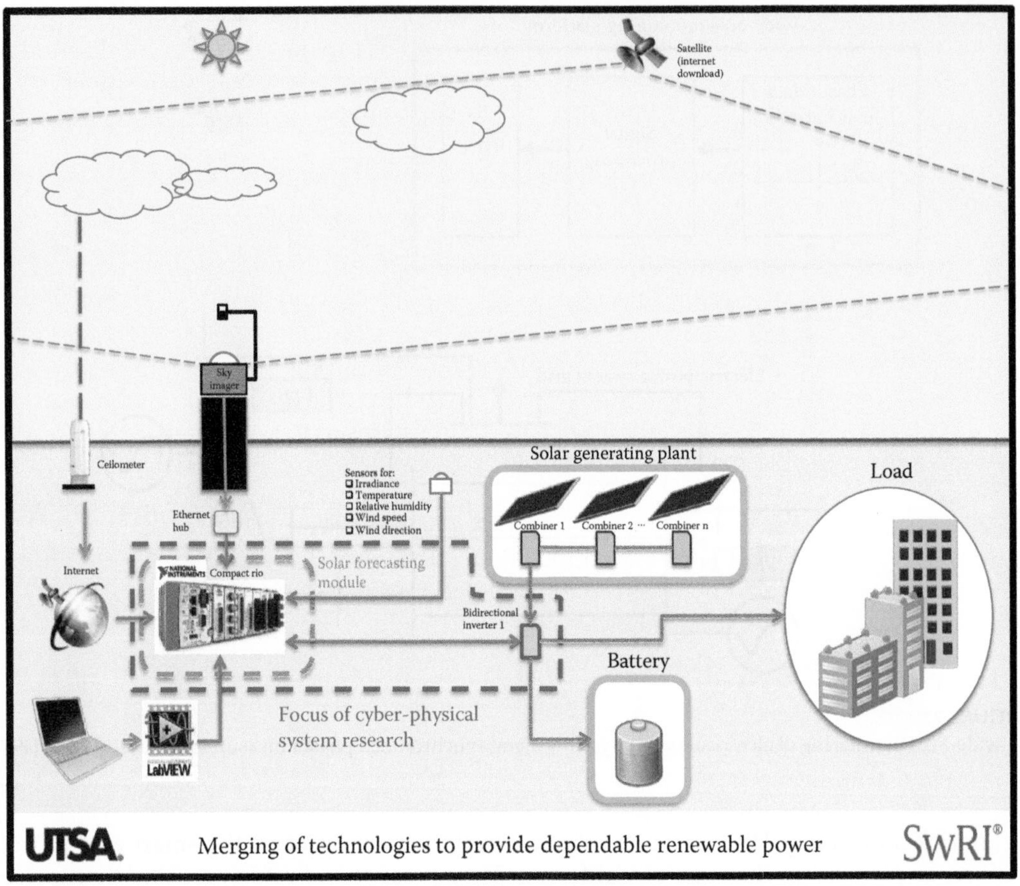

FIGURE 2.10
A typical renewable microgrid as a SoS.

developing countries expand their economies and become more energy intensive. Clean forms of energy are needed to support sustainable global economic growth while mitigating impacts on air quality and the potential effects of greenhouse gas emissions. The United States' growing dependence on foreign sources of energy threatens its national security. As a nation, the authors assert that, we must work to reduce our dependence on foreign sources of energy in a manner that is affordable and preserves environmental quality. Figure 2.11 shows the existing petroleum-based transportation SoS [51].

2.4.2.10 Sustainable Environmental Management from a System of Systems Engineering Perspective

Hipel et al. [52] provide a rich range of decision tools from the field of SE for addressing complex environmental SoS problems in order to obtain sustainable, fair, and responsible solutions to satisfy as much as possible the value systems of stakeholders, including the natural environment and future generations who are not even present at the bargaining table. To better understand the environmental problem being investigated and thereby eventually reach more informed decisions, the insightful paradigm of a system of systems

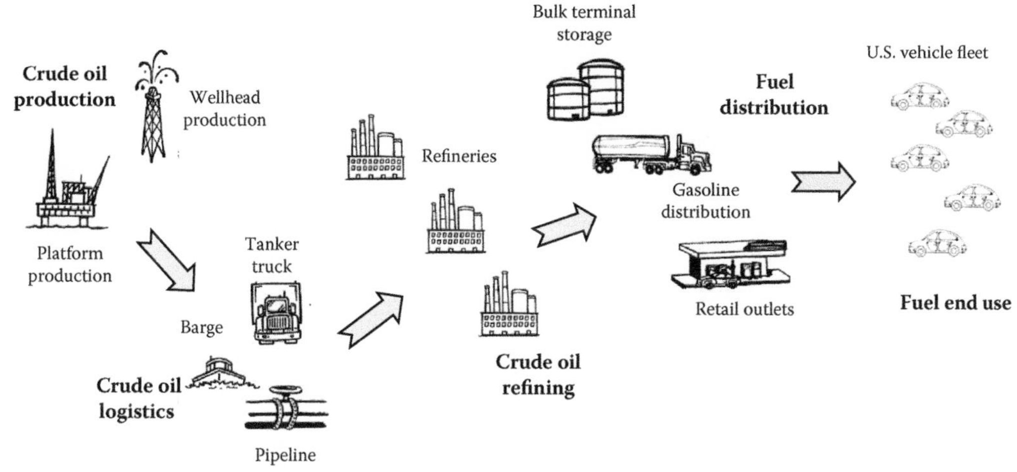

FIGURE 2.11
The existing petroleum-based transportation SoS [51].

can be readily utilized. For example, when developing solutions to global warming problems, one can envision how societal systems, such as agricultural and industrial systems, interact with the atmospheric system of systems, especially at the tropospheric level. The great import of developing a comprehensive toolbox of decision methodologies and techniques is emphasized by pointing out many current pressing environmental issues, such as global warming and its potential adverse effects, and the widespread pollution of our land, water, and air systems of systems. To tackle these large-scale complex systems of systems problems, systems engineering decision techniques that can take into account multiple stakeholders having multiple objectives are explained according to their design and capabilities. To illustrate how systems decision tools can be employed in practice to assist in reaching better decisions to benefit society, different decision tools are applied to three real-world systems of systems environmental problems. Specifically, the Graph Model for Conflict Resolution is applied to the international dispute over the utilization of water in the Aral Sea Basin; a large-scale optimization model founded upon concepts from cooperative game theory, economics, and hydrology is utilized for systematically investigating the fair allocation of scarce water resources among multiple users in the South Saskatchewan River Basin in Western Canada; and multiple criteria decision analysis methods are used to evaluate and compare solutions to handling fluctuating water levels in the five Great Lakes located along the border of Canada and the United States [53]. Figure 2.12 shows the interactions among energy and atmospheric system of systems.

2.4.2.11 Robotic Swarms as a SoS

As another application of SoS, a robotic swarm is considered by Sahin [54]. Here a robotic swarm based on ant colony optimization and artificial immune systems is considered. In the ant colony optimization, the author has developed a multi-agent system model based on the food gathering behaviors of the ants. Similarly, a multi-agent system model is developed based on the human immune system. These multi-agent system models then were tested on the mine detection problem. A modular micro robot is designed to emulate the mine detection problem in a basketball court. The software and hardware components

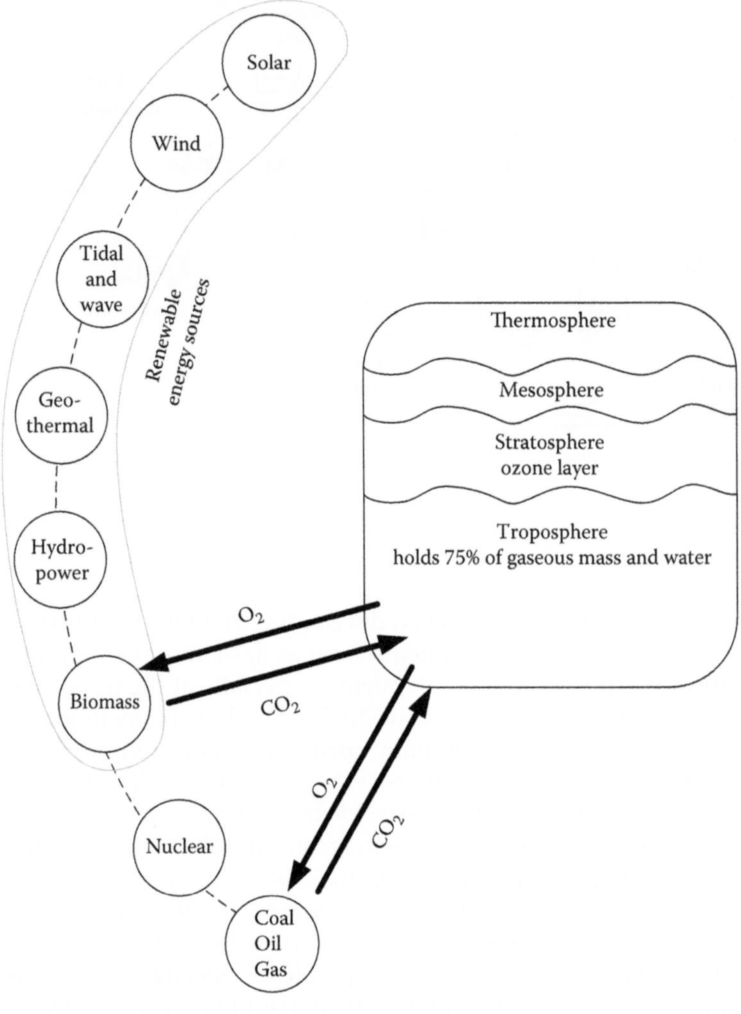

FIGURE 2.12
Interactions among energy and atmospheric system of systems.

of the modular robot are designed to be modular so that robots can be assembled using hot swappable components. An adaptive TDMA communication protocol is developed in order to control connectivity among the swarm robots without the user intervention. Figure 2.13 shows a robotic swarm isolating a mine as a SoS at Rochester Institute of Technology.

2.4.2.12 Transportattion Systems

The National Transportation System (NTS) can be viewed as a collection of layered networks composed by heterogeneous systems for which the Air Transportation System (ATS) and its National Airspace System (NAS) is one part. At present, research on each sector of the NTS is generally conducted independently, with infrequent and/or incomplete

FIGURE 2.13
A robotic swarm isolating a mine as a SoS [54].

consideration of scope dimensions (e.g., multi-modal impacts and policy, societal, and business enterprise influences) and network interactions (e.g., layered dynamics within a scope category). This isolated treatment does not capture the higher level interactions seen at the NTS or ATS architecture level; thus, modifying the transportation system based on limited observations and analyses may not necessarily have the intended effect or impact. A systematic method for modeling these interactions with a system of systems (SoS) approach is essential to the formation of a more complete model and understanding of the ATS, which would ultimately lead to better outcomes from high-consequence decisions in technological, socio-economic, operational, and political policy-making contexts [55]. This is especially vital as decision-makers in both the public and private sector, for example at the inter-agency Joint Planning and Development Office (JPDO) which is charged with transformation of air transportation, are facing problems of increasing complexity and uncertainty in attempting to encourage the evolution of superior transportation architectures. DeLaurentis [55] has addressed this application. Figure 2.14 shows an Entity-Centric Abstraction Model in transportation SoS with two pairs of entity descriptors emerging from the abstraction process: explicit-implicit and endogenous-exogenous [55].

2.4.2.13 Healthcare Systems

Under a 2004 Presidential Order, the U.S. Secretary of Health initiated the development of a National Healthcare Information Network (NHIN), which had the goal of creating a nationwide information system that can build and maintain Electronic Health Records (EHRs) for all citizens by 2014. The NHIN system architecture currently under development will provide a near-real-time heterogeneous integration of disaggregated hospital,

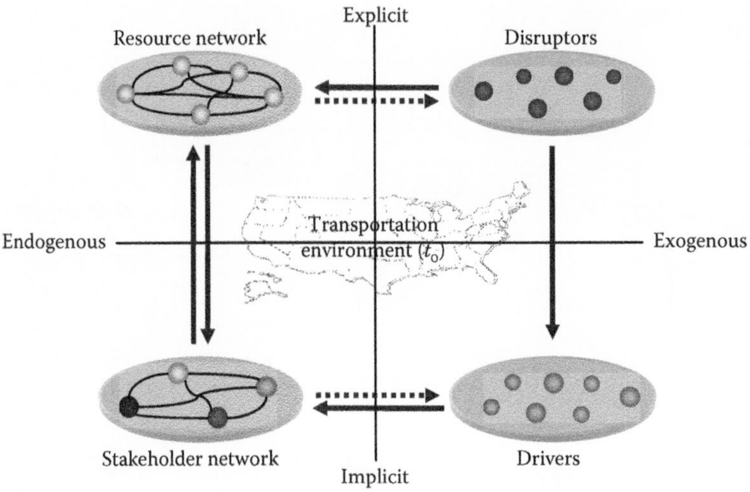

FIGURE 2.14
Entity-Centric Abstraction Model in transportation SoS with two pairs of entity descriptors emerging from the abstraction process: explicit-implicit and endogenous-exogenous [55].

departmental, and physician patient care data, and will assemble and present a complete current EHR to any physician or hospital a patient consults [56]. The NHIN will rely on a network of independent Regional Healthcare Information Organizations (RHIOs) that are being developed and deployed to transform and communicate data from the hundreds of thousands of legacy medical information systems presently used in hospital departments, physician offices, and telemedicine sites into NHIN-specified meta-formats that can be securely relayed and reliably interpreted anywhere in the country. The NHIN "network of networks" will clearly be a very complex SoS, and the performance of the NHIN and RHIOs will directly affect the safety, efficacy, and efficiency of healthcare in the United States. Simulation, modeling, and other appropriate SoSE tools are under development to help ensure reliable and cost-effective planning, configuration, deployment, and management of the heterogeneous, life-critical NHIN and RHIO systems and subsystems [56]. Readers may also see Wickramasinghe et al. [57] on how the healthcare system can be treated as a SoS.

2.4.2.14 Global Earth Observation System of Systems

GEOSS is a global project involving over 60 nations whose purpose is to address the need for timely, quality, long-term, global information as a basis for sound decision making [58]. Its objectives are: (i) improved coordination of strategies and systems for Earth observations to achieve a comprehensive, coordinated, and sustained Earth observation system or systems; (ii) a coordinated effort to involve and assist developing countries in improving and sustaining their contributions to observing systems and their effective utilization of observations and the related technologies; and (iii) the exchange of observations recorded from *in situ*, and open manner with minimum time delay and cost. In GEOSS, the

> SoSE Process provides a complete, detailed, and systematic development approach for engineering systems of systems. Boeing's new architecture-centric, model-based systems engineering process emphasizes concurrent development of the system architecture model and system specifications. The process is applicable to all phases of a system's lifecycle. The SoSE Process is a unified approach for system architecture development that integrates the views of each of a program's participating engineering disciplines into a single system architecture model supporting civil and military domain applications [58].

ICSoS will be another platform for all concerned around the globe to bring the progress and principles of GEOSS to formal discussions and examination on an annual basis.

Shibasaki and Pearlman [58] have presented a detailed description of the GEOSS system, its background, and its objectives and challenges:

> The authors note that the first step is to understand the Earth system—its weather, climate, oceans, atmosphere, water, land, geodynamics, natural resources, ecosystems, and natural and human-induced hazards—is crucial to enhancing human health, safety and welfare, alleviating human suffering including poverty, protecting the global environment, reducing disaster losses, and achieving sustainable development. Observations of the Earth system and the information derived from these observations provide critical inputs for advancing this understanding.
>
> The GEO (Group on Earth Observations), a voluntary partnership of governments and international organizations, was established at the Third Earth Observation Summit in February 2005 to coordinate efforts to build a Global Earth Observation System of Systems, or GEOSS. As of November 2007, GEO's Members include 72 Governments and the European Commission. In addition, 46 intergovernmental, international, and regional organizations with a mandate in Earth observation or related issues have been recognized as Participating Organizations.
>
> The 10-Year Implementation Plan Reference Document of GEOSS (Global Earth Observation System of Systems) states the importance of the Earth observation and the challenges to enhance human and societal welfare. This Implementation Plan, for the period 2005 to 2015, provides a basis for GEO to construct GEOSS. The Plan defines a vision statement for GEOSS, its purpose and scope, and the expected benefits. Prior to its formal establishment, the Ad Hoc GEO (established at the First Earth Observation Summit in July 2003) met as a planning body to develop the GEOSS 10-Year Implementation Plan.
>
> The purpose of GEOSS is to achieve comprehensive, coordinated and sustained observations of the Earth system to meet the need for timely, quality long-term global information as a basis for sound decision making, initially in nine societal benefit are:
>
> 1. Reducing loss of life and property from natural and human-induced disasters;
> 2. Understanding environmental factors affecting human health and well-being;
> 3. Improving management of energy resources;
> 4. Understanding, assessing, predicting, mitigating, and adapting to climate variability and change;

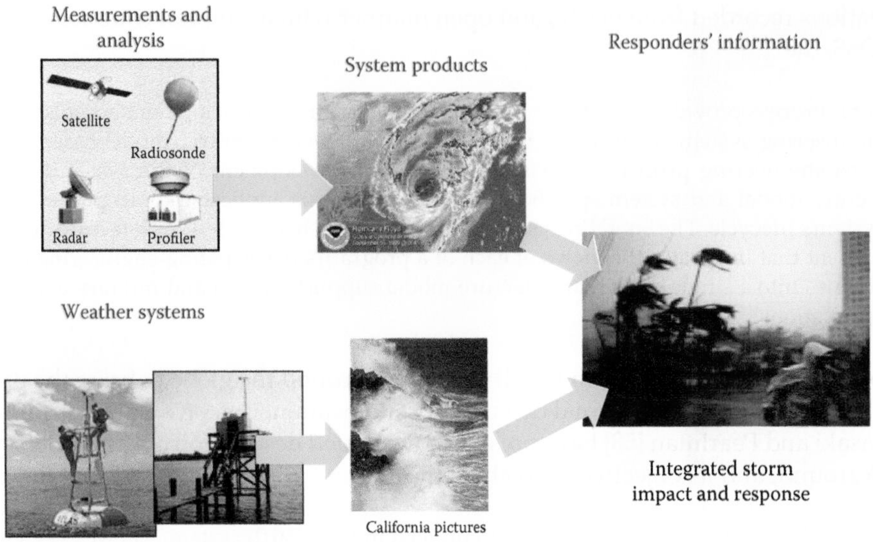

FIGURE 2.15
GEOSS Project Systems in SoS framework. (Courtesy of Pearlman [58].)

5. Improving water resource management through better understanding of the water cycle;
6. Improving weather information, forecasting, and warning;
7. Improving the management and protection of terrestrial, coastal, and marine ecosystems;
8. Supporting sustainable agriculture and combating desertification;
9. Understanding, monitoring, and conserving biodiversity.

Figure 2.15 shows SoS approach in the GEOSS Project.

2.4.2.15 Deepwater Coast Guard Program

One of the earliest realizations of a SoS in the United States is the so-called Deepwater Coast Guard program, shown in Figure 2.16. As seen in the figure, the program takes advantage of all the necessary assets at its disposal, e.g., helicopters, aircrafts, cutters, satellite (GPS), ground station, human, computers, etc.—all systems of the SoS integrated together to react to unforeseen circumstances to secure the coastal borders of the southeastern United States, e.g., the Florida Coast. The Deepwater program is making progress in the development and delivery of mission effective command, control, communications, computers, intelligence, surveillance, and reconnaissance (C4ISR) equipment [59]. The SoS approach, the report goes on, has "improved the operational capabilities of legacy cutters and aircraft, and will provide even more functionality when the next generation of surface and air platforms arrives in service." The key feature of the system is its ability to interoperate among all Coast Guard mission assets and capabilities with those of appropriate authorities both at local and federal levels.

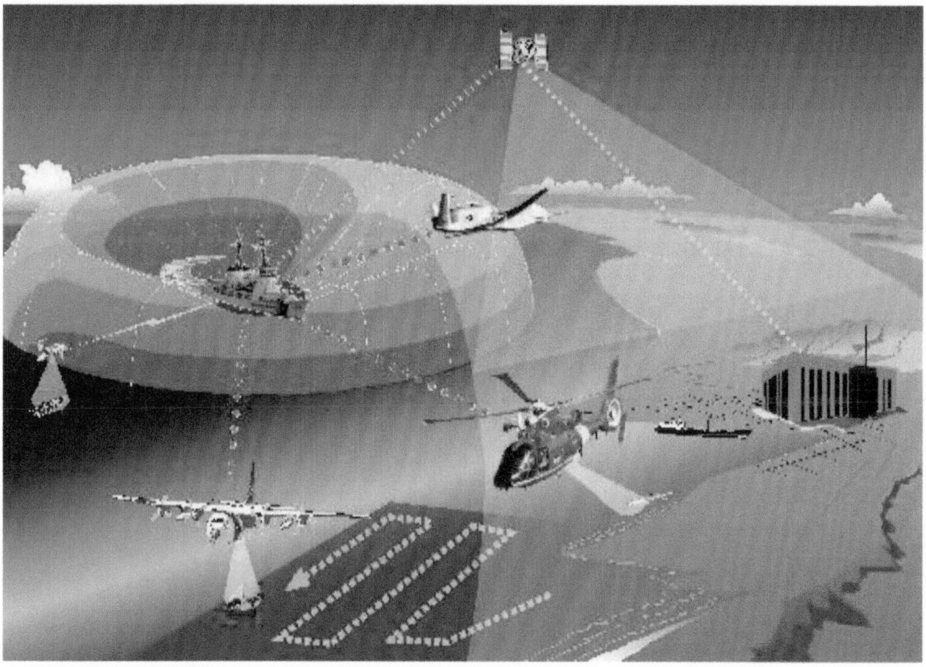

FIGURE 2.16
A security example of a SoS—Deepwater Coast Guard configuration [1].

2.4.2.16 Future Combat Missions

Another national security or defense application of SoS is the future combat mission (FCM). Figure 2.17 shows one of numerous possible configurations of a FCM. The FCM system is

> envisioned to be an ensemble of manned and potentially unmanned combat systems, designed to ensure that the Future Force is strategically responsive and dominant at every point on the spectrum of operations from non-lethal to full scale conflict. FCM will provide a rapidly deployable capability for mounted tactical operations by conducting direct combat, delivering both line-of-sight and beyond-line-of-sight precision munitions, providing variable lethal effect (non-lethal to lethal), performing reconnaissance, and transporting troops. Significant capability enhancements will be achieved by developing multi-functional, multi-mission and modular features for system and component commonality that will allow for multiple state-of-the-art technology options for mission tailoring and performance enhancements. The FCM force will incorporate and exploit information dominance to develop a common, relevant operating picture and achieve battle space situational understanding [60].

See also Dahmann [18] for insights regarding this and other defense applications.

FIGURE 2.17
A defense example of a SoS. (Courtesy of Don Walker, Aerospace Corporation.)

2.4.2.17 National Security

Perhaps one of the most talked-about application areas of SoSE is national security. After many years of discussion of the goals, merits, and attributes of SoS, very few tangible results or solutions have appeared in the national security or other areas of this technology. It is commonly believed that "Systems Engineering tools, methods, and processes are becoming inadequate to perform the tasks needed to realize the systems of systems envisioned for future human endeavors. This is especially becoming evident in evolving national security capabilities realizations for large-scale, complex space and terrestrial military endeavors. Therefore, the development of Systems of Systems Engineering tools, methods and processes is imperative to enable the realization of future national security capabilities" [60]. In most SoSE applications, heterogeneous systems (or communities) are brought together to cooperate for a common good and enhanced robustness and performance. "These communities range in focus from architectures, to lasers, to complex systems, and will eventually cover each area involved in aerospace related national security endeavors. These communities are not developed in isolation in that cross-community interactions on terminology, methods, and processes are done" [1,18,60]. The key is to have these communities work together to guarantee the common goal of making our world a safer place for all. See Dahmann [18] for insights regarding this and other security applications.

2.5 Impacts of System of Systems

In the past two decades or so since the system of systems concept came up, initially in the aerospace and defense industry, a lot of progress has been made. Parallel advances in wireless communication, cyberspace, renewable energy, healthcare, intelligent transportation, driver-less vehicles, artificias intelligence, big data analytics, and machine learning technologies have all contributed to system of systems applications. The impact of system of systems have been great, contributing to a 250 percent increase in the Dow Jones Average in the past decade. What is missing here is the *Engineering of SoS*. Many new theories of areas are depicted in Figure 2.2. System engineering theorists face great challenges in responding to these voids in their field.

2.6 Conclusions

This chapter is written to serve as an introduction to system of systems engineering. The subject matter of this chapter is an unsettled topic in engineering in general and in systems engineering in particular. Attempt has been made to cover as many open questions in both theory and applications of SoS and SoSE. It is the intention that this chapter would be a small beginning of much debate and challenges among and by the readers.

References

1. Jamshidi, M. 2008a (Ed.), *System of Systems Engineering—Innovations for the 21st Century*, Wiley & Sons, Inc., New York, 2009.
2. Jamshidi, M. 2008, *System of Systems Engineering—Principles and Applications*, Taylor & Francis CRC Publishers, Boca Raton, FL, USA, 2008.
3. Crossley, W. A. 2004, "System of Systems: An Introduction of Purdue University Schools of Engineering's Signature Area," *Engineering Systems Symposium*, March 29–31 2004, Tang Center—Wong Auditorium, MIT.
4. Lopez, D. 2006, "Lessons Learned from the Front Lines of the Aerospace," *Proc. of IEEE Iternational Conference on System of Systems Engineering*, Los Angeles, CA, USA, April 2006.
5. Wojcik, L. A., and K. C. Hoffman. 2006, "Systems of Systems Engineering in the Enterprise Context: A Unifying Framework for Dynamics," *Proc. of IEEE International Conference on System of Systems Engineering*, Los Angeles, CA, USA, April 2006.
6. Azarnoosh H., B. Horan, P. Sridhar, A. M. Madni, and M. Jamshidi. 2006, "Towards optimization of real-world robotic-sensor system of systems," in the *Proceedings of World Automation Congress (WAC) 2006*, July 24–26, Budapest, Hungary
7. Abel, A., and S. Sukkarieh. 2006, "The Coordination of Multiple Autonomous Systems using Information Theoretic Political Science Voting Models," *Proc. of IEEE International Conference on System of Systems Engineering*, Los Angeles, CA, USA, April 2006.

8. DiMario, M. J. 2006, "System of Systems Interoperability Types and Characteristics in Joint Command and Control," *Proc. of IEEE International Conference on System of Systems Engineering*, Los Angeles, CA, USA, April 2006.
9. Pearlman, J. 2006, GEOSS-Global Earth Observation System of Systems, Keynote presentation, *2006 IEEE SoSE Conference*, Los Angeles, CA, USA, April 24, 2006.
10. Jamshidi, M., "Theme of the IEEE SMC 2005, Waikoloa, Hawaii, USA," http://ieeesmc2005.unm.edu/.
11. Trans-Atlantic Research and Education Agenda System of Systems, https://www.tareasos.eu/ (accessed April 20, 2017).
12. Clark, J. O., "System of Systems Engineering from a Standards V from a Standards, V-Model and Model, and Dual V-Model Perspective," North Grumman Corp. http://www.ieee-stc.org/proceedings/2009/pdfs/JOC2249.pdf, accessed on February 20, 2017.
13. Wells, G. D., and A. P. Sage. 2008, "Engineering of a System of Systems," *System of Systems Engineering—Innovations for the 21st Century* (M. Jamshidi, Ed.), John Wiley Series on Systems Engineering, New York, 2008.
14. Mittal, S., B. P. Zeigler, J. L. R. Martin, and F. Sahin. 2008, "Modeling and Simulation for Systems of Systems Engineering," *Systems Engineering—Innovations for the 21st Century* (M. Jamshidi, Ed.), John Wiley Series on Systems Engineering, New York, 2008.
15. Pei, R. S., "Systems of Systems Integration (SoSI)—A Smart Way of Acquiring Army C4I2WS Systems," *Proceedings of the Summer Computer Simulation Conference*, pp. 134–139, 2000.
16. Carlock, P. G., and R. E. Fenton, "System of Systems (SoS) Enterprise Systems for Information-Intensive Organizations," *Systems Engineering*, Vol. 4, No. 4, pp. 242–261, 2001.
17. Maeir, M. 1998, "Architecting Principles for Systems-of-Systems," *Systems Engineering*, Vol. 1, No. 4, pp. 267–284.
18. Dahmann, J., "Systems Engineering for Department of Defense Systems of Systems, "*System of Systems Engineering—Innovations for the 21st Century*, Chapter 9 (M. Jamshidi, Ed.), John Wiley Series on Systems Engineering, New York, 2009.
19. Manthorpe, W. H., "The Emerging Joint System of Systems: A Systems Engineering Challenge and Opportunity for APL," *John Hopkins APL Technical Digest*, Vol. 17, No. 3, pp. 305–310, 1996.
20. Azani, C. 2008, "An Open Systems Approach to System of Systems Engineering" *System of Systems Engineering—Innovations for the 21st Century* (M. Jamshidi, Ed.), John Wiley Series on Systems Engineering, New York, 2008.
21. Sage, A. P., and S. M. Biemer. 2007, "Processes for System Family Architecting, Design, and Integration," *IEEE Systems Journal*, ISJ1-1, September, pp. 5–16, 2002
22. De Laurentis, D., C. Dickerson, M. Di Mario, P. Gartz, M. Jamshidi, S. Nahavandi, A. P. Sage, E. Sloane, and D. Walker, "A Case for an International Consortium on System of Systems Engineering," *IEEE Systems Journal*, Volume 1, No. 1, September 2007, pp. 68–72.
23. INCOSE, "Standards of System of Systems Engineering," http://www.incose.org/ChaptersGroups/WorkingGroups/analytic/system-of-systems (accessed February 20, 2017).
24. Stevens Institute of Technology, Body of Knowledge and Curriculum to Advance Systems Engineering (BKCASE), https://www.stevens.edu/news/bkcase-systems-engineering-legacy (accessed February 20, 2017).
25. Pyster, A., D. H. Olwell, T. L. J. Ferris, N. Hutchison, S. Enck, J. F. Anthony, Jr., D. Henry, and A. Squires, "Graduate Reference!Curriculum for Systems Engineering (GRCSE™), version1.0," http://bkcase.org/wp-content/uploads/2014/04/GRCSEv10_Final.pdf (accessed February 20, 2017).
26. Dagli, C. H., and N. Kilicay-Ergin. 2008, "System of Systems Architecting," *System of Systems Engineering—Innovations for the 21st Century*, Chapter 4 (M. Jamshidi, Ed.), John Wiley Series on Systems Engineering, New York, 2008.
27. Sahin, F., M. Jamshidi, and P. Sridhar. 2007, "A Discrete Event XML based Simulation Framework for System of Systems Architectures," *Proc. of the IEEE International Conference on System of Systems*, April 2002.

28. Mittal, S., and J. L. Risco Martin, *"Netcentric System of Systems Engineering with DEVS United Process,"* CRC Press (M. Jamshidi, Series Ed.), 2013.
29. Cloutier, R., M. J. DiMario, and H. W. Polzer. 2008, "Net-Centricity and System of Systems," *System of Systems Engineering—Innovations for the 21st Century* (M. Jamshidi, Ed.), John Wiley Series on Systems Engineering, New York, 2008.
30. Fisher, D. 2006, An Emergent Perspective on Interoperation in Systems of Systems, (CMU/SEI-2006-TR- 003). Pittsburgh, PA: Software Engineering Institute, Carnegie Mellon University, 2006.
31. Keating, C. B. 2008, "Emergence in System of Systems," *System of Systems Engineering—Innovations for the 21st Century* (M. Jamshidi, Ed.), John Wiley Series on Systems Engineering, New York, 2008.
32. Sauser, B., and J. Boardman, "System of Systems management," *System of Systems Engineering—Innovations for the 21st Century,* Chapter 8 (M. Jamshidi, Ed.), John Wiley Series on Systems Engineering, New York, 2008.
33. Jamshidi, M., *"Large-Scale Systems—Modeling, Control and Fuzzy Logic,"* Prentice Hall, Saddle River, NJ, 1997.
34. Nanayakkara, N., F. Sahin, and M. Jamshidi, *"Intelligent Control Systems with an Introduction to System of Systems Engineering,"* Chapter 8, CRC Press, Boca Raton, FL, 2010.
35. Ren, W., and R. W. Beard, *"Distributed Consensus In Multi-Vehicle Cooperative Control,"* Communication and Control Engineering Series, Springer-Verlag, London, 2008.
36. Joordens, M. A., and M. Jamshidi, "Consensus-Based Control of Underwater Robots," *Proc. IEEE SMC Conference,* San Antonio, TX, USA, October 12–14, 2009, pp. 3163–3168.
37. Ren, W., and Y.-C. Cao, "Simulation and experimental study of consensus algorithms for multiple mobile robots and information feedback," Intelligent Automation and Soft Computing, Vol. 14, No. 1, pp. 73–87, 2008.
38. Colandairaj, J., G. Irwin, and W. Scanlon, "Wireless network control systems with QoS-based sampling," *IET Control Theory and Applications,* Vol. 1, No. 1, pp. 430–437, 2007.
39. Office of Secretary of Defense (OSD), "Systems Engineering Guide for System of Systems," Office of Under Secretary of Defense, Washington, D.C., August 2008, http://www.acq.osd.mil/se/docs/SE-Guide-for-SoS.pdf (accessed February 21, 2017).
40. Wilber, F. R. 2007, "A System of Systems Approach to e-Enabling the Commercial Airline Applications from an Airframer's Perspective," Keynote presentation, 2007 *IEEE SoSE* Conference, San Antonio, TX, USA, April 18, 2002.
41. Thissen, W. A. H., and P. M. Herder. 2008, "System of systems perspectives on infrastructures," *System of Systems Engineering—Innovations for the 21st Century,* Chapter 9 (M. Jamshidi, Ed.), John Wiley Series on Systems Engineering, New York, 2008, pp. 257–274.
42. Tien, J. M. 2008, "A System of Systems View of Services" *System of Systems Engineering—Innovations for the 21st Century,* Chapter 13 (M. Jamshidi, Ed.), John Wiley Series on Systems Engineering, New York, 2009, pp. 293–316.
43. Grogan, P. T., and O. L. de Weck, "An Integrated Modeling Framework for Infrastructure System-of-systems Simulation," IEEE Xplore http://ieeexplore.ieee.org/stamp/stamp.jsp?arnumber=6549926 (accessed February 24, 2017).
44. Sridhar, P., A. M. Madni, and M. Jamshidi, "Hierarchical Aggregation and Intelligent Monitoring and Control in Fault-Tolerant Wireless Sensor Networks," *IEEE Systems Journal,* Volume 1, No. 1, September 2007, pp. 38–54.
45. P. Sridhar, A. M. Madni, and M. Jamshidi, "Multi-Criteria Decision Making and Behavior Assignment in Sensor Networks," *IEEE Instrumentation and Measurement Magazine,* Volume 11, No. 1, February 2008, pp. 24–29.
46. Jolly, S. D., and B. Muirhead, "System of Systems Engineering in Space Exploration," *System of Systems Engineering—Innovations for the 21st Century,* Chapter 14 (M. Jamshidi, Ed.), John Wiley Series on Systems Engineering, New York, 2009.
47. Bahsin, K. B., and J. L. Hayden. 2008, "Communication and Navigation Networks in Space System of Systems," *System of Systems Engineering—Innovations for the 21st Century,* Chapter 15 (M. Jamshidi, Ed.), John Wiley Series on Systems Engineering, New York, 2008.

48. Korba, P., and I. A. Hiskins. 2008, "Operation and Control of Electrical Power Systems," *System of Systems Engineering—Innovations for the 21st Century*, Chapter 16 (M. Jamshidi, Ed.), John Wiley Series on Systems Engineering, New York, 2008.
49. Sherick, R., and R. Yinger, "Modernizing the California Grid," *IEEE Power & Energy Magazine*, Vol. 15, No. 2, March/April 2017, pp. 20–28.
50. Y. S. Manjili, "Data analytics-based adaptive forecasting framework for smart energy management in electrical Microgrid," PhD Dissertation, ACE Laboratory, ECE Department, UTSA, San Antonio, TX 2014.
51. Duffy, M., B. Garrett, C. Riley, and D. Sandor, "Future Transportation Fuel System of Systems," *System of Systems Engineering—Innovations for the 21st Century*, Chapter 17 (M. Jamshidi, Ed.), John Wiley Series on Systems Engineering, New York, 2008.
52. Hipel, K., A. Obeidi, L. Fang, and D. M. Kilgour. 2008, "Sustainable Environmental Management from A System Of Systems Engineering Perspective," *System of Systems Engineering—Innovations for the 21st Century*, Chapter 11 (M. Jamshidi, Ed.), John Wiley Series on Systems Engineering, New York, 2008
53. Wang, L., L. Fang, and K. W. Hipel. 2007, "On Achieving Fairness in the Allocation of Scarce Resources: Measurable Principles and Multiple Objective Optimization Approaches," *IEEE Systems Journal*, Volume 1, No. 1, pp. 17–28, 2002.
54. Sahin, F. 2008, "Robotic Swarm as a System of Systems," *System of Systems Engineering—Innovations for the 21st Century*, Chapter 19 (M. Jamshidi, Ed.), John Wiley Series on Systems Engineering, New York, 2008.
55. De Laurentis, D., "Understanding Transportation as a System-of-Systems Problem," *System of Systems Engineering—Innovations for the 21st Century*, Chapter 20 (M. Jamshidi, Ed.), John Wiley Series on Systems Engineering, New York, 2008.
56. Sloane, E. 2006, "Understanding the Emerging National Healthcare IT Infrastructure." *24x7 Magazine*. December 2006.
57. Wickramasinghe, N., S. Chalasani, R. V. Boppana, and A. M. Madni, "Healthcare System of Systems," *System of Systems Engineering—Innovations for the 21st Century* (M. Jamshidi, Ed.), John Wiley Series on Systems Engineering, New York, 2009.
58. Shibasaki, R., and J. Pearlman, "Global Earth Observation System of Systems," *System of Systems Engineering—Innovations for the 21st Century* (M. Jamshidi, Ed.), John Wiley Series on Systems Engineering, New York, 2008.
59. Walker, D. 2007, "Realizing a Corporate SOSE Environment," Keynote presentation, *2007 IEEE SoSE Conference*, San Antonio, TX, USA, April 18, 2002.
60. Wikipedia, "Integrated Deepwater System Program," https://en.wikipedia.org/wiki/Integrated_Deepwater_System_Program (accessed April 4, 2017).

Section II

Theoretical Perspectives

3

DEVS-Based Modeling and Simulation Framework for Emergence in System of Systems

Bernard P. Zeigler

CONTENTS

3.1 Introduction

Emergence as it has been recently treated has both subjective and objective aspects. Objectively, for emergence to be observed, there are changes in the system that surprise the observer. However, such changes may not be significant enough to cause a more fundamental shakeup in understanding. Mittal [1] makes the point that strong emergent behavior results in generation of new knowledge about the system in the form of one or more new abstraction levels and linguistic descriptions, new hierarchical structures and couplings, new component behaviors, and new feedback loops representing previously unperceived complex interactions. Once understood and curated, the behavior returns to the weak form, as it is no longer intriguing, and then can begin to be treated in regularized fashion. Moreover, emergent behavior is likely an inherent feature of any complex system model because abstracting a continuous real-world system (e.g., any complex natural system) to a constructed system model must leave gaps of representation that may diverge in unanticipated directions. Since abstraction is needed to limit the inherently infinite state space to a finite set of tractable and semantically labelled states, Mittal [2] argues that the model's dynamical behavior must account for the elapsed continuous time in the interval between any pair of successive states. This is the case for any computational model whether nominally continuous or discrete. Moreover, high performance computing and big data allow more points to fill the gaps but can never cover the space completely (cf. the mathematics of rationals and reals). Philosophically, following Ashby [3] and Foo and Zeigler [4], the perceived global behavior (holism) of a model might be characterized as: components (reductionism) + interactions (computation) + higher-order effects where the latter can be considered as the source of emergent behaviors [5,6]. In this chapter, we present some features of DEVS that make it the right formalism to use to support the abstraction and observation necessary to deal with emergence in complex systems [7]. We will make the following points:

- DEVS provides the components and couplings for models of complex systems.
- DEVS supports dynamic structure for genuine adaption and evolution.
- Experimental frame supports emergence behavior observation.
- DEVS Markov models support prediction of emergence.
- DEVS enables fundamental emergence modeling.

In the rest of this chapter, we expand on these points, after which we provide some fundamental conditions that may be necessary in order to engineer positive emergence. Finally, we discuss two distinct examples where the application of the concepts and tools discussed here offer clarification and readiness for researcher adoption.

3.2 DEVS Provides the Components and Coupling for Models of Complex Systems

Components and couplings in complex system models must include representation of decision making in natural and artificial environments. DEVS has the universality [8] to represent the discrete (for agent models) and continuous (for natural environments) as well as hybrid (for artificial environments) formalism types needed for adequate complex system model construction. DEVS supports dynamic structure for genuine adaption and evolution. Strong dynamic structure capabilities are needed to specify and flexibly control the changes in components and their coupling to be able to adequately model adaptation, evolution, and emergence in ways that include the possibility of genuine surprise. Recently, a next generation of dynamic structure formalisms has been under development in the DEVS community [9,10,11]. We will briefly review the concepts in the context of an overall framework for modeling and simulation (M&S) based on systems theory and capable of representing existing formulations of dynamic structure [8,12].

3.2.1 Theory of M&S for System of Systems

In systems theory as formulated by Wymore [13,14], systems are defined mathematically and viewed as components to be coupled together to form a higher level system.

Wymore's [14] systems theory mathematically characterizes:

- *Systems* as well-defined mathematical objects characterizing "black boxes" with structure and behavior.
- *Composition of systems*—constituent systems and coupling specification result in a system, called the resultant, with structure and behavior emerging from their interaction.
- *Closure under coupling*—the resultant is a well-defined system just like the original components.

System of Systems (SoS) is a composition of systems, where often component systems have legacy properties, e.g., autonomy, belonging, diversity, and emergence [15]. In this view, a SoS is a system with the distinction that its parts and relationships are gathered together under the forces of legacy (components bring their pre-existing constraints as extant viable systems) and emergence (it is not totally predictable what properties and behavior will emerge.) Here, in Wymore's terms, *coupling* captures certain properties of relevance to coordination, e.g., connectivity, information flow, etc. *Structural and behavioral properties* provide the means to characterize the resulting SoS, such as fragmented, competitive, collaborative, coordinated, etc.

The *Discrete Event Systems Specification* (DEVS) formalism based on systems theory provides a framework and a set of modeling and simulation tools to support Systems concepts in application to SoS engineering [16]. A DEVS model is a system-theoretic concept specifying inputs, states, and outputs, similar to a state machine. Critically different, however, is that it includes a time-advance function that enables it to represent discrete event systems, as well as hybrids with continuous components, in a straightforward platform-neutral manner. DEVS provides a robust formalism for designing systems using event-driven, state-based models in which timing information is explicitly and precisely defined. Hierarchy within DEVS is supported through the specification of atomic and coupled models. Atomic models specify behavior of individual components. Coupled models specify the instances and connections between atomic models and consist of ports, atomic model instances, and port connections. The input and output ports define a model's external interface, through which models (atomic or coupled) can be connected to other models.

Based on Wymore's systems theory, the DEVS formalism mathematically characterizes the following:

- DEVS atomic and coupled models specify Wymore systems.
- Composition of DEVS models-component DEVS and coupling result in a Wymore system, called the resultant, with structure and behavior emerging from their interaction.
- Closure under coupling—the resultant is a well-defined DEVS just like the original components.
- Hierarchical composition—closure of coupling enables the resultant coupled models to become components in larger compositions.

3.2.1.1 Illustrative Example: Turing Machine

To provide a brief primer on DEVS, we illustrate the basic concepts with a well-known concept, the Turing Machine (TM) (see e.g., https://en.wikipedia.org/wiki/Turing_machine). Usually, it is presented in a holistic, unitary manner, but as in Figure 3.1 (top), we can decompose it into two stand-alone independent systems: the TM Control, S1 and the Tape System, S2 [17]. Foo and Zeigler [4] argued that the re-composition of the two parts was an easily understood example of emergence wherein each stand-alone system has very limited power but their composition has universal computation capabilities. Examining this in more depth, the tape system shown in Figure 3.1 (top right) is the dumber of the two—serving a memory with a slave mentality, it gets a symbol (sym) and a move (mv) instruction as input, writes the symbol to the tape square under the head, moves the head according to the instruction, and outputs the symbol found at the new head location. The power of the tape system derives from its physicality—its ability to store and retrieve a potentially infinite amount of data—but this only can be exploited by a device that can properly interface with it. The TM control (Figure 3.1 top left) by contrast has only finite memory but its capacity to make decisions (i.e., use its transition table to jump to a new state and produce state-dependent output) makes it the smarter executive. The composition of the two exhibits "weak emergence" in that the resultant system behavior is of a higher order of complexity than those of the components (logically undecidable versus finitely decidable). The behavior that results can be shown explicitly to be a direct consequence of the component behaviors and their essential feedback coupling—cross-connecting their outputs to inputs as shown by the dashed lines of Figure 3.1 (middle). The resultant is the original concept of Turing, an integral system with canonical computing power (Figure 3.1 bottom). We are going to use this example to discuss the general issues in dealing with such compositions.

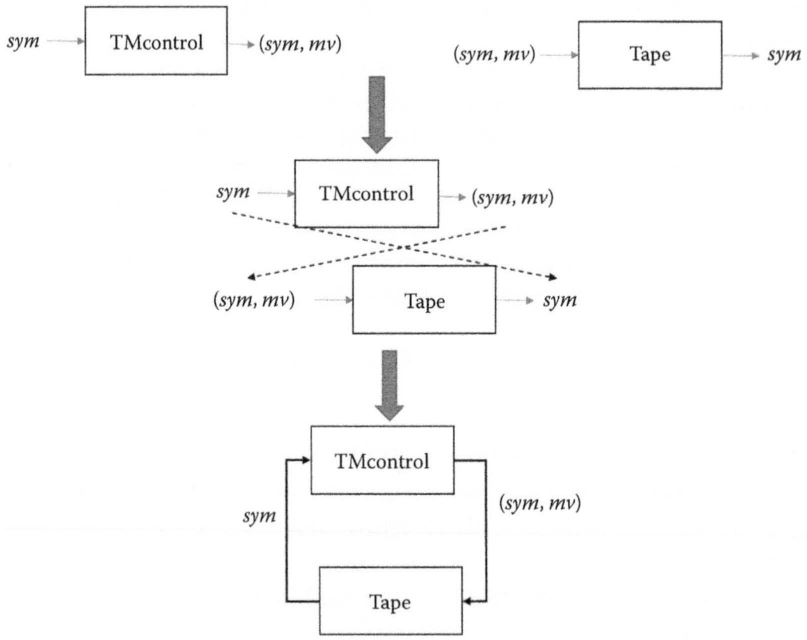

FIGURE 3.1
Composition of components to specify a Turing Machine.

The components in a composition are specified in the DEVS formalism either at the atomic level or recursively, at the coupled model level. To illustrate how DEVS atomic models are described, we provide the tape and control systems as examples.

As presented in [8], a *Discrete Event System Specification* (*DEVS*) is a structure

$$M = \langle X, S, Y, \delta_{int}, \delta_{ext}, \lambda, ta \rangle$$

where

X is the set of input values

S is a set of states

Y is the set of output values

δ_{int}: $S \rightarrow S$ is the *internal transition* function

δ_{ext}: $Q \times X \rightarrow S$ is the *external transition* function

where

$Q = \{(s,e) \mid s \in S, 0 \le e \le ta(s)\}$ is the *total state*

where

e is the *time elapsed* since last transition

λ: $S \rightarrow Y$ is the output function

ta: $S \rightarrow R_0^+$ is the *time advance* function

In the Turing Machine example of Figure 3.1, the tape system and control engine are each *atomic* DEVS. For the tape system, a state is a triple (*tape, pos, mv*) where the tape is an infinite sequence of zeros and ones (symbols), *pos* represents the position of the head, and *mv* is a specification for moving left or right. An external transition accepts a symbol, move pair, writes the symbol in the square of the current head position and stores the move for the subsequent internal transition that executed the specified move. For the control engine, a state is a pair (*st, sym*) where *st* is a control state and *sym* is a stored symbol. An external transition stores the received symbol for subsequent use. An internal transition applies the TM transition table to the (*st, sym*) pair and transitions to the specified control state.

3.2.1.1.1 Tape System

For the tape systems, a state is a triple (*tape, pos, mv*) where $tape : I \rightarrow \{0.1\}$, $pos \in I$, $mv \in \{-1,1\}$ and I is the set of integers; in other words, the tape is an infinite sequence of bits, *pos* represents the position of the head, and *mv* is a specification for moving left or right. An internal transition moves the head as specified; an external transition accepts a symbol, move pair, stores the symbol in the current position of the head and stores the move for subsequent execution. The slot for storing the move also can be null which indicates that a move has taken place.

$$\delta_{int}(tape, pos, mv) = (tape, move(pos, mv), null)$$

$$\delta_{ext}((tape, pos, null), e, (sym, mv)) = (store(tape, pos, sym), pos, mv)$$

$$ta(tape, pos, mv) = 1$$

$$ta(tape, pos, null) = \infty$$

$$\lambda(tape, pos, mv) = getSymbol(tape, pos)$$

where
  $move(pos.mv) = pos + mv$
  $store(tape, pos, sym) = tape'$ where $tape'(pos) = sym$, $tape'(i) = tape(i)$
  $getSymbol(tape, pos) = tape(pos)$

3.2.1.2 TM Control

For the control system, a state is a pair *(st, sym)* where *st* is a control state and *sym* is a stored symbol.

An internal transition applies the TM transition table to the *(st, sym)* pair and transitions to the specified control state. An external transition stores the received symbol for subsequent use.

$$\delta_{int}(st, sym) = (TMState(st, sym), null)$$
$$\delta_{ext}((st, null), e, sym) = (st, sym)$$
$$ta(st, sym) = 1$$
$$ta(st, null) = \infty$$
$$\lambda(st, sym) = TMOutput(st, sym)$$

where
  $TMState(st, sym) = st'$
  $TMOutput(st, sym) = sym'$

3.2.2 Active-Passive Compositions

The interdependent form of the interaction between the TM control and the tape system illustrates a pattern found in numerous information technology and process control systems. In this interaction, each component system alternates between two phases, active and passive. When one system is active the other is passive—only one can be active at any time. The active system does two actions: (1) it sends an input to the passive system that activates it (puts it into the active phase), and (2) it transits to the passive phase to await subsequent re-activation. For example, in the re-composed Turing Machine, the TM control starts a cycle of interaction by sending a symbol and move instruction to the tape system then waiting passively for a new scanned symbol to arrive. The tape system waits passively for the *sym,mv* pair. When it arrives the tape system executes the instruction and sends the symbol now under the head to the waiting control.

Such active-passive compositions provide a class of systems from which we can draw intuition and examples for generalizations about system emergence at the fundamental level. We will employ the modeling and simulation framework based on system theory formulated in [18], especially focusing on its concepts of iterative specification and the Discrete Event Systems Specification (DEVS) formalism. Special cases of memoryless systems and the pattern of active-passive compositions are discussed to exemplify the conditions resulting in ill-definition, deterministic, and non-deterministic as well as probabilistic systems. We provide sufficient conditions, meaningful especially for feedback coupled assemblages, under which iterative system specifications can be composed to create a well-defined resultant and that moreover can be simulated in the DEVS formalism.

TABLE 3.1
Levels of System Specifications

Level	System Specification	Description
4	Coupled	Hierarchical system with coupling specification. System of systems is defined at this level.
3	Atomic	State space with transitions. Behavior and internal structure are specified at this level.
2	I/O Function	State space with defined initial state. Component temporal behavior with respect to initial state is defined at this level.
1	I/O Behavior	Collection of input/output pairs defining temporal behavior is defined at this level.
0	I/O Frame	Defines input/output variables with associated ports over a time base.

3.2.3 Hierarchy of System Specifications

Concepts for organizing models and data for simulation based on systems theory [8,19] and implementable in Model Based Systems Engineering [16] are a necessary background for discussing the *modeling and simulation framework* (MSF). The system specification hierarchy (Table 3.1) provides an orderly way of establishing relationships between system descriptions as well as presenting and working with such relationships. Pairs of system can be related by morphism relations at each level of the hierarchy. A *morphism* is a relation that places elements of system descriptions into correspondence. For example, at the lowest level, two Observation Frames are isomorphic if their inputs, outputs, and time bases, respectively, are identical. In general, the concept of morphism tries to capture similarity between pairs of systems at the same level of specification. Such similarity concepts have to be consistent between levels. When we associate lower level specifications with their respective upper level ones, a morphism holding at the upper level must imply the existence of one at the lower level. The morphisms are set up to satisfy these constraints. The most fundamental morphism, called *homomorphism*, resides at the state transition level.

3.2.4 M&S Framework (MSF)

MSF presents entities and relationships of a model and its simulation as background. The basic entities of the framework are: source system, model, simulator, and experimental frame. As illustrated in Figure 3.2, the basic entities in M&S are the actual system (the "Source

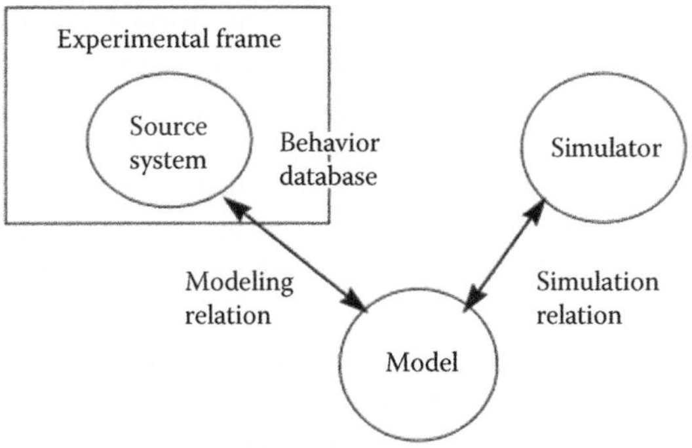

FIGURE 3.2
M&S framework entities and their relationships.

System"), the "Model," and the mechanism for executing the Model (a "Simulator") when the model generates a description of events over time. It is important to understand the relationship between the Model and the Source System (the "Modeling Relation") and the relationship between the Model and the Simulator (the "Simulation Relation").

The adequacy of the model must be judged with respect to the context of use, which includes the domain of input values, the range of output values, and the intent of the user. The experimental frame was originally introduced to operationalize this contextual dependence of adequacy as an object in a status equal to real system, model, and simulator. In the simulation theoretic usage we employ here, the MSF separates models from simulators as entities that can be conceptually manipulated independently and then combined in a relation which defines correct simulation. In addition, the Experimental Frame (EF) defines a particular experimentation process for model input, state, and outcome measurements in accordance with specific analysis objectives. The EF formally recognizes that the intended use (IU) of a model is a fundamental determinant of its validity with respect to the source system. Modular reuse, validity, and executability of simulation compositions are common aspirations among enterprises regularly relying on M&S of System of Systems (SoS) throughout their lifecycles. Such enterprises invest significantly not only in development and experimentation, but also in verification and validation (V&V). The MSF helps clarify many of the issues involved in such activities.

3.2.4.1 The Entities of the Framework

The *source system* is the real or virtual environment viewed as a *source of observable data*, in the form of time-indexed trajectories of variables. The data that has been gathered from observing or otherwise experimenting with a system is called the *system behavior database*. This concept of system is a specification at level 0 and its database is a specification at levels 1 and 2. It is at this level that the conceptual states and I/O trajectories are identified. This data is viewed or acquired through *experimental frames* of interest to the modeler. In *data rich* environments, such data is abundant from prior experimentation or can easily be obtained from measurements. In contrast, *data poor* environments offer meager amounts of historical data or low-quality data (whose representativeness of the system of interest is questionable). The modeling process can direct the acquisition of data to those areas that have the highest impact on the intended uses of the M&S.

In its most general guise, a *model* is a system specification at any of the levels of the hierarchy. However, in the traditional context of M&S, the system specification is usually done at levels 3 and 4. Thus the most common concept of a simulation model is that it is a set of instructions, rules, equations, or constraints for generating I/O behavior. In other words, we write a model with a state transition and output generation mechanisms (level 3) to accept input trajectories and generate output trajectories depending on its initial state setting. Level 3 models usually have a state-space. Such models form the basic components in more complex models that are constructed by coupling them together to form a level 4 specification. The definition of model in terms of system specifications has the advantages of a sound mathematical foundation and a definite semantics that everyone can understand in unambiguous fashion.

Table 3.2 characterizes the level of system specification that typically describes the entities. The level of specification is an important feature for distinguishing between the entities, which is often confounded in practice. Based on this framework, the basic issues and problems encountered in performing M&S activities can be better understood and coherent solutions developed.

TABLE 3.2

Defining the Basic Entities in M&S and Their Usual Levels of Specification

Basic Entity	Definition	Related System Specification Level
Source system	Real or artificial source of data	Known at Level 0
Behavior database	Collection of gathered data	Observed at Level 1 and 2
Experimental frame	Specifies the condition under which the system is observed or experimented with	Constructed at Level 3 and 4
Model	Instructions/algorithms for generating data	Constructed at Level 3 and 4
Simulator	Computational device for generating behavior of the model	Constructed at Level 4

3.3 DEVS Supports Dynamic Structure for Genuine Adaptation and Evolution

So far, the hierarchy of system specifications (Table 3.2) conveys no indication about how changes can be made to the various elements contained therein. The level at which to focus such change is clearly coupled model level at which components and their couplings are specified. Having one such coupled model, we can perform natural operations on its specification, namely, adding or removing components or couplings. Modification of the components themselves calls on the atomic level of specification where component models are described in the basic DEVS formalism. Within this framework, the question then becomes how to initiate such transformations and ensure they are working as intended especially given that such changes must occur while the simulation is executing. The basic idea is that a change in structure is a replacement of one DEVS model by another one, with the latter starting from a state related to the last state of the former.

Recently, Muzy and Zeigler [12] provided a generic definition of dynamic structure system that preserves the DEVS closure under coupling and includes existing formulations [18,20]. The motivation for the generalization was to free up the control of structure change from a single point to allow multiple sources that might be active in a distributed simulation. In this formulation, a coupling is a static relationship between model components that relies on collaboration between components to achieve dynamic changes in the connectivity of their shared couplings. This is a kind of a client-server or a peer-to-peer message communication/synchronization method. In contrast, Park [10] defined a universal coupling specification which is different in that a coupling is a dynamic component that can perform permutation by itself based on its own constraints and requirements. In this approach, component models do not need to possess any additional logic to handle coupling changes. This allows separation of the dynamic structural and coupling permutation logic from the model behaviors unrelated to coupling.

In [10], Park provides a number of types of coupling restructurings inspired by microbiological systems modeling that are potentially applicable to general modeling of emergence. In [11], Steiniger defines a concept of intensional coupling specification distinct from the explicit (extensional) coupling they specify. Biological cellular level modeling similarly motivated this development. We pursue some of the implications of dynamic structure transformation for emergence modeling later with examples in Section 3.7.

3.4 Experimental Frame Supports Emergence Behavior Observation

3.4.1 Positive and Negative Emergence

In a fundamental characterization of emergence, Mittal and Rainey [21] contrast *positive* emergence which fulfils the SoS purpose and keeps the constituent systems operational and healthy in their optimum performance ranges from *negative* emergence which does not fulfill SoS purposes and manifests undesired behaviors such as load hot-swapping, cascaded failures. Their point is that with appropriate concepts, emergence can be harnessed, i.e., emergent behaviors can be controlled and designed by system engineers.

3.4.1.1 Negative Emergence

Negative emergence was clarified in a little recognized report by Mogul [22] which presented a starting point of taxonomies—of emergent *misbehaviors* and their *causes*. The context is software, particularly computer operating systems, but is suggestive generally.

Mogul focuses on system-level misbehavior that does not include bugs, component failures, or "obviously" poor designs such as insufficient resource provision or not using best practice methods, techniques, or algorithms. The causes are "built in" to the system, inherent in its design or implementation—they are features not bugs. Of course, a component failure could trigger a manifestation of such a system-level design failure.

Some misbehaviors are formulated as:

- *Unanticipated resource contention*: Contention for resources that results in wasteful switching overhead such as thrashing in operating systems.
- *Unwanted synchronization or periodicity*: Component systems that were supposed to interact in ergodic unsynchronized fashion fall into synchronized periodic lock step.
- *Deadlock/livelock*: Components cannot move because of circular dependency (deadlock) or get in a loop while still active but where throughput decreases (livelock.) Livelock differs from deadlock in that it is input dependent and can be relieved under reduced load.
- *Phase change*: The behavior of a system changes radically from desired normal modes to undesired failure modes when component behaviors exceed critical thresholds. For example, power drops cause networks to become sparsely connected.

Some causes Mogul enumerates are:

- *Unexpected resource sharing*: Unplanned resource usage patterns arise.
- *Massive scale*: Large numbers of (possibly simple) components interact to give rise to complex global behavior.
- *Decentralized control*: Distributed systems that lack central controls can suffer from incomplete knowledge and delayed information.
- *Unexpected inputs or loads*: Small inputs can excite resonant frequencies and loads can exceed expected limits.

TABLE 3.3
Some Misbehaviors and Causes in Negative Emergence

Misbehaviors/Causes	Unexpected Resource Sharing	Massive Scale	Decentralized Control	Unexpected Inputs or Loads
Unanticipated resource contention	x		x	x
Unwanted synchronization or periodicity		x	x	x
Deadlock/livelock	x		x	x
Phase change		x	x	x

Mittal and Rainey [21] define Emergence Behavior Observers (EBO) that can observe system state transitions and are tuned to catch segments potentially of interest and to record snapshots into memory. A snapshot is defined as an information-set comprising components, and their states, taken at specified moments for specified durations. The concept of Experimental Frame (EF) is a general one which separates the model from the conditions under which it is observed and executed [8] and can include the more specific EBO concept.

We can start associating misbehaviors with causes, singly or in combination, as in Table 3.3:

Developing a comprehensive list of well-defined misbehavior characterizations and their possible causes can support mechanisms for detecting misbehaviors and curing systems of them (akin to debugging). Likewise, were we able to characterize types of positive emergence or emergence in general, we would be able to implement EBO (to be discussed next [23,24]) in infrastructures to offer services for querying component-states and interactions in real time from executing simulations.

3.4.2 Emergence of Topics in Twitter

Emergence in social networks has been defined as the widespread adoption of a norm or convention such as use of a particular taxonomy to categorize topics by using hashtags or certain keywords. Taxonomies created by users are emergent processes whereby users collaborate to categorize topics or things using freely chosen keywords [25]. More relevant is the definition concerning the appearance and popularity of topics when a community adopts a particular means of expression by adopting a hashtag or keywords.

Rather than engage in a lengthy unpacking of such mystery-laden concepts as "irreducible" and "complex," we note that in our architecture there is no mention of reducibility since we are not concerned with the emergence of properties but rather of a new system from existing system components. Further there is no mention of the number and complexity of component systems in the primeval ecology from which the SoS emerges. Thus there is no necessary reliance on large numbers of people, such as in social networks, to collaborate for a topic to emerge. In contrast with large social networks, we consider that a small collection of physicians could form an Integrated Practice Unit (IPU) as an instance of a positive emergence [26]. This brings out the conditions required by the architecture that the components *are initially not bonded to the end state SoS and that they join together to form this system* by relinquishing an existing mode of economic survival in *favor of the new regime the emergent system fosters*. In the case of the IPU, physicians initially owning their own practice band together to form the IPU and give up their own self-directed treatment of patients to conform to the needs of the larger collective with the benefit of increased cost effectiveness and financial viability. None of this is intended to exclude the case of large numbers of component systems or cardinalities of their states as typically being characteristic of the primordial SoS. Indeed, this is typically the case, and drives us toward simulation rather than analysis to understand how the conditions of the architecture can be realized.

3.4.3 Emergence Monitoring and Detection

Although emergence of topics in Twitter is a different kind of emergence, the study of intensity dynamics suggests a parallel to the need for EBOs to detect the emergence in the SoS context. As described by Mittal and Rainey [21], EBOs offer services for querying component-states and interactions in real time from executing simulations that give a heads-up on impending emergence. Figure 3.3 summarizes the methodology employed by Birdsey et al. [25] to develop a user behavior model and employ it to do "what if" simulations of the user community in emergence of topics. The top of Figure 3.3 casts this methodology into a design approach for EBOs. Here we construct a dynamic model of the primordial SoS with the intent to employ it to predict where critical conditions might be developing to implicate emergence of a new SoS. We assume that the primordial SoS has not been well modeled or that such models exist but are too resource-demanding to run in real time. In either case, we employ (big) data-driven techniques to generate dynamic partial models targeted to aspects of emergence of interest and which run fast enough to extrapolate ahead from observed trajectories.

In their ground-breaking paper, Mittal and Rainey [21] integrate the EBO concept into support for sustaining an emergent behavior through control policies. They offer six cybernetic principles to transform EBO outputs into a model that can be understood and scrutinized by humans. These principles are directly applicable to the bottom of Figure 3.3

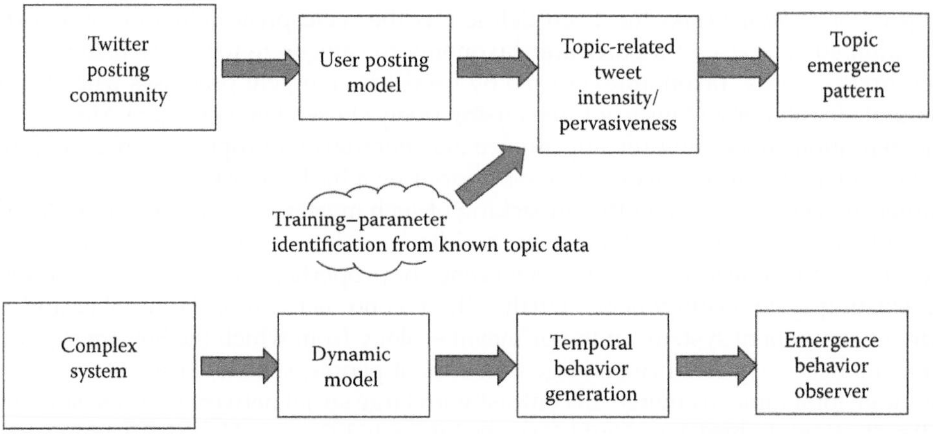

FIGURE 3.3
Detecting topic emergence as a guide to observation of emergence.

suggesting how its Twitter-inspired methodology can be fleshed out and included within a full-capability cloud-based simulation environment.

In the next section we elaborate on the processes in Figure 3.3 to employ Emergence Behavior Observers for emergence prediction.

3.5 DEVS Markov Models Support Prediction of Emergence

Time series of snapshots can be compiled into DEVS Markov models [27] that offer more explanatory and predictive power than the raw inputs. Such compilation can proceed in real time, giving human observers of the simulation a heads-up on imminent events. Figure 3.4 illustrates the role of EBOs and Markov models in real-time emergence monitoring. We start with a coupled model whose components are conceptually partitioned into blocks, and each component's outputs are observed by an experimental frame, called the local EF. This frame samples its block's activity over a (moving) time interval by counting the changes in values of the outputs of the sampled components over the interval. The EF outputs an abstraction of the activity of the block at the end of each interval, e.g., as measured by the scale: zero, low, medium, high, based on a series of increasing thresholds for the total activity. The vector of these abstracted local states forms an abstraction of the global state of the system and is observed by global EBO. By compiling these state

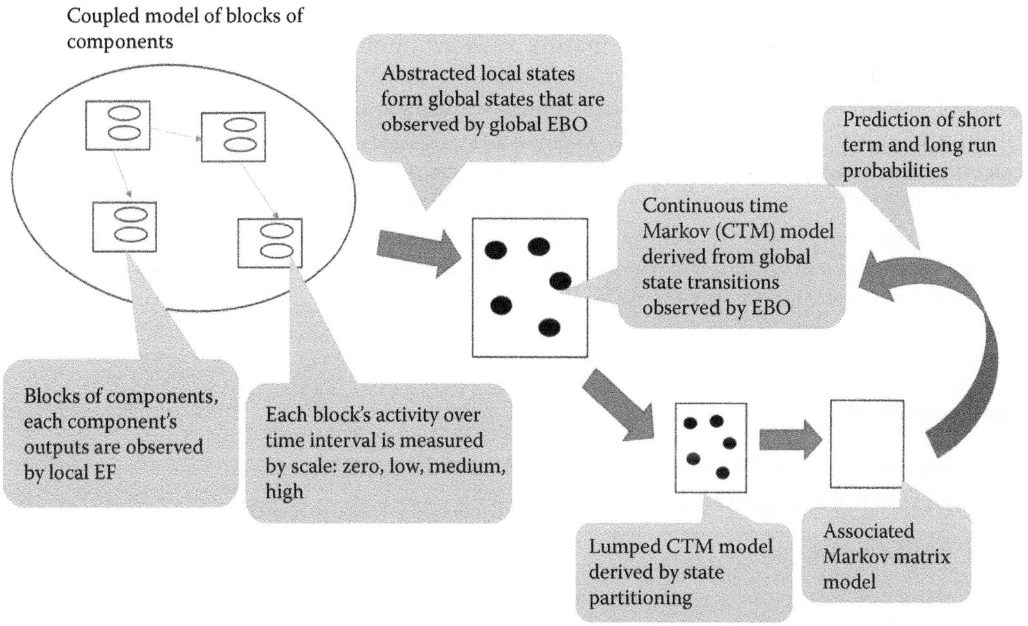

FIGURE 3.4
The role of Emergence Behavior Observation and Markov models in real-time emergence monitoring.

transitions, a Continuous Time Markov (CTM) model [27] is automatically derived. This model is a dynamic structure model as its structure is modifiable depending on the incoming sequence of state labels. Although more compact than the original system, this model may still be too large to comprehend. So it is possible to further reduce this model to a lumped CTM Model derived by state partitioning. The associated Markov chain matrix model can be employed online to provide predictions of short term and long run probability statistics. This enables human observers to understand the underlying cause of behavior deviations and predict when and where they might recur.

3.6 DEVS Enables Fundamental Emergence Modeling

Emergence has been modeled at the fundamental level in terms of formation of new language elements from interaction of components whose behavior does not individually manifest such elements [5,6]. Recently, we have formulated conditions under which compositions of component systems form a well-defined system of systems at the fundamental systems theory level [17]. The formulation states what defines a well-defined composition and sufficient conditions guaranteeing such a result. This formulation offers insight into exemplars that can be found in special cases such as differential equation and discrete event systems [8]. The hierarchy of system specifications forms the framework in which the conditions are stated. At the coupled system level, two requirements were given that can be stated informally as for any global state of a coupled system specification: (1) the system can leave this state, i.e., there is at least one trajectory defined that starts from the state, and (2) the trajectory evolves over time without getting stuck at a point in time. Considered for every global state, these conditions determine whether the resultant of the specification is a well-defined system and if so, whether it is non-deterministic or deterministic. The formalism of iterative specifications, shown to be behaviorally equivalent to the Discrete Event System Specification (DEVS) formalism, supports definitions and proofs of these conditions.

In the following section, we show how implications can be drawn at the fundamental level of existence where the emergence of a system from an assemblage of components can be characterized. We focus on systems with feedback coupling where existence and uniqueness of solutions is problematic.

3.7 Conditions for Positive Emergence and Examples of DEVS Emergence Modeling

3.7.1 Positive Emergence

If negative emergence is an unintended consequence of SoS design then positive emergence should be a deliberate result of design that promotes it. For such an approach we suggest the following tri-layered architectural framework (Figure 3.5).

- *System of systems ecology*: The systems that will become component systems of the SoS already exist as viable autonomous entities in an ecology; however, left unperturbed they would not emerge into the SoS under consideration.

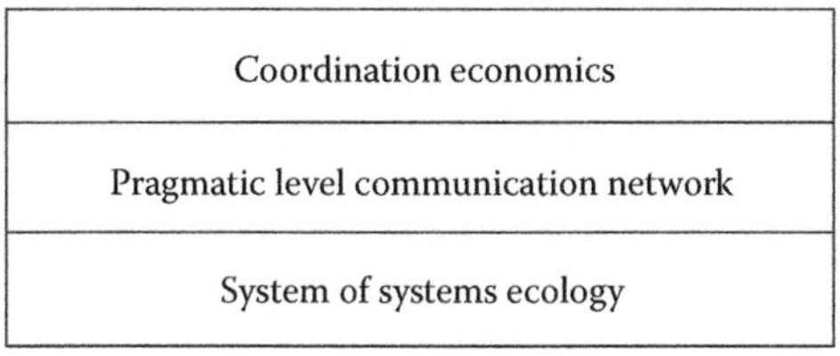

FIGURE 3.5
Tri-layered architecture for positive emergence.

- *Network supporting pragmatic level of communication*: The ability to communicate among putative component systems of the SoS, not only at a technical level but at a level that supports the coordination needed at the next level (this is the "pragmatic" level versus the underlying syntactic and semantic layers [19]).
- *Coordination economics*: A compound term we introduce to refer to (1) the *coordination* required to enable the components to interact in a manner that allows emergence of the SoS with its own purposes, and (2) the economic conditions that enable emergence—the collective benefit that the SoS affords versus the cost to individuals in their own sustainability to contribute to the SoS objectives.

As an analogy, consider an ant colony as a model of a SoS. Here individual ants, members of the colony, are the putative components of the SoS. Prior to evolving collaborative foraging, ant species were part of a natural *ecology*, capable of surviving and reproducing—though not as robustly as in colonies. As the analogue of the second layer, *network supporting pragmatic level of communication*, we point to the communication enabled by the hormone ants secrete while moving. Ants drop pheromone which can be detected by other ants. Trails form from pheromone deposits that accumulate and persist along high production paths and are subsequently followed to reduce search time for food sources. This mechanism of indirect coordination in which the environment is altered by an action that stimulates a next action is called stigmergy. Here it illustrates the *coordination* needed to enable a SoS to emerge, i.e., a colony as a viable autonomous entity with its own survival imperatives. The *economic* cost of pheromone production capability must be overcome by increased survivability (fitness) afforded to individual ants by the colony.

3.7.2 Emergence in National Healthcare Systems

Healthcare in the US can be considered to be in a state characterized by the lowest layer of the emergence conditions stack. It is a system of autonomous systems (consumers, providers, payers) able to survive although at a very high cost to the national treasury. We employ the tri-layered architecture (Figure 3.5) to suggest how healthcare might be reformed to emerge as a SoS which achieves much higher healthcare value, i.e., the same or better outcomes at significantly lower cost [26]. For the *network supporting pragmatic level of communication*, we consider an information technology infrastructure that supports authorized access and exchange of all varieties of health-related information (e.g., patient treatment records and data on treatment results). Such an infrastructure does not exist today—although the internet technology (web, cloud, etc.) provides technical interoperability among hospital and other information systems, common computerized processing of documents at the pragmatic level seems quite far off.

A modicum of such ability to exchange health information is at least necessary to support *coordination* of health-related services needed to improve effectiveness and efficiency and to focus resources where they are most needed. Value-based pricing and purchasing will enable government and insurers to pay for quality of outcomes achieved rather than the services that may or may not have contributed to such outcomes. Determination of value as quality of treatment per dollar of cost and transparency to consumers and payers will enable competition and market forces to weed out poor performers and allow better performers to emerge into a high quality/low cost SoS.

3.7.3 Emergence of Language Capabilities in Human Evolution

Speech acts involve humans speaking to one another in order to change their knowledge or behavior or to otherwise affect them. Questions concern the evolution of motivations and mechanisms. According to Everett [28], humans' interest in interacting with fellow humans differs from that of other apes for reasons such as (1) the discovery of fire and management of fire technology; (2) improving the efficiency of securing and preparing food (accelerated by fire (cooking); and (3) emerging sense of community. Evolution had to prepare early hominids to develop instances of multi-component channels [29,30] in order to communicate with one another, laying the foundation for language, the ultimate communicative tool. The media for conveying messages could have been whistles, humming, or physical signing predating the use of the modern vocal apparatus to produce modern speech sounds. An early speech act may have been spontaneous grunting and pointing vigorously to a serendipitously appearing prey. "Theory of mind"—recognizing others as having the same propensities as oneself—may be prerequisite to shared symbols in messages. The range of employed speech acts is governed by general principles by which all members of the culture recognize that a speech act of a certain kind has taken place. Thus culture sets the channel symbol size and response of senders and receivers. While syntax concerns message formation, and semantics concerns message meaning, pragmatics concerns a listener's ability to understand another's intended meaning. Mutual comprehension requires speaking the same language with a shared ontology. Standard languages and principles for computer agent to agent communications are being researched [31].

Our approach to emergence of language starts with this background, but formulates the problem as an instance of a set of components that must be coupled to form a system with the ability to communicate that does not exist initially. From this perspective, agent communication research assumes existing linguistic competence while evolution speculates on how it could have developed but not as an instance of emergence from a formal M&S perspective.

Assume that both agents have the basic components—motivation and theory-of-mind basis for communication and Shannon audio-visual mechanisms to create signs for encoding/decoding messages. They must set up a channel—first two uni-lateral channels then perhaps one bi-lateral channel. Example: A wants to inform B: "wild boar there, grab it." A needs to get B to attend to this audio-visually encoded message, thus setting up the channel whereby B needs to decode this message and interpret it as information and urge to action. The situation might be depicted as in Figure 3.6 a) where a uni-directional channel emerges from one agent to another. DEVS enables formal modeling—and subsequent simulation—of this dynamic as depicted in Figure 3.6 b) where ports provide potential points of input and output information/energy flow among component models and couplings from specific output ports to input ports determine the actual flow paths.

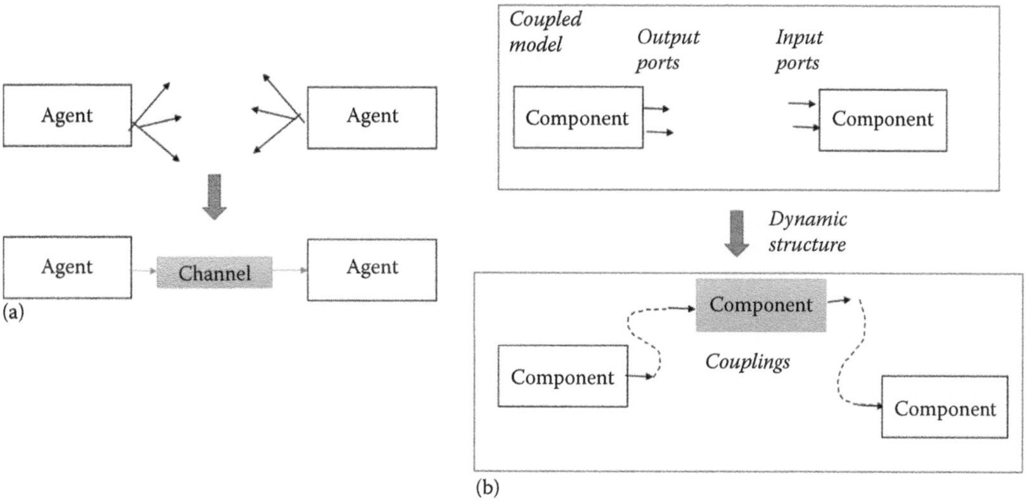

FIGURE 3.6
Hypothetical DEVS modeling of emergence of language capabilities in human evolution. (a) addition of channel for agent-to-agent communication, (b) addition of such a component with dynamic structure.

The addition of a new component and associated couplings is mediated by dynamic structure transformation that can be induced when a model satisfies the requisite conditions.

The next step in emergence of language might have been the addition of a second channel allowing bi-directional communication (or the same channel with duplex capacity). Conceptually, this would be a second instance of the process in Figure 3.6b). However, the establishment of a disciple or protocol for when to speak and when to listen may be more novel and problematic. The problem might be suggested by talking with a friend on the phone (perhaps a successful instance) or pundits from opposite sides of the political spectrum attempting to talk over each other (a negative instance). We outline two perspectives—application domain and system modeling—and show how they may converge to provide a solution.

From the domain perspective, we can recognize that two Shannon components are needed: thinking (e.g., encoding, decoding, interpreting, etc.) and production/transduction (e.g., vocalizing, auditioning, etc.). Limited cognitive capabilities would bound the number and level of activities that could command attention simultaneously (single-tasking more likely than multi-tasking). Thus agents would likely be in either speaking or listening mode, not both. Conversational interaction might ensue as the talker finishes and the listener now has motivation to speak, e.g., "Let's grab it." The discipline of alternation between speaking and listening might then become institutionalized.

From the system modeling perspective, we refer to the problem that components with cyclic (looped)—as opposed to acyclic—coupling face in forming a well-defined system. We see that the passive/active alternation interaction for a cross-coupled pair can provide the conditions for a well-defined system. Identifying listening with passive (although mental components are active, they do not produce output) and speaking with active (output production) we see that alternation between speaking and listening would satisfy the conditions for establishing a well-defined coupled system from a pair of hominid agents.

It is instructive to place the examples of true emergence—national healthcare system and evolutionary language emergence—in the same framework in the form of Table 3.4 along with the process of topic emergence in Twitter [25]. The latter case is informative

TABLE 3.4

Comparison of Conditions for Emergence in Twitter, Healthcare, and Language Emergence

	Healthcare in US	Twitter Topic Dynamics	Language Emergence
Component autonomous systems	Patients, providers, payers	Twitter users (micro-bloggers)	Hominids in small groups
Component interactions	Payers pay for services provided to patients	Tweet messages exchanged among participants	Information exchange for basic survival
Basis for economic survival	Mainly fee-for-service	Independent of posts	Fire management, food provisioning, community bonding
Communication platform at technical levels	Snail mail, phone service, Internet, electronic records	Users access Twitter through the website interface, SMS or mobile device application	Basic Shannon Communication Capabilities = sound production, gesturing, theory of mind, etc.
Communication platform at pragmatic level	Health information exchange (not yet established nationwide)	Online social networking service is a platform to build social relations among users who share similar interests, activities, backgrounds or real-life connections, enables users to exchange 140-character "tweets"	Association of meaning and intended actions with speech acts
Coordination required	Care delivery chains, pathway-based coordination	Platform provides topic naming (via hashtag), following and retweeting mechanisms	Alternation between speaking and listening—one speaker at a time half duplex protocol
Favorable economic conditions	Value-based purchasing, fee-for-performance	N/A (see text)	Greatly accelerated survival potential in employing organized food searches, fire-management technology hand-off

since many of the features and conditions that have been identified in the former cases also apply. However, one that seems glaringly absent is the last one in the table—favorable economic conditions seem not to be a driving force in topic emergence—although indications are that emergence of a business model featuring such a driving force may be imminent.

3.8 Conclusion and Perspective

We have demonstrated the ability of DEVS to provide the concepts and formalisms needed for modeling and simulation of emergent behavior. DEVS has a number of features that enable the representation of systems components and couplings that can change dynamically as needed for genuine adaptation and evolution. We indicated how DEVS Markov models can support prediction of emergence derived from observation of complex system behavior. Finally, we provided some fundamental conditions that may be necessary in order to engineer positive emergence and two distinct examples where the application of the concepts and tools discussed here offer clarification and are ready for researcher adoption.

References

1. Mittal, S. Emergence in stigmergic and complex adaptive systems: A formal discrete event systems perspective. Cognitive Systems Research 21 (2013), 22 39. Stigmergy in the Human Domain.
2. Mittal, S., and Zeigler, B. P. Context and attention in activity-based intelligent systems. ITM Web of Conferences 3 (2014), 03001.
3. Ashby, W. An Introduction to Cybernetics. University Paperbacks, Methuen, London, 1964.
4. Foo, N. Y., and Zeigler, B. P. Emergence and computation. International Journal of General Systems 10, 2–3 (1985), 163 168.
5. Kubik, A. Toward a formalization of emergence. Artificial Life 9, 1 (2003), 41 65.
6. Szabo, C., and Teo, Y. M. Formalization of weak emergence in multiagent systems. ACM Trans. Model. Comput. Simul. 26, 1 (Sept. 2015), 6:1 6:25.
7. Zeigler, B. P., and Muzy, A. Some modeling and simulation perspectives on emergence in system-of-systems, Proceedings of the Modeling and Simulation of Complexity in Intelligent, Adaptive and Autonomous Systems 2016 (MSCIAAS 2016) and Space Simulation for Planetary Space Exploration (SPACE 2016) Pasadena, California, SCS, April 03–06, 2016.
8. Zeigler, B., Praehofer, H., and Kim, T. Theory of Modeling and Simulation: Integrating Discrete Event and Continuous Complex Dynamic Systems. Academic Press, Boston, 2000.
9. Ören, T. I., and Yilmaz, L. Awareness-based couplings of intelligent agents and other advanced coupling concepts for m&s. In SIMULTECH 2015—Proceedings of the 5th International Conference on Simulation and Modeling Methodologies, Technologies and Applications, Colmar, Alsace, France, 21–23 July 2015. (2015), 3 12.
10. Park, S., and Hunt, C. Coupling permutation and model migration based on dynamic and adaptive coupling mechanisms. In Agent Directed Simulation Conference/Spring Simulation Multiconference (Von Braun Center Huntsville, Alabama, USA, 2006), 6 15.
11. Steiniger, A., and Uhrmacher, A. M. Intensional couplings in variable-structure models: An exploration based on multilevel-devs. ACM Trans. Model. Comput. Simul. 26, 2 (Jan. 2016), 9:1 9:27.
12. Muzy, A., and Zeigler, B. P. Specification of dynamic structure discrete event systems using single point encapsulated control functions. International Journal of Modeling, Simulation, and Scientific Computing 05, 03 (2014), 1450012.
13. Wymore, A. W. A Mathematical Theory of Systems Engineering: The Elements. New York: John Wiley & Sons, 1967.
14. Ören, T. I., and Zeigler, B. P. (2012b). System Theoretic Foundations of Modeling and Simulation: A Historic Perspective and the Legacy of A. Wayne Wymore, Simulation, 88(9):1033–1046.
15. Boardman, J., and Sauser, B. 2006. System of Systems—The meaning of "of." IEEE/SMC International Conference on System of Systems Engineering. Los Angeles, CA: IEEE.
16. Mittal, S., and Risco-Martín, J. L. 2012. DEVS Net-Centric System of Systems Engineering with DEVS Unified Process. CRC-Taylor & Francis Series on System of Systems Engineering.
17. Zeigler, B. P., and Muzy, A. Iterative Specification of Input/Output Dynamic Systems: Emergence at the Fundamental Level, Systems 2016, 4(4), 34, Special Issue "Second Generation General System Theory: Perspectives in Philosophy and Approaches in Complex Systems," doi:10.3390/systems4040034.
18. Uhrmacher, A. M. Dynamic structures in modeling and simulation: A reflective approach. ACM Trans. Model. Comput. Simul. 11, 2 (Apr. 2001), 206 232.

19. Zeigler, B. P., and Hammonds, P. Modeling and Simulation-Based Data Engineering: Introducing Pragmatics into Ontologies for Net-Centric Information Exchange, Academic Press, Boston, 2007.
20. Barros, F. J. Towards a theory of continuous ow models. International Journal of General Systems 31, 1 (2002), 29 40.
21. Mittal, S., and Rainey, L. Harnessing emergence: The control and design of emergent behavior in system of systems engineering. In Proceedings of the Conference on Summer Computer Simulation, SummerSim '15, Society for Computer Simulation International (San Diego, CA, USA, 2015), 1 10.
22. Mogul, J. C. (2005, December 22). Emergent (Mis)behavior vs. Complex Software Systems. Retrieved from HP Labs Tech Reports: http://www.hpl.hp.com/techreports/2006/HPL-2006-2.pdf.
23. Zeigler, B. P. A note on promoting positive emergence and managing negative emergence in systems of systems. The Journal of Defense Modeling and Simulation: Applications, Methodology, Technology, pp. 133–136, doi 10.1177/1548512915620580,13,1, http://dms.sagepub.com/content/13/1/133.abstract 2016.
24. Zeigler, B. P. Contrasting Emergence: In Systems of Systems and in Social Networks, JDMS 2016.
25. Birdsey, L., Szabo, C., and Teo, Y. M. Twitter Knows: Understanding the Emergence of Topics in Social Networks, Proceedings of the 2015 Winter Simulation Conference, L. Yilmaz, W. K. V. Chan, I. Moon, T. M. K. Roeder, C. Macal, and M. D. Rossetti, Eds. 2015.
26. Zeigler, B. P., Discrete Event System Specification Framework for Self-Improving Healthcare Service Systems, IEEE Systems Journal Volume: PP Issue: 99.
27. Zeigler, B. P., Nutaro, J. J., and Seo, C. Combining DEVS and model-checking: Concepts and tools for integrating simulation and analysis. J. Process Modeling and Simulation. Special Issue on: "New Advances in Simulation and Process Modelling: Integrating New Technologies and Methodologies to Enlarge Simulation Capabilities," 12(1), pp. 2–15 https://doi.org/10.1504/IJSPM.2017.082781.
28. Everett, D. L. Language: The Cultural Tool, Vintage (December 11, 2012) New York, NY.
29. Zeigler, B. P. Emergence of Human Language: A DEVS-Based Systems Approach, SpringSim, Virginia Beach, VA, April 5, 2017.
30. Shannon, C. E., and Weaver, W. The Mathematical Theory of Communication. Univ of Illinois Press, 1949. ISBN 0-252-72548-4.
31. Chopra, A. K., Artikis, A., Bentahar, J., Colombetti, M. et al., Research Directions in Agent Communication, TIST, vol. V, No. N, July 2012, Pages 1–27. 2016.

4

Sources for Emergence and Development of System of Systems

John M. Colombi

CONTENTS

4.1 Introduction

This chapter examines a biology-inspired model, with extension, to explain man-made systems evolution and emergence within a system of systems (SoS). Most of the material in this chapter was collected, discussed, modelled, and debated over the last several years in an Advanced Topics in System Architecture graduate course (SENG 740) at the Air Force Institute of Technology (AFIT). It was my intent to take systems engineering students beyond their first system architecture course that introduced frameworks, structured and object-oriented methodologies, modelling languages, and process or physics-based executable architecture. This advanced course would target architecturally relevant topics such as system of systems engineering and management, network-centric warfare, interoperability, network dynamics, and graph theory. Effectively, my students and I were attempting to understand and model the material force structure of the U.S. Air Force as a complex adaptive system, together with its evolution and emergence.

A case is made that the concepts of punctuated equilibrium, self-organized criticality (SOC), and the fitness of evolving biological systems within a natural ecosystem apply equally to the variety of interconnected man-made (Defense) systems within the military environment. The Bak-Sneppen (B-S) model is described and simulated and applied to several plausible underlying network structures and replacement (generation) rules. Next, the concept of measuring interoperability between systems is presented. This presentation allows the aggregation of the simple B-S evolutionary model with varying strengths of interdependence between systems (interoperability). In summary, man-made systems within a system of systems should evolve into a self-organized state of fitness that is dependent on varying network structures, modernization rules, and strength of relationships.

4.1.1 Emergence

Numerous definitions of emergence exist within the literature. Standish defines emergence as "the concept of some new phenomenon arising in a system that wasn't in the system's specification to start with" [1]. Hitchins defines it as "the bringing together of different parts in such a way that their dynamic interactions result in properties, capabilities, and behaviors of the whole that are not exclusively attributable to either of the rationally separable parts" [2]. Halloy and Deneubourg similarly state that emergence is "a collective behavior that is not explicitly programmed in each individual but emerges at the level of the group from the numerous interactions between these individuals that only follow local rules" [3].

The definitions of emergence allow for both *planned* emergent properties and future, *unplanned* emergent properties. System engineers deal with planned emergent properties regularly when they realize system-level requirements of a complex system. For example, in the design of an aircraft, a primary requirement might be range in order to successfully fly from point A to point B. This property is emergent due to the fact that the ability to fly over a particular distance cannot be assigned to any particular subcomponent of the aircraft, such as the wing, engine, landing gear, radar, pilot, etc. It only can be assigned to the overall aircraft itself, making it an emergent property. In fact, Abbot affirms that "systems engineering may usefully be understood as the design and development of systems that have desired emergent properties" [4].

However, as one considers today's increasingly complex and interconnected systems and systems of systems, the need for engineering principles to analyze unplanned emergent behaviors is very evident. This is the more common interpretation of Standish's definition—the unplanned phenomenon, that was not part of the design specification. One infamous example of an unplanned emergent behavior was the Northeast blackout of 2003. This series of failures was one of the largest electric grid failures of the time, affecting nearly 10 million people across several U.S. states and Ontario, Canada. A surge of electricity within western New York and Canada resulted in a cascading set of failures. Individual electric sub-grids attempted to protect themselves as the surge tripped protective relays. Further emergent effects from the blackout included looting, a spike in deaths due to heat, thousands of stranded commuters, congested and blocked phone lines, and revenue loss from lost employment.

Other, more positive, emergent behaviors can be the result of mass collaboration [5]. When Jimmy Wales developed his first, online encyclopedia (called Nupedia.com), it was not very successful [6]. Articles took months to get released through a peer-reviewed academic, top-down, centralized, seven-step review process. Nupedia produced 24 articles in several months [5]. When Wales followed with Wikipedia.com, emergence took place on a massive scale. Users could cooperatively produce, review, and edit articles. Compared to Nupedia.com, Wikipedia grew practically without bound. It produced 200 articles the first month, and 18,000 the first year. It now boasts approximately 40 million articles in more than 250 languages. It would have been hard to foresee the magnitude of the emergent capabilities at the start, especially without trust from academic review. Paper encyclopedias have become nearly extinct in the current searchable online digital environment. Similarly, the increase in digital customizable online content has challenged the printed newspaper business model, as online shopping is pressuring the brick-and-mortar retail model. The underlying relationships between systems and services, people and information technology, and their coevolution must be understood.

4.1.2 Fitness and Evolution

The measure of fitness is an important concept in evolutionary models and their application to complex systems. Using Bak-Sneppen's definition [7], fitness reflects the ability and likelihood for a species to survive or mutate. This value can be considered a barrier to mutation and thus reflects the stability of a species. Higher fitness values reflect a species that is more fit in the ecosystem and will be more stable, or less likely to (need to) evolve. Likewise, less fit species have a higher likelihood to evolve or mutate. Their barrier to evolution (finding a more preferred state of fitness) is lower. Fitness is often the basis for selection decisions and the key criterion driving evolutionary dynamics. In addition, since evolutionary systems interact within a dynamic coevolutionary environment, the system fitness will change in time. Perhaps fitness metrics should arise from measuring the system response to interactions with its environment. This idea is expressed by Harman and Clark [8]; any metric used to optimize systems can be considered a good fitness function, provided it has a large search space, low computational complexity, approximate continuity, and an absence of known optimal solutions.

Engineered systems are often designed for cost, schedule, performance, and/or risk. But within a system of systems, other factors may be more appropriate. Maier suggests a design heuristic for these collaborative systems, which he terms "stable intermediate forms" [9]. As the SoS evolves, "the systems within it must achieve intermediate configurations, that are technically, economically, and/or politically stable." These could be considerations for modelling and evaluating an engineered system's fitness within the SoS. Perhaps a system's technical fitness is based on its technology maturity or widespread technology use/adoption. A system is economically fit if it maintains affordable development/acquisition costs, production costs, operating and support or warranty estimates, and/or is profitable. Politically, fitness may be tied to local employment, county or state revenues, or contributions.

4.1.3 Self-Organized Criticality

Self-organized criticality (SOC) is a phenomenon of dynamic systems in which the system components, over time, self-organize and tend toward some critical point or critical state (sometimes called an attractor) which arises from local interactions of components.

This mechanism is considered to explain a wide number of phenomena in physics, chemistry, geology, economics, and social sciences.

A related concept is that of "avalanches," which are ubiquitous and related to self-organized critical states. Consider mountains (in a stable state), until some amount of snow, ice, rock, or mud slides down a face. One can find the notion of a sudden event such as an avalanche (pulling a system out of the stable state) in biology, economics, and geology, with the system returning to a stable state after the event. Of interest, often, is the size of the avalanche, both spatially and temporally—how big (size) and how long it lasts, which is a macrolevel emergent property, caused by the local interactions and detailed dynamics of the overall system.

In DoD systems, "avalanches" in capability need to be considered when upgrading systems. An upgrade to one system has the potential of causing a chain reaction across several systems, making them less capable, less interoperable, and less fit. This is especially true for data-oriented systems. One system that controls a shared database could impact several other applications which assume and rely on its authoritative data and schema.

Assuming self-organized criticality could be a possible mechanism to explain evolutionary and coevolutionary behavior of man-made systems within an SoS, a number of interesting questions can be posed. What are the rules and behaviors that guide man-made systems? How are systems connected and related, both from a technical perspective (interfaces) and also regarding influence, shared utility, or resource dependencies? What are the attributes that describe fitness, both locally and globally? Lastly, and most importantly, toward what critical state is the SoS self-organizing?

4.1.4 System of Systems

Within the Systems Engineering community, there has been great progress in describing, characterizing, and guiding the engineering efforts of system of systems. Many researchers refer to Maier's definition that identifies two main SoS characteristics: "A system-of-systems is an assemblage of components which individually may be regarded as systems, and which possesses two additional properties: operational independence of the components (and) managerial independence of the components" [9]. Interestingly, these characteristics were reduced from his 1996 paper identifying three other characteristics: "evolutionary development, emergent behavior, and geographical distribution" [10]. A variety of authors have refined these characteristics [11,12]. For example, Boardman and Sauser [11] suggest five characteristics of an SoS, easily remembered as "ABCDE," which are Autonomy, Belonging, Connectivity, Diversity, and Emergence. These are strongly related characteristics attributed to biological systems. Biological systems are autonomous, with instincts to survive and procreate. While they don't exhibit shared purpose for belonging, they often share a belonging to the environment or ecosystem. They are highly connected in complex direct and indirect ways, often exhibiting shared dynamics (like predator-prey and food-chain relationships). Biological systems, due to environmental pressures, have resulted in a diverse set of species, in which it is often difficult to precisely characterize subtle evolutionary differences. Lastly, biological systems, through mutations, demonstrate evolution, which leads to emergent behavior.

The increasing complexity and networked capabilities of modern defense acquisition has led to a corresponding increase in the dependencies and interdependencies of acquisition development and production programs [13]. In the current environment, a program may have major impacts on, or be impacted by, other programs based on these dependencies. A program may depend on other programs to ensure funding, develop technology, provide support and resources, or even justify its continued existence. In these cases, the program is dependent upon the other program(s). In cases where two or more programs both depend on each other for their continued functioning, the programs are interdependent.

Various definitions of dependency exist and they differ with the type of dependency. The Merriam-Webster Dictionary defines dependency as "the quality or state of being influenced or determined by or subject to another." Thus, adapting this definition to program dependency results in the quality or state of a program being influenced by or subject to another program.

Consider dependency as a one-way characteristic. Essentially, program A depends on program B, but B does not necessarily depend on A. A clear example of a system-level dependency is exhibited by the Global Positioning System (GPS) satellite constellation. A variety of GPS-aware vehicles use and depend on GPS signals. Some unmanned quadcopters used by hobbyists depend on GPS exclusively and will implement a fail-safe and return to the launch point (RTL) if the signal is lost. Most manned vehicles will switch to a less precise Inertial Navigation System (INS) when GPS is lost. But GPS does not exhibit any technical dependency on any of the vehicles. One could debate whether the GPS constellation would be required if there were no GPS receivers and users. No demand would impact the necessity of the supply, which is a more political and socio-technical relationship.

Program dependencies may manifest for many reasons, aside from technical interface and direct data exchanges. Typically, dependencies are caused by access to program funding, resources (contractor key personnel), support (shared acquisition staff, shared financial management staff, or shared contracting staff), or if program requirements and/or schedule can be influenced by another program.

Factors that affect dependencies may be grouped into two broad categories: direct and indirect. Direct factors represent a relation between two programs with no intermediaries between them. The programs are directly linked to each other, physically or logically, such as with a communication network. Indirect factors are measured between two programs which are connected by their ties to a common third party or entity. An example would be two programs which rely on the same prime contractor, and compete for key technical personnel. While the programs or systems may be largely unrelated, they are still both related to a common organization, and so a level of interdependency may exist between them.

Research by Brown, Flowe, and Hamel has demonstrated that increasing program interdependencies is correlated to schedule problems [13]. Their results provide convincing evidence to suggest that, when considering acquisition performance as an indicator of program "fitness," more interconnected programs are statistically more likely to encounter programmatic problems than their single system counterparts. Another study by the Software Engineering Institute (SEI) attempted to capture and graph program dependencies for major Defense programs [14]. This type of analysis is difficult to capture completely, with so many possible indirect dependencies. The report provides an example of 571 major programs with 1174 documented direct relationships. This results in an average degree of 4.11 dependencies per program. Visually, in Figure 4.1, one notices a majority of the programs (on the edge of the graph) have few links, and fewer in the middle are more connected. This is generally a key characteristic of scale-free graphs.

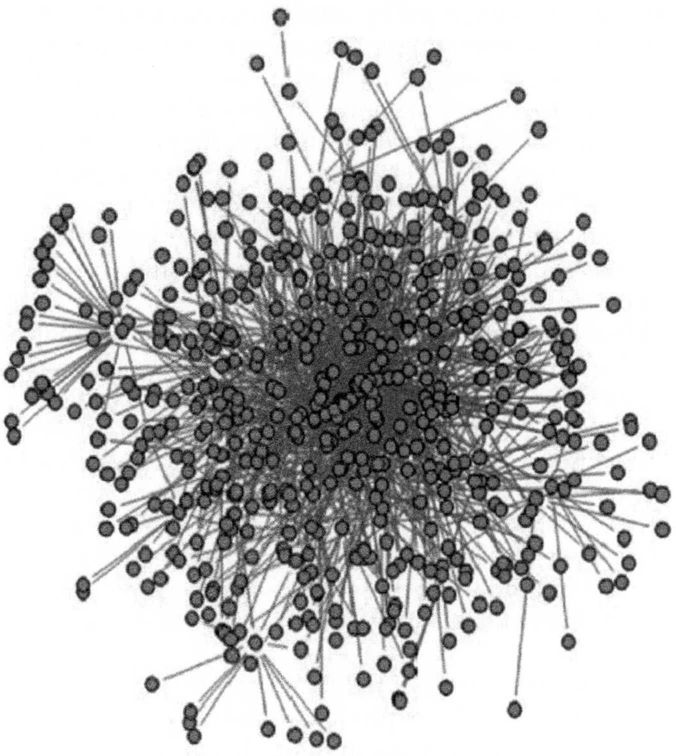

FIGURE 4.1
Network graph of Defense program interdependencies (modified from [16]).

4.2 Making the Case: Aerospace Examples

4.2.1 C-47 Case

The military aircraft C-47 "SkyTrain" was a modified civilian Douglas DC-3 design, and was produced in huge numbers (over 10,000) during World War II. See Figure 4.2. Like the DC-3, designed to satisfy early aviation requirements for a comfortable sleeper transport aircraft, with significant reliability and range requirements at the time [15,16], the C-47 was used for transport of troops, cargo, and wounded soldiers. The escalation of transport requirements due to the war and the reliable design of the DC-3 made the aircraft very valuable. This point is significant; environmental factors can greatly and dynamically affect a system's fitness or a systems fitness barrier.

All man-made technological systems decrease in utility and value over time. For the C-47, this would be due to advances in technology, advances in aeronautical and engine design, and the changing post-war environment. That said, the commercial variants of these aircraft surprisingly find continued use in niche markets such as aerial spraying, freight and skydiver transport, passenger service, and sightseeing. Perhaps this is attributed to the system architecture—the C-47 (and its DC-3 foundation) introduced the hollow fuselage and dual cockpit design; both facilitated many variants and modifications.

FIGURE 4.2
C-47 transport aircraft originally for U.S. Army Air Corps (U.S. Air Force photo).

This case highlights the concepts of fitness and fitness barrier. The DC-3 is considered one of the most successful commercial aircrafts and started the era of profitable airline designs. It was highly valued (fit) for both commercial and military use. Generally, fitness conveys survival and stability. But the overwhelming number of updates and modifications to the C-47 should be noted; nearly 80 variants of the C-47 were designated during and after World War II. Here are just a few examples:

- C-47A, a modified C-47 with a 24-volt electrical system.
- RC-47A, a C-47A modified with photographic reconnaissance equipment.
- C-47B, a C-47 modified with upgraded engines and extra fuel capacity.
- AC-47D, a C-47 modified with side-firing machine guns.
- EC-47D, a C-47D modified with equipment for the Early Warning mission.
- C-47E, a modified cargo variant with space for 27–28 passengers or 18–24 litters.
- C-117A Skytrooper, a C-47B with 24-seat airline-type interior for staff transport use.
- XCG-17, one C-47 tested as a 40-seat troop glider with engines removed.

This fact implies that the C-47 system, or its design, had a lower fitness barrier, which made it relatively easy to evolve, while also being highly fit for the environment. Like all aircraft, the C-47 was eventually surpassed by later generation aerospace systems, the Douglas C-54 (also a commercial variant) and then the C-124 Globemaster II. Man-made systems are assumed to eventually retire and cease their useful life, due to mechanical wear-out, structural fatigue, or end-of-service life, as well as changing environments, requirements, priorities, and technologies.

This case demonstrates the constant evolutionary nature of man-made systems. The scale of change for our man-made systems, according to some futurists, has been increasing exponentially [17]. Consider not just the growth of integrated circuits (Moore's Law), but internet users, speed of computers, speed of communications, memory size, human genome mapping and sequencing, etc. On a scale much faster than that of biological evolution, the complete set of systems, say across all Defense systems, will see constant modifications or replacements. These dynamic changes occur at all scales—on the entire system, on a major segment, on subsystems within the system, on individual parts or suppliers, or on the continual maintenance of software. The combination of a system's fitness (from various perspectives—technical, economical, and/or political) and fitness barrier (ability to evolve or be replaced) will determine system life.

4.2.2 F-117 Case

The aircraft which operationally pioneered stealth technology was the F-117 "Nighthawk." This highly faceted fighter aircraft had its inaugural flight in 1981 and was retired in 2008. A total of only 65 aircraft were built. Shown in Figure 4.3, the single-seat aircraft featured a unique faceted body, specifically designed to scatter radar energy, making the aircraft less prone (invisible) to enemy forces.

FIGURE 4.3
F-117 stealth fighter aircraft (U.S. Air Force photo).

Operationally, the F-117 was publicly acknowledged in 1988 at the time it was first put in combat during the 1989 U.S. military actions in Panama. The aircraft saw remarkable success during the Gulf War in 1991, flying approximately 1,300 sorties and attacking 1,600 high-value targets. This suggests high fitness during that timeframe.

Some claim the F-117 was limited in missions (air-to-ground) and had limited munition storage. It was suggested the system was costly to maintain because of the radar absorbing materials (RAM) painted onto the aircraft shell. Still others claimed the F-117 was not well integrated with other military aircraft. These claims suggest that the overall fitness, including operational considerations, may have been rapidly changing after its Gulf War success. The unique aircraft outer mold-line, giving the aircraft its primary stealth abilities, together with limited internal munition space and no external mounts could have limited the system's ability to evolve like the C-47.

Other less-stealth and older aircraft at the time continued in the U.S. Air Force inventory, such as the F-16 and F-15. These fighter aircraft were designed in the 1960s and produced in larger numbers throughout the 1970s, and they continue to support present day Air Force missions. The F-117 was supposedly replaced by the F-22 Raptor, the fifth-generation fighter which entered service in 2005. A system's fitness and fitness barrier are influenced by many factors. Perhaps stealth technology and adversary radar capabilities quickly changed the DoD ecosystem and newer, more stealthy (low-observable) systems like the F-22, B-2, and the Joint Strike Fighter (F-35) were considered more fit, in one or more critical attributes. This example demonstrates the complex, dynamic, and often hard to quantity, fitness landscape for military weapon systems.

4.2.3 B-52 Case

The last aircraft we examine, different in mission from the cargo hauling capabilities of the C-47 or the air superiority capabilities of the F-117, is the B-52 "Stratofortress." Shown in Figure 4.4, this 1940s World War II bomber is still flying in the U.S. Air Force inventory. While developed as a long-range, heavy bomber, the B-52 can now perform a wide variety of missions, as well as carry a wide variety of both nuclear and conventional ordnance. "In a conventional conflict, the B-52 can perform strategic attack, close-air support, air interdiction, offensive counter-air and maritime operations" [18].

From an SoS perspective, it is interesting to ponder why this system continues to survive and be fit in the modern DoD ecosystem. It came into the Air Force inventory in 1952, built to carry nuclear weapons for Cold War deterrence missions. The political, operational, and especially technological environment has changed drastically since 1952. The B-52 continues operationally despite new bomber aircraft like the B-1 Lancer and the B-2 Spirit. Perhaps this is another example of low fitness barrier to mutation (modernization). The podded engines hanging below the wings have allowed the engines to be upgraded several times. Sources [19] describe upgrades over the last 50 years in communications, avionics, structure, defense systems, munition bay modifications, types of munitions, targeting systems, pilot cockpit systems, etc.

During Desert Storm in 1990-91, nearly 40 years after entering Air Force service, the B-52 was still a preferred aircraft, delivering 40 percent of all the weapons. Current projections suggest the B-52 fleet life could extend to 2040. The lessons learned for lifecycle design should be extensive, in addition to the SoS foundations for evolution and emergence.

FIGURE 4.4
B-52 Stratofortress has survived in the collection of Defense systems for over 60 years.

4.2.4 Case Insights and Integration

These short cases were provided to set a foundation for a model that can be used to understand and predict the evolution of man-made systems. First, all military systems exist within the cycles and patterns of both major warfare (World War II, Korean War, Vietnam War, Cold War, the Gulf Wars, War on Terrorism) and smaller military actions (Grenada, Libya, Panama, Somalia, Haiti, Yugoslavia, etc). Second, in addition to these patterns of warfare, there is also the pattern of Defense spending which greatly impacts new system development and modernization; this second factor may be related to external forces due to patterns of economic prosperity. Third and last, there are internal military organizational and leadership effects. All these effects contribute to the actual or perceived fitness of a system to stay operationally viable.

Consider the interdependence of the three types of systems highlighted in the case studies—fighter, bomber, cargo. These systems, within the Defense force structure, provide "complementary and mutually enabling" capabilities [20]. For example, fighters (like F-117) support air superiority and/or force application and can strike the adversary's air bases and air defense systems. These actions suppress threats on friendly aircraft, such as bombers (like B-52) and/or cargo/mobility aircraft (like C-47). Likewise, bombers also reciprocate and suppress command and control, communications, and air base operations that allow fighters and cargo aircraft to be more effective. Another example is fighter aircraft, providing airspace protection for air refuelling operations, then using the exact services of air refuelling to stay flying within the area of operations. These systems exemplify the synergies (dependencies) across one or more military system of systems.

4.3 Bak-Sneppen Evolutionary Model

For use in a graduate course, the Bak-Sneppen model provides a very simple model of some features describing biological evolution, and clearly captures complex concepts of fitness, punctuated equilibrium (avalanches), and self-organized criticality (SOC). These concepts can be easily demonstrated by students in MATLAB® or Python. However, for systems engineers, it becomes a model to promote discussion on possible analogies to system of systems, man-made system evolution, engineered emergence, and system fitness.

4.3.1 Simulation

In the basic Bak-Sneppen (B-S) model, each species *i* is represented by a single fitness barrier value, B_i. The model for this chapter uses a 1-dimension linear structure, with wrap-around. So, every *N* species is connected to exactly 2 other neighbor nodes—effectively a 2-regular graph, as shown in Figure 4.5.

The two nearest neighbors represent simple interrelation between species within an ecosystem. To initialize the graph, each species is assigned a random fitness barrier from a uniform distribution between 0.0 and 1.0. At each epoch, an evolutionary rule is applied that replaces the weakest (lowest fitness barrier) species with a new random fitness, also chosen from a uniform distribution. This would be analogous to a random mutation or a random replacement. In addition to the weakest species, the two nearest neighbors are also replaced, as the fitness of a species is coupled to other species in the ecosystem. Bak and

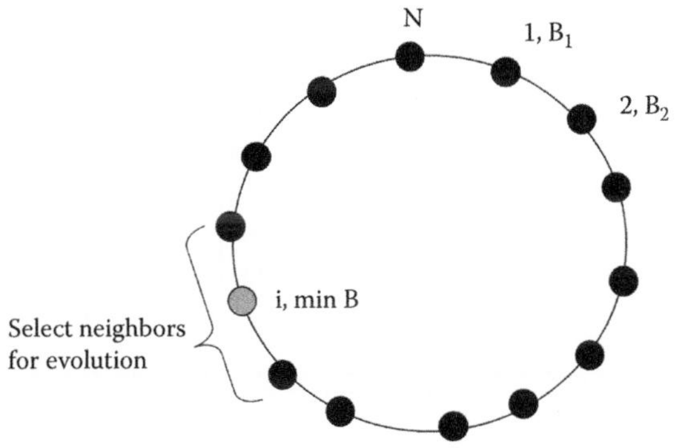

FIGURE 4.5
B-S model with N species. Mutation (or replacement) will happen for weakest species with the minimum fitness barrier.

Sneppen state that "when species make an adaptive move, it changes the fitness landscape of its neighbors." This rule is applied indefinitely.

After a transient phase, the system converges to a stable critical state, where all fitness values are uniformly distributed greater than some critical fitness barrier, $B_i > B_c$. Figure 4.6 shows the initial set of 5000 species and their self-organization with a critical barrier emerging at ~0.67. Figure 4.7 plots the minimum fitness of the species selected for mutation/replacement. This is the instantaneous critical barrier.

Note the "jagged" top edge in Figure 4.7 once the ecosystem reaches the stable state. While it is a simple model, Bak-Sneppen demonstrates SOC with punctuated equilibrium. That edge demonstrates avalanches where the fitness of one or several species has randomly fallen below the critical barrier. This is typical of many complex dynamic systems where some internal or external force affects the fitness of one species, which impacts its neighbors, which then further impacts the neighbors of those neighbors. Eventually, the system returns to a stable state. Figure 4.8 depicts the avalanches (highlighting which species are affected) over many epochs.

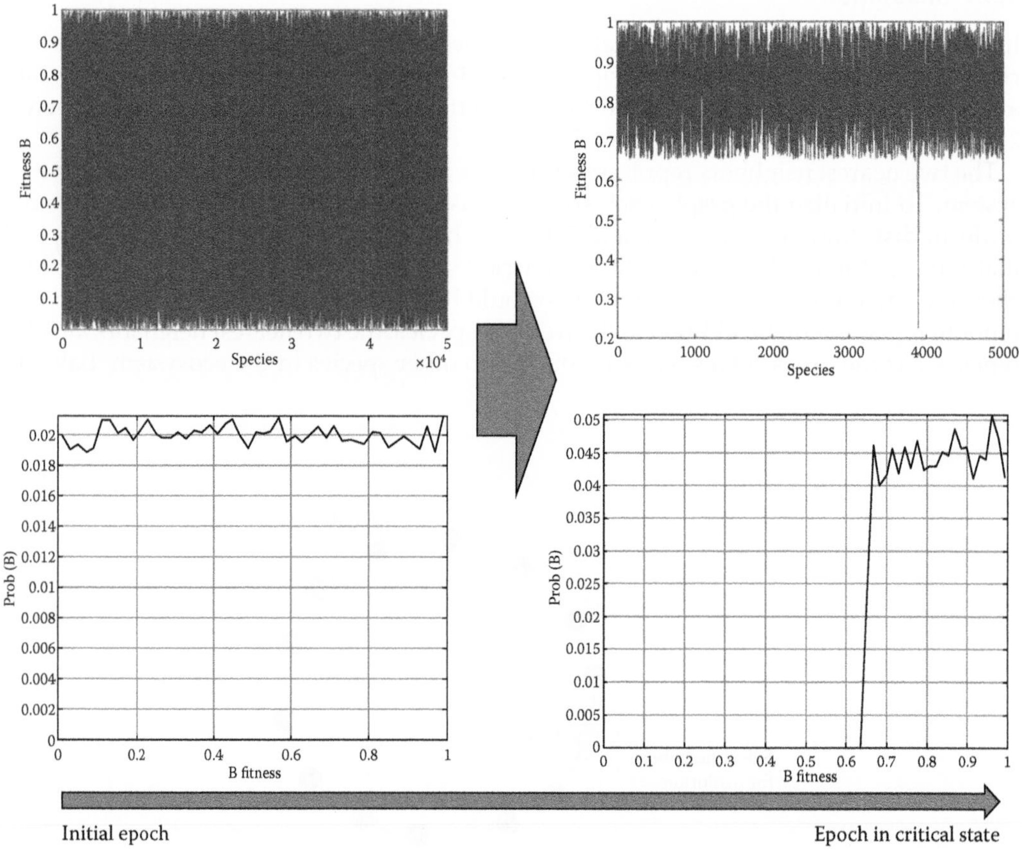

FIGURE 4.6
Bak-Sneppen model starts with Uniform fitness barriers ($0 \leq B_i \leq 1$) across 5000 species in the initial epoch. After convergence, the fitness is still Uniform but a critical barrier emerges ($B_c \sim 0.67$).

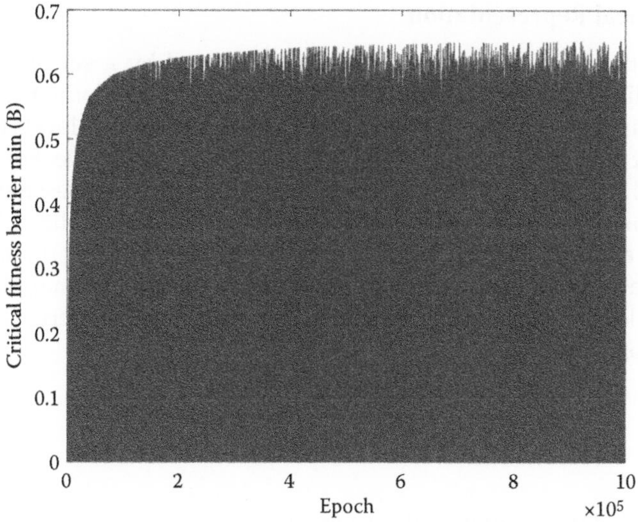

FIGURE 4.7
Evolution/coevolution of the 5000 species critical barrier to approximately ~0.67 fitness, with punctuated equilibrium occurring in the critical state.

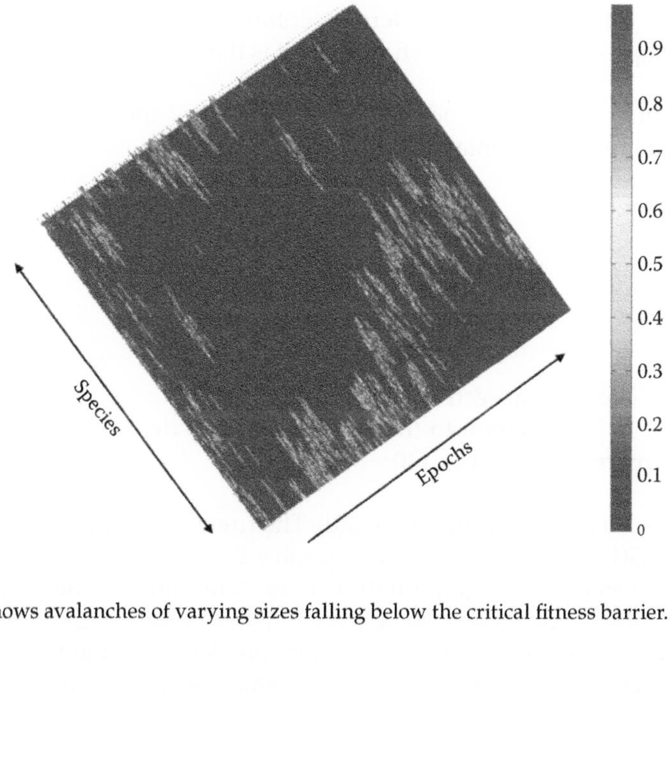

FIGURE 4.8
Dynamic model shows avalanches of varying sizes falling below the critical fitness barrier.

4.3.2 Mathematical Representation

Mathematically, these dynamic networks can be defined by a set of nodes (systems or species), a set of links or edges (relationships), an incidence mapping function that relates each edge to a pair of source-destination nodes, and a set of micro-level rules or behaviors. There is also a set of one or more characters that describes each node. For example, instead of a single overall fitness value between 0 and 1, multiple fitness assessments, such as technical, economic, and political, could be captured for each node. The rules are iteratively and consistently applied at each epoch. One can also assume that the environment may feedback information and control, which affect nodes, edges, the mapping, the character values, or the rule(s). A dynamic network G is first defined, structurally, as the 3-tuple extending notation from [21].

$$G = [S, L, f], \text{a basic graph} \tag{4.1}$$

where

$S = \{s_1, s_2, \ldots, s_N\}$ is the set of N nodes (systems)

$L = \{l_1, l_2, \ldots, l_M\}$ is the set of M links (edges or relations)

f: $L \rightarrow S \times S$, maps links to nodes (source and destination)

One can then a set of attributes describing each node, extending G to

$$G = [S, L, f, X, \Sigma] \tag{4.2}$$

where

$X = \{x_1, x_2, \ldots, x_K\}$ is the set of K characters describing the nodes,

Σ: $\sigma \rightarrow S \times K$, maps systems to their characters; it is the set of node instantiations.

Lastly, an update is made to introduce evolutionary rules and behaviors to capture network dynamics and environmental effects.

$$G(\text{t}) = [S(t), L(t), f(t), X, \Sigma(t)] \tag{4.3}$$

$$G(t+1) = R[G(t); E(t)] \tag{4.4}$$

where

the graph $G(t)$ is captured at each epoch, t,

the graph $G(t + 1)$, its evolution to the next timestep, is driven by a set of rules R and external environment forces, E(t).

The graph will start at some initial state, $G(0)$. The rules will be applied to transition $G(0)$ to $G(1)$, then from $G(1)$ to $G(2)$, and so on. The graph will either converge to a final state, and remain within that state, or the graph will diverge. Alternatively, the graph may demonstrate stochastic convergence where some distribution or other statistic of a random variable eventually converges. This notation in Equations 4.1–4.4 equally applies to network dynamics in social science, physical science, economics, geology, or biology.

System of Systems are described by a time-variant graph, $G(t)$, made up of a diverse variety of systems, operationally and managerially independent. Links between systems can be either technical interfaces or represent relationships/influences between systems or programs. Links map directionally from source to destination. Systems can be described by attributes, such as fitness, or other interoperability measures, and the changing environmental conditions will influence the systems and their fitness, or fitness barriers, differently. The systems evolve through modernization or replacement. The system of systems will not continue to operate unfit systems, or will choose to modernize those with a low fitness barrier. Lastly, an evolutionary rule describes the common managerial behaviors to replace or modernize systems.

4.3.3 Varying the SoS Evolution Rules

Often when they are introduced to the Bak-Sneppen model, students ask, "What if the links don't affect fitness?" or "Can't we make the replacement system always 'better' than before?" The model is so simple that these are easy questions to simulate, while promoting critical thinking regarding man-made systems analogies. First, if one only replaces the weakest species as originally, but without coevolving the neighbors, the entire system converges monotonically to perfection—all species tend to 1.0. This is due to the fact that each epoch will either maintain the current state (a particular species is weakest and will stay weakest), or the system improves (weak species get better than the new weakest). See how the critical barrier slowly converges toward 1.0 in Figure 4.9 a).

Likewise, if the weakest species is replaced with a "better species," but the effects of coevolution with nearest neighbors still exist, the result is the same as the original Bak-Sneppen. The rationale is that man-made systems should always improve and become more fit. As shown in Figure 4.9 (right), self-organized criticality will again evolve to a critical fitness barrier level of ~0.67. Here, the "better species" replacement for a weak

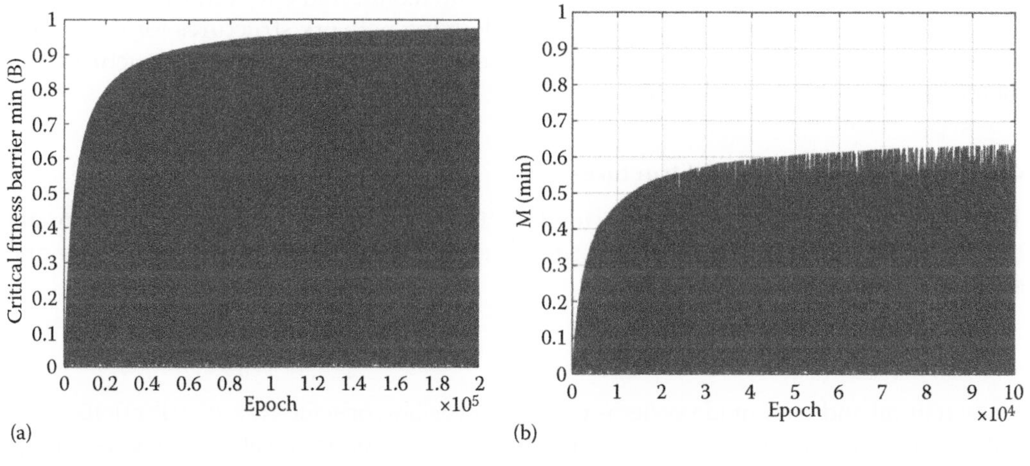

FIGURE 4.9
Evolutionary rule changes. (a) No coevolution, barrier height converges (slowly) to 1.0. (b) Improved replacement with coevolution. Same as original with critical barrier ~0.67.

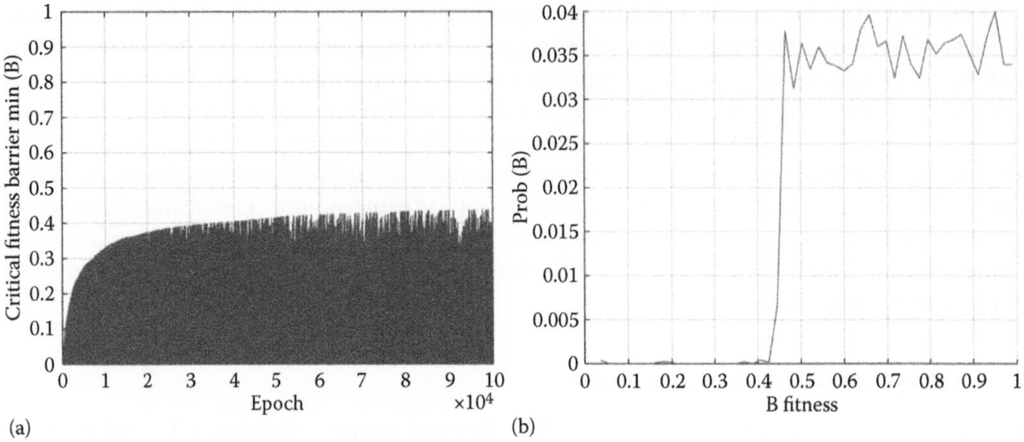

FIGURE 4.10
Evolutionary rule changes. Coevolve the weakest species and its four nearest neighbors. (a) Barrier converges to a lower value ~0.45. (b) The state distribution of the fitness is still Uniform.

(minimum) species with fitness of B_{min} will have the new random fitness drawn from a Uniform distribution with bounds (B_{min} and 1.0), thus, randomly more fit.

One more question that comes from students is, "What if each species were more connected and had more nearest neighbors?" The initial hypothesis is lowering of the critical barrier in the steady state, due to coevolutionary effects from more neighbors. Effectively, there is a greater chance that if a species is strong, one of its weak neighbors will pull its fitness lower. As expected, Figure 4.10 shows the lowering of the barrier to ~0.45. Thus, the connectedness of our man-made systems could have consequences on their overall fitness. Both Defense and commercial industries are making larger expenditures in cyber risk assessments, white hat testing, and resilient designs; perhaps this is a visible result of attempting to make systems more fit in a highly interconnected environment.

Regular graphs are not representative of underlying network structures for man-made systems of systems. The next sections examine other, more appropriate, structures and their effect on network dynamics.

4.3.4 Varying SoS Network Structures

The first change to the Bak-Sneppen model will be to change the link distribution and the mapping of links to nodes. This is addressed in the next section using scale-free and small-world graphs.

4.3.4.1 Scale-Free (Power-Law) SoS

Many natural and man-made systems exhibit power-law, or scale-free, distributions of the links, including biological food webs, neuron networks, communications systems, energy grid networks, and world-wide web hyperlinks [22]. From a graph theory perspective, these networks are described by two properties: large clustering coefficients (which captures high local neighbor interactions) and a small diameter (maximum of the shortest distances between every pair of nodes in the network). Scale-free networks are characterized by a power-law distribution, such that the probability that a node in the graph is connected or linked to k other nodes is defined by $Prob(k) \sim k^{-\alpha}$, where generally $\alpha \sim 2\text{–}3$. For example, Albert et al. [22] mapped one .edu domain and found α to be 2.1–2.45.

One proposed mechanism that explains scale-free networks is preferential attachment [23]. Barabasi and Albert propose that as a network grows in size (nodes), each new node will attach to the existing network with probabilities based on the existing degrees of the nodes. In other words, a new node will prefer (tend) to connect to highly connected node. This creates a network with many nodes having a few connections, and few nodes (called "hubs") that have many connections. See Figure 4.11 for a typical power-law distribution (left), which is often checked for a linear regression on a log-log plot (right).

Using the method of Barabasi, a scale-free network of 5000 nodes is created, iteratively grown from 5 initial nodes, then used with the Bak-Sneppen evolutionary model. Other researchers have examined this behavior [24]. The resulting graph has the same number of average links and total links as the original 2-regular graphs used in all previous examples. A smaller 50 node example graph is provided in Figure 4.12a, as it is difficult to show the link structure across 5000 nodes.

For this experiment, coevolution will still occur, using *all* neighbors of the weakest (minimum) species. At some times during evolution, a node will have a well-connected hub as a neighbor. At other times, a hub itself may become the weakest species and force mutation for all its many neighbors. Each of the nodes is again initialized using Uniform random fitness values between 0 and 1.0. Figure 4.12c shows the evolution of the critical barrier and the final fitness values (b) as well as the distribution of fitness values (d) in steady state.

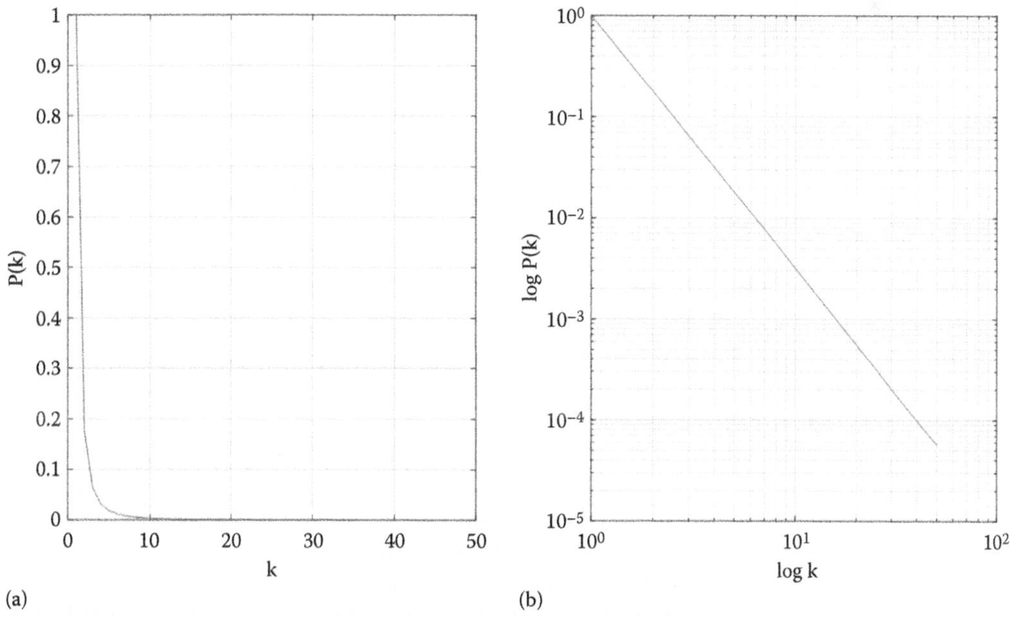

FIGURE 4.11
(a) Scale-free networks are governed by a power-law distribution of number of links, k. (b) Log-log graph is the typical depiction for power-law networks, clearly showing the linear relationship.

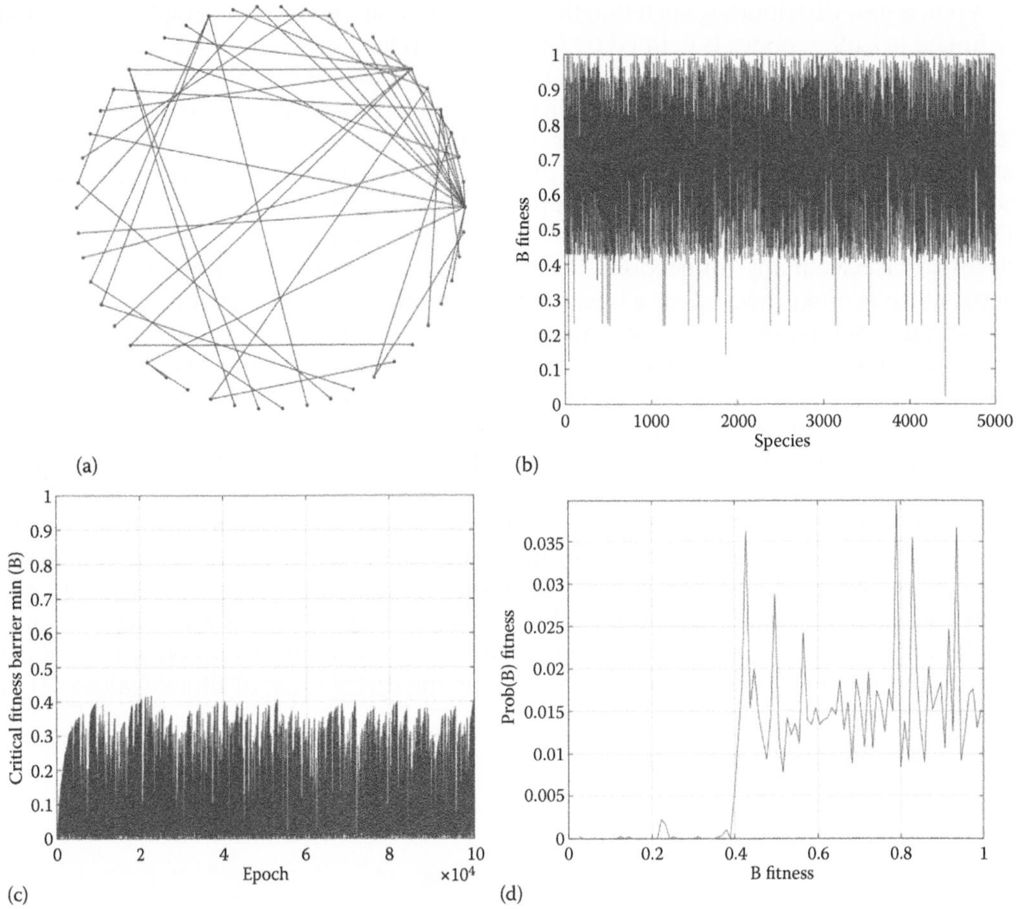

FIGURE 4.12
(a) Example scale-free 50 node network, created by iterative addition of nodes using preferential attachment. (b) Species fitness in steady state. (c) Evolution of the critical barrier occurs quickly and stays low ~0.40. (d) Fitness uniform distribution above the barrier.

4.3.4.2 Small-World SoS

Another class of networks that may also be appropriate for underlying Bak-Sneppen dynamics is a small-world network. Like scale-free networks, many biological, technological, and social networks can be modeled as a small-world, which is a network that has properties somewhere between a regular network (high diameter and characteristics path length), and a random network (high clustering coefficient and low path length). Watts and Strogatz [25] named these types of networks after the small-world phenomenon (i.e., six degrees of separation) attributed to Milgram [26]. This type of network makes large populations feel small, due to the relative small path lengths.

Dynamical models which are based on small-world networks show one interesting property—fast signal-propagation. For example, an infectious disease will spread more quickly across a small-world network than across a comparable regular graph. This could have analogous consequences for man-made systems within a system of systems. Changes to one system's fitness, for any reason, may quickly influence the entire SoS, spreading through local neighbors and simultaneously influencing more remote systems.

For this next example, the Watts-Strogatz random rewiring procedure is used to create the underlying SoS structure [25]. This procedure creates a small-world by starting with a k-regular graph (Figure 4.13 a), then for each link, it uses a probability, p, to randomly rewire the link to a random node. Figure 4.13 b-d) shows the effects for a small 50 node graph, as the rewiring probability is increased from $p = 0$ (no change to the regular graph) toward $p = 1$ (completely random graph).

Results were unexpected after running the Bak-Sneppen evolutionary rule for a 5000 node scale-free network. See Figure 4.14 where the critical barrier is checked and the fitness values plotted at steady state.

The density of the fitness values in steady state are also less uniform, with probability spikes that vary. This may be due to the number and size of hubs within each of the 5000 node simulated system of systems. Moreno and Vasquez suggest that the barrier threshold should decrease as N (number of nodes) increases on a scale-free network [24]. Simulated results confirm this behavior. The threshold barrier is ~0.55 for a scale-free network with

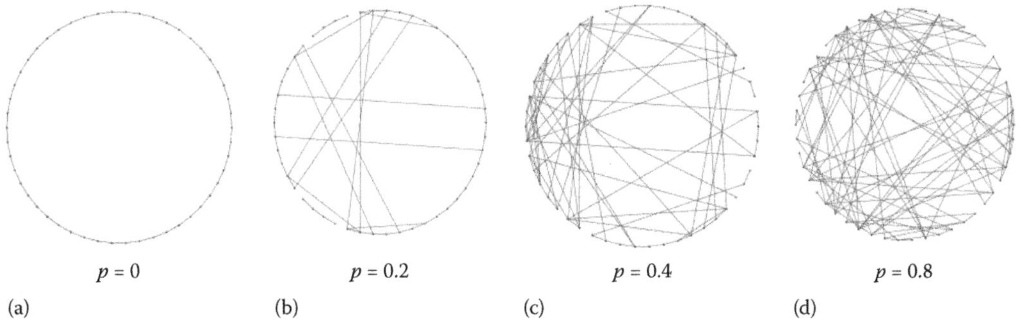

FIGURE 4.13
Application of Watts-Strogatz rewiring method to create small-world graphs with increasing amounts of rewiring. (a) Baseline 50 node regular graph. (b) Baseline after rewiring ($p = 0.2$). Shortcuts begin to emerge. (c) Baseline after rewiring ($p = 0.4$). Characteristic path length is further lowered. (d) Baseline after rewiring ($p = 0.8$) resembling a random graph.

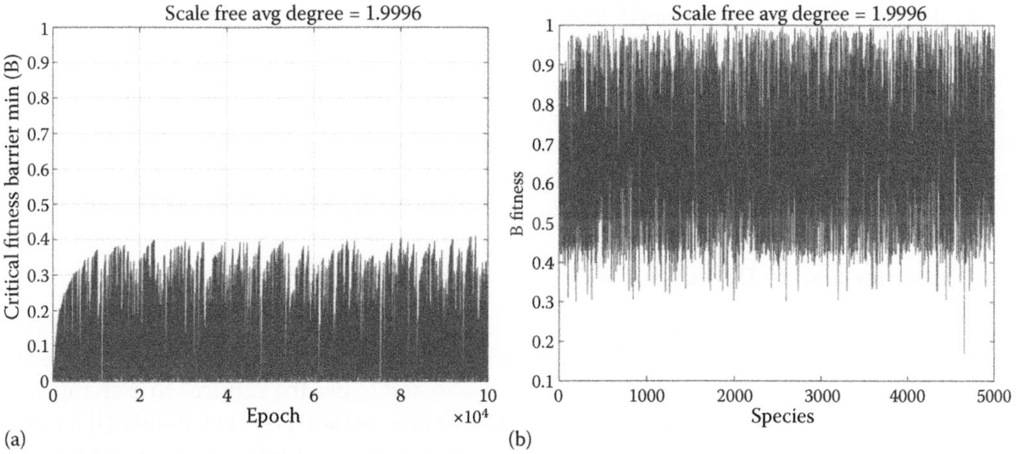

FIGURE 4.14
(a) Scale-free network shows rapid convergence to a critical state, with a barrier ~0.42, with significant punctuated equilibrium behavior. (b) Fitness values in steady state y, with critical barrier less defined than past models.

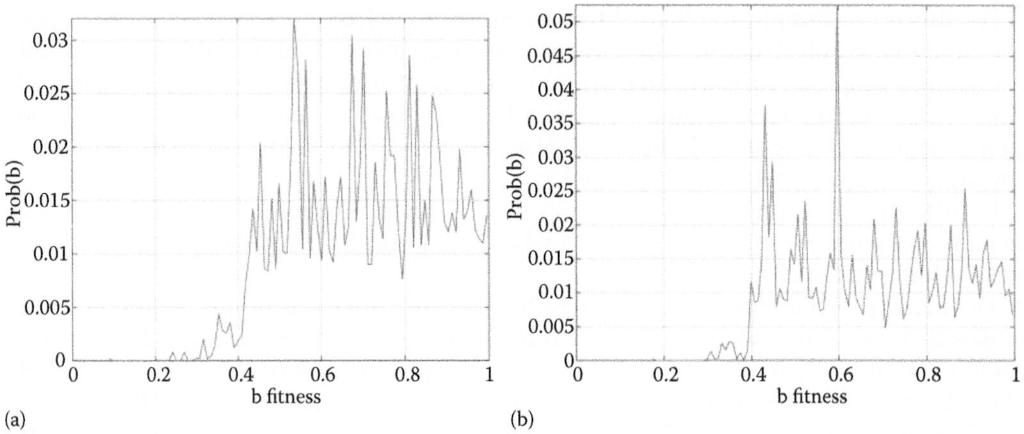

FIGURE 4.15
Scale-free graph structure changes the steady state distribution of fitness values, while still self-organizing to a critical barrier of ~0.42. (a) Scale-free graph with a critical barrier of ~0.42, but less uniform fitness density in steady state. (b) Another trial of a scale-free graph with similar barrier, but highly variable fitness density.

50 nodes, drops to ~0.46 for 1000 nodes, then to ~0.42 for 5000 nodes, and finally 0.38 for 10,000 nodes. Again, this is due to the existence of hubs. When a hub is selected for mutation/replacement, it coevolves lots of neighbors. Likewise, a neighbor of a hub is likely to be selected for mutation/replacement. If a neighbor is selected, it mutates the hub, which leaves the hub more vulnerable to selection. The cycle continues, resulting in a lower barrier threshold. Figure 4.15 depicts two different trials of the steady state fitness distribution, each providing a critical fitness barrier of approximately ~0.42.

4.4 Interoperability

Generally, systems engineers need to ensure interoperability across external interfaces, such as between systems within an enterprise or a system of systems (SoS). While there may be a vast array of internal interfaces, those are generally under the design control of the system engineer. Within design and operation of military systems, the external interfaces have historically been a challenge. There are often many complexities and/or constraints, different contractors, different time scales, inclusion of both new and old technologies, all of which must work together across Service organizational boundaries.

4.4.1 Interoperability Definition

One of the earliest definitions of interoperability [27] from DoD Directive 4630.5 (first published in 1977) is one of the most popular: "the ability of systems, units, or forces to provide services to and accept services from other systems, units, or forces and to use the services so exchanged to enable them to operate effectively together." This definition makes a few key points: (1) it infers that interoperation occurs between many types of entities (e.g., systems, units, or forces), (2) it describes interoperability as a relationship between these entities, (3) it implies that interoperation is a directional relationship between a "provider" and an "acceptor," and (4) it explains the reason why interoperability matters, so as to enable more effective operation.

4.4.2 Interoperability Types and Measurement

Over the last two decades, dozens of papers have been published specifically on interoperability measurement. One can classify each as a leveling or non-leveling method. Leveling interoperability assessment methods are largely based upon the software capability maturity model (CMM) and represent maturity by thresholds of increasing interoperability capability. Non-leveling interoperability assessment methods are a much more diverse group and, as a whole, generally pre-date the leveling methods. They also are specialized to a particular type of system or interoperability (technical, communications, data, etc.).

Work by Ford, Colombi et al. proposed a general method to measure interoperability using one or more characters that describe the relationship between systems [28].

Assume a set of N systems $S = \{s_1, s_2, \ldots, sN\}$ across the SoS, described by a set of k characters $X = \{x_1, x_2, \ldots, x_k\}$. These characters can be highly varied, such as morphological, physiological, interfacial, ecological, and can be discrete or stochastic [29], unidirectional or bidirectional, or can be restricted to Boolean. Ideally, the set of characters chosen are natural and diagnostic, and related to the type of interoperability measurement that is to be undertaken. The characters could be any mixed set of Boolean, rational, or reflect levels.

Once systems, their interoperability characters, and the values of those characters have been identified, then specific systems can be instantiated with characters and their values. In [28], the authors suggest a modified Minkowski similarity measure to quantify interoperability when the characters reflect levels. Therefore, given any number of factors or characters that express the degree to which systems interoperate (and/or provide services to one another), one can evaluate their interoperability measurement. In the next section, this measurement is used to weight the local fitness values in the Bak-Sneppen evolutionary model.

4.5 SoS Evolution with Interoperability

Combining the concepts from the last few sections, one can use interoperability measurement to explicitly and directionally modulate the fitness values of systems across the SoS. Interoperability, then, is the extent to which systems interact. This approach introduces aspects of local and global fitness across the SoS.

4.5.1 Local and Global Fitness

Several authors have suggested complex dynamic systems behaviors are associated with varying scales. In researching SoS evolutions, Davendralingam and DeLaurentis [30] remind the reader that "typical engineering efforts focus on locally incremental developments and do not explicitly consider their effects within the larger context of the original SoS architecture." Thus, local modernization, together with system interdependencies and the inherent technical and programmatic uncertainties of acquisition, make for unknown global impacts. Further, Holland introduced a taxonomy of four types of emergence [31] that differ in their local and global effects. These four types are classified as:

- Type 1 Nominal: local system to global SoS feed-forward emergence effects
- Type 2 Moderated: global SoS to local system (top-down) feedback mechanisms, either negative (stable) or positive (unstable)

- Type 3 Multiple: global SoS to local system with both positive and negative feedback
- Type 4 Evolutionary: local evolutionary changes in rules, behaviors, and attributes due to local and/or global influences

To address these different scales, the Bak-Sneppen model can be extended with both local and global fitness. One researcher has examined this extension in economics, where business firms are linked by "mutual business interests" [32].

4.5.2 Bak-Sneppen with Interoperability Measurement

Consider a global fitness landscape which takes the current (local) fitness barrier from the Bak-Sneppen model and adds weighted fitness effects from the local neighbors. Bartolozzi et al. drew inspiration from economics, as the dynamics of different firms are correlated [32]. Their modification "takes into account the feedback from the environment on the single [system]." As before, assume a one-dimensional array of N nodes, with wraparound, and only interaction from nearest neighbors. Initialize the set of nodes to have a Uniform random local fitness attribute. Also, initialize a set of link weights which reflects the strength of the relationships between nodes. These could be considered an interoperability measurement. Define a global fitness F_i for species i,

$$F_i = \lambda_{i-1,i}B_{i-1} + \lambda_{i+1,i}B_{i+1} + B_i \tag{4.5}$$

where

B_i is the local (self) fitness of node i,
B_{i-1} and B_{i+1} are the local (self) fitness of i's two neighbors, and
$\lambda_{i-1,i}$ and $\lambda_{i+1,i}$ are the directional weights toward node i, reflecting fitness contribution.

The Bak-Sneppen evolution rule is now changed to mutate the *global* weakest species. This is accomplished by first redrawing a Uniform random value for the local fitness, $B_{min.}$ Since this new model explicitly captures the strength of the relationships between systems, if the weakest system evolves or is replaced, then all four relationships' values with the nearest neighbors (two in and two out) must also be redrawn. Thus, when species k is selected as the globally weakest, the evolutionary rule will redraw the following five parameters:

B_k: local fitness

$\lambda_{k-1,k}$ and $\lambda_{k+1,k}$: nearest neighbor contribution of their local fitness toward species k

$\lambda_{k,k-1}$ and $\lambda_{k,k+1}$: species k contribution of its local fitness toward nearest neighbors

Figure 4.16 shows the evolution of the global critical fitness barrier. Recall that fitness is the weighted combination of three local fitness values, and thus varies between 0 and 3.0.

In Figures 4.17 and 4.18, both the local and global fitness values and their distributions are shown respectively. These have been shaped uniquely by the addition of interoperability measurement. There is continued self-organized criticality, with a tendency toward 1.0 for the local fitness and a clear critical barrier of ~1.7 for global fitness.

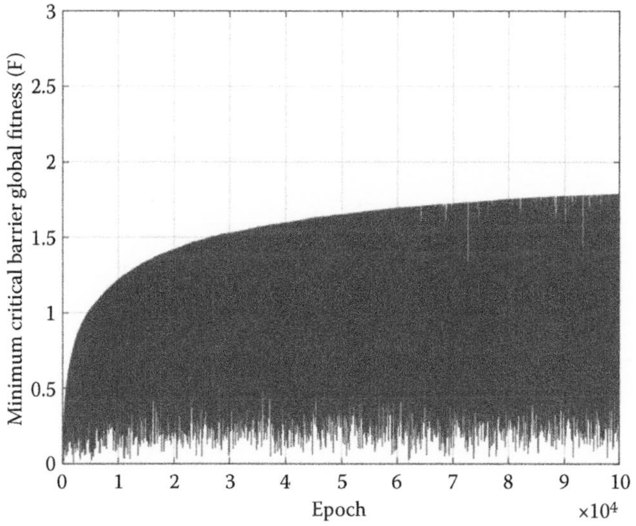

FIGURE 4.16
Evolution of critical global barrier. The graph evolves to a global fitness below 2.0, with relatively few periods of punctuated equilibrium.

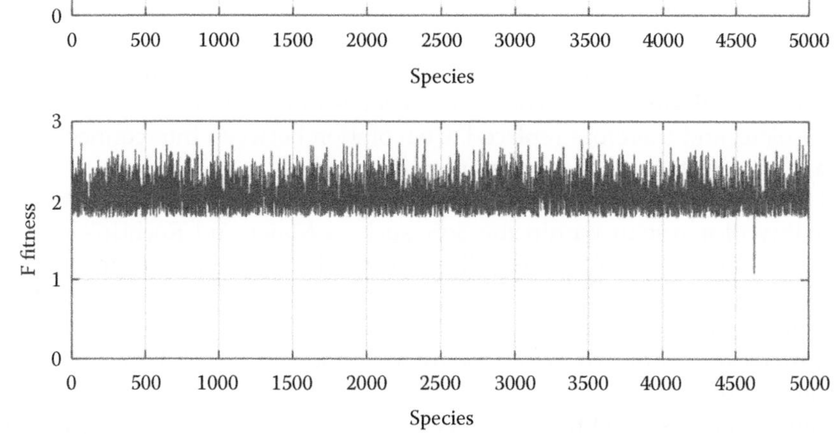

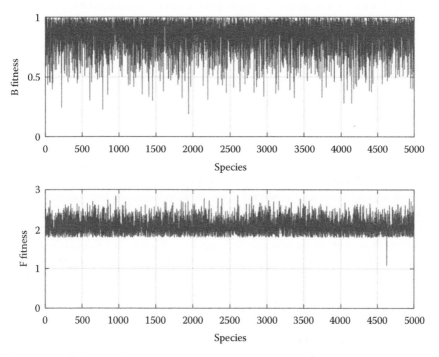

FIGURE 4.17
Local (B) fitness (top) and global (F) fitness (bottom) in steady state.

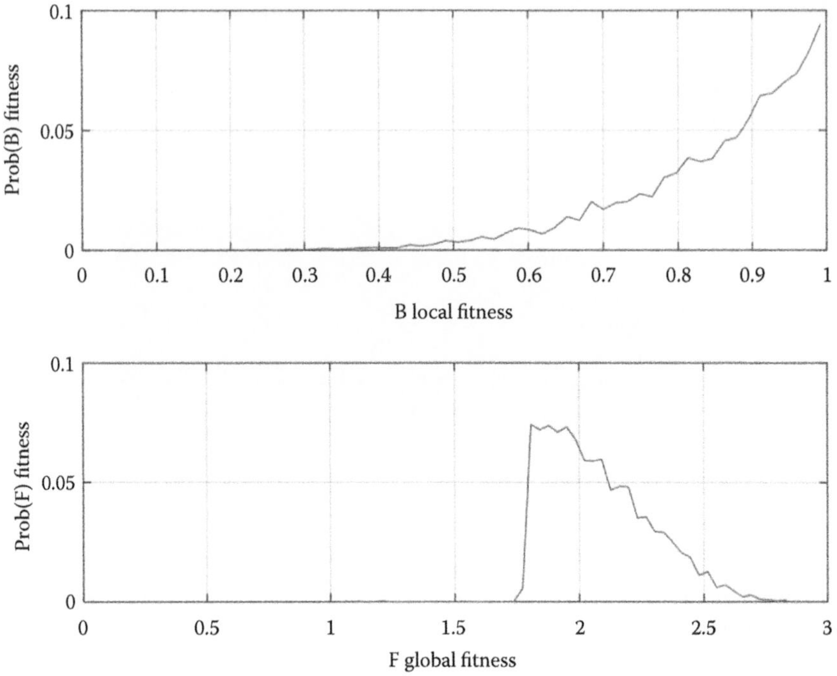

FIGURE 4.18
Global fitness (F) evolves to a clear barrier ~1.7, but drops off linearly. Local fitness (B) no longer exhibits a clear barrier value, but tends away from 0.

4.6 Discussion and Conclusions

This chapter suggests that biological models of evolution can instruct and inform the evolution of man-made systems within a system of systems. Engineered systems evolve through their lifecycle, being created (often for a specific purpose), operated, and often updated, then disposed. There exist similarities to biologic system lifecycles over generations of mutations. Systems must continue to evolve and be modernized, if they can be, or become obsolete, and therefore replaced. Coevolution between interconnected systems is the primary source of emergence for man-made systems.

Several authors have proposed SoS metrics and attributes. Some of these attributes relate to the stability of a system within the SoS, such as Maier and Rechtin's technical, economically and political "stable intermediate forms" [15]. Attributes regarding stability are related to fitness; per Bak-Sneppen [7], "the stability of each species is characterized by a barrier height separating its local fitness" from others. Pragmatic rules for evolution and the structure of system or programmatic interdependencies will drive dynamic changes to fitness and stability.

This chapter demonstrated the effects of Bak-Sneppen dynamics on regular, scale-free (power-law), and small-world graphs, as well as graphs with explicit interdependency/interoperability measurements. Varying graph structures can weaken the entire SoS fitness landscape (lower fitness barrier), thereby impacting the amount and frequency of avalanches, and can transform the steady state fitness landscape and distribution.

System and program interdependencies are found across commercial and Defense environments. Program have direct and indirect relationships and influence during acquisition, and the resulting systems have different direct and indirect relationships within the operational SoS. Evolutionary models of system dynamics may provide great insight for studying the sources of emergence affecting system of systems.

References

1. Standish RK. On complexity and emergence. Complexity International. 2002.
2. Hitchins D. Emergence, Hierarchy, Complexity, Architecture [Internet]. Systems Approach. 2016 [cited 2017 Jan 15]. p. 1–10. Available from: http://systems.hitchins.net/systems/systems-approach/emergenceetc.pdf
3. Halloy, J.; Deneubourg JL. From biological to artifical complex systems [Internet]. Bruxelles, Belgium; 2004. Available from: https://cordis.europa.eu/pub/ist/docs/directorate_d/st-ds/halloy.pdf
4. Abbott R. Putting Complex Systems to Work. Work. 2006;34.
5. Tapscott D, Williams AD. Wikinomics. How mass collaboration changes everything. 2006. 320 p.
6. Matt C. Fail study: Jimmy Wales and Nupedia [Internet]. Wired. 2011 [cited 2017 Jun 1]. Available from: http://www.wired.co.uk/article/fail-study-jimmy-wales
7. Bak P, Sneppen K. Punctuated equilibrium and criticality in a simple model of evolution. Phys Rev Lett. 1993;71(24):4083–6.
8. Harman M, Clark J. Metrics are fitness functions too. In: Proceedings, 10th International Software Metrics Symposium. 2004. p. 58–69.
9. Maier MW. Architecting Principles for Systems-of-Systems. Syst Eng J Int Counc Syst Eng. 1999;2(2).
10. Maier M. Architecting principles of systems-of-systems. In: 6th Annual Symposium for the International Council of Systems Engineering. Boston, MA; 1996. p. 565–73.
11. Boardman J, Sauser B. System of Systems - the meaning of Smc. 2006;(April):1–6.
12. Gorod A, Sauser B, Boardman J. System-of-Systems Engineering Management: A Review of Modern History and a Path Forward. IEEE Syst J. 2008;2(4):484–99.
13. Brown MM, Flowe RM, Hamel SP. The Acquisition of Joint Programs: The Implications of Interdependencies. CROSSTALK J Def Softw Eng. 2007;20–4.
14. Flowe RM, Kasunic M, Brown MM, Hardin III PL, Mccurley J, Zubrow D et al. Programmatic and Constructive Interdependence: Emerging Insights and Predictive Indicators of Development Resource Demand [Internet]. Pittsburgh, PA; 2010 [cited 2017 May 13]. Report No.: CMU/SEI-2010-TR-024. Available from: http://www.sei.cmu.edu
15. Maier MW, Rechtin E. The Art of Systems Architecting. 3rd ed. Boca Raton, FL: CRC Press; 2009.
16. Raymond AE. The Well-tempered Aircraft. In: Wilbur Wright Memorial Lecture. London: Royal Aeronautical Society; 1951.
17. Kurzweil R. The Law of Accelerating Returns [Internet]. 2001 [cited 2017 Mar 1]. Available from: http://www.kurzweilai.net/the-law-of-accelerating-returns
18. "US Air Force." B-52 Stratofortress [Internet]. US Air Force. 2016 [cited 2017 May 1]. Available from: http://www.af.mil/About-Us/Fact-Sheets/Display/Article/104465/b-52-stratofortress/
19. World Heritage Encyclopedia. B-52G [Internet]. World Heritage Encyclopedia. [cited 2017 May 1]. Available from: http://self.gutenberg.org/articles/eng/B-52G
20. Murphy EM. Complex Adaptive Systems and the Development of Force Structures for the United States Sir Force. Maxwell AFB, AL; 2014.
21. Lewis T. Network Science: Theory and Applications. Hoboken, NJ: Wiley & Sons, Inc.; 2009.

22. Albert R, Jeong H, Barabasi A-L. Diameter of the World Wide Web. Nature. 1999;130.
23. Barabasi, Albert-Laszlo ;Albert R. Emergence of Scaling in Random Networks. Sci [Internet]. 1999 [cited 2017 May 13];286(15). Available from: www.sciencemag.org
24. Moreno Y, Vazquez A. The Bak-Sneppen model on scale-free networks. Europhys Lett [Internet]. 2002;57(5):765–71. Available from: papers3://publication/uuid/6E322D0D-B8CA-41A6-94CD-515BDBCE2DCD
25. Watts DJ, Strogatz SH. Collective dynamics of "small-world" networks. Nature [Internet]. 1998;393(6684):440–2. Available from: http://www.ncbi.nlm.nih.gov/pubmed/9623998
26. Milgram S. The Small-World Problem. Psychol Today [Internet]. 1967;1(1):61–7. Available from: http://snap.stanford.edu/class/cs224w-readings/milgram67smallworld.pdf
27. Ford TC, Colombi JM, Graham SR, Jacques DR. Survey on Interoperability Measurement. In: Proceedings, 12th Annual International Command and Control Research and Technology Symposium (ICCRST). Newport, RI; 2007.
28. Ford TC, Colombi JM, Jacques DR, Graham SR. A General Method of Measuring Interoperability and Describing Its Impact on Operational Effectiveness. J Def Model Simul Appl Methodol Technol. 2009;6(1).
29. Chalyvidis CE, Ogden JA, Johnson AW. Using supply chain interoperability as a measure of supply chain performance. Supply Chain Forum [Internet]. 2013;14(3):52–73. Available from: http://www.scopus.com/inward/record.url?eid=2-s2.0-84921860924&partnerID=tZOtx3y1
30. Davendralingam N, DeLaurentis D. An analytic portfolio approach to system of systems evolutions. In: Procedia Computer Science. 2014.
31. Holland OT. Taxonomy for the modeling and simulation of emergent behavior systems. … 2007 spring Simul multiconference-Volume 2 [Internet]. 2007;1:28–35. Available from: http://dl.acm.org/citation.cfm?id=1404684
32. Bartolozzi M, Leinweber DB, Thomas AW. Symbiosis in the Bak-Sneppen model for biological evolution with economic applications. Phys A Stat Mech its Appl. 2006;365(2):499–508.

5

Leveraging Deterministic Chaos to Mitigate Combinatorial Explosions

Josef Schaff

CONTENTS

5.1 Introduction

This chapter focuses on emergence from deterministic chaos, which is a specific subset of complexity and emergence, where it is possible to instantiate and control the resulting emergent behavior. This chapter does not provide a complete overview of complexity science. Instead, selected aspects of complexity science are described to the level needed in order to understand how to implement this control of emergent complexity. Following the background development, an example is given for solving a computationally hard ad-hoc networking problem. This solution is also applicable to massive swarms.

Our world consists of many intricate and complex things. Most of nature consists of elements at various sizes and form, which make up larger natural systems. A tree consists of cells that are divided into those that make the bark on the outside, the wood inside, and the leaves as well as flowers. Each of these cells in the tree is composed of many complex parts. This is also true in the animal kingdom. Our bodies are made up of about a trillion cells. Each cell is produced specifically for one of the complex organs (including skin), as well as blood and the skeletal structures of our bodies. Both the trees and human bodies start from a much simpler form and differentiate to create a complex system. While individual cells may die after a few months or so, the body can live for many decades, implying some form of collective stability in complexity. Most of the current engineered systems that we use today are built differently, as we try to minimize complexity while creating systems to perform complex tasks. Even when systems are engineered to a simpler form, when the components of the system are connected, there still are unexpected complex behaviors, i.e., emergence, that seem to go against what we intended.

The simplest complex system is the three-body problem in physics, where three objects interact by [1]. This problem and the compound pendulum (i.e., a rigid pendulum arm with more than one pivot point) are simple systems yet they confounded Isaac Newton [3]. In such systems, parts with predictable behaviors can form unpredictable system behavior. Nature uses complexity to its advantage, so why don't we do the same with our engineered systems? To better understand this, we will look at complex systems with associated emergence.

5.2 Emergent Patterns in Complex Systems—A Brief Overview of Complex Systems

5.2.1 Brief History

The study of complexity science has occurred throughout history. The focus in this chapter will be on the past three decades, with reference to the Santa Fe Institute established by two Nobel Laureates, and the New England Complexity Sciences Institute (NECSI), established by Yaneer Bar-Yam. The Santa Fe Institute was established in the early 1980s by a Nobel Laureate in physics, and another in economics. It was precipitated by interdisciplinary researchers at the Los Alamos National Laboratory who had backgrounds as diverse as physics, cosmology, economics, and biology. All were investigating chaos and doing complex systems research.

NECSI originated within a few years of the start of the Santa Fe Institute, primarily as a product of Bar-Yam's research that he began a few years prior for his doctorate. NECSI is also interdisciplinary but focuses on the systems-level aspects of complexity and emergence, while Santa Fe focuses on complexity, emergence, and the interactions of multi-agent systems [39].

5.2.2 Complexity and Related Definitions

Order can be defined by simple equations, as in Newtonian systems. As we will see later in this chapter, a deterministic chaos equation can be as simple as the Newtonian ones, yet produce emergent behaviors that are repeatable.

Complexity can be defined as stated in title of Waldrup's book—*Complexity: The Emerging Science at the Edge of Order and Chaos* [39]. In order to better understand this, we need to understand the boundaries—order and chaos. Order is self-evident, and can be considered from many different aspects, e.g., entropy of zero, or a set of any consistent elements placed in a regular sequence.

Chaos needs further definition. According to Strogatz, "chaos is aperiodic long-term behavior in a deterministic system that exhibits sensitive dependence on initial conditions." Strogatz then defines aperiodic long-term behavior as follows: "there are [some] trajectories which do not settle down to fixed points, periodic orbits, or quasiperiodic orbits." [43]

Emergence has several definitions that are based upon its context, but for the purpose of this chapter we focus on emergence from complex systems and chaotic dynamics. The definition that fits our purpose best is from Waldrup [39], where emergence is simply the (sometimes unexpected) outcome of complexity science research known as *emergent complexity*, or *"emergence"* for short. We now investigate three known examples of how emergence is related to complexity and chaos.

5.2.3 Predator-Prey, Boids, and Life—Three Examples of Models with Emergence

Predator-prey or Lotka-Volterra equations are nonlinear first order differential equations that have been used to represent where an interdependency exists between the populations of two or more species, e.g., rabbits and foxes. This dependency causes the behaviors of the equation to produce a sinusoidal phase-shifted relationship between predator and prey. Under certain conditions of population growth, it can produce an emergent aspect, where the population of one group varies in a non-uniform aspect to the other group. A predator population may sporadically increase after the prey population increases and decrease when the prey is killed off. In this instance, variation in population sizes may not necessarily be cyclical, but may vary chaotically.

The two populations would be expected to reach equilibrium; however, the emergent reality of this relationship is that it never reaches equilibrium but continuously oscillates, as shown in Figure 5.1. Perhaps this can be better understood by observing a simpler form of population dynamics, known as the logistic equation.

The predator-prey model is taken in its simplest form directly from the logistic equation for population growth, where population size is dependent upon resources available and the rate of reproduction. The logistic equation is represented as:

$$x_{n+1} = r^*x_n(1 - x_n)$$

with x_{n+1} representing the population of the next generation, x_n the population of the current generation, and r representing the growth rate. For values of r between 1 and 3,

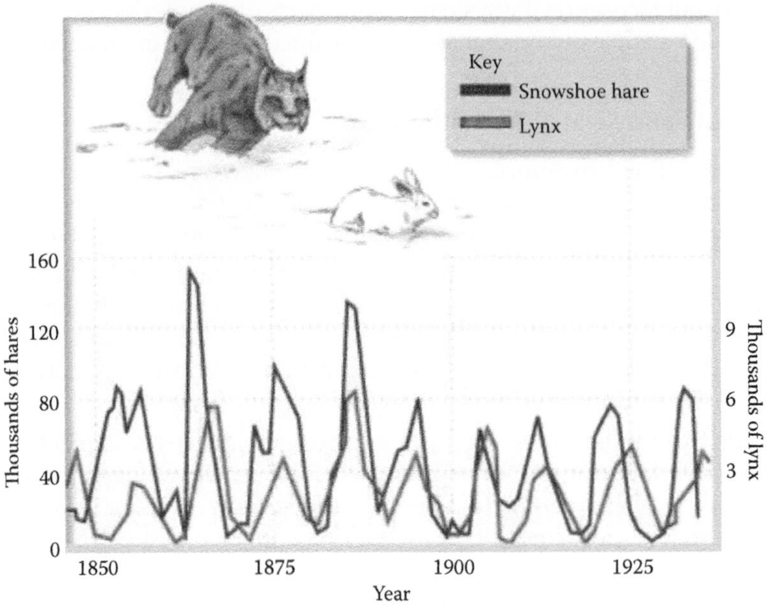

FIGURE 5.1
Lynx vs. hare populations. The pattern is seemingly chaotic, yet cyclical with relationships between both populations evident in the picture.

the population eventually reaches a steady state. When r is greater than 3.3, there is a *period doubling*, or *bifurcation*, that happens frequently for slight increases in r until it reaches 3.57, where chaos ensues. This is best seen in an orbit diagram, which is a representation of a set of points that act as attractors, similar to an orbit of a planet around its attractor, e.g., earth orbiting around the sun. This orbit diagram produces an emergent behavior, which transitions from order to chaos and back again multiple times, as seen in the well-known Feigenbaum (orbit) diagram [7], shown in Figure 5.2. When the bifurcation occurs the first time, the single point attractor is split into two point attractors. Each subsequent bifurcation doubles the number of attractors and subsequent orbit paths. When r (the **X**-axis of left-hand diagram in Figure 5.2) reaches 3.57, the bifurcations (theoretically) have split into an infinite number of attractor points, and chaos is evident in this aperiodic long-term behavior. At several regions of values for r greater than 3.57, order can be seen re-establishing itself. An interesting property of some of these regions, e.g., values of r between 3.84 and 3.85, is that if magnified, it produces a nearly identical Feigenbaum diagram, which shows an attribute of self-similarity at different scales. The right-hand diagram in Figure 5.2 shows this region between 3.84 and 3.85, and the self-similarity to the left-hand side between about 3.3 and 4.0 is evident. This property of self-similarity will reappear in the later sections of this chapter as fractals.

Reynolds' Boids was an artificial life program created 30 years ago that contained rules for behavior. Boids consists of simple agents that follow three simple behaviors: *separation*, or keeping greater than a minimum distance from adjacent boids, *alignment*, which causes all boids to fly in the same aggregate direction, and *cohesion*, which causes the boids to aggregate around a flock's "center of mass." These three behaviors can cause emergent flock behaviors such as chaotic flight paths, or bifurcation (splitting) of the flock, as well as a stable and orderly flock. Another emergent behavior occurs during an obstacle avoidance, where the flock scatters then reunites after the obstacle.

FIGURE 5.2
(a) Example of an emergent static pattern from Conway's Life game. (b) Feigenbaum diagram.

Conway's "Life" game dates from nearly 50 years ago [41], and displayed emergence from simple cellular automata, which followed a few fundamental rules for survival, birth, and death. Life consists of a checkerboard-like set of squares or cells each of which, if filled (i.e., black), denotes a "living" cellular automata. When a cell is empty, it is devoid of any living cellular automata. Each cell is surrounded by eight adjacent (touching) cells. There are only three rules:

1. **Survival**—if a cell has two or three neighbors, then it survives to the next generation (or iteration of the simple state machine).
2. **Death**—if a cell has four or more neighbors, it dies from overcrowding. If it has one or no neighbors, it dies of isolation.
3. **Birth**—if there is an empty cell adjacent to exactly three neighbors, then the cell becomes filled, i.e., a birth, in the next generation.

There are many emergent patterns that result, both static and dynamic, with some of the dynamic patterns oscillating through several states. Below in Figure 5.2a is one example of a static pattern.

Both Boids and Life are simple entities or agents constrained by a few simple rules. Life is a simple state machine, with the entities defined by a binary state of alive/dead. Despite this simplicity complex behaviors are emergent in both systems. The next section focuses on the current conventional methods for managing complexity, their shortfalls, and developing an approach for a general solution.

5.3 Methods for Managing Complexity

5.3.1 Reductionism or Design by Decomposition and the Lego Blocks Mindset for Systems Integration

Design by decomposition is typically used in systems engineering to decompose a composite system into its functional elements. The limitation to this approach appears as unexpected effects in complex systems that may make a seemingly well-designed system non-functional upon completion. According to Bar-Yam, this limitation is due to the interactions and interdependencies of components for complex systems, which may inadvertently have been ignored. These complex systems are therefore also untestable, as it may be infeasible to test all possible combinations of components [31].

The International Council on Systems Engineering (INCOSE) in their vision for the future of large and **complex systems** stated that current practice does not address the increasing number of failures in complex systems, and also that there are factors that are either not consistently applied or are omitted from the model of the System of Systems (SoS) to be created. This would result in **emergent behaviors** in the actual SoS constructed. Additionally, DeRosa et al. discussed autonomous agents as an intrinsic property of complex systems, and the aspect of emergence—i.e., a novel property that was not an attribute of any of the system's subsets, that established some type of quasi-stable state for the SoS [33].

5.3.2 Formal Methods—Correctness and Its Limitations

Formal methods provide a solid foundation for rigorous analysis and trusted behaviors in mission-critical systems. The biggest strength of formal methods is in the direct flow of requirements to engineering models, and the subsequent identification of inconsistencies or incompleteness in the specifications. By starting with well-defined modeling languages, e.g., SysML, the requirements for a detailed system can be used to generate a model of that system. This model can also provide effective test cases from the specifications, ultimately determining the "correctness" of the SoS. There are also tools for the analysis of the system's architectural design, e.g., the Society of Automotive Engineers (SAE) standard, Architecture Analysis and Design Language (AADL).

If formal methods worked for the complete spectrum of all SoS including creative or emergent behaviors, then there would be no need to leverage complexity. Such behaviors would be predicted in entirety by formal methods. Unfortunately, there are limitations to formal methods. Some of these include "correctness" of the specifications or requirements themselves, their implementation, and the non-analytical aspects of creativity with respect to design processes as well as aspects of emergence in machine learning. Because of some of these limitations, formal methods do not scale well to very large projects or SoS, but work best with clearly defined tightly-coupled software. That is one of the reasons a cybersecurity program that incorporates formal methods such as the Defense Advanced Research Projects Agency (DARPA) HACMS (High Assurance Cyber Military Systems) program is based on a micro-kernel (a simple and small-scale operating system foundation).

These limitations in current SoS design and analysis indicate the need for a novel approach that can take advantage of SoS attributes, which manifest themselves as size and complexity increases for large systems. Since we are exploring "new territory," the approach we lay out should be treated as a starting point and not a final, polished product. Our exploration will likely need to assess mathematics that does not yet exist, and is therefore the product of computational, or experimental mathematics [40]. This is not as difficult as it may appear, as the search for new mathematical formulae can be directed by a heuristic approach using machine learning techniques. Experimental mathematics has just begun to flourish due to the rapid increase in computational power resulting from Moore's law, and the variety of processors that originated from advances in computer graphics such as the GPU chip. Many future discoveries in experimental mathematics may result from this new increase in computational power.

5.3.3 Towards a General Solution

At the Institute of Electrical and Electronics Engineers (IEEE) International Systems Conference, SysCon 2008, DeRosa et al. described the type of *problems that would use complexity science,* and the existence of gaps in knowledge. Candidate problem boundaries, such as closed systems, were described, and complexity was contrasted with "complicated" systems. Self-organization and adaptation were defined within their role in complex systems, their causing factors, and how and when these occur in complex systems. Adaptation was described as an iterative process; therefore, it may be modeled by some kind of iterative function [33].

Bar-Yam presented a fundamental view of how complex systems need to have a different engineering approach than the simpler systems. His approach, "Multiscale Analysis," showed where the typical systems engineering approach failed and at what level of complexity, while giving an analytic basis for determining the degree of complexity. He suggested leveraging off of evolutionary engineering, which is an iterative process of building upon numerous engineered products that interact in some way. Using this approach, the existing system is augmented by modifying components while in use, so as to evolve towards a system with desired characteristics [31]. To formulate a better approach to complex systems, we can start with DeRosa et al. and Bar-Yam's concepts. The first step is to look at complexity science, and some of its attributes such as emergent order.

5.4 Overview of Complexity Science—Some SoS Types with Emergence

Complexity science is a relatively new science that has been increasingly studied in the past 50 years, and in general terms describes the dynamics of interacting systems.

Novel coherent properties can result from self-organizing System of Systems (SoS). Collective actions of many entities in a system produces *emergence*. One aspect that has been a topic of interest is the chaotic dynamics that occurs in certain nonlinear systems. The results of such interactions are not easily predictable and produce unexpected or "emergent" behaviors.

The field of complexity science is interdisciplinary, where resulting emergent behaviors have been observed in areas as diverse as economics, social structures, business corporate structures, biological systems, human learning, and engineering models for machine learning, e.g., Neural Networks (NNs). In many instances, the seemingly "accidental" emergent result significantly benefits the systems. In *complex adaptive systems*, learning and the resulting changes in behaviors of the entities that compose the system affect the optimization and sometimes the survival of the complex system.

Historically, systems engineers have attempted to minimize any aspects of emergence, but as Systems of Systems (SoS) become more complex in their interactions with themselves and the environment, this has become increasingly more difficult. The emergent behaviors that result, however, may not be disadvantageous to SoS functionality, but instead may augment a system's capabilities. Since a cyber or military adversary may already know most of our systems, or at least their intended behaviors, one way to surprise an opponent would be to leverage emergence resulting from some complex systems. Ideally, the system's initial behaviors should appear random or chaotic. This would afford the element of surprise when emergent order occurs.

It has been suggested that complexity science should instead be called "Order-Creation Science" [27], since it addresses aspects of emergent behaviors that can form order in disparate systems. Under certain circumstances, chaotic systems can synchronize, providing alternate means of conveying information [25,28]. This order creation starts with aspects of self-similarity, or analogous attributes of the systems or environment. Applications of this order creation could have significant implications for massive swarms and cyber botnets, as well as aggregates of heterogeneous systems in a Battlespace environment.

5.4.1 Analogies, Self-Similarity, and Chaotic Synchronization

Much of human learning is by analogy: When a person sees a new object in his or her environment, he or she may treat it the same as some known object that is considered similar to it. The original is a template for modeling interactions with the new one. More observant individuals then go on to note the differences between the new object and the one previously encountered. This indicates that a large portion of acquired knowledge is based on self-similar relationships and their respective transformations. An additional factor to consider is that the model of the world is based on our perception of it, and this is limited to the relatively narrow range of human senses. Because this self-similar paradigm has been used for our internal world abstractions that we use to define our observations, any model that we have built so far would have been impacted by the similarity paradigm.

Over two decades ago, an experiment was conducted to provide a means of detecting a signal below the environmental noise level, using synchronized chaotic circuits. Pecora and Carroll's research [26] on information embedded below chaotic noise threshold showed that a *similar chaotic* circuit can "decrypt" signal from noise, making an inaudible signal embedded in noise become understandable. The *similar chaotic* circuit can be thought of in the way that a human uses an analogy to understand a difficult concept, by likening it to a known concept.

5.4.2 Emergent Order from Complexity and Chaos

Complex systems may consist of many components associated by structure or just an abstract relationship. The system may be scalable and self-similar at more than one level, and is not described by simple rules or from the fundamental level. Emergence can produce "creative" system behaviors. Examples of applied emergent behaviors can be found in the research field of Artificial Life, which uses emergence generating algorithms. The algorithms of choice include genetic algorithms, neural nets, and cellular automata. A specific applied example is "The Sims," which uses genetic algorithms for its automata.

Emergent behavior has been observed over the past half century in cellular automata (Conway's Life game [12]) as well as in a simple agent-based model at the Naval Postgraduate School that produced swarm organization (the Boids simulation [13]). Further evidence of desirable emergent behaviors appeared in the RoboCup competition. The team with the robots that first realize collectively that they should play with teammates to oppose the actions of the other team's players wins every time. The teaming aspect for multi-agent systems is the emergence of self-organization. The threshold measurement for this organization is referred to as an "avalanche" metric. This metric is applied to player agents on a team to determine when it affects the overall team performance. The specific example teams are organized into the "RoboCup" and "RoboNBA" games. The metric showed that a self-organizing pattern emerged when players on one team would try to pair up with the opponent team's worst player that was un-paired. This resulted in an overall win for the team that had the most success in doing these pairings.

It may be initially surprising that software paradigms can exhibit emergence, especially since software is constructed from simple deterministic statements that execute on the Boolean logic elements of a computer. Such a software paradigm is Conway's mathematical game "Life." Life, described earlier in this chapter, is a simple state transition sequence for cellular automata. The set of resulting next-state transitions becomes an iterative process based on the current state of each cell and its adjoining neighbors. In both Life and Boids, the base elements (cellular automata and agents) are instantiated many times over. Perhaps there is a way to leverage a repetitive pattern, yet still converge to a solution.

5.4.3 Emergence—What Can and Cannot Create It

There are various methods to create complex SoS that produce emergence, yet have results that are controllable. New approaches in computational (experimental) mathematics for multi-agent systems, and deterministic chaos (fractals) are a couple of examples. Emergent behaviors result *not* from stochastic continuous (e.g., thermodynamics) models, but instead from multi-agent interactions (e.g., RoboCup). Moreover, emergent SoS *cannot* be designed by functional decomposition [31]. Emergence results when predictability "collapses" as a sequence progresses (complexity increases). Chaos can result from even small changes, and known initial and intermediate conditions can have unpredictable results, as shown in the logistic equation and resulting bifurcation diagram in Figure 5.2.

An example from nearly a half century ago of a complex system with emergent behavior was Mandelbrot's "transmission line noise." This initially appeared to be random noise, and as such could not be eliminated. Instead, it was found to be a predictable "Cantor Dust"—which now is known as a fractal pattern—and therefore could be cancelled out by generating an inverse pattern [36]. Because of the "fractional" Brownian motion description of the patterns, Mandelbrot later coined the name *"fractal"* for this class of complex emergent patterns. This leads to the question of whether we can discover seemingly complex patterns that have similar properties, much as in the way that a human recognizes complex structures and their relationships [4,23].

5.4.4 Complexity and Attractors—Coalescing Towards Dynamic Stability

The dependence of attractors to impact emergent behavior was shown above in Figure 5.2 by the orbital diagrams associated with the logistic equation. These equations describe many relationships existing in nature, e.g., predator-prey or the impact of resources on population size. To model complex behavior, a good start would be to use nature as a template because this is an already known working system. At this point, then, we go back to the fundamentals of nature to determine the driving factors for why things differentiate the way they do. Most of nature appears to be complex and chaotic; hence, many aspects of nature can be represented by complexity and chaos theory. Narrowing down our analysis to something more tractable, we observe that much of nature has affinities for particular types of behaviors or complex structures that seem to have some sort of regular pattern, possibly converging around some abstract set of attractors. A paradigm based on probability would partially work—it would be a good generalization that bounds the problem without really understanding the nature of it. Alternatively, determining some generic linear equations and then calculating several orders of perturbations could provide the hope of capturing some of the emergent behaviors. Specific analytical definitions, however, are much better. Instead of treating all of nature as a random distribution of permutations, consider some permutations to have a "weighted" value, or specifically, the ones where a local minimum exists would be more likely to happen than those elsewhere. In many instances, there can be a regular pattern or topology to these minima, or *basins of attraction*. Each of these minima would in turn form a pattern as to how entities that are attracted to it cluster, and these distributed minima would likely have similar cluster topologies. As large groups of minima cluster together, a larger scale version of the cluster topology may be found, thus uncovering a self-similar scaling.

These local minima, or attractors, become a topological tool that can be used to describe either overlapping behaviors or shared similar structures between heterogeneous entities. To define a set of these minima and their respective synergistic effects on any interaction, we must first preserve their topology. Subsequently, by reproducing the same topology but with a slight transformation, we would have a similar effect. In order to preserve the kind of effect that a particular topology depicts, we would need to preserve its relative topology.

One of the best ways to preserve the kind of effect that a particular topology exhibits is by using geometric transformations of the original topology. The *affine transformation* preserves the relative shape as well as accounting for collinearity (i.e., sequential positions of points on a line), dilation, and rotations of objects. Examples of affine transformations are the changing of a large n-sided polygon into a smaller n-sided polygon, or a rectangle into a parallelogram. The parallelogram still maintains four vertices and sides as well as the parallel nature of opposite sides. Using this transformation, a geometric shape could be rotated and skewed but still retain its relative shape (e.g., square-rectangle, triangle). This would help to create a class of self-similar shapes. One class of affine transformations is the contractive affine. Given a set of attractors, this transformation would, after a number of iterations, cause a series of diffuse points to coalesce towards the attractors. The nature of the contractive affine set is to always coalesce; thus, there is no risk of it diffusing (i.e., it will always form clusters around an attractor).

5.4.5 Self-Organizing Systems and Entropy

Civilizations, economies, evolution [14], and biological organisms, as well as our cognitive thought process, are dependent on complexity. Most body functions exhibit complex behavior, e.g., the fractal pattern of heartbeat [18], polarization of ionic channels in membranes, etc. These are so crucial to the optimal functioning of biological systems that when the electrical activity pattern of the heart's pacemaker cells becomes *less* complex, it indicates a potential heart problem. Artificial biological systems also exhibit this behavior, which can be seen in Neural Networks, genetic algorithms, Boolean nets [14], and cellular automata [15].

Cellular automata theory offers a number of examples of self-organizing systems. One of the most common examples of this is Conway's game of Life, mentioned previously. The automata are defined by simple rules of interaction (much simpler than in real-world situations) and are capable of generating many modes ranging from static to cyclic, and even chaotic system states. This begs the question as to whether a system that starts in an arbitrary (random) state can end up producing a *repeatable* highly ordered end state. Does this violate the second law of thermodynamics—are not the rules regarding entropy always true? Mostly true would be more accurate, and dependent upon the semantics. A very simple example of this can be produced using Boolean nets. Kauffman [14], who has been investigating Boolean nets for several decades, found that Boolean nodes that have less than four connections always result in a stable ordered end state, whereas Boolean nodes with four or more connections may result in a chaotic state. Let's step aside from the semantic argument as to whether entropy rules are violated and consider that some random systems can end up in a stable, highly ordered state.

Conceptually, the easiest way to envision this is to consider the example of crystal formations, where an amorphous liquid loses energy to form a highly ordered structure—a crystal. This is clearly due to the nature of the system—the interaction of the molecular forces inherent in the crystallizing substance. The crystal is clearly the lowest energy formation of the substance. A similar analogy can be applied to stable states in cellular automata, which are closely allied with the Boolean nets previously mentioned and other random systems with selected constraints. This is also addressed by Wolfram [15], as is the seeming violation of rules on entropy. These stable states result when we approach a minimum (either local or absolute) on a particular mapping of points, whereby the mapping of the points can vary only slightly within the constraints of the minima. These "energy mappings" can be used to show convergence to a solution.

Approximately 25 years ago, a physicist named Per Bak co-authored a paper that coined the term "self-organizing criticality," after studying how organization emerges from disorder [16]. The stability resulting from complexity has been investigated by Glass and Mackey [17], and in collaboration with others [18] were able to show the chaotic aspect of cardiac rhythms. A particularly interesting result was the conclusion that the less chaotic the cardiac waveform, the greater the chance of a massive cardiac failure, resulting in an indicator "tool" for potential problems. (The optimal cardiac waveform had a particular fractal aspect to it.)

The network packet flow patterns through major internet hubs provide another example of *emergent self-organization*, which we can perhaps leverage in some way. Typically, packet traffic was modeled as a stochastic distribution through major network nodes, since the packets were assumed to maintain their randomness in distribution. The analytical tool of choice was a Poisson variant (e.g., Gaussian) distribution of the packets. When that model was applied, the results displayed were significantly different than those observed.

The observed packet flows had the emergent property of self-organizing into topological structures called fractals [5,29]. Fractals have been successful paradigms for real-world representations and have been observed in numerous phenomena ranging from the nano-scale transitions of magnetic patterns in metals [21] and network packet traffic flowing through the largest internet hubs, to the neural ring patterns such as in the pacemaker cells of the heart.

5.5 Some Complex Problems

Now let's look at two problems and their respective solution, which can be derived using complexity-based emergence. This is just an outline of the problems and one solution that will be described in the next few sections of this chapter. This will be used as the example and one of its respective solutions for the rest of the chapter.

5.5.1 Application Focus

Problem 1: Consistent robot behaviors within a world model: Cognitive robotics incorporates the behaviors of intelligent agents within an abstraction of reality known as a *world model*. Multi-agent systems create challenges for desired behaviors within a planned environment due in part to the problem of translating and using symbolic reasoning for world abstractions [28]. These world abstractions are needed by the agents to successfully share a *world model*.

Problem 2: Spatially distributed large dynamic networks: Loss of edge node communications—The Congressional Research Report [9] showed scaling limitations for large numbers of networked nodes needed for battlespace. Calculations result in a combinatorial explosion from massive numbers of route calculations. To increase availability and resiliency in network-centric clouds and swarms, ad-hoc nodes must rapidly self-organize using shared topology data. Additionally, topology can affect the network failures and success of cyber offense and defense.

Solution needed: Both of these problems have issues with shared information. Problem 1 has the issue of translating the world into actions for the distributed agents to produce their desired emergent collective behaviors that can efficiently accomplish a mission. Problem 2 has the issue of sharing topology information between dynamic nodes on massive networks, so that rapid self-organization can take place, providing network resiliency. The emergent macroscopic (i.e., multi-agent world) perspective can address both problems 1 and 2, for computationally intractable solutions—e.g., the large-scale (10,000+ nodes) self-organization of ad-hoc networks and adaptive swarms. A method needs to be defined to accomplish this solution.

Dependencies: Adaptable open architectures, possibly integrating machine learning with agent models. Need for agent-based modeling with emergence for SoS.

Limitations of existing methods: Current systems engineering is limited in its approach to SoS to be able to predict novel/emergent behaviors in any consistent way. System elements need to be more adaptable, loosely coupled, and create a dynamically interoperable environment. Emergent SoS cannot be designed by functional decomposition.

5.6 Large Ad-Hoc Network Routing Issues and a Possible Path to a Solution

Some limitations for large-scale system of systems may be caused by emergent phenomena that are common in complex systems [31]. Moffat [24] discussed the potential impact of complexity science upon large network-centric topological environments. While these emergent behaviors of complex systems may cause issues if not constrained, they may also present a solution to emergent network topologies. Since the topologically distributed nodes need to share both local and distant information, as described by Bar-Yam, it would be prudent to observe how nature creates self-organization. One method used by nature is called stigmergy [8], or the sharing from sparse yet dense information [37], and the impact of different topologies upon these behaviors specifically for distributed network nodes [32].

The internet and smaller network topologies have been well studied to understand their positive attributes and potential weaknesses [6], as described by Kleinberg et al. as resilience to random failures while being vulnerable to attacks targeted at a few critical nodes [35]. Additionally, this provides a rationale for developing a resilient topology, unlike the internet [34], while *mitigating unwanted emergent behaviors* due to the magnitude of the number of nodes [31]. The mathematical descriptions of complex networks [38], their construction [31], and the emergent order [16] have helped to direct us to look for a mathematical topology for resilient ad-hoc networks. Since complexity cannot be eliminated, it should instead be leveraged for its emergent order [16]. Self-similar patterns like fractals may provide such a solution [36], while the randomness of nodes entering and leaving the swarm or battlespace would lead us to look at randomly generated fractals [19].

5.6.1 From Chaos to Iterative Order

5.6.1.1 Applying Fractals and the Iterated Function System (IFS)

Many real-world phenomena are self-similar (i.e., a smaller section of the whole can be scaled iteratively to represent the whole). Examples include trees, broccoli, cauliflower, and their respective branches; planets with multiple moons that can scale to a solar system with multiple planets; coastlines that can scale both upward and downward; and many other phenomena. Many phenomena can be represented by recursive generating functions, or Iterated Function Systems (IFS), which typically generate fractals [5,19,20]. The IFS generates a fractal shape by constructing many copies of itself and overlaying these to produce the fractal pattern. The copies are affine transformations of the original shape, and are typically contractive-affine, or a reduced size copy of the original. This IFS is the result of a set of contractive affine elements, by use of what is called the *Hutchinson operator*, which converges to a unique attractor [22]. The overlays of these affine transformations are mathematically the union of these functionally transformed shapes, which create a fractal topology.

These generating functions can be iterated to the desired degree of behavioral fidelity with respect to real-world observations. This means that although the true fractal may be infinite, we can take the IFS computation to almost any arbitrary level and still see some of the resulting emergent pattern (similar to a hologram, where each small element of the hologram still contains enough critical attributes to reconstruct the complete picture).

5.6.1.2 Static or Dynamic?

The above approach will give us a static picture, but what about dynamical systems in general? That is indeed the nature of our world—dynamic. What if we consider another approach, i.e., to use an adaptive, stand-alone IFS to generate unique self-organizing sets of patterns, and let these evolve into a representation of an ordered structure? This dynamic IFS would adapt, morph, and distort the fractal to include new points as they are acquired (e.g., entities could join or leave swarms or ad-hoc networks). Because fractals are self-similar, their elementary parts can each represent an aspect of the real world. Such a "rubber" fractal can stretch and adapt, and by its nature coalesce or contract to a stable set of patterns. The vertices on this fractal would be related by analogy to real-world parameters. This produces a paradigm that allows a machine to behave more like the dynamics of natural phenomena.

5.6.1.3 The Dynamic IFS—What Is It?

The dynamic IFS described below is just one of possibly infinitely many new or unknown mathematical functions that, instead of being derived from formal proofs and existing mathematical functions, is a product of research using experimental mathematics. In this case, it is derived from one aspect of Barnsley's "chaos game" [19], or Markov Chain Monte Carlo algorithm, and is a simple linear algorithm that takes minimal computational power to *generate a repeatable and predictable complex system*. The algorithm is noted for its high accuracy coupled with low memory usage [30]. This specific algorithm creates an adaptive fractal, whose shape is formed from many points in self-similar or repeating patterns, analogous to the way that tree branches are similar to a twig at the smallest scale, and trunks at the largest end.

This dynamic IFS is created from points that topologically represent the positions of nodes or simple agents. The positions of these nodes are mapped to the IFS equation, which is dependent upon several parameters, specifically a small subset of initial points or vertices that become attractors (e.g., targets or waypoints for mobile entities), and a contractive scaling factor. When the scaling factor approaches either +1 or –1, the system displays what appears to be a random scattering of node points. When the scaling factor approaches 0, the points coalesce in clusters around an arbitrary attractor, covering all the attractor vertices. These emergent clusters provide the topology needed for point-to-point node communication links, as well as a choice of inter-cluster routing topologies.

This solves several problems, including those associated with a topologically distributed network or swarm. It also solves some of the currently intractable distributed information problems by creating *emergent solutions* through sharing of information. This also *leverages the inherent complexity* in the SoS, i.e., the iteration of each function in the IFS represents each system, while the total IFS represents the SoS. A stigmergic solution results when the small amount of data that is information-dense is shared among all nodes—in this case, the vertex positions and the scaling factor value. By each node knowing this sparse amount of data, each can reconstruct the entire topology of the SoS.

This has narrowed down our approach to using a dynamic IFS, which produces an emergent solution that depends upon a relatively small change in one or more variables. The solution would also need to represent the random changes in a large ad-hoc network. Finally, the results would need to be repeatable each time that the dependent parameters are set to the same values. This should be achievable through a deterministic chaos solution, which is the domain where many fractals exist. Now we need to find a specific solution that can be applied to the problem.

5.6.2 Introduction to the NPPR Algorithm

Now we are ready to discuss an application of the dynamic IFS on the routing problem of ad-hoc network, which we call NPPR algorithm. First, it is important to realize that there would be many possible techniques and mathematical relationships (some of which may as yet be unknown) that could solve this problem and this solution is only one example that uses emergence within a complex system. This specific solution to the large-scale network problem is based on a random set of points representing nodes generated with certain constraints. Points can be at *arbitrary locations* on the surface and *still allow the solutions to converge in any random order.* This is a characteristic of a self-organizing system. The system is called a *non-predetermined parametric random iterated function system* (NPPR IFS). Like many IFSs, iterating through will generate highly ordered patterns in many instances, while some still appear to be truly random sets. The nature of contractive affine objects is that they will always converge if a particular value is <1. Figure 5.3 shows the

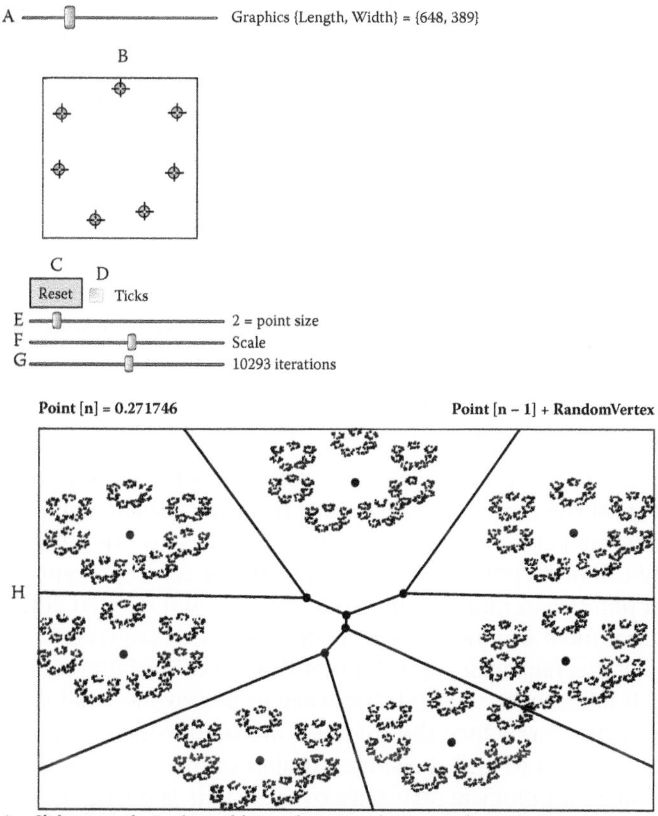

A = Slider controls size (# pixels) in node-points plotting window, at bottom.
B = Hubs topology map, used to drag-and-drop a hub relative to others, or create hubs.
C = Resets diagram to a default 3-vertex, 0.5 scale for equilateral Sierpinski gasket.
D = Checkbox that toggles display of horizontal and vertical axes.
E = Slider for number of pixels selected to represent each node plotted.
F = Scale slider for the NPPR parameter (floating point multiplier).
G = Slider for the total number of points (nodes) to plot.
H = Lines indicate Voronoi partitions, for cluster observation guidance.
I = Nodes plotted using formula at top of window. Center points correspond to hubs.

FIGURE 5.3
A screenshot of the tool for displaying NPPR parametrics.

computer screen for a program that displays the topologies produced by the NPPR function, and allows the parameters to be manipulated for producing the various topologies. The formula is also displayed on the screen, with the denominator in the generating function always >1, meaning that for a function f(n) = a/n where a ≤ n, f(n) will always be <1. The code is simple, with two lines defining the function's primitive for generating the patterns from a random number generator. This system starts randomly with only a few data points, and due to its dynamical nature, it adapts to new points outside the realm of the initial system. Because it is an IFS, that is, a contractive affine transformation, it converges to a solution set. Since only a few points of information are needed to define complex patterns, the system can make estimates from sparse information, which may appear as though it has intuitive abilities. By adding networked sensor data into the system, the NPPR IFS could modify (adapt) itself to a dynamically changing environment or changing threat behavior. Networked inter-agent communication would allow shared knowledge to enhance and update each agent's knowledge map [22], promoting a consistent shared knowledge of the networked environment.

Because the system is contractive affine, it will still encompass the previously discovered data points and contain future data points within its boundaries. Some of the emergent patterns produced by this algorithm are provided the figures at the end of this chapter. There are a tremendous number of as-yet-unknown patterns that the algorithm could produce.

5.6.2.1 NPPR Adaptive Fractal—How It Works, and What Is Displayed by the Computer Tool

Like the simple point-slope equation for line, the deterministic chaos equation is given as:

$$\mathbf{X(n) = M * X(n-1) + Z}$$

X(n − 1) is the *current point*, **X(n)** is the *next point*, **M** is the scale parameter, which controls the distance that the *next point is generated* from the *current point*, and **Z** corresponds to one of the randomly selected "vertices." These vertices are the set of initial points that constrain all network node points, and for our example can represent network hubs. The number of initial points can range from three to over 100 if needed, although experimental results show that most unique emergent behaviors are observed with 12 or less initial points. The specific **Z** *is randomly selected* out of this set of initial points. The value of M is $0 < |M| < 1$, and it takes experimental discovery to determine what values of **M** give the desired emergent behaviors. Once these values are determined, any random run of this equation with the selected **M** and the same number of initial points in similar topologies will produce the same topological structure overall. This is particularly interesting, since changing the random number seed produces almost no visible change to the topologies generated, despite the fact that all the points are newly generated, and justifies this mathematical function belonging to the area of deterministic chaos. Both variables **M** and **Z** share interdependencies that affect the overall network topologies, including thresholds for clustering and implicit mappings to certain cluster elements.

The algorithm has "tunable" values that control the repeatable transition from order to chaos. This is controlled by the variable **M** coupled with the initial set of vertices. The resulting data points, in addition to producing a topological map for network routing, may also be used to create several categories of information. The information may result from sensor data that consists of several features to be associated with feature categories and can be represented by topological relationships. If several sensors at different locations detect something from a common location, but categorize it differently, the sensors' data maps into multiple interrelated categories. The "topological map" would contain the sensor positions and data and correlate the relationships between them while preserving the differences. This allows shared information to define distinct yet common features. A small amount of data transferred between sensor elements could be used to share the global *picture* for the sensor nodes, i.e., a stigmergic relationship.

Nature uses a similar stigmergic approach when ants and other social insects mark a location that is of interest. As other ants pass the marked location, they too mark it, and a sufficient number of marks cause all ants to follow that trail of landmarks. This small chunk of information can become extremely powerful and is analogous to the sparse data elements used for defining behaviors in the NPPR algorithm.

For the investigation of the NPPR algorithm, a tool was built in Mathematica® to display parametric changes as they are made dynamically by the user. This tool is shown in Figure 5.3 with definitions of labels A-I displayed below it. The areas to focus on within this tool are B, which is the topological layout of the initial points, F, which corresponds to **M** in the equation, and H, which is the box in the lower half of the picture. This box contains all the points plotted from the equation as well as Voronoi lines, which are independently validating the partitioning of clusters onto separate tiles. The points in B can be dragged around the small box; points can be added or deleted so as to increase or decrease the number and positions of the initial points. These initial points form attractors around which clusters may coalesce.

F uses a slider that changes the value of the floating-point number **M**, visible within the equation in the center of the screenshot. The Voronoi lines in H are created internally using the Voronoi cluster analysis tool embedded in Mathematica®. A more mathematically detailed description of the NPPR algorithm is given in Appendix A.

Comparing the two pictures in Figure 5.4, one can see the relationship between the initial points, which can also be applied to a topological representation for data analytics consisting of five categories, with each attractor/vertex representing a category. As described previously, the top left (smaller) rectangle in each picture contains five points that represent the initial categories (in this example, for data analytics) that are spatially separated. The set of clustered points in the large rectangle in each picture represents the data. The left picture shows symmetric clusters resulting from a symmetric layout of categories. When the lower left category is moved so that it is closer to some of the other categories, the clustering data points change to match the relative category positions.

To see the impact of parameter **M** on the topology, in Figure 5.5, M is chosen as a value greater than 0.5, and the resulting pattern for five initial points is random scattering, which would be the default pattern guess anyone would make for a simple function with the random aspect of NPPR. Now **M** is changed to around 0.33 in Figure 5.6, and the resulting pattern resembles the classic video game of Space Invaders. Other example patterns are included at the end of this chapter.

From above, we can see that the emergent pattern of clusters in Figure 5.6 resulted from the seemingly random cloud of points in Figure 5.5. Therefore, in this case for 5 vertices, changing the scaling parameter **M** from about 0.5 or more to about 0.33 causes the order to emerge. The lines that divide the pictures into tile-like sections are the result of applying a Voronoi function mapping to the points on the screen. The Voronoi tessellations provide an unbiased method of determining which points map to a particular cluster. This avoids the human errors or biases that would result from estimating just where a cluster begins

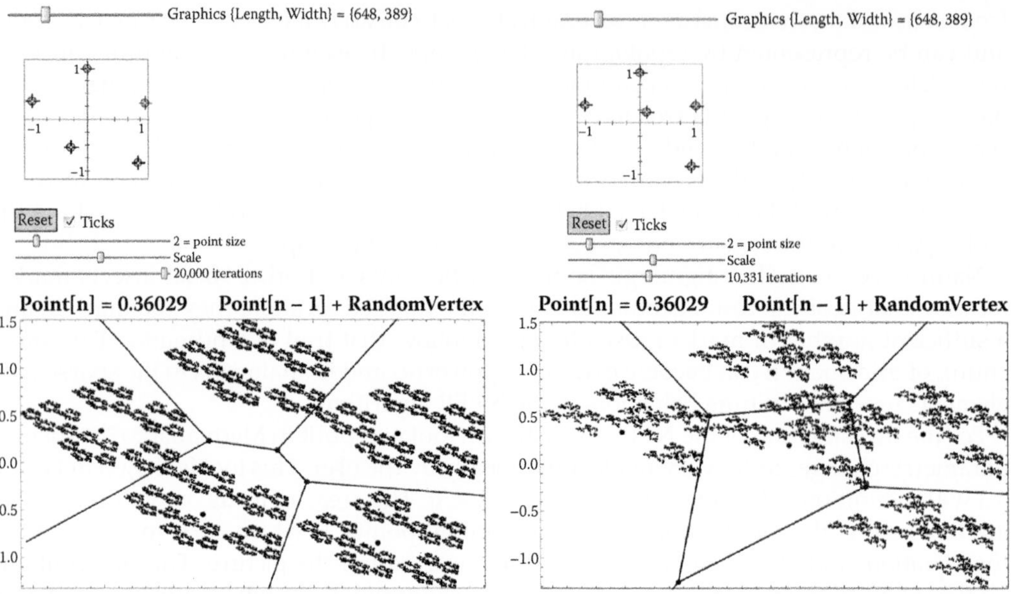

FIGURE 5.4
How data may self-organize topologically for a data analytics problem.

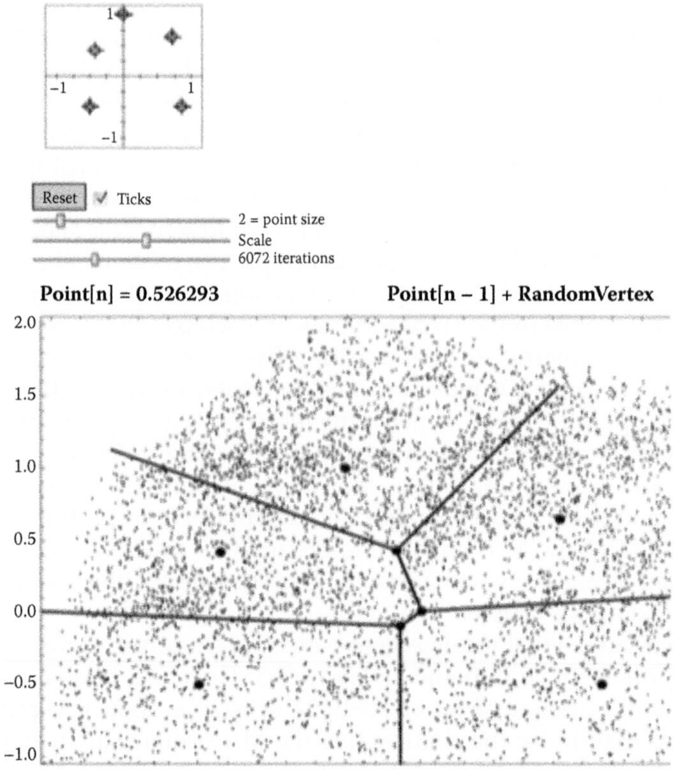

FIGURE 5.5
Random 5-vertex NPPR for M > 0.5.

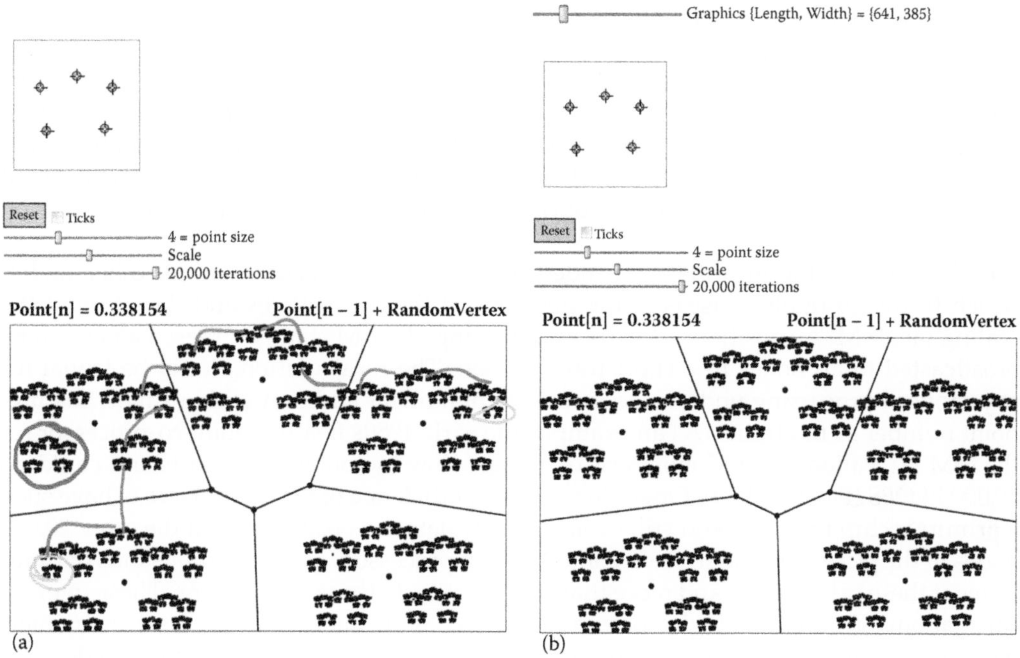

FIGURE 5.6
(a) Routes drawn between two nodes for 5-vertex structured topology. (b) Highly structured topology around vertices for M value approximately 0.33.

and ends, or the illusion of a cluster being at a location, i.e., pareidolia. The Voronoi is not infallible, as occasionally some parts of a cluster will appear on the line between two tessellations. It provides a general metric for when clusters are readily separable.

5.6.2.2 Applying NPPR to Solve the Ad-Hoc Routing Problem

We now have enough information to provide a solution to our large ad-hoc network routing problem. Using the previous highly structured topology shown in Figure 5.6, we now add routing, and define a communication distance. This result is shown in the marked-up version of 5.6, labeled as Figure 5.6a.

The five major clusters in the picture, i.e., one major cluster per Voronoi tile, is divided into five sub-clusters, one of which is circled in dark gray, which are again divided into sub-sub-clusters, and onwards due to the self-similar nature of the IFS. Let's assume that the window with the clusters is a two-dimensional map of a region within which these points are located. Let's assume that each one of the smallest points on the screen corresponds to an ad-hoc radio within an agent or autonomous vehicle (each point is not explicitly visible since this run of the tool plotted 20,000 unique points, many of which are tightly clustered). For the sub-cluster circled in dark gray, let's assume that the circle's diameter is the typical range of a point-to-point communication between two nodes. Finally, two nodes have been selected (actually for ease of viewing, two sub-sub-clusters were selected) and are circled in light gray. For the purpose of showing the routing, the light gray circled nodes were selected at diagonally opposite ends of the window.

Since each node would be capable of calculating the general layout of all the other nodes, it would be able to determine which direction to forward its message for the next leg, or "hop" of its route. The emergent self-similarity makes it easier since despite the nodes being randomly iterated, the overall topology will remain self-similar, at least for certain values of the scaling parameter **M**. If each node knows its range limitations, i.e., the diameter of the dark gray circle, then it would broadcast to the sub-sub-clusters at the edge of its range and in the appropriate direction in order to reach the destination node.

Each node has the information needed to calculate the positions of all other nodes, which in this example consists of the positions of the five vertices and the value of the scaling factor **M**. As positions of vertices are changed, or **M** changes, these values can be broadcasted to all nodes in a short transmission. The recalculation would be trivial for any modern processing node, since it consists of a linear addition with a single floating-point number as multiplier. As an example, the early 1980s floating point co-processor for the IBM PC, called the 8087 chip, was extremely slow by today's standards and could do 50,000 FLOPS (Floating Point Operations per second). If one of the nodes were to have such a primitive chip then it would still be able to calculate the positions of all the other 20,000 nodes in the picture in under one second. Modern processors are many thousands of times faster at these calculations and more than sufficient to do these calculations. These internal calculations of node topologies provide all nodes with the ability to create their internal world model that is consistent with the overall topology, hence the ability to derive a shared world model. This imparts a shared understanding for communication between the nodes, even when parameters dynamically change.

5.6.2.3 Walking through the Point-to-Point Network Connection

As we walk through the route from the source node to the destination, it would be easiest to number the clusters and sub-clusters in some order. Arbitrarily, let's go clockwise within each cluster and sub-cluster, starting with the lower left, and ending with the lower right. So, for the gray circled nodes in the lower left, we can refer to these cluster elements by three numbers (cluster, sub, and sub-sub) as 1,2,1. The gray arcs represent the paths for each hop within the range limit of each node. The approximate sequence of hops from the light gray circled nodes in 1,2,1 to the destination light gray circled nodes in 4,4,5 is as follows:

1,2,1 to 1,3,3 to 2,5,1, to 2,4,{3,4 or 5} to 3,2,1 to 3,3,2 to 3,4 {1 or 2} to 4,2,2 to 4,3,2 to 4,4 {3,4,5}.

The braces represent the approximate hops, since all the nodes within braces are in range, so if some are unavailable, others nodes can take their place.

While the exact routes chosen may not be optimal every time, this type of routing topology provides a measure of resilience for both edge-nodes and more centrally located ones, as the communication routes can work even if some sub-clusters and sub-sub-clusters lose up to half of their nodes. This has a clear advantage over current ad-hoc routing protocols, e.g., AODV (Ad-hoc On-Demand Distance Vector Routing), which suffers from large latency time in route discovery. Instead, this approach using NPPR consists of calculating just a few route permutations to find an efficient, albeit not optimal, route.

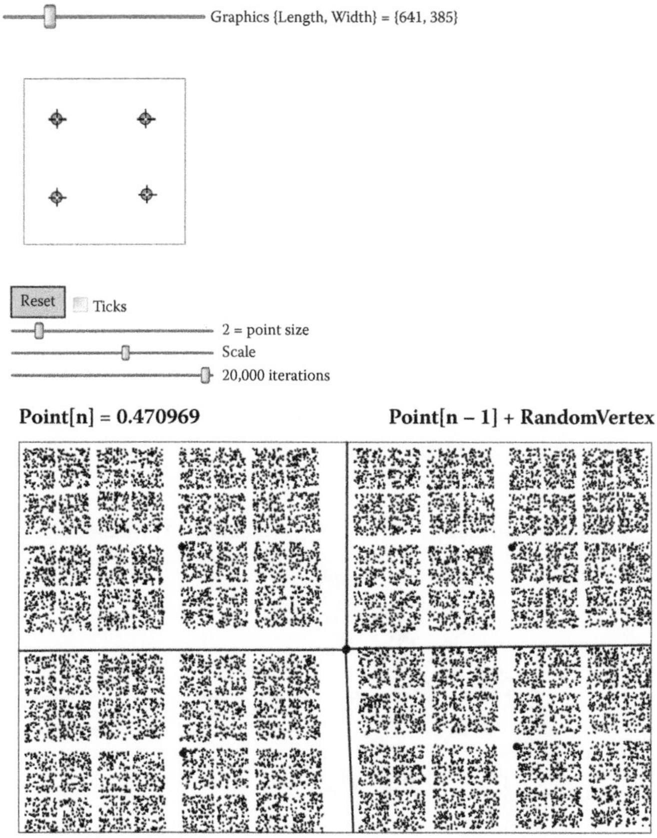

FIGURE 5.7
4-vertex stone-cubes. M = 0.47.

This topology can also apply to the emergent organization of massive swarms, since a change in one parameter can cause the entire swarm of autonomous vehicles to either produce emergent flocking behaviors, or scatter into a random pattern. The swarm application can also provide a single-point control for many vehicles whose objective is to arrive at a few destinations (i.e., the vertices can represent destinations).

The emergent patterns that appear from this simple equation can be surprisingly different, and are dependent upon the number of vertices as well as the scaling multiplier. Figures 5.7 through 5.12 show some of the patterns that emerge by varying the positions and number of vertices.

5.7 Conclusion and Future Work

Emergent behavior is generally treated as a liability, even though emergence is effectively used in nature (e.g., crystallization, flocks of birds, etc.). Current methods of addressing SoS do not take into account the emergent behavior that occurs in large-scale systems.

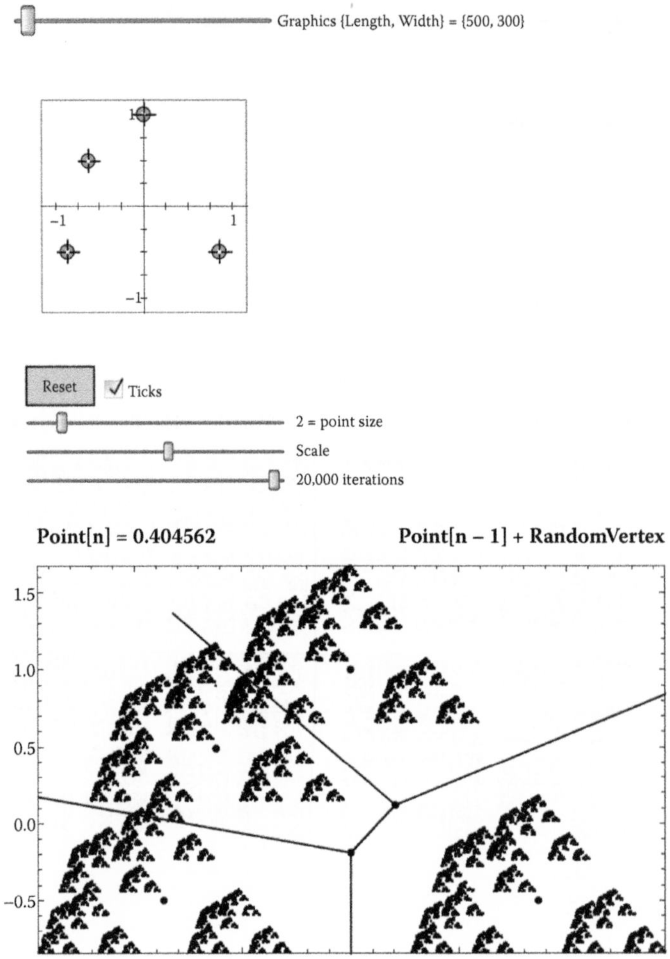

FIGURE 5.8
4-vertex oblique. M = 0.39–0.41.

Generally, these SoS will exhibit a wide range of complex behaviors. These range from well-ordered predicted behaviors to completely chaotic ones.

Several problems of scale have not yet been successfully solved, such as large swarm behaviors and the routing dynamics of large-scale ad-hoc communication networks. A first principle derivation of how to address these SoS problems leads to various examples from nature, distributed systems, and cellular automata. In order to better understand these examples, we need to consider the aspects of randomness and order at the fringe of order and chaos. This boundary contains *deterministic chaos* that can be computed, is *repeatable*, but is not the predicted or anticipated SoS behavior, and therefore exhibits *emergence*. Because we need to be able to compute the patterns effectively to "trigger" emergent behavior such as entities coalescing around a series of points, we use contractive affine

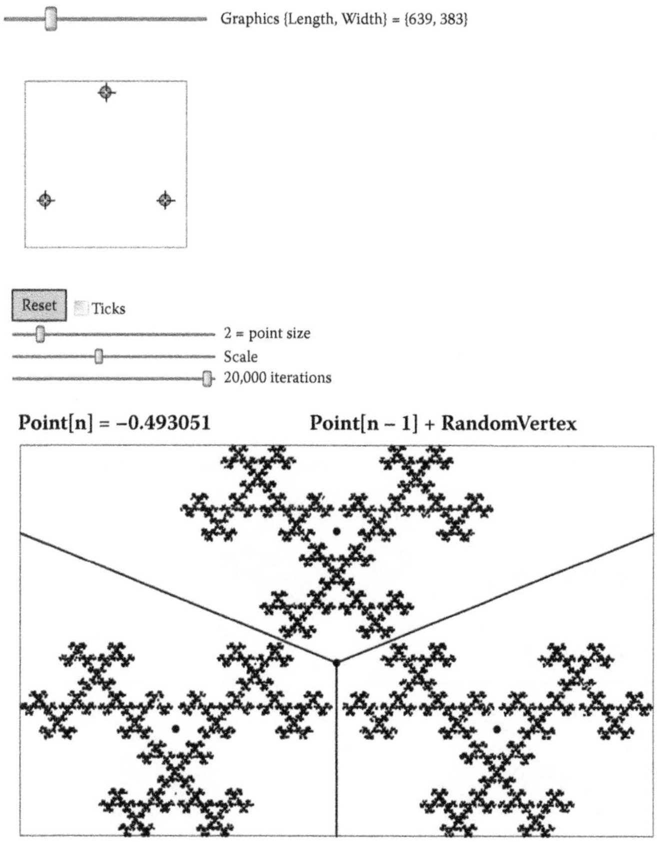

FIGURE 5.9
3-vertex inverse-Sierpinski. M = minus-0.5.

set topologies. The IFS for such sets produces *fractals,* and the algorithm described and applied here includes the dynamical nature of the SoS. Using selected constraints, random iterated points converge to ordered patterns and dynamically adapt to changes in the environment. Such emergent topologies can provide solutions to massive distributed agent problems.

The example of an IFS given in this chapter is but one of many ways to leverage emergence within complex systems to produce desired behaviors. This may be further expanded to sets of fractal attractors, called superfractals [30], and function in an analogous manner to the principle of superposition for differential equations, commonly used in quantum physics. There are also many as-yet-undiscovered methods within mathematics that may not be derivable or result from theorems or proofs. The field of experimental mathematics is likely to have many yet-to-be-discovered equations with emergent properties, some of which may solve currently intractable problems.

While it is beyond the scope of this chapter to advise on how to approach experimental mathematics, some starting points for investigation would be the deterministic chaos used

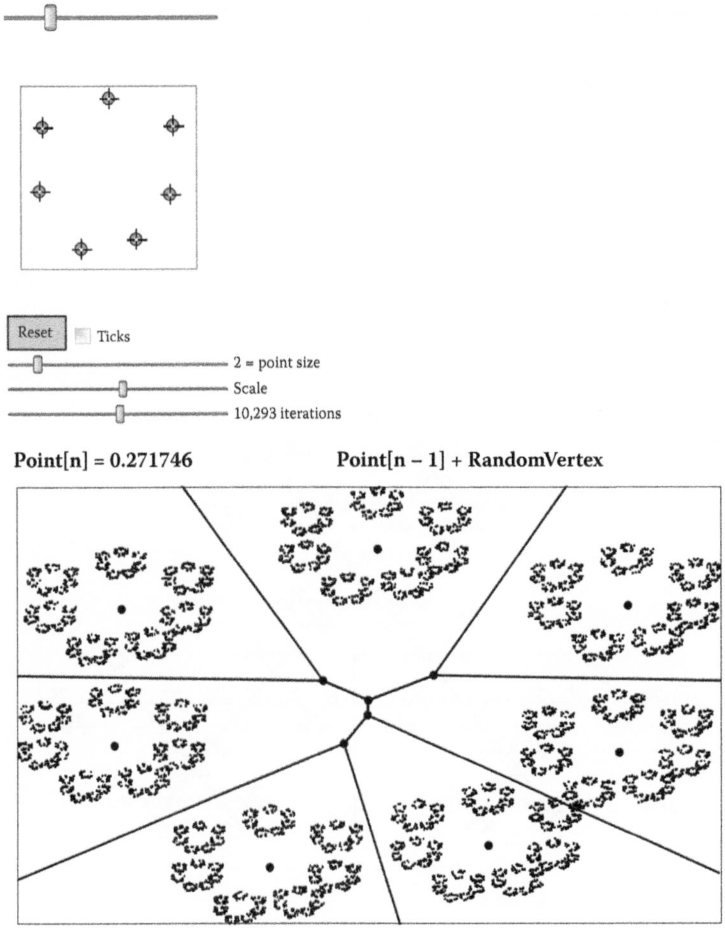

FIGURE 5.10
7-vertex ring-clusters. M = 0.23-0.30.

in this chapter, specifically leveraging off of self-similarity. Another approach would be to use machine learning (ML) algorithms that work by analogy, e.g., support vector machines and radial basis neural nets. Unsupervised ML algorithms may also provide a way to discover new mathematical algorithms, specifically when a deep learning is applied with known constraints.

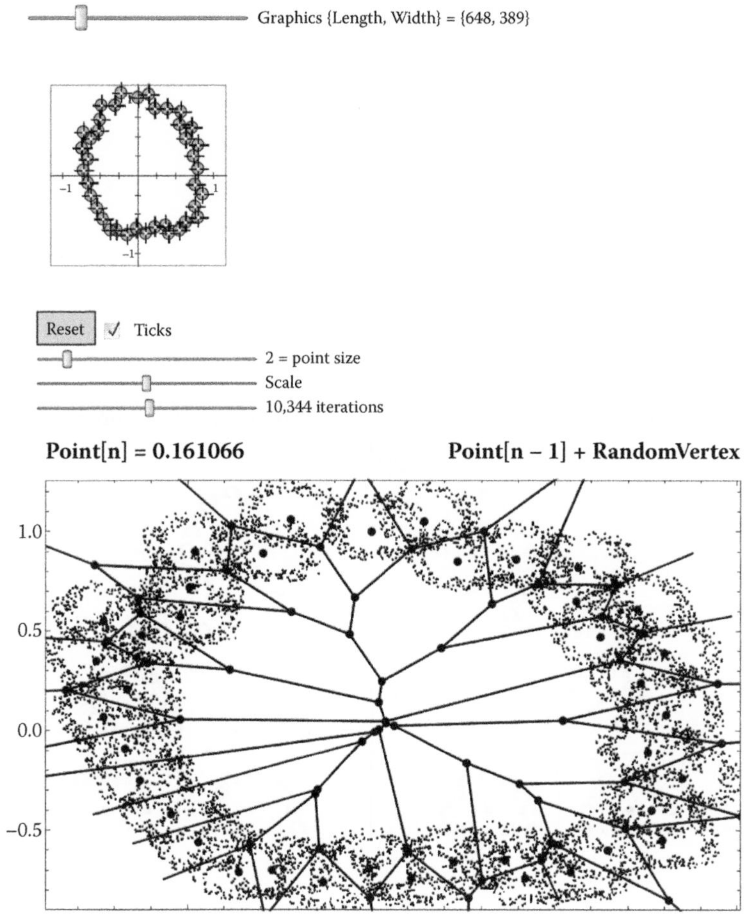

FIGURE 5.11
40-vertex ring-clusters. M = 0.16-0.17.

It would be interesting to watch the interactive dynamics of a network-centric "society" of agents constructed from this mathematical paradigm. Perhaps the lessons learned from such observation could be used to develop novel economic and political paradigms as well. This alternate approach to large-scale SoS solves some of the issues mentioned in the beginning of this chapter.

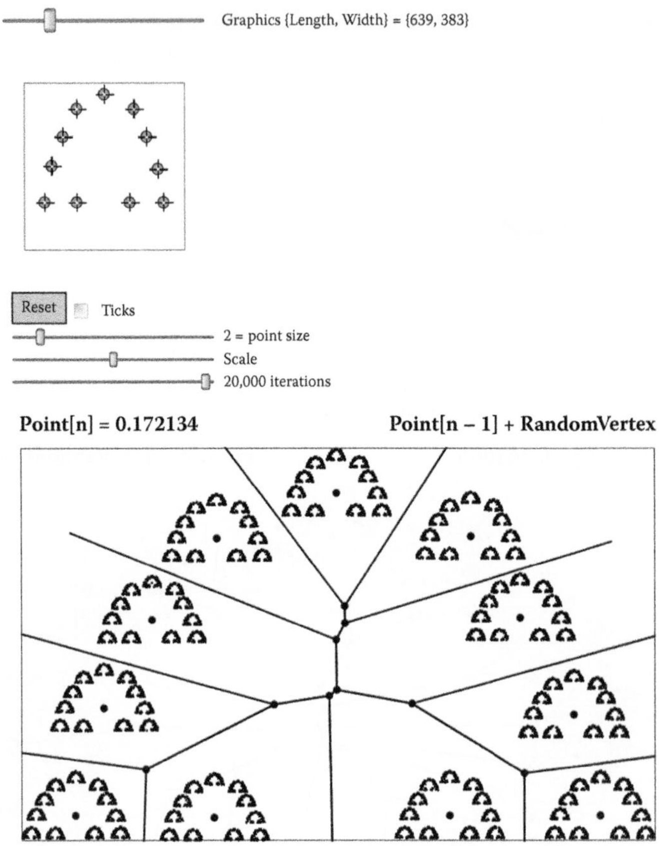

FIGURE 5.12
11-vertex bell. M = 0.17.

References

1. F. Diacu, "The solution of the n-body problem," *The Mathematical Intelligencer,* vol. 18, no. 3, pp. 66–70, 1996.
2. Office of the Deputy Under Secretary of Defense for Acquisition and Technology, "Systems and software engineering. Systems engineering guide for systems of systems, version 1.0," ODUSD(A&T) SSE, Washington, DC, 2008.
3. "Classical mechanics." [Online]. http://en.wikipedia.org/wiki/Newtonian_mechanics.
4. J. Schaff, "Chaos/complexity: Causing software failures and creating an intelligent adversary pilot," presented at the Joint Avionics Weapons Systems, Sensors and Simulation Symposium, San Diego, CA, Jul. 2001.
5. K. Park and W. Willinger, "The internet as a large-scale complex system," in *Santa Fe Institute Studies on the Sciences of Complexity,* New York: Oxford University Press, 2005.
6. D. Graham-Rowe, "Sprawling systems teeter on IT chaos," *New Scientist,* Nov. 27, 2004.
7. "Bifurcation diagram." [Online]. http://en.wikipedia. org/wiki/Bifurcation_diagram.
8. "Stygmergy." [Online]. Available at http://en.wikipedia.org/wiki/Stygmergy.
9. Library of Congress, Congressional Research Service. (2007, updated). *Network centric operations: background and oversight issues for Congress.* (Report Order Code RL32411). Retrieved on January 2, 2015 from http://www.fas.org/sgp/crs/natsec/RL32411.pdf.

10. "Ludwig Boltzmann." [Online]. http://en.wikipedia. org/wiki/Ludwig_Boltzmann.
11. Königlich Preussische Akademie der Wissenschaften zu Berlin, *Monatsberichte der Königlich Preussischen Akademie der Wissenschaften zu Berlin*, 1872.
12. "Conway's game of life." [Online]. http://en.wikipedia.org/wiki/Conway_game_of_life.
13. C. Reynolds, "Flocks, herds and schools: A distributed behavioral model," SIGGRAPH '87: *Proceedings of the 14th Annual Conference on Computer Graphics and Interactive Techniques*, pp. 25–34. Association for Computing Machinery. doi:10.1145/37401.37406. ISBN 0-89791-227-6.
14. S. Kauffman, *At Home in the Universe*. New York: Oxford University Press, 1995.
15. S. Wolfram, *A New Kind of Science*. Champaign, IL: Wolfram Media Inc., 2002.
16. P. Bak, C. Tang, and K. Wiesenfeld, "Self-organized criticality: An explanation of 1/f noise," *Physical Review Letters*, vol. 59, pp. 381–384, 1987.
17. L. Glass and M. Mackey, *From Clocks to Chaos*. Princeton, NJ: Princeton University Press, 1988.
18. A. L. Goldberger, V. Bhargava, B. J. West, and A. J. Mandell, "On a mechanism of cardiac electrical stability. The fractal hypothesis," *Biophysics Journal*, vol. 48, no. 3, pp. 525–528, 1985.
19. M. F. Barnsley, *Fractals Everywhere*. San Diego, CA: Academic Press, 1988.
20. S. Demko, S. Hodges, and B. Naylor, "Construction of fractal objects with IFS," *Computer Graphics*, vol. 19, no. 3, pp. 271–278, Jul. 1985.
21. D. Aurongzeb, "Self-assembly of fractal nanowires and stripe magnetic domain on stretchable substrate," *Applied Physics Letters*, vol. 89, no. 12, pp. 123–128, 2006.
22. H. Peitgen, H. Jurgens, and D. Saupe, *Chaos and Fractals: New Frontiers of Science*. Springer-Verlag, 1992.
23. J. Schaff, "Creating human-like behavior in a UAV: A counter-terrorism model," in *Proc. Fourteenth Annual Software Technology Conference*, May 2002.
24. J. Moffat, *Complexity Theory and Network Centric Warfare*. Washington, DC: DOD Command and Control Research Program (CCRP), 2003.
25. N. Hall, Ed., *Exploring Chaos*. New York: W. W. Norton and Co. Inc., 1993.
26. L. M. Pecora and T. L. Carroll, "Synchronization of chaotic systems," *Chaos* 25, 097611 (2015); doi: 10.1063/1.4917383.
27. B. McKelvey, "Complexity science as order-creation science: New theory, new method," E:CO Issue vol. 6 no. 4, pp. 2-27, 2004.
28. B. Hu and J. Liu, "Exploring and Exploiting Complex Behavior in Self-Organizing Multi-Agent Systems." *Proceedings of 2005 IEEE International Workshop on VLSI Design and Video Technology*, USA, pp. 137–142, 2005. doi: 10.1109/IWVDVT.2005.1504570.
29. V. Paxson and S. Floyd, "Wide area traffic: The failure of poisson modeling." IEEE/ACM Transactions on Networking, 3(3), 226–244, 1995.
30. M. F. Barnsley and A. Vince, "The chaos game on a general iterated function system." *Ergodic theory and dynamical systems*, vol. 31, no. 4, pp. 1073–1079, 2010. doi: http://dx.doi.org/10.1017/S0143385710000428.
31. Y. Bar-Yam, About Engineering Complex Systems: Multiscale Analysis and Evolutionary Engineering, 2005. Retrieved from New England Complex Systems Institute website: http://necsi.edu/research/multiscale/ESOA04.pdf.
32. D. Dagon, G. Guofei, C. P. Lee, and L. Wenke. *A taxonomy of botnet structures*. Computer Security Applications Conference, 2007. ACSAC 2007. pp. 325–339. doi: 10.1109/ACSAC.2007.44. Retrieved on 08/09/13 from: http://ieeexplore.ieee.org/stamp/stamp.jsp?tp=&arnumber=4413000&isnumber=4412960.
33. J. DeRosa, A. Grisogono, A. Ryan, and D. Norman. (2008, April). *A Research Agenda for the Engineering of Complex Systems*. Paper presented at SysCon 2008—IEEE International Systems Conference, Montreal, Canada.
34. S. Dorogovtsev and J. Mendes, "The shortest path to complex networks." *American Journal of Medical Genetics*, vol. 71, pp. 47–53, 2004.
35. J. Kleinberg, M. Sandler, and A. Slivkins, "Network failure detection and graph connectivity." *SODA '04 Proceedings of the Fifteenth Annual ACM-SIAM Symposium on Discrete Algorithms*, Philadelphia, pp. 76–85, 2004. Society for Industrial and Applied Mathematics.

36. B. Mandelbrot and J. Van Ness, "Fractional brownian motions, fractional noises and applications," *SIAM Review* vol. 10, no. 4, pp. 422–437, 1968.
37. B. Masoumi and M. R. Meybodi, "Speeding up learning automata based multi agent systems using the concepts of stigmergy and entropy." *Expert Systems with Applications*, vol. 38, no. 7, pp. 8105–8118, 2011. doi: 10.1016/j.ewsa.2010.12.152.
38. B. Shargel, H. Sayama, I. Epstein, and Y. Bar-Yam, "Optimization of robustness and connectivity in complex networks," *Physical Review Letters*, vol. 90, no. 6, 2003. 068701-1-4. doi:10.1103/PhysRevLett.90.068701.
39. M. M. Waldrop, *Complexity: The Emerging Science at the Edge of Order and Chaos*. Simon and Schuster, 1993.
40. D. Bailey, J. Borwein, N. Calkin, R. Girgensohn, D. R. Luke, and V. Moll, *Experimental Mathematics in Action*. Wellesley, MA: A.K. Peters, Ltd, 2007.
41. M. Gardner, Martin (Oct. 1970). "Mathematical Games—The fantastic combinations of John Conway's new solitaire game 'life,'" *Scientific American*. 223: 120–123. ISBN 0-89454-001-7.
42. R. A. Brooks, "Intelligence Without Reason." IJCAI '91 *Proceedings of the 12th International Joint Conference on Artificial Intelligence*, Sydney, vol. 1, pp. 569–595, 1991.
43. S. H. Strogatz, *Nonlinear Dynamics and Chaos*. Perseus Books Publishing, 1994.

6

Phenomenological and Ontological Models for Predicting Emergence

Gary O. Langford

CONTENTS

6.1 Emergence—Through Lenses of Ontology and Phenomenology

It is difficult to live outside the milieu of perception, however critical and inspired our thinking (Merleau-Ponty 1992). The very notion of theory and fact rely on observation, measurement, and interpretation as perceived and explained by observers. Observers look for patterns and derive their explanations from a foundation of consistency and personal well-being and self-worth. The objective of the observer is to distinguish provenance, relate differing perspectives, interpret order, and predict upcoming events. The study of the observer's sense of what appears to be is called phenomenology. Husserlian phenomenology (Bernet, Kern, Marbach 1993) allows for both subjectivity and objectivity (van Atten and Kennedy 2003).

Thinking critically through accepted propositions and formal methods challenges phenomenology to accept only that which is not falsifiable. However in the most exacting interpretation, ontology imposes absolute meaning through consistency of logic, continuity of method, applicability across disciplines, scalability from the micro to macro, and capaciousness. A model of ontology provides "an explicit specification of a conceptualization" (Gruber 1993), that promotes clear and specific communications, allows for the reuse of artifacts, and describes the objects and processes (including relations and sequence of relations of variables that are causal) (Garud and Kumaraswamy 1995), causality (Sowa 2000; Casini 2009), allowable interactions through transfers of EMMI (Langford 2012, 2017), and the effects of partitioning and managing boundary conditions. From the perspective of ontology, the universe can be formalized devoid of cognitive and social biases (Bäck 2011).

A primary task of ontology is to bridge gaps in knowledge and in how knowledge is used, and to provide theories of how the world works compared to the model of how the world works (Sowa 1999). Knowledge about the world can be represented by rules, abstractions of relations, interactions on many scales, and mechanisms that make things happen (Schiemenz 2002; von Bertalanffy 1968; Holland 1986). When such is merged into a proper framework, the ontology represented in the frames of that framework and the rules that drive the nexuses of the combinatorial intersections of frames provide a disciplined hermeneutic approach to exploring the actions and events within the aggregations, systems, and system of systems. The error analysis is facilitated by using the Leśniewski formalisms, the ontological framework to structure and organize the work, and the taxonomy developed by Langford (Langford 2017). The results of that error analysis are ongoing. To that end, a simple 10 × 10 matrix rule set (typical of a small system) predicted a small number of black swan emergent behaviors over the systems lifecycle.

A gap exists between the meaning and interpretation of what is observed when objects interact and the reality of the results of interaction. The gap has many edges, each forged from founding pillars of philosophy. One edge of the gap is the works of Hume positioned in regards to ethics (Socrates and Plato). Lewes forges another edge within Husserlian phenomenology. On the opposite side of the Hume-Lewes edges situates a combination of Aristotelian metaphysics and Leśniewskian ontology. The formative edge that defines the gap in interpreting emergence is explicated through the works of Stanislaw Leśniewski and his general theory of objects. However, the goal is not to close the gap or to reconcile the gap as to in any way detract from the well-reasoned pillars of philosophy. Rather, it is the existence of the gap that is important to help determine a path to explore the relation between interaction and emergence. Regardless of the pillar of philosophy chosen, the gap between the other pillars must remain distinct, consistent and plausible, and distinguishable. Each of the pillars will imbue emergence with some aspects of interaction. This research reconciles the edges of the gap through an integrative framework (Langford 2012) that organizes and structures interactions between objects. It has been from the lack of an integrative framework by which to structure an ontology for systems and system of systems that our comprehension of interactions and emergence has not progressed sufficiently to advance the knowledge of how to manage and predict emergence. From the view of model-based systems engineering and model-based system of systems engineering, the integrative framework is an essential means to formulating valid models.

6.1.1 Purpose

The purpose of this chapter is to develop insight into predicting emergence, without which effective modeling of a system of systems fails to account for consequences of missing emergent functions and emergent processes. The phenomenological interpretation of emergence and its role in modeling emergence are examined in contrast to the ontological interpretation of emergence for system of systems. Ontology is then developed for systems and system of systems, from which emergence is examined and discussed.

6.1.2 Brief Introduction to Emergence

At the instant of every beginning, there is interaction—whether it be turning on a light switch, a bird's flight, or the universe expanding. And when a sufficiency of electrons flow in an appropriate mixture of gas, through a conductor or semi-conductor, or are emitted spontaneously from algae or plants, interactions release photons. Light belongs not to a preceding occurrence, cause, or event, but rather results from the interaction and continues as such until a subsequent interaction. Similarly, the bird wings interact with molecules in the air, without which flying is not possible; and the expanding universe creates galaxies, planets, and life through a myriad of interactions. It is often the case with systems that sustainment of interactions engenders reproducible effects that can be observed after initiation. Sustainability requires energy exchange and therefore a constancy of emergence.

When photons are emitted from plants, light bulbs, and stars, light is emergent, i.e., a consequence of interactions. Flight is emergent from the interaction of wings with air, so provocative and engaging that humans took centuries to discover and apply the subtleties of such emergence. Molecular masses interact through gravitational, electromagnetic, and nuclear forces to produce nebulae, stars, and planets—all emergent from those interactions.

Emergence can be bold, yet remain mysterious; and otherwise, show mundane effects seemingly quite ordinary and unremarkable. However, linking emergence with complexity dodges the nature of emergence. And, associating emergence with causal events imposes a strong emphasis on identifying mechanisms that give rise to, force, or sustain actions that link one process to another process. Our appreciation for the nature of emergence has evolved down the road of history. To that end, the remarks of Hume related the cause with the effect in the sense that, "Now such as the parts are, such is the whole" (Hume 1839, p. 39), without acknowledging the accord between the cause and effect. Through that thinking, Hume argued that the existence of an object and our beliefs about that object did not arise from the parts of the object, but rather in the manner that we conceived the object. The notion of emergence was unacceptable by Hume plausibilities, conceived contraveneous to the idea that is anything is more than its parts.

According to Hume, our individual conceptions about an object are the same, i.e., no one has superior knowledge. If one could have superior knowledge it would imply a greater understanding of the whole part relationships, require explanation which might be different from a demonstration of the whole part relationships, and therefore constitute proof that a causal mechanism existed that could not be demonstrated. To that end, Hume (a pantheist) cites with certitude his eight common-sense general rules to establish a cause-effect relation.

1. "The cause and effect must be contiguous in space and time" (restated as the premise of contiguity).
2. "The cause must be prior to the effect" (restated as the premise of temporal proximality).

3. "There must be a constant union betwixt the cause and the effect" (restated as the premise of propinquity).
4. "The same cause always produces the same effect, and the same effect never arises but from the same cause" (restated as the premise of invariance).
5. "Like effects imply like causes …", and "… where several different objects produce the same effect, it must be by means of some quality, which we discover to be common amongst them" (restated as the premise of commonalities).
6. "The difference in the effects of two resembling objects must proceed from that particular, in which they differ"; "… irregularity proceeds from some difference in the causes" (restated as the premise of irregularity).
7. "When any object encreases or diminishes with the encrease or diminution of its cause, it is to be regarded as a compounded effect, derived from the union of the several different effects, which arise from the several different parts of the cause. The absence or presence of one part of the cause is here supposed to be always attended with the absence or presence of a proportionable part of the effect. This constant conjunction sufficiently proves, that the one part is the cause of the other" (restated as the premise of proportionateness).
8. "… an object, which exists for any time in its full perfection without any effect, is not the sole cause of that effect, but requires to be assisted by some other principle, which may forward its influence and operation. For as like effects necessarily follow from like causes, and in a contiguous time and place, their separation for a moment shews, that these causes are not compleat ones" (restated as the premise of separability).

Hume attempted to substantiate his claims that the whole is equal to the mere sum of its parts, in essence promoting what later has been termed "causal realism." Interpreting Hume through modern insights purports that causal reasoning is built on causal mechanisms that link what is demonstrated to be the cause with what we observe as the effect. Since there is no imaginative way to conceive or explain that which is different from a demonstration, a causal mechanism is the sequence of events that are justified by preconditions that must be observed. The Hume rules stipulated that the only way to determine cause and effect was by associating a pattern of events that could be interpreted according to his eight rules.

In particular, rule seven (based on the premise of proportionateness) inspired George Lewes (a British philosopher), who in 1875 countered that causes were related to two factors—resultants and emergents. Resultants can be traced directly to causes, whereas emergents are obscured from investigative work. Lewes stated:

> The great and obvious unlikeness of a product to any one of its factors, of the effect to any one of its causal moments,—unlikeness which is seen to be necessary,—has led to the fallacy that the product is unlike the combination of its factors, the effect unlike the cause, which is the single term for all the co-operants; and this simply because we select one of the factors to represent the whole, and are also in the habit of regarding the whole of the enumerated factors independently of their combination.

Said another way, the whole is equal to greater than the sum of its parts—in notable contrast to Hume's view.

Resultants derive from the interactions of physical objects—traceable in their actions and components, because they are "co-operant processes" (Lewes 1875). We observe resultants, recognizing (with slight differences from Hume) their contiguity, proximality, propinquity, invariance, irregularity, commonalities, and separability. However, the premise of proportionality is upended by observations of natural and artifactual systems. In contrast, emergents derive from the inability to "trace the steps of the process, so as to see in the product the mode of operation of each factor" (Lewes 1875). Emergents and resultants did not strike down the cause and effect representations of Hume, but effectively modified the list of eight to accommodate emergents—by replacing rule seven.

This chapter builds on the portrayal of emergence as expressed in 1875 (Lewes 1875) by generalizing the notions of resultants and emergents as the totality of actions that result from "... interactions of concrete elements" (Lewes 1875). Furthering the exposure of the essence of emergence, the actions resulting from interactions were codified and formalized by Stanislaw Leśniewski's "Foundations of the General Theory of Sets" (Leśniewski 1924). The works of Lewes and Leśniewski illuminate the nature and phenomena of emergence, which this chapter develops the foundation from which to define systems and system of systems. The basic ideas of emergence can be properly shown as one of four formal mereotopology foundations for systems and one of the five mereotopology foundations for a system of systems (Leśniewski 1988).

6.2 Expounding on Emergence

Evoking fascination, delight, and scourge, the observed effects and consequences of interactions between objects that are observed at some point during the product's lifecycle result in emergents. To believe that emergents are unusual would seem to be an oddity in nature. Why and when emergence occurs when interactions arise would either place restrictions on the types of interactions or on the objects themselves. What about an object that would precipitate something that would violate one or more of the eight principles? If the focus was only on the seventh principle (as was that of Lewes), then the right kind of interaction might act as a catalyst if the structures and thresholds of the object were "ripe" to enact emergent behavior(s). Perhaps the correct mix of structures and thresholds to stimulate actions within an object are a necessary condition for an interaction to enact emergent behavior(s).

For argument's sake, assume that a particular mix of structures and thresholds is required for an interaction to enable emergence. Is there a dividing line between emergence and no emergence for such mixtures and thresholds? Or is there some emergence that results and then more emergence results due to an increase in the intensity or geometric area of the interaction? At the molecular level, we observe discrete thresholds of energy for electronic transitions, below which atoms are stable while above which atoms become unstable. Energy build up (for example, heat imparted by photons) increases atomic vibrations as photons are absorbed. This kind of electronic excitation increases the size of the molecule, but the mass remains the same as before energy build up. As the energy increases, the molecule will be unable to expand and may dissociate, causing an electron to be expelled or the molecule to come apart. The structure of the molecule changes due to interaction with photons. Should we say that the mechanisms of excitation and dissociation cause the emergent behaviors of expansion and destruction of water molecules that are heated? Were we not to know anything about steam rising from a pot of boiling water heated by high intensity laser light, we might share our ignorance and not think of steam as emergence. With a bit more knowledge, the observers might appreciate the workings of molecular chemistry and thermodynamics in sufficiency to explain what others might deem as emergence. This *gedanken* experiment with water molecules illustrates the dividing line between what should be called emergence and what should not is determined by interaction. That which is gaseous hydrogen and that which is gaseous oxygen gas is not by simple summation, liquid water (Berenda 1951). The relations between interaction and the complexity of the objects or the actions of the objects does not distinguish emergence as more than the consequence of interaction. All interactions result in emergence, i.e., something more than the mere sum of actions of the parts. Heat from friction, however commonplace, must have been thought of as an amazing discovery, whose importance is lost in history. And then, there was the discovery of fire from two sticks of wood. Whether the experiment is with "simple" phenomenology, such as rubbing two hands together to generate heat, or the most complex situations (the study of consciousness at all levels of living things; or a vexing computer-driven communications system of systems integration), the observer may be confounded by behaviors unknown, unexpected, and indescribable. Emergence can now be investigated, defined, and its types determined.

The eight principles espoused by Hume and the contribution made by Lewes in relating the actions and consequences of interaction with resultants and emergents moved the scholarship along. The topic of emergence has ratcheted slowly since Lewes' work, progressing to a test that depends on an observer's decision to include behaviors that "surprise" (Ronald, Sipper, and Capcarrère 1999). The element of surprise was noted in this publication by Ronald, Sipper, and Capcarrère as a common theme in case studies involving artificial life. These authors caution researchers in the fields of artificial life to establish standards by which the term emergence should be used with regards to the observer's surprise, so as to not "… devalue its significance, and bring work centered on emergence into disrepute." Further, Ronald, Sipper, and Capcarrère pose five comments within the context of emergence and ontology. Their five comments regarding emergence and ontology are listed below, after which a brief discussion is warranted for each comment:

- Not an object within a context—"like a stone in your pocket"
- Not a particular behavior—"known to occur at some moment in time"
- Not a delimited category—"like the stones on some particular beach"
- Define an all-inclusive category—defining the ontology for emergence
- Observer's choice—to include or exclude behaviors

Not an object within a context: Context is the situation or framework (Aerts, Broekaert, and Gabora 2003) in which two objects interact (Langford 2012). An object without context does not account for boundaries. Boundaries, in part, determine what interactions are possible for an object that is within another object. Those interactions may proceed as prescribed by the boundary conditions of the bounding object. The emergence that occurs with an object within the boundaries of another object results from interactions between the two objects—which might result from physical contact (assuming the objects are not interactive through electromagnetic fields, e.g., thermal absorption or radiative emissions). The context of all objects is an essential consideration for ontology and emergence.

Not a particular behavior: Behaviors are defined by the use of an operational definition (Kerlinger and Lee 2000) that specifies objects and functions in ways that are measurable. Behaviors result because functions either do or do not exist (Langford 2012). Additionally, behaviors result because objects do or do not exists and objects are interacting or are not interacting (Langford 2012). Objects that can anticipate interactions due to an awareness of context may by the nature of their boundary conditions react to the existence of an object (friendly, threat, unknown) or the absence of an object (known but unseen; or expected and unseen). Behaviors change when we expect something and it happens, not expect something and it happens, or expect something and it does not happen. It is possible, but rare, there would be situations that change behaviors when we are not expecting something and it does not happen. The element of timing of an expected or unexpected event is subject to the context and the type of interactions that could occur, regardless of the expectations. Any deviation from the expected timing of an event might be cause for investigation to determine whether the expected mechanism was operable or inoperable. Similarly, objects that interact produce functions that may or may not have utility in a given context and therefore may or may not be useful (Taguchi 1986; Ariew 2002). Behaviors change because object is available to use, not available to use, or whose availability is unknown but expected. The particulars of any and all behaviors are essential considerations for ontology and emergence, as behaviors change because of objects and functions or lack thereof.

Not a delimited category and ask, how do we define the bounds of emergence, i.e., where, when, if, how, why, what occurs? Emergence appears everywhere with and without pattern—always resulting from interactions between objects. Perhaps a particular beach has very coarse sand (1.0–2.0 mm diameter) which causes the flow of water back from the wave surge to "bark," i.e., an emergent from the interaction of fast flowing water over rough sand (or rocks). Imagine the surprise when a person takes off the earbuds and hears barking sand. No category is without emergence as long as there are at least two objects present. The water rushing over the sand moves the sand which creates sound. The person who hears the sound refers to the beach as "barking sands" in text messages sent worldwide. Emergence is found with each of the interactions—physical (at the boundary of water interacting with sand to generate pressure waves in the air), functional (at the boundary of the air pressure wave and the person's ear drum "to hear 'barking sounds'"), behavioral (to send text message), and so forth from person to person.

How do we define the bounds of emergence, i.e., where, when, if, how, why, what occurs? Emergence appears localized in the physical/intellectual domain, more widespread in the functional domain, and limited only by the spread of behaviors.

Then, define an all-inclusive category: The ontology may not be one category, but rather many categories. The kinds of categories that encompass emergent behaviors is determinable by the theory and the evaluative framework by which theory is brought into practice. The role of theory is to organize, explain, and predict actions and events. Theory inspires the application and use of frameworks from which to observe, explain, and interpret empirical data and qualitative aspects that take on significance for the practice that indurates the applicability of the theory's use. The link between theory and its use is stringently enforced to maintain contiguity of boundaries (in the comprehensive senses of quantum mechanics, classical mechanics, and relativistic physics), thereby including those objects that are contiguous (within an appropriate space and time juxtaposition). There underlies the importance of this comment by Ronald, Sipper, and Capcarrère. The condition of boundary contiguity ensures that appropriate partitioning of objects is achieved so that all conceptualized as within the category is indeed within the category.

Observer's choice: The intent of their publication was to justify the naming of an observation that surprised the observers. The reasons are many, but in the main the primary one is there has not been a sufficient grounding in theory, ontology, and method to tease the essence of emergence from the flutter of ideas that rely on empty platitudes to find footholds in the discussion. When a behavior is unexpected, unknown, or surprising we have taken to using the word emergent—the now commonplace word that is as empty in value as it is in worth. It neither improves our understanding of a given situation nor decreases our losses due to our ignorance. If surprised by what is observed, observers who are less ignorant may identify more resultants and fewer emergents than observers with less knowledge. Therefore, it is the progression of knowledge across the field, domain, and discipline that helps observers define the thresholds for identifying resultants and detecting emergents rather than an observer's choice. This reaffirms that the test of ignorance dominates the determination of what are and what are not emergents. This determination is far from the intent of either Hume or Lewes; and indeed, only applicable should the observer influence the interactions taking place between the observed objects.

6.3 Exploring Emergence

To explore emergence, there should be a starting point with suitable provenience, a course of conduct, a Gladstone bag, and a statement of inquiry. The beginning of this research journey to explore the nature of emergence focuses on Hume's common-sense general rules as moderated by Lewes' characterization of emergence arising from interacting objects. The goal of the research is to answer the question that looms because of the knowledge gap—the difference between the meaning and interpretation of what is observed when objects interact. The research question is then: What about the nature of interaction causes resultants and emergents? While related questions abound, the aim of this research is to explore the nature of interaction from the perspectives of theory, approach, method, framework, ontology, and taxonomy. Fulfilling the needs and requirements for these perspectives stuffs the Gladstone bag with the appropriate set of tools, techniques, knowledge, and sensitivities necessary to explore various paths on which the researcher follows a course of conduct in search of an answer to the research question.

There are various approaches for exploring resultants and emergents. The portrayal that the observer's surprise correlated with emergence (Ronald, Sipper, and Capcarrère 1999) applied phenomenological thinking, i.e., the structures of consciousness (Smith 2013), to bear on the subject of emergence. There, the test for emergence was cast as it appeared experientially to the observer—phenomenologically sound evocation. To differentiate between the ethical interpretation (Hume 1839), the phenomenological interpretation of emergence (Ronald, Sipper, and Capcarrère 1999), and the logical explanation based on the causal association of interactions and emergence (Lewes 1875) suggests that another pillar of philosophy may illuminate the conceptualization of emergence and possibly a different discourse. Ethics premises a notion of moral excellence, without evidence; logic determines if evidence has been found, but does not go about finding it; and phenomenology associates observations with emotions, yet with no assurance that emotions bear relation with causal actions. Departing from the monist view taken by emergentists (McLaughlin 1992) and the view of nonreductive physicalism (Beckermann, Flohr, and Kim 1992) (Bechtel and Richardson 1992), this research explores emergence through the bifocals of ontology and fundamental mechanisms. Ontology (the broader view) is the study of what is, through broad, all-encompassing categories that capture causation and interaction. Within the ontological structures, fundamental mechanisms form the causal process engines that convert inputs of Energy, Matter, Material wealth (e.g., money), and Information (EMMI) into outputs of EMMI.

Through a progression of activities, albeit sometimes seemingly unstructured and asynchronous, physical/intellectual objects are provided with the capacity* to do work. Process builds capacity through mechanisms. Mechanisms provide the means by which objects and processes change. EMMI is the capacity to do work. Process builds capacity, EMMI is the capacity. The rate that process builds capacity refers to the replenishment of EMMI that is used, i.e., the quicker an object has the ability to do a particular amount of work in the manner that is typical for the circumstances. The amount of work that can be done is determined by the amount of EMMI available to be used. EMMI is the capacity. Therefore, interaction, i.e., the enactment of mechanisms, results in a change in capacity, capability, and ability.† At the fundamental level, the mechanistic view is based on interactions. A broader discussion across various issues embodied in processes, approaches, models, perspectives, and relevant theory brings out the fundamental mechanisms from which interaction operates on objects and processes. The scope of this chapter is not to explicate these deeper issues, but rather to realize that at the core of all interactions regardless of scale (micro-, meso-, or macro-), fundamental mechanisms constitute and sustain all physical/intellectual objects and their aggregate structures through endogenous and exogenous interactions. Fundamental mechanisms cause change by changing observables within ontology (i.e., the structure of being) and changing the appearance of what we observe (phenomenology).

The value of a process is characterized by a set of measures of accomplishment (through comparison with other processes). Accomplishment is a combination of resultants and emergents from interaction—driven by process. One of the measures of a process is effectiveness. Measures of effectiveness are beyond the scope of this chapter.

6.4 Paths of Ontological Investigation

The paths of research—that is, the approach—based on ontological thought are diverse and mostly well-trodden (Emmeche, Køppe, and Stjernfelt 1997; Ferreira and Tejeda, Jenny 2011; Honour and Valerdi 2006; Madni 2006), including taxonomy (Rueger 2000; Bedau 2002; Fromm 2005; Holland 2007); representations of ontologies (Uschold and Gruninger 1996; Bar-Yam 2004); simulation (Bedau 2002; Beurier, Simonin, and Ferber 2002; Bonabeau 2002; Chen, Nagl, and Clack 2008); and prediction (Heylighen 1989; Crutchfield 1994; Shalizi and Crutchfield 2001).

* Capacity is the potential to do work, i.e., object has the potential EMMI to do work; capability is the fitness to do an amount of work and do it to the appropriate level of quality, i.e., object has the minimum threshold of ability with acceptable quality to do the required work; and ability is the requisite aptitude (the actual knowledge and skills), i.e., object has the minimum knowledge and skills to do the required work.

† Note that the vulnerability of an object refers to the capability to resist and recover from the impact of a hazard. Therefore, a vulnerability can be construed to be a weakness that can result in a loss (Langford 2012) of capability. A measure of disruptions to an object is its vulnerability. Whether a vulnerability is exploitable is referred to as susceptibility.

6.5 Research Approach

The approach used in this research was hermeneutic (Ricoeur 1979) as furthered by including an interpretive perspective to include emergence, such as "breakdowns" at the cessation of interaction due to faulty design (Winograd 1995, 1996, 1997). This approach was premised on mechanisms that operate within objects—in keeping with the formal mereology of objects and processes (Leśniewski 1988), where mereology is the abstract study of whole-part relationships. The specifics of abstraction, that restricted the details of the activities comprising the mechanisms, and scope that clarified the extent of the investigation, are described briefly.

6.6 Foundational Mereological Structure

The objective of this approach was to investigate emergence through a theoretic interpretive framework that parses interaction according to ontology of objects and processes. This framework is of ontological origin and intent. The particular framing in two ontological domains grew out of the ideas in Stanislaw Leśniewski's "Foundations of the General Theory of Sets" (Leśniewski 1988). A formal mereotopology foundation was developed for a general structuring of systems and system of systems. Through an expository framework of object-related and process-related concepts all interactions can be categorized. In the most general appreciation of epistemology, such a mereotopology framework that codifies the most comprehensive ontology represents all knowledge (Sowa 2000). The integrative framework was first introduced in the context of a general theory of integration (Langford 2012, 2013). Then, the integrative framework was applied to the notional ontology for systems and system of systems for developing the rules of taxonomy and to demonstrate "... flexible dimensionalities and proper atomic form to allow for integration according to the rules of part-whole mereology so as to capture all relationships between structures, delineate all processes, and stipulate every interaction between objects in both event-based and time-based contexts" (Langford and Langford 2017). This research expands the integrative framework to an interpretive framework that juxtaposes the ontology of interaction and the ontology of causation.

6.6.1 Foundational Mereological Structure

Within the logic of objects and processes (Leśniewski 1984; Boyd 1999; Goodman 1978; Dori 2002; Radder 2006; Hacking 2007; Osorio, Dori, and Sussman 2011) two frames of dimensionality are constructed (Langford 2012). The essential structure of the interpretive integrative framework is the division into mutually exclusive categories: one based on objects and their interactions, and the other based on processes and their causal mechanisms. The frame for objects has three dimensions—physical/intellectual objects, interactions of physical/intellectual objects (i.e., functions), and behavioral consequences of objects or the lack of objects, and behavioral consequences due to functions or lack of functions. The frame for processes has three dimensions—cognitive activity (cogitating), mechanisms (mechanizing), and modeling (i.e., the process of representing cognitive activities and mechanisms). The object frame and the process frame are ontologically orthogonal, i.e., formally independent as shown through predicate calculus (Leśniewski 1988). Langford (2012, 2013) interpreted Leśniewski's formal relationships between objects and processes by formulating an interpretive integrative framework that integrated the three dimensions of process along the abscissa axis with that of the three dimensions of objects along the ordinate axis. Therefore, the three dimensions of objects map to each of the three dimensions of processes, and vice versa. The intersecting frames of object dimensions and process dimensions form the interpretive integrative framework. This framework of causal interactions and causal processes captures the totality of all objects, processes, and their actions (Langford 2016).

The formal ontology of objects and processes captures all the results of interaction within two schemas (objects and processes). Leśniewski's contribution is for a first-order theory of a binary predicate that has been shown to be a complete descriptive foundation for complete Boolean algebras and whose extensions afford it the likelihood of being the theoretical basis of set theory. And, while there is not now considerable interest in extending Leśniewskian mereology to be a new foundational formal theory for mathematics, that was the intent of Leśniewski—cut short by his untimely death. The formal structure and the mathematical model that underpins the integrative interpretive framework formulate the complete causal essence of the universe, the world, nature, and all that is built—regardless of complexity. That causal essence captures both resultants and emergents as a consequence of the allowable rules for interaction.

6.6.1.1 Ontology of Interaction: The Object Frame

Objects are entities that are things or represent things physical. Objects are of obvious interest to every living thing, and as such, objects have been accommodated, considered, or discussed since life's beginning. Before Parmenides and Plato, to Einstein (Einstein 1950), Gödel (Gödel 1951 [1995]), and after (Korman 2011), objects continue to reveal different aspects of their existence and importance (Langford 2012). Objects have location in space (Everett III 1957) and are measurable in that position space (Lyre 1995), and are representable by ideas that are the agents of physical/intellectual things. Objects are both physical and intellectual. Physical objects are made up of parts that are also physical objects; therefore objects are divisible. An object is an object regardless of the number of objects that comprise the parts of an object. Physical and intellectual objects make up the world and all within it, the solar systems, and the universe. The cognitive representation of an object is an object.

Some objects are artifactual and are designed and made to have intended functions. Not all objects have intended functions. But, if an object has no function, then it is not interactive with any other object. The teleological discourse is broadly extended to artifactual and natural objects. Objects are absolutes (Martin 1988). Physical/intellectual objects, people, and ideas are representative of objects. As absolutes, objects are viewed independent of other objects. An object's independence results from its individual nature that is differentiable from other objects. Each object is bounded, i.e., separated from other objects.

Objects exist within metric spaces. A metric space is a set of objects where each object is identifiable by the notion of "distance" between objects. Mereotopology represents objects as extended parts in its space, akin to locale theory (Johnstone 2001). Although Leśniewski did not incorporate the notion of individual points within his mereotopology, the means to add points (and therefore the concept of distance) would be to define points as objects with no proper parts. Then, distance could be construed to indicate the degree of separation.

That distance might be measurable with reference to a standard (e.g., meter), by ratio (e.g., the quantitative relation between two amounts showing the number of times one value contains or is contained with the other), or by a generalized distance that results from the correlation of variables (e.g., "... a single measure of the degree of divergence in the mean values of different characteristics of a population ..." Mahalanobis 1936; Taguchi and Jugulum 2002). The boundedness of an object and the distance between objects can be measured objectively (relative to a standard) and subjectively (relative to self). Multidimensional scaling is a technique of exposing the essential dimensions and distances between objects that are inspired by subjective judgments. In this manner, objective or subjective measures can determine the distances. This emphasis on dimensionality is facilitated by interpreting the similarities, dissimilarities, or proximities of objects as patterns (which have scalability in terms of distance). Distance is a measure of difference(s) between objects as scaled by the relative juxtaposition of objects (e.g., patterns) and the absolute limits of distance imposed by the separation(s) of the objects. With objects and distances incorporated into mereological formulations, the Leśniewski formal theory for whole part relations is well suited for modeling, designing, building, and sustaining systems and system of systems. The interpretive integrative framework provides the object frame from which to organize physical objects, interactions of those objects, and behaviors that result from interactions or lack of interactions.

6.6.1.1.1 Object Boundaries

All objects are bounded. Boundedness is perfected by three types of boundaries that are the limits of physical objects. These limits are expressed as the physical limits of the object, the results of the uses of the object (functional space), and the consequences of either having the object or the function or not having the object or the function (behavioral domains). The physical limits of an object have continuity of parts that comprise the limit of an object. That continuity is expressed as the molecular components of the objects, commonly referred to as the surface of the object. At the micro-scale, the object's surface will have unbounded molecular components that may be in proximity but not strongly bonded. The micro-scale propinquity is beyond the scope of this discussion. However, the meso-scale issues derive from the context and environmental situation in which the object's physical, functional, and behavioral boundaries are considered. If the temperature and pressure is such that the physical surface is reactive with its environment, then significant quantities of the object's molecular material may be associated with the object, but neither strongly bonded nor in close proximity to the bulk of the object's material. The meso-scale propinquity is also beyond the scope of this discussion. In the classical physical sense, the boundary at the macro-scale is as determined to be the surface, i.e., the commonly referred to surface of an object that can be seen and touched.

When objects interact with other objects, the effects are determined by the condition of connectivity. Physically touching the surface of an object by another object (macro-scale connectivity) results in an exchange of EMMI. The matter-to-matter contact between two objects enacts a transfer of energy, matter (perhaps material wealth, e.g., money), or information. The physical boundaries of the two objects are made observable both by the lack of exchange of EMMI as well as the transfer of EMMI.

Definition: An object is a bounded set of material properties and traits whose structure is comprised of constituent objects and mechanisms. An object is or represents a tangible, material thing. Objects have influence. Objects have boundaries. Objects can change. Objects have properties. Objects have traits. Objects have attributes. Objects can be produced. Objects can interact with other objects. Objects interact with other objects through energy, matter, material wealth, and information (EMMI). EMMI expresses the interactions between objects.

6.6.1.1.2 *Object Boundary Conditions*

An object is influenced by its sensitivities to conditions. For example, conditions that define the mediation of exchange of EMMI across a boundary are termed *boundary conditions*. A convenient means of describing objects is to characterize their intrinsic properties, situational traits, and attributes. A condition is the circumstances that encompass an object; the factors that affect the manner and ways in which the object interacts; the situation in which the object operates; or the terms under which an object behaves (Langford 2012, p. 42). Citing Norman Swartz, "A condition *A* is said to be *necessary* for a condition *B*, if (and only if) the falsity (/nonexistence/non-occurrence) [as the case may be] of *A* guarantees (or brings about) the falsity (/nonexistence/non-occurrence) of *B*" (Swartz 1997). Many conditions (e.g., circumstances) may apply. An object is influenced by its sensitivities to these conditions. Boundary conditions mediate the flow of EMMI across interfaces at boundaries.

6.6.1.1.3 *Object Interaction*

Objects interact. Objects interact with other objects through the exchange of EMMI. A sampling from the many authors who have discussed interactions based on various inputs and outputs for functions is presented here: energy and matter (Sage and Armstrong 2000); energy and information (Morris and Pinto 2004); energy, material, and information (Wieringa 1996; Oliver, Kelliher, and Keegan 1997; Kossiakoff 2003); and energy, matter, and information (Miller 1978; Bornemann and Wenzel 2006; White 2007; Edwards 2009; Tan, Hao, and Yang 2003; Wells and Sage 2009). However, the expression of energy, matter, material wealth, and information (EMMI) (Langford 2012) captures both the physical domain of inputs and outputs as well as the intellectual domain of inputs and outputs, according to the general expression of objects as physical and intellectual. Therefore, EMMI is the means of expressing interaction for the purpose of integration (Langford 2012, 2013) and emergence (Langford 2017).

Interaction leads to constraints on the two objects (Langford 2013). Limitations (boundaries) are given by the domain of the interactions, whereas, constraints are a structural property of the interactions.

> Therefore, interactions cause a change in boundaries, which in turn changes the constraints on the objects, which results in emergence.

6.6.1.1.4 *Object Constraints*

A constraint exists due to the conditions that govern within the context of objects or processes. These conditions are specified by the amount of EMMI that is available for an object or process. The mechanism, which drives the physical manifestations of an object, acts according to the capacity, the needs of the mechanism(s), and the kinds and amounts of EMMI. The object's output EMMI is a transformation of the input EMMI and the effects of EMMI on the object's mechanism. In all cases of interaction, there is loss of EMMI. Similarly, the conditions under which a process is enabled and driven by input EMMI through its procedures and mechanistic behaviors and then carried out by objects are likewise constraining for the process. As well, the processes interact to constrain processes that result in objects being constrained if those objects also interact with the constrained process. An object or process under such conditions is said to be constrained. One constrained process within a network of constrained processes can constrain the whole of the network, depending on the architecture and design of that network.

Constraints change the boundaries and boundary conditions of objects. Changes in boundaries or boundary conditions precipitate emergence. The amount of change in the constraints and therefore the amount of change in the boundary conditions leads to non-linear changes that are observed as emergence. There are threshold effects that limit the amount of emergence that is observable. In other words, a small, imperceptible change in boundary conditions may not be observable, but the cumulative effects of small changes in boundary conditions may exceed a threshold after which the effects can cause subsequent interactions to be observable, but without traceability to the precipitating sub-threshold effects. Interacting entities are subject to limitations, conditions, and constraints (Langford 2012).

Boundary conditions are either satisfied by incident EMMI or not satisfied. If satisfied, then the EMMI becomes an input to the object. If not satisfied, the incident EMMI is either not accepted or in some fashion enters the physical boundary of the object. The object may be disrupted or destroyed if its boundary conditions are not satisfied. As with constraints, a limit imposes a boundary and changing a limit changes a boundary and perhaps a boundary condition.

6.6.1.1.5 Objects in Context

Premised on a state-context-property theory (Aerts and Gabora 2005), objects exhibit various traits based on context. While objects may be categorized according to any number of priorities or measures, a rule-based categorization process is inconsistent with the results of modern empirical and qualitative research (Rosch 1978, 1983, 1999). Observing objects within various contexts, from moving frames of reference (e.g., velocity) to changes due to processes, the result of contexts is often noticed as a change in traits of an object. Relating observable traits to an internal mechanism of an object introduces the concept of "state." The conception of "state" is a level of detail unnecessary for this chapter and is therefore beyond its scope. However, without adopting the concept of "state," as is done by Aerts and Gabora (Aerts and Gabora 2005), objects exist in context(s). It is that context that imbues objects with various traits that we observe and measure. As the context changes, the measurements may change. The characterization of objects as perhaps exhibiting different observables by changes in their traits is not meant to imply that context is only the mechanism that is correlated with traits. Changes that are internal to the object also may result in a host of different observables. This research posits that objects may exhibit various traits due to context, and context may change due to its interactions with the object. Similarly, the object may change its context. For example, moving an object with mass m may result in changes in mass as a function of velocity v, and in addition, space-time may be changed by its interactions with the object.

If certain traits are in some way related to or identifiable as properties in a given context, then those traits might in fact be emergent properties. This formulation of emergent behaviors is consistent with Lewes' definition and current interpretations as expressed in systems engineering literature (Polani 2006; Fromm 2005; Valerdi et al. 2011). Rather than ascribe emergent behaviors as an attribute which is intrinsic to objects, that context and objects are operands on each other through interaction offers additional insight into the context by which emergence arises. Systems integration creates contexts within which objects interact and it is that context coupled with the object from which emergence arises. In a general sense, integration may be merely an object with its context. The impact of context is observed "... as an element of complexity ... ," "... in the case of adversary-created emergence, changes in enemy tactics serve to adapt or interact with our efforts to ..." interpret those (Cabana et al. 2006). In these interpretations of observed emergence, "... the principle that whole entities exhibit properties which are meaningful only when attributed to the whole, not to its parts ..." is discussed in (Hitchins 1992, p. 10); and (Weiss and Glanville 2002).

The discussion of objects in context relates to emergence through precipitant events arising out of objects as they change contexts. This change is shown to be a testable hypothesis, with the result that there exists a threshold for emergence that can be defined and is observable (Langford 2013). Whether there is actually a gradient of change that is at some point versus one event that results in a "tipping point" (Gladwell 2002) is identified as a topic for further research, suggesting a transition from an element (as an object that is changed due to a process) based on a threshold(s) that represents some resistance to the transition to from one object to a new object.

6.6.1.1.6 Functions

The exchange of EMMI creates a function that did not exist before the interaction between two objects, notionally based on Gottlob Frege's ontology of properties (concepts as part of realism) and relations of objects, i.e., functions (Frege 1979, p. 177). It is the interaction that provides the use of one object by another. Without interaction, there is no access to another object and therefore no means to use that other object. The minimum requirement for an object to use another object is interaction sufficient to provide for that use.

A function is the action realized when objects interact (Langford 2012). A function is measurable by performances, referred to as functional performances. Functions arise due to the enactments of mechanisms of at least two objects that are interacting. Functions have inputs of EMMI and outputs of EMMI. More formally, Object O and Object P create a function f if and only if:

a. Object O interacts with Object P.
b. Function f is the set of all actions that are a consequence of Object O interacting with Object P.
c. An action is defined as the release or receipt of EMMI.
d. EMMI represents energy, matter, material wealth, and information. EMMI activates mechanisms.
e. Object O is changed by its mechanism(s) because of an interaction with Object P, or Object P is changed by its mechanism(s) because of an interaction with Object O, or both Object O and Object P are changed by their respective mechanism(s) because of an interaction.
f. Mechanism is set of rules and logic constrained by context and environment that govern the transformation of EMMI by objects.
g. Mechanisms are the means or methods by which objects change, i.e., the execution of activities and acts.
h. A change is defined as the difference between an object before interaction and the same object after interaction.

An example of interaction to produce a function with emergence is to use a carpenter's hammer to nail a 2 x 4 brace and brackets to provide restraints on an installed, gas-fired water heater. The purpose of the project is to provide bracing of the water heater to the structural wood-framed wall to secure the water heater in case of an earthquake. This situation is commonplace in a residential setting within an earthquake prone area. The function of "to secure (water heater)" is in response to the problem of preventing a gas-line attached to the water heater from becoming dislodged during an earthquake and causing a fire that results in loss of life or property due to inadequate response from overwhelmed emergency services. A hammer, nails, and bracing materials are needed to secure a water heater to a structural wall. Procedurally, the gas should be turned off before working in the area surrounding the water heater, then turned on once the task of securing the water heated is completed. Buffo, the homeowner's pet dog, stays with the tradesperson from the start to finish of these activities at the home. The interaction between the tradesperson and the hammer with a mass of 473 grams to pick up, swing, and strike nail, produces the function of "use (hammer)." The subfunction of "F: to strike (nail)" begins the interaction between the hammer and the nail. Since functions are measurable and testable, the functional performance specification for "FP: to use (hammer)" is an 850 +1100/-730 Newton force to perform the subfunction of "F: to strike (nail)." Buffo places his mouth over the gas shut-off valve and cracks it open slightly after the tradesperson leaves the work area to respond to a distressing text message. When the tradesperson returns, he finds Buffo waging his tail, quite content that he has helped. Thinking about the text message, the tradesperson slams the hammer onto a 16-penny nail (posed to secure a strap to the wall), clipping the edge of the nail, then ricocheting the nail point into the nearby metal corrugated connector pipe, puncturing the gas line. Gas begins to leak into the immediate area. Without realizing the damage and still reeling from the text message, the tradesperson jams another nail into the same nail-point site and winds up, taking quick aim, then punishes the nail with a thrashing blow. A spark flies from the contact point at the physical boundary of the hammer and the nail. Neither the tradesperson nor Buffo are harmed by the explosion or during the writing of this example. The resultant function of "F: to nail" was enacted with less functional performance than required to complete the intended task. The functional performance was inadequate and "out of spec"; the boundary conditions were not satisfied; yet the function of "to use" was enacted. The interactions between the tradesperson and the hammer and then two combinations of the tradesperson, hammer, nail resulted in the additional functions of "F: to fly (nail)"; "F: to puncture (pipe)"; "F: to create spark"; "F: to ignite gas"; and "F: to knock bejezes out of the cognitive processing of stakeholders (tradesperson and Buffo)." The function of "F: to fly (nail)" results from interaction between tradesperson-hammer-nail, as measured at 2 meters/second; the function of "F: to puncture (pipe)" results from the interaction between nail-pipe, as measured at 150 Newton force to penetrate gas pipe; the function of "F: to create spark" results from interaction between tradesperson-hammer-nail, as measured at 0.05 Kilojoule; the function of "F: ignite gas" results from the interaction between spark and mixture of gas and air, as measured at 5–15% natural gas concentration in a volume of air with 0.33 Millijoules of spark energy; and the function of "to shock cognitive processing" as measured by blast overpressure of 700-1400 kilograms per square meter. All of the interactions in this example show emergence, some intended, some not. Regardless of intention, the results of interaction are the enabling of functions, i.e., uses that may or may not be wanted or used.

If two objects interact such that one or both of the objects change, then a function is derived as a trait of the relation or behaviors that is evoked because of the interaction. A function is a set of actions that result when two objects interact. A function is measured by performance of the bi-object's properties or traits, either individually or pairwise.

Functions arise due to the enactments of the mechanisms of at least two objects. Intended functions require proper EMMI.

6.6.1.1.7 Proper Part

A proper part is a trait of an object that is comprised of multiple objects, all of which are essential to the actions of the composite object. A person with a scoop of delicious French vanilla ice cream in a sugar cone (together termed the ice cream object) in her/his mouth (a person–ice cream object) has an improper part: the ice cream object, i.e., the ice cream object is not intrinsic to the person.

The differentiation between improper part with proper parts is that the whole relation of the object is proper. All aspects that are intrinsically the object are proper. An improper part is not intrinsic to the object. An example of proper and improper parts is described using the example of humans whose inputs are improper. Once input EMMI is within a person, if the use of that input EMMI results in sustainment of the person, then that use is proper and the results of internal mechanisms become a proper part of the person. Every part of an object is either a part that is proper or a part that is improper. For example, a system can have both proper and improper parts. A proper part is any object that is essential and fundamental to that object. An improper part is the waste products of the system. The mechanisms that separate waste products from used products are intrinsic—the waste products that become outputs of the system are improper parts. Systems have both proper and improper parts, and consequently a system of systems has both proper and improper parts.

6.6.1.2 Ontology of Causation: The Process Frame

A process is a partially ordered set of activities, use of tools, and practices set down in a structured order, that are carried out within limitations, subject to constraints to provide a result that is intended to satisfy a need (adapted from (Humphrey 1989) and referring to the work of (Lonchamp 1993)). Processes can be measured relative to other processes and improved relative to themselves (Goldberg et al. 1994). In the main, processes are sets of activities, of which the make-up and sequencing is often accommodated by the participants' skills, resource availability, and supporting infrastructure. There is no set of required acts, activities, or procedures; no absolute sequencing; and no fixed or unalterable juxtaposition of work.

A process is a systematic pattern, a coordinated set of procedures, tasks, activities, or acts that convert inputs of EMMI into outputs of EMMI. Process is an amalgamation of cogitation, acts, activities, procedures, and developing models or invoking representations of concepts. In other words, all things necessary to convert an input into an output are accomplish by process. The basic unit of a process is an act—a single factor that can be evaluated in isolation. An act is a single step in a series of steps that when combined are recognizable as an activity. Activities are orchestrated by procedures to form process.

Processes rely on inputs and incur losses to achieve outputs. Processes are measurable with objective measures such as cost, number of people, and the amount of loss to achieve certain results. Processes are comparable to other processes, but on a subjective basis, i.e., without a standard or reference. Consequently, it is quite difficult to say that one process is better than another.

Processes are of two kinds: those that are continuous and expressive within the postulates of four-dimensional space-time, and those that are discrete. Discrete processes exist within boundaries, but with piece-wise continuous expressions that concatenate interrupted segments of activities into the semblance of continuity. At the fundamental level, all processes are discrete. In this regard, the use of process in this chapter is generalized to include both continuous and discrete kinds.

Process is a metaphor—a core metaphor (Martin 1988, p. 38). A core metaphor is a set of metaphors linked by association that extends to all activities that make up or support the commonly shared metaphor. In this regard, process can be considered an enclosure (Martin 1988, p. 1), or as an environment (Martin 1988, p. 2). That processes have inputs and outputs corresponds with the concept of enclosure by which the wrapper delineates both process and the contexts of process as the process environment.

There are noticeable differences between processes, in general, with some consuming more labor, some requiring greater resources, some costing more money. In isolation, the differences can be viewed as objective measures. Since one process differs from another by its make-up of activities and its environment, any comparison may have little meaning since their basis may be quite different.

An activity is the means by which a mechanism applies force to an object. Processes have inputs of EMMI and outputs of EMMI. Process provides for the enactment of a function, i.e., the capacity to do work. Every function has multiple measurable performances associated with the perspectives of stakeholders. To achieve various levels of functional performances, processes must be enabled by EMMI to enable mechanisms. Processes derive from mechanisms as the capacity to do work on objects. Note: excess capacity is a trade-space (Emerson 2011), by which the design and architecture for systems and system of systems is a primary concern. Tradeoff studies create an objective mechanism (Daniels, Werner, and Bahill 2001). Objective mechanisms are those that are evaluated either by whether or not formal rules have been followed (i.e., process) or by measurements relative to a standard unit of measure, e.g., rule of length, time, mass, energy, power, velocity, or temperature (i.e., function).

6.6.1.2.1 Mechanism

A mechanism is defined as the set of processes (procedures, activities, and acts—whether a constituency or acting individually) that operates in the context of forces. More formally, Machamer, Darden, and Carver explain that "Mechanisms are entities and activities organised such that they are productive of regular changes from start or set-up to finish or termination conditions" (Machamer, Darden, and Carver 2000). Additionally, Machamer, Darden, and Carver state that two distinct kinds of building blocks make up a mechanism—entities and activities. Mechanisms "force" objects or processes to be conducted. Mechanisms are carried out by objects that interact with objects and processes (Gabora 1995); and by objects, systems, and processes, all interacting in various combinations with each other (NASA 2004).

Mechanisms explain (Illari and Williamson 2011) how things change. Change is caused by the input of EMMI into an object or parts of an object. Without an input of EMMI, there is no change. Change means that an object is different because of its use of EMMI. That difference is observable from one instance in time to another instance in time. An event is the result of the enactment of a mechanism by input EMMI which is transformed into output EMMI. The output EMMI is termed functional performance, i.e., the performance that a user derives from applying or using the function. Functions occur at the physical interface of or between two objects. Behaviors occur due to the existence or non-existence of object(s) and due to function(s) or no function(s). Every enactment of a mechanism results in a change, the results of which are an event.

EMMI stimulate the enactments of mechanism(s) that exert(s) force(s) that influence(s) some or all of the structures that comprise the object or its parts. The structures (Reed 2008), e.g., mechanical, electrical, and chemical activities and procedures within an object, give way to the influences of the forces derived from EMMI. Structures may give way to these influences depending on their susceptibility to such influence. When structures give way, they can have an impact on other structures, influencing these other structures depending on the coupling and cohesion between the structures. Structures that give way in turn may influence other structures to give way. The properties of structures (those intrinsic characteristics that have mechanistic and process resilience, i.e., relaxation and restoring action) have a semblance of reliability if their variations in performance about a central nominal value stay within a range of upper- and lower-boundedness.

Objects have structures with mechanical characteristics that result in giving way to influences from EMMI. Mechanisms can be characterized by an enabling space, a transacting space, and an outcome space (Trockel 1999). The output EMMI is parsed for convenience into two components: one describable as performance (functional performance), the other indicated as losses of EMMI. Functional performance is measurable (with appropriate instruments with sufficient accuracy and precision). Output EMMI may be uni- or multi-variant. Therefore, there can be multi performances for a single function.

Each object transforms the input EMMI into an output EMMI by the process of changing the input through the actions of the object's mechanism(s). The mechanism is a form of force, whether mechanical, chemical, or electrical. Force is defined as the influences of EMMI on objects—no influence on an object, no force, and no change. The test for influence is determined by the net of power (i.e., work done) on an object as observed by the output EMMIs of that object; changes in the object's properties, traits, or attributes; or other changes in boundary, boundary conditions, physical issues, and functional or behavioral issues (Kocsis 2008).

An object sends EMMI (e.g., releases, sets free, or give ups) through the process(es) of an object's internal mechanisms. Similarly, an object receives EMMI (e.g., accepts, collects, or takes in) through the process(es) of an object's internal mechanisms. An interaction is defined by identifying the sending and receiving objects. James Lake argued that "functional congruence must exist between phenomena that underlie a specified symptom pattern and phenomena operationalized as the mechanism of action ..." (Lake 2007).

6.6.1.2.2 Force and Energy

Processes are akin to force (Feynman 1970). Processes enable mechanisms to do work. For an object to do work, processes must build the capacity to do work by enabling mechanisms within an object to accept input EMMI and to release EMMI. EMMI is the capacity to do work. Work may be observable by the behaviors of the objects that are responsive to interactions. The interaction (the input to the mechanism) facilitates work.

The broader definition of "force" captures the general notion of overcoming resistance to change, i.e., the enactment of processes to "P: take-in (EMMI)" and "P: to release (EMMI)."

That change in the object due to the interaction also results in a change in the object's energy. The energy represents the capacity of an object; and a change in capacity is indicated by the inputs and outputs of EMMI. Interaction results in a change in capacity with process being responsible for building or reducing capacity.

The empirical (in contrast to the theoretical) form of interaction takes place through processes (forces) that act to enable mechanisms. A metaphorical interpretation of interaction is the exchange of capacity that results in interactions of objects. It is likely that the local environments have received EMMI lost from the interactions, for example, the absorption of heat resultant from the physical heating of the atmosphere along the path of transmissions of an electromagnetic beam; or the reduction in the interpretable information due to a reduced signal-to-noise ratio caused by dissipation of energy over distance transverse and longitudinal to the primary motion. Therefore, the aggregate consequence of two-object interactions is not the simple sum of one action (from the sending object with that of the action of the receiving object). The objects may be different, including their internal mechanisms, their actions, and their losses, and the resulting changes they undergo are different than each other's changes.

Mechanisms are the means by which objects and processes change. Change means that an object or process is different at one instance from the previous or next instance. Change is precipitated by EMMI. If there is no change in an object, there is no energy loss.

6.6.1.2.3 Cognition

Cognition is the acts and activities carried out to think, reason, perceive, meditate, and learn. The result of these activities is to form the psychological make-up of being. Applying cognition to processes, the activities span the gamut from thinking (e.g., about thinking), formulating (e.g., the thought patterns to detect proximate events), determining (e.g., how to trade off approaches and methods), determining (e.g., how to identify the best course of action), laying out (e.g., how to carry out planning), reasoning (e.g., through the kinds of issues), meditating (e.g., to calm the soul); rationalizing (e.g., the efficacy of a set of procedures); considering (e.g., feasibility of a decision), to give some examples. The kinds and types of thinking are embodied in the process ontology that determines all that follows in the causal frame.

In greater detail (albeit without defining terms), cognition—the process of thinking, reasoning, perceiving, meditating, learning—can be decomposed into conceptional processes:

To cogitate (think, reason, perceive, meditate, learn)

- 1.0 To apply mental model
  - 1.1 To build mental model
  - 1.2 To search mental model
- 2.0 To reconcile
  - 2.1 Self-biases
  - 2.2 Habits
- 3.0 To interpret
  - 3.1 Distinguish patterns
  - 3.2 Compare patterns to standards
- 4.0 Develop cognitive structures
  - 4.1 For context
  - 4.2 For meaning
- 5.0 Distinguish between changes
  - 5.1 Identify proper changes
  - 5.2 Identify improper changes

- 6.0 To recognize
  - 6.1 To reason information
    - 6.1.1 To reject
    - 6.1.2 To accept
    - 6.1.3 To consider
  - 6.2 To contrive patterns
    - 6.2.1 To compare
    - 6.2.2 To identify repetition
  - 6.3 To match patterns
    - 6.3.1 To rotate
    - 6.3.2 To scale
    - 6.3.3 To reconfigure
  - 6.4 To remember data
    - 6.4.1 To retain
    - 6.4.2 To combine
    - 6.4.3 To replace
    - 6.4.4 To modify
    - 6.4.5 To delete
  - 6.5 To remember information
    - 6.5.1 To retain
    - 6.5.2 To combine
    - 6.5.3 To replace
    - 6.5.4 To modify
    - 6.5.5 To delete
  - 6.6 Integrate linguistic and non-linguistic information
  - 6.7 Evaluate error
    - 6.7.1 Determine error
    - 6.7.2 Identify problems due to error
  - 6.8 Correlate information with sensory inputs
  - 6.9 Superposition subconscious with conscious information
  - 6.10 To disjoint two concepts
  - 6.11 To compare two concepts
  - 6.12 Detect anomaly

Learning is the change in knowledge. Extending knowledge into new areas of thought by extrapolation, interpolation, and conjecture reveals the self-logical structures invented to remove, to control, or to accept intrinsic biases based in part within the self-belief system. The use of predicate logic permits reasoning to be based on properties and relations of individual physical/intellectual objects. To change cognition to allow for thinking in the manner of the rules accepted by a formal model (based on predicate calculus), Leśniewski extended and validated a general mathematical model to explore and explain the relations of wholes and parts. In the parlance of this research, the relations of wholes and parts is that of notasystems (not-a-systems) (Langford 2012), systems, and system of systems. From the perspective of the person thinking in terms of systems, the goal of improving cognition is to relegate self to serve an ever-improving mental model, perhaps in the vein of Leśniewski's transformative mereology.

6.6.1.2.4 *Modeling and Representing*

To model objects (the process) is to imitate or represent the dynamic relations between objects, the interactions of objects, the consequences of interactions, and the behaviors of objects on a scale that is convenient for imitating causal properties, traits, and attributes of objects. To model processes (the process) is to imitate or represent the dynamic relations of objects regarding causal cognition, causal procedures, and causal mechanisms. The enactments of modeling or representing objects and process are by abstracting, granularizing, reduction, and partitioning.

6.6.1.2.5 *Abstracting*

Abstracting is the process of isolating, integrating, and differentiating. Abstracting isolates a broadly stated concept that embodies the key aspects of data and information found in the details of that concept. Abstracting extracts all the essential expressions of objects and processes by aggregating and summarizing the details. Abstracting blurs the details yet retains the essences of those details. Integration further isolates detailed concepts and by forming a new structure that is enlightened with emergents—aggregations with additional similarities and essences of the simple sums of the constituent parts. Then, abstraction differentiates the general idea from the formative parts, at the same time highlighting emergent concepts that arise from integration. These three processes define the process "to abstract." Defining a word through its decomposition is a convenient means of knowing more about the word, detailing how to use the word, and applying the word appropriately in any specific domain or for particular use. A process can be conceived as capturing a level of abstraction that carries with it the conceptualization of all its included activities. The process of abstracting involves isolating concepts (concepts are themselves abstractions), encapsulating emergence from the integration of a set of concepts, and then summarizing the emergence to the level of detail necessary to capture the artifacts needed for the model or representation.

The result(s) achieved when these activities are accomplished are not only expected to be greater than any one activity but they are also presumed to be greater than the simple sum total of the tasked roles and assignments. For activities, a meaningful (yet indirect) measure of one activity versus another activity is an evaluation of the relative differences between results for the two activities. In essence, the series of acts and activities taken to achieve an outcome, distinct from that of an unorganized approach, is termed as a process.

6.6.1.2.6 *Granularizing*

Granularizing is the process of differentiating and separating objects or processes into a heterarchical or hierarchical conceptualization based on the relations between objects or processes. That which is within the limits of each separation is termed a granule. Granularity is interpretable at a level of abstraction, but not necessarily a different level of abstraction in a descriptive phenomenology (Keets 2008). From a modeling perspective, granules are a natural form of separating objects or processes according to a set of local context—much like grounded theory derives theory from situationally localized sets of data. There is no standard way to granularize processes, broadly stated to include cognition, procedures, acts, activities, and modeling and representing same.

6.6.1.2.7 Reduction

Reduction is the process of relating the whole to its parts without emergence. Parts in isolation simply aggregate as sums and partition according to fractions. Reductionism is an approach that ignores the relationships due to interactions; in other words, reduction implies all can be said to be unity—a consistency of interpretation that derives all things from the same building blocks, deprived of the consequences of interaction. While reduction does not contain reality, except in the most parochial sense where it finds favor with phenomenologists who express their observations within their self-made domain. Thinking systems, systems engineering, and other engineering fields provide significance leverage in dealing with the reductionism with "substantial method proportions of its methodology design to address problems involving ambiguity and uncertainty" (Ferris, Cook, and Honour 2005).

6.6.1.2.8 Partitioning and Modularizing

Partitioning is the process of dividing the parameter space to satisfy a set of requirements. Those requirements are stipulated according to the objectives of the partitioning. Consider the requirements of the space to be based on the principle of partitioning, stated as: Partitioning of objects creates tractable solutions to solve, if and only if boundary contiguity is achieved. Tractability implies outcomes of partitioning that result in solutions are easy to control or influence if all adjacent partitions (in a geometrical sense) touch each other's boundaries such that no partition overlaps another partition and no partition does not adjoin an adjacent partition (underlap). The conditions of no overlaps and no underlaps describes the expected actions and decompositions of objects, functions, and behaviors. From a requirement's perspective, no object shall unintendedly pierce the physical surface of another object and cause degradation of either object. To do so creates emergence whose consequences are unintended (no overlap—nail piercing a gas pipe). An object shall interact physically in the intended manner with the object for which interaction is required (no underlap—retaining nut not placed on bolt). Overlapping or underlapping physical/intellectual objects create emergence. The requirement for a function is that two objects interact in the manner prescribed. An interaction that fails to exchange a sufficiency of EMMI or EMMI that is unsuitable to create the intended function is an underlap—not enough force was applied or force was applied incorrectly to close the door on the submersible vehicle. The meaning of underlap is that there may be functions missing that are necessary to provide all required functions and performances indicated by the abstractions higher in the functional hierarchy. Emergence occurred and submersible sank and was unrecoverable from the ocean trench. An interaction between two objects shall not include EMMI that is unintendedly in excess of that needed to create the intended function—irradiating a cancer patient with several beams of electron energy that cross paths inadvertently and thermally burn a patient rather than result in the ameliorative dosage (overlap). The meaning of overlap in the functional domain is that the functions are not necessarily exclusive, perhaps having two causal actions determining functional performance, control, or behaviors. Overlapping or underlapping functions create emergence. In the behavioral domain, overlap means the cause for one effect may be uncorrelated with the cause for another effect, or that a causal action or event is missing. Buffo the dog receives a tasty treat for moving his largest prime rib bone from the well-trafficked hallway (behaviors of treat-giver overlap with Buffo's behaviors). Underlapping behaviors are illustrated if Buffo expects another treat for placing his largest prime rib bone on the out-of-the-way, dimly lit basement stairs. Overlapping or underlapping behaviors create emergence.

Emergence can be thought of as a symptom, merely a consequence of the traits of an object undergoing interaction or not participating in the interaction that is intended. Also, if boundary conditions change (as with behaviors), emergence may lead to a lack of reciprocity in those emergent traits, by which the constituent parts cannot be reconstituted back to their original traits and properties. Buffo's behaviors may no longer elicit treats.

Partitioning of processes can be driven by the need to minimize the amount of energy expended in managing industrial processes (Bedeaux, Standaert, Hemmes, and Kjelstrup 1999), to construct multiple decision procedures (Finner and Strassburger 2002), to engineering large software projects (Gannon, Hamlet, and Mills 1987), Requirements Determinations (Hevner, Mills 1995), to highlight a few. A fundamental process in systems engineering is to partition physical/intellectual objects, functions, behaviors, and processes into a hierarchical decomposition or a heterarchically modular structure. Partitioning is the method used by Leśniewski and this research. For an in-depth discussion of partitioning, refer to Langford 2012.

To be a part (i.e., a partitioned element), a boundedness must differentiate one part from that of other part. A whole means to have a boundary that differentiates one whole from another whole. The Principle of Partitioning indicates what is necessary to create a tractable solution, i.e., that of maintaining boundary contiguity—the adjoinment in spatial or temporal contexts. The discreteness property of parts and wholes coupled with their mechanisms that result in interactions between the parts is the recipe for emergence. Changing any one or all of the physical, functional, or behavioral boundaries between parts affects the whole, as seen in both the resultants and emergents. A part before a boundary change can indicate a different part after that change. A change in the boundary of the whole also can indicate a different whole, each part potentially affected by the change in the whole. Partitioning is a means of changing boundaries of a part within the whole, or as a means of changing boundaries of the whole. Since boundaries are, in effect, constraints, emergence can be said to either change constraints or to invoke constraints. The difficulty faced during integration of parts into a whole (systems) or wholes into a whole (system of systems) is not knowing the full complement of consequences (emergence) of abstracting, granularizing, partitioning, modularizing, or reduction. For this reason, the process of partitioning has the most leverage in orchestrating control over emergence, i.e., minimizing the deleterious effects of emergence. The bounding by partitioning is a method of controlling interaction, integration is a method of managing the boundaries to build a system or system of systems.

6.7 Interpretive Integrative Framework

The philosophical interpretation of Leśniewski's writings on the mereology of objects and processes is that many exist, one exists, or none can exist. The study of things that exist or may exist is the subject of ontology. From the vista of thinking in terms of notasystems, systems, and system of systems, the challenge is to detect, identify, evaluate, and manage emergence. There are three causal actions that govern putting things together. First, interactions between physical/intellectual objects create uses; second, all interactions result in a loss of EMMI; and third, all interactions cause emergence. While the expectation will be that interactions can be tamed and managed, there may be predictable interactions with unforeseeable consequences and unpredictable interactions with unforeseeable consequences. Any action taken to aggregate physical/intellectual objects into part-whole relation assumes interactions within an abstract problem-solutions domain—a formal compromise of requirements, needs, and wants with insufficiencies of knowledge and no essential relations to context. This research indicates there are two separate and distinct ontologies which entwine to fabricate the abstract problem-solutions domain for all interactions. The process frame encapsulates and expresses the ontology of causation: cognition, mechanisms, and modeling/representing. Causality can be considered by ontology—entities and activities (Machamer, Darden, and Craver 2000); type—system, statistical, mechanism, and algorithmic (Doreian 2001); the existence of ontological emergence (Humphreys 1997; Atlan 1998; Kim 2005; Christensen and Hooker 2001; and Campbell 2009). Evidence for causation is pluralistic (Russo and Williamson 2007). The object frame describes succinctly the ontology of interaction from exchange of EMMI between objects, the functions that arise from those interactions, to the behaviors that follow as a result or effect. These frames are configured to intersect so that each ontological element captures the nine concomitance synergies between the actions derived from causation and interaction, as shown in the sections below.

6.7.1 Cognition AND Object Frame

Cognition AND Behavior

Cognition AND objects interacting (Function)

Cognition AND physical/Intellectual Objects

Examples of Cognition include: planning, evaluating, assessing, considering, and contemplating.

6.7.2 Mechanism AND Object Frame

Mechanism AND Behavior

Mechanism AND objects interacting (Function)

Mechanism AND physical/Intellectual Objects

Examples of Mechanism include: perceived statistical structures, mechanical structures with two building blocks—entities and activities.

6.7.3 Modeling/Representing AND Object Frame

Modeling/Representing AND Behavior

Modeling/Representing AND objects interacting (Function)

Modeling/Representing AND physical/Intellectual Objects

Examples of Modeling/Representing include: imitating, abstracting, summarizing, simplyfying

6.7.4 Object AND Process Frame

Physical/Intellectual Objects AND Cognition

Physical/Intellectual Objects AND Procedures

Physical/Intellectual Objects AND Modeling/Representing

Examples of Physical/Intellectual Objects include: atoms, molecules, insects, rocks, systems, people, system of systems, planets, stars, galaxies, and universe.

6.7.5 Object Interaction (Function) AND Process Frame

Objects Interaction (Function) AND Cognition

Objects Interaction (Function) AND Procedures

Objects Interaction (Function) AND Modeling/Representing

Examples of Objects Interaction (Function) include: perceived

6.7.6 Object Behaving (Behaviors) AND Process Frame

Physical/Intellectual Objects Behaving AND Cognition

Physical/Intellectual Objects Behaving AND Procedures

Physical/Intellectual Objects Behaving AND Modeling/Representing

Examples of Object Behaving (Behaviors) include: perceived

The general nature of examining all exchanges of EMMI is a daunting assignment at any level, and often a more complete relation between what initiates mechanisms and the causal effects of mechanisms is found at the micro-level (Engelhart 2001). It is demonstrated in this chapter that the causal effects and causal conditions are exposed at all levels of interaction, aggregation, and integration. Importantly, the aggregations of micro-level mechanisms into meso- and macro-level mechanisms are also traceable to triggering events which result in the causal effects following their causal conditions. The interpretive integrative framework with intersecting frames of objects and processes expose interactions between consequences of interactions between objects in the form of new objects, new functions, new behaviors, new cognition, new mechanisms, and new models/representations. All of these emergents are expressed in the interpretive integrative framework.

Interactions are the threads that bind objects together in the fabric of integration. When EMMI impinges on an object, the object reacts, whether at the micro-, meso-, or macro-level. A result of that input EMMI is to stimulate a mechanism(s) to effect change in the object by controlled or uncontrolled process(es). EMMI impinging on an object always results in a loss of EMMI by that object. For every action, there is a loss of EMMI. Objects that interact with other objects cause processes within those objects to react to inputs and outputs of EMMI. Some of those reactions of those objects may be desirable, while others may be disruptive or extremely harmful.

The basis of the interpretive integrative framework is a model of the two ontologies which represent all interactions and all causation. The approach taken with this research was to build the permanent structures of ontology and scaffoldings to allow for endurants (objects that exist as notasystems at all times), perdurants (objects that exist as ProtaSystems (an aggregation that is transitional between NotaSystem and a System), systems, or system of systems—unfolding temporally or in phases), and stages (instantaneous parts of perdurants) (Bittner, Donnelly, Smith 2000).

Formulating and building a framework is a means to view the object ontology juxtaposed with the process ontology, applying the same characterization of mereology as Leśniewski posed. The ontological frame that objectifies objects interacts with the ontological frame that subjectifies processes. These two frames, when appropriately laid out, present a comprehensible means to anchor a general foundation for all interactions, aggregations, and integrations in all disciplines, fields, and domains. According to the rules within the framework (which is in effect the taxonomy (Langford 2017)), the influence of interactions on emergents can be interpreted in the sense of the classes of emergence (simple, weak, strong, and spooky).

The scope and applicability of a framework determines the completeness and fitness for use. Leśniewski's formal model for mereology of the object and process ontology is broadly applicable, complete, and organized to identify and explain how emergence arises. The validity of an ontology is determined by its applicability, completeness, and organization to identify and explain how emergence arises (Brooks 1972).

In particular, the interpretive integrative framework (Figure 6.1) incorporates the particulars related to the problem, question, or competition for which the framework is adapted. The physical/intellectual objects, functions, and behaviors (comprising the ontology of interaction) and the cognition, mechanisms, and representing (or modeling) (comprises the ontology of causation) are fully integrated into the specifics, by design. IF the interpretive integrative framework was set up to design, architect, and integrate a system of systems,

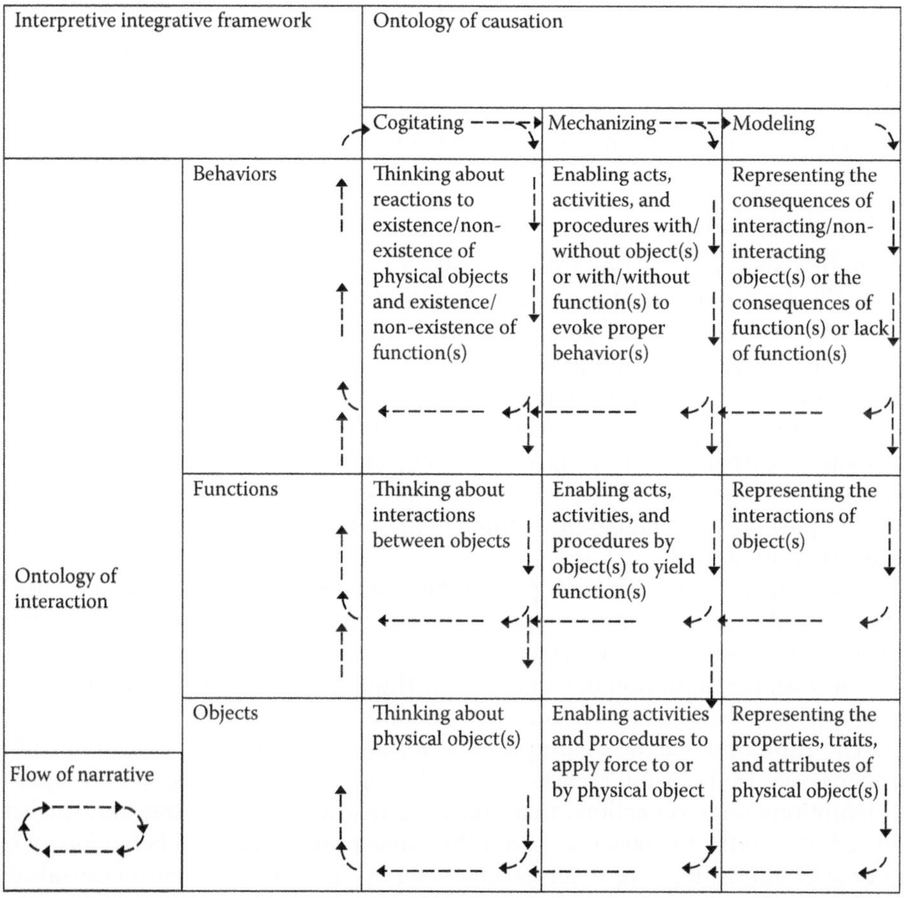

FIGURE 6.1
The interpretive integrative framework. (Adapted from Langford 2012.)

then the objects would represent the physical/intellectual objects of the solution, including parts, test equipment, facilities, people—ALL objects required to carry out the tasks. The management functions would derive from the interactions between people, between people and objects, and between objects and objects. The objects and functions of management would result in the various behaviors that would be found on such a program, project, or task. The planning (process) would take place in the domain frame due to the physical/intellectual objects, e.g., people, physical/intellectual objects, and facilities through cognition (thinking about planning and the ramifications of planning), through procedures (working through the procedures that should be used), and through modeling or representing that thinking in the planning process according to a model of best practices (for example). The moment a plan is formulated in the causation frame, the plan becomes an object and participates in the object domain through interactions with people and other objects (such as for a person to store the plan). Here, F: "to store" (plan) is a function with a performance target value of 2 minutes (To Be Resolved, TBR) with a +/– variance about that performance target of 1 minute (TBR). If there is no reason to ascribe a functional performance and therefore no need to test that performance, then a decision is made to relegate P: "to store" (plan) as a process. In this manner, key functions are identified and brought forward as measures which will be tracked and possibly used to build metrics from which to monitor work.

The process frame of causality orchestrates the interactions between objects. The moment at which an object is born in either its parts or whole, that object becomes an element in the object frame of causality. The interaction frame that constitutes half of the interpretive integrative framework construes the ontology of interaction as the mereological sum of its three dimensions—objects, functions, and behaviors.

6.7.7 Mereological Sum

Informally, the mereological sum *S* is the set of *s* if and only if every part of *s* has a common part with some *s*. *S* and *s* are objects. The concept of the set of *s* expresses that every object *s* is a part of *S*. The concept of "part" is referred to as the "proper part." Every part of an object is either a part that is proper or improper.

And, a class of *s* is a set of all *s*. The concept of the class of objects *s* expresses that every object *S* can have object *T* as a part where *T* is a part of *S*, and as such, some part of *T* is a part of some *s*. Further, an object *s* is a class of objects *s*. And, if *S* is a class of objects *s* and *T* is a class of objects *s*, then object *S* is object *T*. The concept of a part of object *T* expresses *T* and every part of *T*. A mereological sum *S* is the set of *s* consisting of one or more *s*, but not necessarily all of the *s*.

Therefore, surprise is an improper part of the intensional definition of emergence. If the definition included the caveat of surprise, the notion that emergence might improve or degrade expected behaviors would be totally dependent on the perspective of an observer. In stark contrast, the definition of emergence without the caveat of surprise has at its core the assumption that the observer is independent of the actions of interactions between objects. The stipulation of the assumption of objectivity neither impinges nor influences the properties or traits of the objects before, during, or after interactions. Proper intensional definitions seize on actions that attach correlative or causal mechanisms to what is observed. Pointedly, the observer is not the cause of the change in behaviors of objects during or after interaction. The essential characteristic is that the actions of causality occur through interaction of the observed objects and not the influences of the observer.

The ontology of interactions describes the three categories of causal interactions that are allowed. Physical/intellectual objects interact to create functions. Behaviors result from physical/intellectual objects or lack of physical/intellectual objects and from functions and lack of functions. The causation model formulates the other half of the interpretive integrative framework. This causation frame consolidates the ontology of causation with causal processes.

The substance of this formal ontology of integration is nominally equivalent to the genres "entity" and "activity" (Machamer 2004). In Machamer's book, entity and activity are differentiated (and carrying with them all the historically significant encumbrances that these two words harbor) as a workable ontology. Applying Machamer's formalism, objects have mechanisms that in themselves are objects and acts/activities, a blending of objects and processes. Objects have properties (intrinsic to their being that object), traits (the combination of the object's properties and the object's environment due to interaction(s)), and attributes (that which is associated with the object, but neither intrinsic nor situational, e.g., environmental).

Machamer's phenomenologically-based entity and activity model is in keeping with the formalisms of Leśniewskian mereology: objects exist as objects; the constituents (parts) of objects are objects; and compositions, agglomerations, and combinations of objects are objects. Further, no object has any property that is not a property of an object. Leśniewski presented the idea of a theory of relations for parts and whole (i.e., a mereology) (Simon 1987) as objects.

Leśniewski's development of the logical theory of relations between objects shows that relations do not depend on recognizing objects as sets of points (Srzednicki 1988), but rather as either distinct entities (objects) or by domains that embody objects. Leśniewski's formulation is extensible to points, although Leśniewski did not seem to have that intention. Processes, on the other hand, derive from mechanisms as resources, or the capacity to do work on objects through mechanisms.

Interaction is typified by the exchange of EMMI between the two objects. The significance of this perspective is based in part on a philosophical foundation that a possible result of interaction is integration, which is describable as many, one, or does not exist—a blending and restatement of the concepts expressed by Stanislav Leśniewski (Henry 1972). The strict interpretation of this descriptive union of many exist, one exists, or cannot exist emphasizes integration as a result of interaction. Therefore, as with interactions, integrations also have emergents. This interpretation contrasts with Parmenides philosophy of "one exists, does not exist, cannot exist"—which implies unity, i.e., without emergence. Since all that we do involves interactions, emergence is part of every endeavor, domain, discipline, and field. In the context of system of systems, emergence is at "… the core space of engineering" (Ferris 2017).

An object is completely defined by the interactions or potential interactions of its parts, not merely by the existence of its parts—interacting or not interacting. Parts alone do not define anything except that there are parts. Interaction between the parts creates all. Interaction between the parts creates functions that provide uses for other objects, whether expected or not.

6.8 Phenomenological Perspective

The term *emergent*, in reference to unintended effects caused by integration, is attributed to George Henry Lewes (Lewes 1875). The seven uses of the word *emergence* by Lewes

(Lewes 1875) clearly and unequivocally shows it was not the aim of Lewes' intensional definition to convey or promote the notion of surprise that might result from a difference between expected behaviors or consequences and what is actually observed observing interaction(s). However, in 1999, Edmund Ronald and Moshe Sipper and Mathieu Capcarrère (Ronald, Sipper, and Caparrère 1999) appended the notion of "surprise" onto the Lewes definition to form a "test" for emergence. Based on a sampling of phenomena whereby the researchers expressed some form of surprise of their results, the notion of "surprise" has since flourished in a vacuum brought on by the seemingly ill-defined word of emergence. In spite of the excellent article by John Holland (Holland 1998) in which he pointed out that unexpectedness (surprise) was not a proper part of the way emergence occurs, the Ronald, Sipper, and Caparrère article attempted to avoid criticism of their approach, which relied on a limited set of examples in which surprise was a dominant reaction to results, by restricting the test for emergence to artifactual things. Ronald, Sipper, and Caparrère acknowledged Holland's point that "surprise" was inappropriate for emergence in nature. Their admission was that Holland's words might better describe emergence. Yet, the result of the article by Ronald, Sipper, and Caparrère was to define a test for emergence (which included surprise by the observer) which they purported was substantiated by phenomenological logic. Therefore, by gaining knowledge about the effects of interaction we should find the term *emergent* used less and less for what we have come to understand, and then appropriately applied to integrations of immense complexity caused by the effects of flummoxing interactions. This definition of emergence which incorporates surprise poses a major problem for communicating ideas clearly. What is emergent from one observer's perspective, however, may well be understood differently by another observer, and therefore not emergent from the second observer's perspective. The test proposed by Ronald, Sipper, and Caparrère for emergence is effectively a test for ignorance or stupidity with regard to the difference between those who recognize why something happens and those who are surprised. This portrayal is not meant to be a harsh appraisal of the test for emergence as described by Ronald, Sipper, and Caparrère. Rather, advancing science is more than merely pointing out that emergence ensues from interaction of ideas without the notion of surprise. Instead, the opportunity is to now focus on emergence as results from object to object interaction as an essential and worthy introduction for the necessity of ontology frames. The original Lewes' definition is retained, the factor of surprise is rejected. From a phenomenological perspective, emergence refers to the changes caused by combining objects through interactions of EMMI, whether unobserved, unexpected, unforeseen, or unanticipated.

6.8.1 Phenomenological Perspective of Emergence

As a natural course in refining and amending definitions, academics, practitioners, and writers nuance definitions to improve communications and merit greater accuracy and precision in describing observed behaviors and effects. Since the topic of emergence will lead new ideas into previously used words, certain liberties are taken to either strengthen or weaken, add to or delete, clarify or generalize commonly used words. That is the situation with the term *phenomenology*.

Traditionally, phenomenology is the science and systematic study and description of human experience (Stanford 2013). However, in the most general sense, phenomenology is about naturally-cognitive and artificially-cognitive existentialist beings determining meaning in their existence. In the spirit of Edmund Husserl [1859-1938] and Jean-Paul Sartre [1905-1980], as interpreted by their contemporary colleague Maurice Merleau-Ponty

[1908-1961], the development of the phenomenological account for emergence requires a blend of thought leading to causality is some mix of objective and subjective perspectives that are otherwise premised on entities and activities.

Since the intent of this chapter is to build an appreciation for emergence based on a rigorous approach rather than a superficial discernment based on bias, emotion, and hearsay, the broadly defined description of phenomenology must be reconciled with the need to separate the observer as an instrument of reporting from the feelings the observer may have about the reporting. The aim is to objectify separately the observer's descriptions of what things are from how the observer feels about how things are or should be. The result is examination and judgment based on observables in which every question is posed to have only one right answer. The observer is said to be uninfluenced by emotion or prejudice. In the language of mereology, the observer must describe what is in the boundaries of the observed. Therefore, the observer's emotions are seen as outside of the context of the observed.

6.8.2 Kinds of Phenomenology

Recognizing that the discipline of phenomenology traditionally encompasses both observer's experiences relative to context of the observed, we distinguish between transeunt phenomenology which is defined as the systematic study of objects as they appear to the objective self; and immanent phenomenology which is the systematic study of reactions to transeunt phenomenology and its results as they appear to the causal self. Transeunt phenomenology focuses on how things appear to be correlated within the set of events observed, which is to say that transeunt phenomenology concerns only things causal or correlative to events observed. Within the context of the observed, which is outside of how the observer feels about the events observed, transeunt phenomenology offers reason, logic, differentia, and rationale to describe what is observed. Outside of the context of the observed, immanent phenomenology deals with that which is inside the context of how the observer feels about the things observed, yet strictly not bleeding through into the realm of transeunt phenomenology.

6.8.3 Proper Part and Improper Parts

Emergence is a whole-part property created by objects that interact. Interactions distribute Energy, Matter, Material wealth (money), and Information (EMMI) to all proper parts within the structure of objects. Sometimes, EMMI is also distributed to improper parts, e.g., to cancer cells. What ensues from that interaction can be a proper part and an improper part depending on the boundary conditions that govern the interaction between the objects. A proper part is a part that is intrinsically natural to the whole after interaction, i.e., integrated and interoperable with the whole. A whole depends on its proper parts for existence (Tahko and Lowe 2015). For example, a fox's paw is a proper part of the fox, whereas the air the fox breathes is comprised of proper and improper parts. As the fox interacts properly with the air during inhalation, the processes to oxygenate blood are enacted. Oxygen and other air molecules not used during the oxygenation processes are expelled properly during exhalation. In other words, the breathable air must be integrated and made interoperable with the appropriate fox mechanisms that satisfy various boundary conditions. Parts are proper if and only if they interact with a whole to be used by the whole to enable a function of the whole. Improper parts are those parts that do not interact with the whole to enable a function of the whole. A splinter that pierces the fox's skin and becomes lodged in the fox's paw is an improper part. The fox function of 'to respond' to foreign object embedded in paw may result in the splinter being dislodged and removed, or broken off and encapsulated in a membrane that is generated by processes in the fox. In either of these two splinter-events (removal or embedding), the splinter remains the splinter (an improper part). If the lodged splinter causes an infection in the fox paw, the splinter is still a splinter, an improper part. The body responses to the improper part may disrupt the cellular structure and cause a reaction. If the splinter is eventually dissolved by fox biological mechanisms, the resultant usefulness of those dissolved products become proper parts, or else they are removed by various processes. Objects that are not used during interaction are improper parts of the whole. And, waste products that are by-products of processes of the whole are also improper parts of the fox. Proper and improper parts are objects. The body deals with improper parts in such a way as to remove them or in some manner accommodate them. Proper part means "part of but not identical to" (Parsons 2012, interpreting Simon 1987).

In the parlance of interaction, formative wholes are the objects before interaction. Formative objects have whole-part properties. A new whole-part property may be either a whole-part property that is emergent during interactions of the objects or arises as a consequence after the formative objects have interacted. Therefore, during interaction there may be a whole-part property that is different than the whole-part property after interaction. These whole-part properties are different from the whole-part property of each of the objects before interaction begins. The whole-part property that occurs during interaction is a proper part of the objects during interaction; and the whole-part property that ensues after interaction is a proper part of the objects after interaction. Improper parts are either unused during interaction or expelled by means of whole-part processes. An outcome of action of interaction, i.e., emergence, is the functionalities of the resultant whole are different from the individual functions of the formative wholes. Through interactions (both internal and external to the whole), change occurs and may be observable (Langford 2012). Change is noted by differences in proper parts before and after interaction.

6.8.4 Phenomenological Perspective of Emergence without Surprise

Focusing on "... the effects of integration," and eliminating the emotional expression of surprise which sometimes ensues when the results of interaction deviate from the observer's intent, significantly helps one remain unbiased as a researcher of emergence. The progress toward developing a logical, unbiased presentation of emergence is accelerated without subterfuge. Causal intension concerns the logically necessary conditions that apply qualitatively to defined terminology. Intensional definitions that are too broad in their scope mean a short list of conditions. Definitions that are said to have restrictive extensional structure are narrowly scoped with a long list of categories stipulated. Logic dictates that causal intension cannot be applied rigorously; that restrictive extensional definitions should be refined to the higher level of intension; and that an operational definition needs to be developed to simplify, characterize, and clarify use to yield an academically rigorous, reproducible result (Swartz 1997; Copi, Cohen, and McMahon 2016).

After experiencing and observing a phenomenon, the intensional definition begins to knit the logically necessary conditions that apply to the word being defined. Without a classically intensional definition, any observed emergent phenomenon may be ill-expressed and inefficiently communicated. Classifying emergence according to the intensional approach first differentiates the behaviors of objects before interaction from that of behaviors of the same objects due to or after interaction; and second, distinguishes between individual objects and their known properties and traits before interaction from the phenomenon that is observed during or after interaction. These two differences stipulated in genus and differentia give rise to the intensional definition of emergence. The essential characteristic of the phenomenon that is captured in the intensional definition of emergence is that a change in behaviors of objects before and during (after) interaction may be observed. However, the rule of equivalence is broken when the definition includes more or less specification than required. This situation is the case with the historical and recently promulgated definition of emergence. The rule of equivalence is a check that follows from the inclusion principle—only that which is necessary is included, all else, not.

Rule of Equivalence: an object x is a mereological sum of the group W if and only if every W is part of x and every x is compatible with some W.

The rule of equivalence is built on the definition of a mereological sum, which means that because every object is subordinate to itself, no class of objects is not subordinate to itself. A mereological sum is not the numerical result of a mathematical process, but rather the imbuement of properties of objects with spatially or temporally continuous traits with mixed kinds of things and stuff. According to Dr. Ferris, "The word 'stuff' is deliberately used with its Jacobean era definition as including both the material of which things are made and the things themselves" (Martin and Ferris 2008).

Phenomenology is most fundamental to development of cognitive beings within the world that is a system of systems. Cognitive beings observe, demonstrate action (whether by design or by intention), and deal with consequences. Such is the plight of interacting object to object.

To gain insight into the phenomenological basis for the concept of emergence, appropriate theory, approach, and method must be derived from the phenomena.

6.9 The Gladstone Bag Equipped for Ontological Exploration and Investigation

The optimal path for research is when the researcher takes the shortest period to answer the research question, i.e., the quest. Optimal implies the research is directed, by knowing which branch points along the path offer the shortest path to the answer. The researcher must determine to start the journey, filling the satchel with the appropriate set of techniques and tools, the correct formulation of theory, an approach that will provide a sufficiency of guideposts, and a method that will expose the relevant and avoid the fool's errand. The art of research is to recognize which discoveries shorten the period or open up greater questions that serve to mature the destination or the contents of the research satchel. Which jewels to pick up and place in the satchel and which to leave for another trek are continuous decisions along the path. Herein lies the paradox of research—pathfinding is as much the task as is filling the contents of the satchel. If either the path or the satchel is insufficient for the journey, perhaps a lesser treasure is to be found. From this allegory, each researcher assumes his satchel will mature or at least be useful for the path chosen. Each researcher attempts to find her optimal path which may or may not coincide with the importance of the information discovered.

For this particular research, the inspiration was recognizing the gap between the meaning and interpretation of what is observed when objects interact and the reality of the results of interaction. The work of Leśniewski, the review by Herbert Simon 50 years later

(Simon 1987), and Machamer (Machamer 2004) related the fundamentals of interaction as a mereology of objects and processes. A means of translating that mereology into two frames of ontology (interaction and causality) was devised (Langford 2012) and revised in this chapter. A taxonomy was proposed (Langford 2017) that built on the interpretive integrative framework. The interpretive integrative framework communicates the partitioned artifacts and their interfaces between the object-function-behavior ontological frame and the cogitating, mechanizing, modeling ontological frame to facilitate building, analyzing, and predicting notasystems, systems, and system of systems. The contribution of a formal, yet general, ontological framework is to standardize the language of emergence. The contribution of this research is to show that a fundamental result of interacting objects is as claimed by Lewes, resultants and emergents.

Resultants and emergents having simple mereological sums of objects, i.e., the results of mechanically combining simple objects (string strung securely and tautered tightly between the ends of a long straight stick, e.g., a bow to strum as a musical instrument or to launch arrows).

A simple mereological sum of a set *P* is an object such that everything which overlaps with it also overlaps with something in *P*, and everything that overlaps with something in *P* also overlaps with *P*. Resultants and emergents having weak mereological sums of objects, i.e., the results of mechanically combining objects to achieve weak functional dependencies, e.g., building and hanging a ballistic pendulum with two string-lengths from an elevated structural beam, resulting in circulatory motion in one direction that then slows and stops before reversing to retrograde motion—each motion being a cycle in grade/retrograde until frictional damping reduces no motion in the Earth-Structure reference frame. Resultants and emergents having strong mereological sums of objects, i.e., the results of mechanically and electrically combining objects to achieve strong functional dependencies, e.g., constructing an autonomous passenger vehicle. Resultants and emergents having spooky mereological sums of objects, i.e., the results of mechanically combing objects to achieve spooky functional dependencies, e.g., configuring, integrating, and making interoperable a system of systems that incorporates a smart highway with autonomous passenger vehicles, autonomous freight trucks, high-speed autonomous rail transportation merged with pedestrian and bicycle traffic.

Emergents are ontological sums of objects driven by causal processes that move objects or parts of objects. The gap between what is observed to take place with interacting objects and what actually results from interacting objects is represented by the mereological summations of the products of interaction. The gap is lowest when the resultants form Notasystems. The gap is highest when there are both resultants, expected emergents and unexpected emergents. A measure of the gap is the level of knowledge of the observer.

6.10 Conclusion

In this chapter the predictability of emergence was discussed from the standpoint of a formal mereology of objects and processes. Formulating two frames of ontology—one for causation (process) and one for interaction (object) extends the formal mereology of Stanislaw Leśniewski to inspire the construction of an ontologically-driven interpretive integrative framework by which to explain what emergence is and why it exists. The mechanics of predicting, identifying, detailing, and explaining emergence is then set forth by reviewing the planned interactions for resultants and emergents; and by discerning the spurious interactions that are likely, possible, and impossible unless certain conditions are satisfied. Likely interactions are unplanned, but given extreme or foreseeable circumstances would happen. Possible interactions are unintended, yet could happen in spite of controls and mechanisms to the contrary. Impossible interactions are akin to black swan events and are at the confluence of an undeterminably small likelihood of occurrence (notably rare), but harboring devastating effects. Through the ontological frame of interactions, "likely" means that the objects are within the system or system of systems and their connectivity is established and known. "Possible" means the connectivity may be through agency relationships in the form of object-to-object-object-object and the mediation of interactions between these objects is controlled with various limits and constraints (which are deemed adequate to prevent emergence). "Impossible" means that the number of influences removed from each object is great, the likelihood of occurrence of a causal interaction is undeterminably small, but the consequences of such a confluence of causal actions is dire. All three kinds of interactions are detectable within the structures of objects and processes—by hand, a formidable task; by machine modeling and simulation, tractable in the first order. The black swan events may be premised on spurious or ad-hoc interconnectivity with a sufficiency of coupling and cohesion to cause "impossible" emergence. Emergence was discussed as simply the result of interaction(s) between physical objects whose mechanisms were the forcing event that resulted in consequences. The system of systems engineer is required to be most attentive to emergents facilitated by system-level and system of system–level interactions that change boundaries and boundary conditions for both systems and system of systems.

Given the persistent interpretations afforded various observations and perceptions of the world, phenomenologists challenge the view of the formal ontological calculus by usually being first to weigh in with reasonable and rational logic that explains. The objective of this chapter was not to lessen the value of phenomenology, which often has been responsible for early and plausible explanations for what we observed. Instead, this chapter compared but means to predict emergence using phenomenological thinking with that of the fundamental, formal method of expressing system artifacts in terms of their causal events as the impetus for developing a different understanding of that which we do not yet know. Closer attention by systems engineering ontologists is warranted when there can be great harm done by emergent behaviors. Rather, the aim of this research was to explore the nature of interaction through the interpretive integrated framework of ontology so that systems and systems of systems needs and requirements can be tamed through the use of the Gladstone bag with appropriate tools, techniques, knowledge, and sensitivities. The tools were laid out, the techniques were summarized, and the sensitivities to systemic behaviors were discussed. The sufficiency of this chapter to predict emergence is shown; however, the analysis of error that assumes this formal approach is a work in progress, to be reported later.

References

Aerts, D., Broekaert, J., and Gabora, L. (Eds.). 2003. *A case for applying an abstracted quantum formalism to cognition. Mind in Interaction.* Amsterdam, John Benjamins.

Aerts, D., and Gabora, L. 2005. "A Theory of Concepts and Their Combinations I," Kybernetes 34(1/2), pp. 167–191.

Atlan, H. 1998. Intentional Self-Organization. Emergence and reduction: Towards a Physical Theory of Intentionality. *Thesis Eleven*, 52(1): 5–34.

Bäck, E., Social and Cognitive Biases in Large Group Decision Settings, ISBN 978-91-7447-320-9. Printed in Sweden by US-AB, Stockholm 2011 Distributor: Department of Psychology, Stockholm University.

Bar-Yam, Y. 2004. "A Mathematical Theory of Strong Multiscale Variety," Complexity, vol. 9:15–24.

Bechtel, W., Richardson, R. C. 1992. Emergent Phenomena and Complex Systems. In A. Beckermann, H. Flohr, J. Kim (Eds.) *Emergence or Reduction? Essays on the Prospects of Nonreductive Physicalism*. (257–288). Berlin: Walter de Gruyter.

Beckermann, A., Flohr, H., and Kim, J., eds. 1992. *Emergence or Reduction?* Berlin: Walter de Gruyter.

Bedau, M.A. 2002. "Downward Causation and the Autonomy of Weak Emergence," Principia, vol. 6: 5–50.

Bedeaux, D., Standaert, F., Hemmes, K., and Kjelstrup, S. 1999. Optimization of Processes by Equipartition, J. Non-Equilibrium Thermodynamics, Number 825, Vol. 24: 242–259. Leiden Institute of Chemistry, Gorlaeus Laboratoria, Leiden, The Netherlands, 2 Delft University of Technology, Delft, The Netherlands, 3 Institute of Physical Chemistry, Norwegian University of Science and Technology, Trondheim, Norway.

Berenda, C. 1953. "On Emergence and Prediction," Journal of Philosophy, vol. 50(9): 269–274.

Bernet, R., Kern, I., Eduard, and Marbach, E. 1993. *An Introduction to Husserlian Phenomenology*. Evanston: Northwestern University Press.

Beurier, G., Simonin, O., and Ferber, J. 2002. "Model and Simulation of Multilevel Emergence," Proceedings 2nd IEEE International Symposium on Signal Processing and Information Technology, pp. 231–236.

Bittner, T., Donnelly, M., and Smith, B. "Endurants and perdurants in directly depicting ontologies," 2004, AI Communications, 13: 247–258.

Bonabeau, E. 2002. "Predicting the Unpredictable," Harvard Business Review, March.

Bornemann, F., Wenzel, S., 2006. "Managing Compatibility Throughout the Product Life Cycle of Embedded Systems—Definition and Application of an Effective Process to Control Compatibility." INCOSE 2006—16th Annual International Symposium Proceedings: Systems Engineering: Shining Light on the Tough Issues, Toulouse, International Council on Systems Engineering (INCOSE).

Boyd, R. 1999. Homeostasis, species, and higher taxa. In R. A. Wilson (Ed.), *Species*. New Interdisciplinary Essays: 141–185, Cambridge: MIT Press.

Brooks, H. 1972. A framework for science and technology policy. *IEEE Workshop on National Goals, Science Policy, and Technology Assessment*. Warrenton, VA.

Cabana, K.A., Boiney, L.G., Lesch, R.J., Berube, C.D., Loren, L.A., O'Brien, L.B., Bonaceto, C.A., Singh, H., Anapol, R.L. 2006. "Enterprise Research and Development: Agile Functionality for Decision Superiority," Volume 9, Enterprise Systems Engineering Theory and Practice, The MITRE Corporation, Bedford, MA, Report Number MP05B0000043, February.

Campbell, R. 2009. A Process-Based Model for an Interactive Ontology. Synthese 166: 453–477.

Carlos, O., Dori, D., and Sussman, J. 2011. "COIM: An object-process based method for analyzing architectures of complex, interconnected, large-scale socio-technical systems," in Systems Engineering, vol. 14: no. 3, pp. 364–382.

Casini, L. 2009. "Social Mechanisms: What Social Sciences Can Learn from Natural Selection," http://www.kent.ac.uk/secl/philosophy/jw/2009/macits/abstracts.pdf (accessed 9 June 2012).

Chen, C., Nagl, S.B., and Clack, C.D. 2008. "A method for validating and discovering association between multi-level emergent behaviours in agent-based simulations." Lecture Notes in Artificial Intelligence, vol. 4953:1–10.

Christensen, W.D., and Hooker, C.A. 2001. Self-Directed Agents. In McIntosh, J.S. (Ed.), *Naturalism, Evolution, and Intentionality*. Canadian Journal of Philosophy, Supplementary Volume 27:19–52. http://www.kli.ac.at/personal/christensen/SelfDirected_agents.pdf.

Copi, I.M. 2014. *Introduction to Logic*. (Cohen, C., and McMahon, K. (Eds.). 14th ed. New York: Routledge.

Crutchfield, J.P. 1994. "The Calculi of Emergence: Computation, Dynamics and Induction," Physica D, vol. 75:11–54.

Dalkin, S.M., Greenhalgh, J., Jones, D., Cunningham, B., Lhussier, M. 2015. "What's in a mechanism? development of a key concept in realist evaluation," in Implementation Science, Vol. 10 no. 49. https://implementationscience.biomedcentral.com/articles/10.1186/s13012-015-0237-x.

Daniels, J., Werner, P.W., and Bahill, A.T. 2001. "Quantitative Methods for Tradeoff Analyses," Systems Engineering 4(3): 190–212.

Donnelly, M. 2011. "Using Mereological Principles to Support Metaphysics," Philosophical Quarterly 61(243), pp. 225–246.

Doreian, P. 2001. *Causality in Social Network Analysis, Sociological Methods & Research*, Vol. 30, No. 1, 81–114 (2001) University of Pittsburgh: SAGE Publications.

Dori, D. 2002. *Object-process methodology: A holistic systems approach*, Springer-Verlag, Berlin.

Edwards, M. 2009. "The Impact of Technical Regulation on the Technical Integrity of Complex Engineered Systems," Proceedings of the INCOSE International Symposium. Singapore, International Council on Systems Engineering.

Einstein, A. 1950. *The Meaning of Relativity*. Princeton: Princeton University Press.

Emerson, R.F. 2011. "Infrastructure Systems Characteristics and Implications," CSER 2011, Redondo Beach, University of Southern California.

Emmeche, C., Køppe, S., Stjernfelt, F., Levels, Emergence, and Three Versions of Downward Causation, in 'Emergence' in Downward Causation, ed. P.B. Andersen, C. Emmerche, N.O. Finnemann & P.V. Christiansen (Aarhus University Press, 2000).

Engelhart, L.K. 2001. Wholeness and the Rational Structure of Inquiring Systems, Diss., March, Saybrook Graduate School and Research Center, San Francisco.

Everett III, H. 1957. "Relative State" Formulation of Quantum Mechanics, Review of Modern Physics 29(3):454–462.

Ferreira, S. and Tejeda, J. "An Evolving Understanding for Predicting Emergent Properties," 2011 IEEE International Systems Conference, 2011.

Ferris, T. June 2017. "Emergence as a Subject of Research, Research Methods, and Engineering Knowledge and Practice." Paper presented at the 12th International Conference on System of Systems Engineering (SoSE), Waikoloa, Hawaii.

Ferris, T.L.J., Cook, S.C., Honour, E.C. 2005. Towards a structure for systems engineering research, Proceedings of the 15th INCOSE Annual International Symposium—Systems Engineering: Bridging Industry, Government and Academia, Rochester, New York, 10–14 July, paper 6.1.1.

Feynman, R. 1970. *The Feynman Lectures on Physics*. 3 Volume Set, Addison-Wesley Longman.

Finner, H., and Strassburger, K. 2002. The partitioning principle: A powerful tool in multiple decision theory, The Annals of Statistics, Vol. 30, No. 4, 1194–1213.

Frege, G. 1979. "Posthumous Writings," Hermes, Kambartel, F., Kaulbach, F. (Eds.), translated by Long, P., and White, R., Blackwell, Oxford.

Fromm, J. 2005. "Ten Questions about Emergence." from http://arxiv.org/abs/nlin/0509049.

Gabora, L. 1995. *Meme and Variations: A Computational Model of Cultural Evolution*, SFI Studies in the Sciences of Complexity, Lecture Volume VI, Addison-Wesley.

Gannon, J.D., Hamlet, R.G., and Mills, H.D. 1987. Theory of Modules. IEEE Transactions on Software Engineering, Vol. SE-13, no. 7.

Garud, R., and Kumaraswamy, A. 1995. "Technological and Organizational Designs for Realizing Economies of Substitution." Strategic Management Journal 16: 93–110.

Gladwell, M. 2002. *The Tipping Point: How Little Things Can Make a Big Difference*. New York, Little, Brown and Company.

Gödel, K. 1951 (1995). Collected Works, Volume III. Oxford: Oxford University Press.

Goldberg, B.E., Everhart, K., Stevens, R., Babbitt III, N., Clemens, P., and Stout, L. 1994. System Engineering "Toolbox" for Design-Oriented Engineers. Marshall Space Flight Center, Alabama, National Aeronautics and Space Administration (NASA): 306.

Goodman, N. 1978. *Ways of Worldmaking*. Indianapolis: Hackett Publishing Co.

Gruber, T. 2007. "What Is an Ontology?," http://www-ksl.stanford.edu/kst/what-is-an-ontology.html (accessed 20 January 2010).
Hacking, I. 2007. Natural kinds: Rosy dawn, scholastic twilight. Royal Institute of Philosophy Supplements, vol. 61: 203–240.
Henry, D.P. 1972. *Medieval Logic and Metaphysics.* London: Hutchinson & Co. Ltd.
Hevner, A.R., and Mills, H. 1995. Box-structured requirements determination methods, Decision Support Systems 13: 223–239. North-Holland.
Heylighen, F. 1989. "Self-organization, Emergence and the Architecture of Complexity," Proceedings of the 1st European Conference on System Science, pp. 23–32.
Hitchins, D.K., 1992. *Putting Systems to Work.* Chichester, New York: John Wiley & Sons.
Holland, O. 2007. "Taxonomy of the Modeling and Simulation of Emergent Behavior Systems," Proceedings of the 2007 Spring Simulation Multiconference, vol. 2.
Holland, J.H., Holyoak, K.J., Nisbett, R.E., and Thagard, P.R. 1986. *Induction: Processes of Inference, Learning, and Discovery.* Cambridge, MA: The MIT Press.
Honour, E.C. and Valerdi, R., "Advancing an Ontology for Systems Engineering to Allow Consistent Measurement", 2006 Conference on Systems Engineering Research, Los Angeles, CA, 2006.
Hume, D. 1839. *Treatise of Human Nature.* Clarendon Press, University of Oxford, London.
Humphrey, W.S. 1989. *Managing the Software Process.* Reading: Addison-Wesley.
Humphreys, P. 1997. How Properties Emerge. Philosophy of Science, 64: 1–17.
Illari, P., and Williamson, J. 2011. "Mechanisms Are Real and Local. Causalities in the Sciences". Illari, P., Russo, F., and Williamson, J. (Eds.). Oxford: Oxford University Press.
Johnstone, P.T. 2001. "Elements of the History of Locale Theory," *Handbook of the History of General Topology,* Vol. 3, Dordrecht: Kluwer, 835–851.
Keets, C.M. 2008. "A Formal Theory of Granularity: Toward Enhancing Biological and Applied Life Sciences Information Systems with Granularity." Computer Science Department. Free University of Bozen-Bolzano, Bozen-Bolzano. Ph.D. diss.: 268.
Kerlinger, F.N., and Lee, H.B. 2000. *Foundations of Behavioral Research.* 4th Edition. Belmont, California, Centage Learning.
Kim, J. 2005. *Physicalism, or Something Near Enough.* Princeton: Princeton University Press.
Klepper, S. 1996. "Entry, exit, growth, and innovation over the product life cycle," in The American Economic Review, vol. 86, no.3, pp. 562–583.
Kocsis, J.G. 2008. Determining Success for the Naval Systems Engineering Resource Center. Department of Systems Engineering. Monterey, CA, United States Naval Postgraduate School. M.S.: 101.
Korman, D.Z. 2011. "Ordinary Objects." Stanford Encyclopedia of Philosophy, published online by Stanford University, http://plato.stanford.edu/archives/spr2007/entries/ordinary-objects/. (accessed 13 August 2011).
Kossiakoff, A., and Sweet, W.N., 2003. *Systems Engineering Principles and Practice.* Hoboken, New Jersey: John Wiley & Sons.
Lake, J. 2007. *Textbook of Integrative Mental Health Care.* New York: Thieme Medical Publishers, Inc.
Langford, G.O. 2012. *Engineering Systems Integration: Theory, Metrics, and Methods.* Boca Raton: CRC Press, Francis & Taylor.
Langford, G.O. 2013. Toward a General Theory of Systems Integration. Defence and Systems Institute, School of Electrical and Information Engineering, Ph.D. diss., University of South Australia, Adelaide, Australia.
Langford, G.O. 2016. Maintenance Scheduling Using Systems Engineering Integration. 26th Annual INCOSE International Symposium (IS 2016), International Council on Systems Engineering, Edinburgh, Scotland, July 18–21.
Langford, G.O., and Langford, T.S.E. 2017. The making of a system of systems: Ontology reveals the true nature of emergence. IEEE SoSE Conference, Hawaii.
Langford, G.O. 2017. Verification of requirements: System of systems theory, framework, formalisms, validity. 27th Annual INCOSE International Symposium (IS 2017), International Council on Systems Engineering, Adelaide, Australia, July 15–20.

Leśniewski, S. 1916. "Foundations of the General Theory of Sets, I." In Surma, S. et al. (Eds.), Stanislaw Leśniewski: Collected Works [Nijhoff International Philosophy Series, 44] Dordrecht: Kluwer Academic Publishers, 1992, pp. 129–173.
Leśniewski, S. 1988. Leśniewski's Lecture Notes in Logic. Srzednicki J.T.J., and Stachniak, J. (Eds.). Dordrecht: Kluwer.
Lewes, G.H. 1875. *Problems of Life and Mind*. Boston: James R. Osgood and Company.
Lonchamp, J. 1993. A Structured Conceptual and Terminological Framework for Software Process Engineering. In Proceedings of the 2nd International Conference on the Software Process (ICSP 2), Berlin, Germany, IEEE Computer Society Press.
Lupaşc, A., Lupaşc, I., and Negoescu, G. 2010. "The role of ontologies for designing accounting information systems," in The Annals of "Dunarea de Jos" University of Galati Fascicle I—2010. Economics and Applied Informatics. Years XVI—No 1.
Lyre, H. 1995. "The Quantum Theory of Ur-Objects as a Theory of Information." International Journal of Theoretical Physics 34(8): 1541–1552.
Machamer, P., Darden, L., and Carver, C.F. 2000. "Thinking About Mechanisms," Philosophy of Science 67(1): 1–25.
Machamer, P. 2004. "Activities and Causation: The Metaphysics and Epistemology of Mechanism." International Studies in the Philosophy of Science 18(1): 27–39.
Madni, A. M., "The Intellectual Content of Systems Engineering: A Definitional Hurdle or something More?", INCOSE Insight, October, pp. 21–23, 2006.
Mahalanobis, P.C. 1936. "On the Generalized Distance in Statistics," Proceedings, National Institute of Science of India, 2: 49–55.
Martin, J.H. 1988. A Computational Theory of Metaphor. Electrical Engineering Computer Science Department, Berkeley, University of California. Ph.D.: 224.
Martin, J., and Ferris, T. 2008. "On the Various Conceptualizations of Systems and Their Impact on the Practice of Systems Engineering." INCOSE.
McCarthy, W. 1982. "The REA accounting model: A generalized framework for accounting systems in a shared data environment," in The Accounting Review, vol. LVII, no. 3 July.
McLaughlin, B. P. 1992. The Rise and Fall of British Emergentism. In A. Beckermann, H. Flohr, J. Kim (Eds.) *Emergence or Reduction? Essays on the Prospects of Nonreductive Physicalism*. (49–93). Berlin: Walter de Gruyter.
Merleau-Ponty, M. 1992. *Phenomenology of Perception*. London: Routledge.
Miller, J.G. 1978. *Living Systems*. New York, McGraw-Hill Book Company, Inc.
Morris, P.W.G., and Pinto, J.K. (Eds.). 2004. *The Wiley Guide to Managing Projects*. New Jersey, John Wiley & Sons, Inc.
NASA. 2004. National Airspace System: System Engineering Manual, version 3.0, 30 September, Washington, D.C. (http://www.faa.gov/about/office_org/headquarters_offices/ato/service_units /operations/sys engsaf/seman/ (accessed 14 December 2008).
Oliver, D.W., Kelliher, T.P., and Keegan, J.G. 1997. *Engineering Complex Systems with Models and Objects*. New York: McGraw-Hill.
Osorio, C., Dori, D., and Sussman, J. 2011. "COIM: An Object-Process Based Method for Analyzing Architectures of Complex, Interconnected, Large-Scale Socio-technical Systems," Systems Engineering 14(3): 364–382.
Parson, J. 2012. "The many primitives of mereology," in Mereology and Location, Kleinschmidt, S. (Ed.). Oxford: Oxford University Press.
Polani, D. 2006. "Emergence, Intrinsic Structure of Information, and Agenthood," International Conference on Complex Systems (ICCS 2006): Boston, MA.
Pawson, R. 1989. *A Measure for Measures: A Manifesto for Empirical Sociology*. London, New York: Routledge.
Radder, H. 2006. *The World Observed/The World Conceived*. Pittsburgh: University of Pittsburgh Press.
Reed, I. 2008. "Justifying Sociological Knowledge: From Realism to Interpretation." Sociological Theory 26(2): 101–129.

Ricoeur, P. 1979. The model of the text: Meaningful action considered as a text. In Rabinow, P., and Sullivan, W. (Eds.), *Interpretive Social Science*. Berkeley, CA: University of California Press.
Ronald, E.M.A., Sipper, M., and Capcarrère, M.S. 1999. "Design, Observation, Surprise! A Test of Emergence," Artificial Life, Summer, Vol. 5, No. 3: 225–239.
Rosch, E. 1978. "Principles of Categorization," in *Cognition and Categorization*, pp. 133–159, Rosch, E., and B. Lloyd, B. (Eds.), Lawrence Erlbaum, Hillsdale, NJ.
Rosch, E. 1983. "Prototype classification and logical classification: The two systems," in *New Trends in Conceptual Representation: Challenges to Piaget's Theory?*, pp. 133–159, Scholnick, E.K. (Ed.) Lawrence Erlbaum, Hillsdale, NJ.
Rosch, E. 1999, "Reclaiming Concepts," Journal of Consciousness Studies, 6: 61–78.
Rueger, A. 2000. "Robust Supervenience and Emergence." Philosophy of Science 67(3): 466–489.
Rueschemeyer, D. 2009. *Usable Theory: Analytic Tools for Social and Political Research*. Princeton: Princeton University Press.
Russo, F., and Williamson, J. 2007. "Interpreting Causality in the Health Sciences," International Studies in the Philosophy of Science. 21(2): 157–170.
Sage, A. P., and Armstrong, J.E. 2000. Introduction to Systems Engineering. New York, John Wiley & Sons, Inc.
Schiemenz, B. 2002. Managing Complexity by Recursion. European Meeting on Cybernetics and Systems Research. Vienna, Austria.
Seel, R. 2000. "New Insights on Organizational Change," in *Organisation and People*, vol. 7, no 2, pp. 2–9.
Shalizi, C.R., and Crutchfield, J.P. 2001. "Computational mechanics: Patterns and prediction, structure and simplicity," Journal of Statistical Physics 2001:104:817–879.
Simons, P. 1987. *Parts: A Study in Ontology*. Oxford: Clarendon Press.
Smith, D. 2009. "Mereology Without Weak Supplementation," Australasian Journal of Philosophy, 87(3), pp. 505–511.
Smith, D. 2016. "Phenomenology," The Stanford Encyclopedia of Philosophy (Winter 2016 Edition), Zalta, E. N. (Ed.), https://plato.stanford.edu/archives/win2016/entries/phenomenology/ (accessed 18 May 2017).
Sowa, J.F. 1999. "Signs, Processes, and Language Games Foundations for Ontology." Invited Lecture, International Conference on the Challenge of Pragmatic Process Philosophy, University of Nijmegen, http://www.jfsowa.com/pubs/signproc.pdf (accessed 12 February 2017).
Sowa, J.F. 2000. *Knowledge Representation: Logical, Philosophical, and Computational Foundations*. Pacific Grove, CA: Books/Cole.
Swartz, N. 1997. "Definitions, Dictionaries, and Meanings," Philosophia Vol. 3:2–3:167–178 April-July 1973, http://www.sfu.ca/~swartz/definition.htm (accessed 8 November 2016).
Taguchi, G., and Jugulum, R. 2002. *The Mahalanobis-Taguchi Strategy*. New York: John Wiley & Sons.
Tahko, T.E., and Lowe, E.J. 2016. "Ontological Dependence," The Stanford Encyclopedia of Philosophy (Winter 2016 Edition), Zalta, E.N. (Ed.), https://plato.stanford.edu/archives/win2016/entries/dependence-ontological/ (accessed 19 February 2017).
Tan, H.B.K., Hao, L., and Yang, Y. 2003. On formalizations of the whole-part relationship in the Unified Modeling Language. IEEE Transactions on Software Engineering, 29(11): 1054–1055.
Torgerson, W.S. 1967. *Theory and Methods of Scaling*. New York: John Wiley & Sons, Inc.
Trockel, W. 1999. "Integrating the Nash Program into Mechanism Theory, Economics," Working Paper 787, Department of Economics. University of California, Los Angeles, and Bielefeld University.
Uschold, M., and Gruninger, M. 1996. "Ontologies: principles, methods and applications," Knowledge Engineering Review, 11(2): 93–136
Valerdi, R., Friedman, G., Marticello, G.D. 2011. "Diseconomies of Scale in Systems Engineering," Proceedings Conference on Systems Engineering Research (CSER), Redondo Beach, Ca, University of Southern California.
van Atten, M., and Kennedy, J. 2003. On the Philosophical Development of Kurt Gödel, Bulletin of Symbolic Logic, 9(4): 425–476. Reprinted in *Kurt Gödel: Essays for His Centennial*, Feferman, S., Parsons, C., and Simpson, S.G. (Eds.), Cambridge, UK: Cambridge University Press.

Varzi, A. 2008. "Universalism entails extensionalism," Analysis 69, pp. 599–604.

von Bertalanffy, L. 1968. *General System Theory: Foundations, Development, Applications*. New York: George Braziller.

Weiss, F., and Glanville, D. 2002. "The Need for Descriptions of a System's Dynamic Behavior on Projects Involving the Integration of Large and Complex Systems," Systems Engineering, Test and Evaluation Conference, Sydney, Australia, October.

Wells, G.D., and Sage, A.P. 2009. *Engineering of a System of Systems*. John Wiley & Sons, Inc., Hoboken.

White, D.R. 2007. Innovation in the Context of Networks, Hierarchies, and Cohesion. Dordrecht, Springer Methodos Series.

Wieringa, R.J. 1996. *Requirements Engineering: Frameworks for Understanding*. John Wiley & Sons, Inc., Amsterdam.

Winograd, T. 1997. Categories, disciplines, and social coordination. In Friedman, B. (Ed.), *Human Values and the Design of Computer Technology*, pp. 107–114. Cambridge, UK: Cambridge University Press.

Winograd, T. 1996. *Bringing Design to Software*. New York: ACM Press.

Winograd, T. 1995. Heidegger and the design of computer systems. In Feenberg, A., and Hannay, A. (Eds.), *Technology and the Politics of Knowledge*, pp. 108–127. Indianapolis: Indiana University Press.

Zafirovski, M. 2005. "Social exchange theory under scrutiny: A positive critique of its economic-behaviorist formulations," Electronic Journal of Sociology, vol. 2, pp. 1–40.

7

System of Systems Process Model

Gary O. Langford

CONTENTS

7.1 Introduction

This chapter introduces a metaprocess model for developing, enhancing, and sustaining lifecycles of complex systems of systems. This metamodel is a heterarchical conglomerate of domains of artifacts for constituent systems and corresponding domains of artifacts for the system of systems (Figure 7.1).

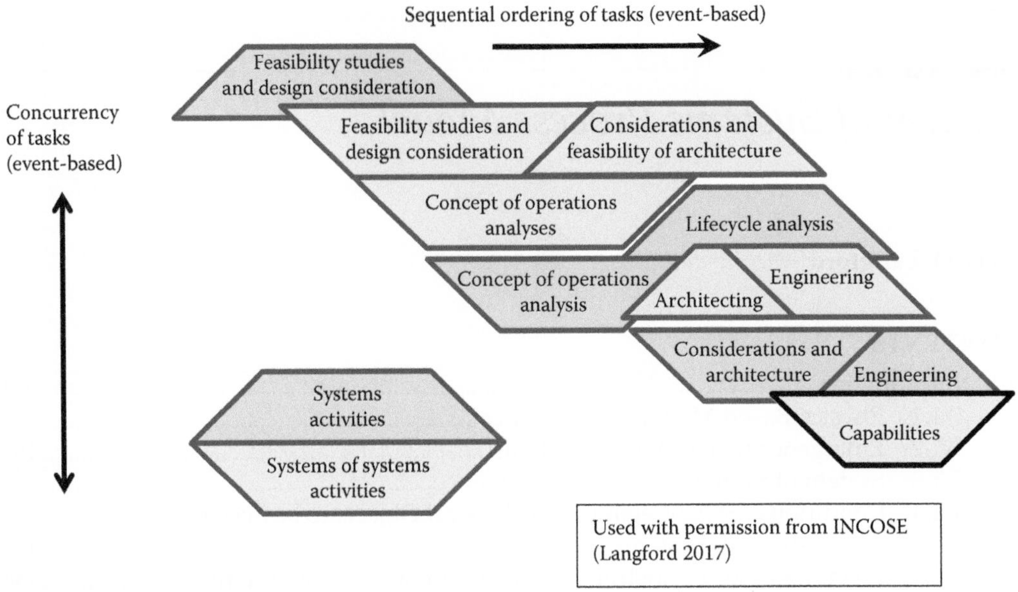

FIGURE 7.1
The metamodel for lifecycle work with standard systems engineering process models.

Whereas the traditional systems engineering process models in part rely on a hierarchical structuring of requirements, functions, processes, performance, and risk, the metamodel presumes that the structuring of artifacts needs to accommodate both hierarchies as well as other structures. The metaprocess model represents a nearly flat abstraction at the same level of decomposition across the categories in contrast to layering of successively greater details in hierarchical fashion. The logic and reasons for supplying artifacts, regardless of the structures within the receiving process model, builds on the rooted concept that systems of systems have part-whole relationships that neither supersede nor supplant their constituent systems. Rather, the five key characteristics for systems of systems (Maier 1998)—operational independence of constituent systems, managerial independence of constituent systems, evolutionary development, geographical distribution, and emergent behavior—provide a convenient means to distinguish systems of systems from systems. There are harmonious and suitable relations between constituent systems in a system of systems. Consequently, the artifacts from constituent systems of the system of systems need to neither harm nor be harmed during their lifecycle participation in the system of systems.

Several approaches to formulating requirements for systems of systems have been developed. Notably, data modeling and various modeling languages have been devised to structure artifacts for or to support the formulation and management of requirements, e.g., object management, the unified-modeling-language (UML), object-oriented methods, rational unified processes, object-process methodologies, eXtensible Markup Language (XML) metadata-interchange specification, and System Markup Language (SysML), to mention a few of the methods and tools. Additionally, formal methods and mathematical approaches and analytical techniques are used to verify requirements and address conflicts with requirements.

In contrast to all previous approaches, methods, and tools to develop and manage requirements, this chapter introduces an ontological framework premised on a physics-based model of emergence. The discussion follows a strict mathematical formulation following the formal logic of objects and processes laid down by Leśniewski (1988). The chapter begins with a physics-based model of emergence and then incorporates emergence in the lifecycle of a system of systems; followed by the strict conditions for aggregations of objects, process models, ontology, principles, and essential foundational definitions. The details of how to formalize the artifacts for requirements is detailed in the Appendix (A.7: Sequencing of Domain Artifacts).

7.1.1 Physics-Based Model of Emergence

Underpinning this research is the development of a physics-based model of emergence that is consistent with the characterization of emergence by Maier (1998). The commonly used terminology of objects as physical objects and processes as mechanisms and activities are developed here according to their counterparts in physics. Process is force (Feynman 1970) and function is energy (Langford 2012, 2013). Process, expressed through mechanisms, applies force that influences and moves physical objects. When objects interact, they show emergence (as a result of the interaction) (Langford 2012, 2013, 2018). Interaction does not infer only physical contact. Contact can be made by energy, matter (physical contact), material wealth (money—which may or may not be physical), and information—which can be conveyed through energy, matter, or material wealth. Interacting objects produce a function (Langford 2012, 2017). Without interaction there is no system of systems—no interaction, no function. Function results from the action of one object with another object. The physics of interaction used within this chapter is presented within the domain of classical Newtonian physics. The particulars of this discussion are extendable and scalable to relativistic physics and quantum mechanics. The selection of the appropriate physics is determined by the scale of the system of systems. For optical computing in a system of systems, the quantum domain is descriptive of the effects of interaction; whereas in nuclear medicine that involves interactions between electrons and human tissue, the domain of classical physics is based on everyday typical macro-scale observations.

7.1.2 Emergence in a System of Systems

Emergence in a system of systems derives from the interaction between its constituent systems. A system of systems exists only when its constituent systems interact. That interaction is typically when the constituent systems exchange energy, matter, material wealth, or information (EMMI). Atypical interaction occurs when the behaviors of the constituent systems change in anticipation of corporeal interaction, but as yet the exchange of EMMI has not yet commenced. Without interaction, there is no system of systems. With interaction, the system of systems exists. If an entity only exists when two constituent entities interact, the resultant and emergent are definable according to Lewes (Lewes 1875). Lewes stated that the resultant was the original of two constituent entities (which had no change after interaction), whereas the emergent would result in the system of systems. In the original context of Lewes' portrayal of emergence, the system of systems is the emergent and the constituent systems remain the same before and after interaction. From a fundamental perspective, the emergence from interacting entities is an entity, specifically function(s) (Langford 2012). Therefore, systems of systems functionalities are emergent functions that either did not exist within the constituent systems and emerged from the agglomeration or existed in one or more of the constituent systems and extended their functionalities to other constituent systems. In either case, the emergent function might exhibit change in functional performance or quality. In artifactual systems, the emergent is typically a change in performance or quality. In natural constituent systems, the emergent is typically a new function that is not expressed in any of the constituent systems.

7.1.3 System of Systems Lifecycle Emergence

Systems of systems emergence, therefore, is related to the lifecycle of the system of systems. Before the system of systems comes into existence, the system of systems has no emergence. Once the first two constituent systems begin to interact, emergence transpires and the system of systems becomes a force greater than any of the constituent systems. When the last two constituent systems disengage from interaction, the system of systems ceases to provide enhanced functional capability to any of the constituent systems.

7.1.4 Conditions for Aggregations of Physical Objects to be Systems and Systems of Systems

There are four conditions to determine if an agglomeration of objects constitutes a system and five conditions that determine if an agglomeration of systems constitutes a system of systems. The four conditions to be a system are metastability, internal agility, external adaptability, and non-reciprocal emergence; whereas the five conditions to be a system of systems are metastability, internal agility, external adaptability, and non-reciprocal emergence—to sustain its systemic state, in addition to the system having reciprocal emergence with the system of systems (Langford 2012, 2013, 2015, 2017). All four conditions for a system must be satisfied for an agglomeration of parts to be a system. And, all five conditions for a system of systems must be satisfied simultaneously for a system of systems to exist.

Each constituent system must retain its ability to be dynamically stable. Dynamic stability means that each system must change in response to internal and external contexts. Changing the capability of individual physical objects or aggregations of individual objects to accommodate changing contexts means that the various aggregations have control mechanisms that respond to context and attempt to revert the actions of the aggregations to the stability of various target values that are sustainable, i.e., revert to initial states after change (Langford 2012). The actions of dynamic stability create a metastable condition that drives feedback—the first condition for a system. In this regard, emergence for each constituent system is reciprocal. Reciprocal means like-kind response by an object with EMMI to input from another object. For example, gift for gift, or responses to questions. Reciprocal does not mean the same for same, but rather appropriate response to sustain or complete interaction.

The condition of metastability requires reciprocal emergence or the system changes irreparably. Metastability is a condition for sustaining systemic behaviors that means the system's state may change from one to another, but in all cases the systemic behaviors are retained regardless of the end result of the change. An example of reciprocal emergence is a battery with limited capacity that is drawn down by an electrical load. If only one battery exists, then after the power depletion, the load either becomes inoperable or draws needed power from another source. During the lifecycle of the battery, the load is sustained. Rechargeable batteries cycle their electrical outputs and can continue to sustain their loads as long as recharging occurs. For example, the solar electromagnetic flux bathes a point at the equator on the rotating Earth for approximately 12 hours each day. Twelve hours of daylight and then 12 hours of darkness change the production of CO_2 (daylight hours) and O_2 (dark hours), recharging the atmosphere.

The second condition is internal agility to move EMMI as necessary to sustain capability. Movement of EMMI means interactions, and interactions means emergence. As an agglomeration of objects encounters interactions from outside of its internal interactions, if some of its internal mechanisms are operable to support the inherent internal needs of individual objects or aggregations of internal objects, a condition for a system is met. The sustained agility of the internal capabilities to move EMMI is a necessary condition for a system. Emergence, albeit emergence that results in a loss of EMMI, is necessary to sustain internal agility. To sustain emergent functionality in a system, EMMI must be drawn down from internal capacity, and at some point, must be either replaced or drawn down with resultant changes in the systemic behaviors that change the system from one state to another that is consistent with the metastability condition. Internal agility to move EMMI is a necessary condition that must be met for a system to exist (Langford 2012).

The third condition for a system is to have external adaptability. Interactions with entities external to the system may require the system to expend EMMI to respond, to hide, to defend, to adapt, or to accept (for example). External adaptability requires that the condition for internal agility be satisfied, as well as the condition for metastability (Langford 2012).

The fourth condition for a system is to demonstrate non-reciprocal emergence. Non-reciprocal emergence means that once the system parts have interacted, that interaction is non-reversible or non-reciprocal. There is no return interaction of EMMI from the receiving object that is like-kind, i.e., non-reciprocal. Unlike reciprocal emergence that is required for metastability, a system requires non-reciprocal emergence to change the individual parts to the whole that exhibits properties, traits, and attributes that are irreversibly different (Langford 2012). An example of non-reciprocal emergence for artifactual systems is changes in procedures to extract additional capability out of existing physical objects, changes in behaviors to use existing objects differently than before, e.g., to improve sustainability. The non-reciprocal emergents are enforced by rules, for example, enacted by people. For living systems, an example is a seed germinating into a plant. All four conditions must be satisfied for a system.

A system of systems must respect all system-ness of its constituent systems and do no harm to irreparably change them. The "do not harm" to its constituent systems is the first condition for all systems of systems. The other four conditions for systems of systems are that each constituent system must simultaneously satisfy the four conditions for systems; the agglomeration of constituent systems must be metastable; the agglomeration of constituent systems must have internal agility to move EMMI between the constituent systems; and the agglomeration of constituent systems must have external adaptability to behave in a systems manner with the other constituent systems. The emergence that results from the interactions of the constituent systems must be that of a system without the requirement to damage any of the constituent systems.

For systems of systems, no constituent system shall interact with another constituent system to harm each other or any other system of systems. For artifactual and natural systems, the emergent functions that enable system of systems will most certainly require each constituent system to expend EMMI. Each system that participates in a system of systems will experience a loss for its lifecycle participation with other constituent systems. For artifactual systems of systems, the decision to participate must be made by each constituent system based on the relative worth of that participation. Under most circumstances, it is self-defeating for a system to participate as a constituent system to its complete detriment and demise. However, one constituent system may be comfortable with sacrificing itself for the good of the whole and therefore join the system of systems with that as a possible outcome for its participation.

7.1.5 Metamodel for Predicting Emergence in Systems of Systems

The materiality of the five conditions for systems of systems versus the reciprocal requirement of systems to retain autonomous behavior is a typical culprit in misunderstanding the nature of emergence that makes up the system of systems. The metamodel (MULE) is presented which enables appropriate artifacts in proper temporal sequence to derive the necessary emergents for lifecycle planning and lifecycle building of both constituent systems and their system and systems. See the Appendix (A.7: Sequencing of Domain Artifacts) for detailed discussions as to how to lay out the sequencing of artifacts from both the system of systems and for each of the constituent systems.

This metamodel encompasses the results from *multiple* states of existence and localizations of functions in space (*ubieties*) that reconcile differences (*liaise*) to achieve non-local *expressions* of capabilities—*m*ulti-*u*bieties *l*iaise *e*xpressions (MULE). MULE provides properly sequenced artifacts that result from multiple interactions between the various domains of constituent systems and the system of systems. The traditional systems engineering process model, with the requisite artifacts and emergents structured by MULE, is conceptually appropriate for lifecycle work on a system of systems. Without such structure, the traditional systems engineering process model is subject to abject failure.

7.1.6 Traditional Systems Engineering Process Models

The traditional systems engineering process models do not adequately develop and provide artifacts from constituent systems and system of systems domains in correspondence with the characteristics of the systems of systems. Therefore, structuring the artifacts from constituent systems as subservient or servile due to hierarchical thinking destroys the value of the system of systems.

7.1.6.1 Hierarchy

With few exceptions, the systems engineering process models are hierarchically configured and therefore apropos for systems, many of which are modeled after the concept of central control. That is not to indicate that systems are centrally controlled, rather that the systems engineering process model represents distributed processing in a hierarchical manner. This structuring of artifacts in a hierarchical style (Holland 2007) should not be cast upon the artifacts for a system of systems or the systems engineering process model, as hierarchical formatting will induce emergent behaviors that will counter the native characteristics desired in the systems of systems.

7.1.6.2 Context

Further, the traditional systems engineering process model drives the requisite inputs necessary to carry out the stages, phases, or steps predetermined by best practices within that hierarchical conviction. Therefore, the method used by the systems engineers following the tenets of the traditional systems engineering process model will consider the input artifacts as not having or resulting in any particular preferential order with little regard for the importance of sequencing as a primary means to influence design, architecture, and operations. Intermingling artifacts from the constituent systems domain with those artifacts from the system of systems domain without proper sequencing is akin to disruptive syntax that obscures or minimizes meaning. The issue is context for the artifacts. Context means that the circumstances, conditions, and relations that form the setting for an artifact are inextricably linked through the result of a measurement of a variable that depends on another measurement of a different variable. The syntactic function of context is to categorize and classify, i.e., ontology and taxonomy, respectively. Therefore, the context of relations between constituent systems and the system of systems enables adequate communication (Langford and Ferris 2014) that is improved by proper sequencing. Context with proper sequencing results in convergence to the appropriate meaning of the relationship between a domain artifact from the constituent system and a concomitant domain artifact from the system of systems. Proper parts are those parts that satisfy three self-evident criteria: first, nothing is a proper part of itself (a proper part is part of but not identical to the individual or whole); second, if A is a proper part of B then B is not a part of A; third, if A is a part of B and B is a part of C then A is a part of C. To explain using an example: the minimum situation within a set of situations is the situation that does not have proper parts that are also in the set of all situations. The minimum situation is the one which contains just enough parts to be the minimal of all properties, traits, and attributes such that nothing can be removed and still be considered within the set of situations. Further, a proper part may not be present in the minimum situation.

7.1.6.3 Sequencing

Proper sequencing is the sequence that satisfy three self-evident criteria: no entity in sequence is a proper part of itself (a proper sequence is one in which no one entity is repeated one after the other in sequence and in which an entity in sequence is not identical to the sequence of entities); second, if X is a proper sequence within a sequence Y then the Y sequence is not a part of X; third, if X is a proper sequence within Y and Y is a proper sequence within Z then X is a proper sequence of Z. And, alternatively, if X is a proper sequence of Y, then there is some sequence W that is a part of sequence Y and disjoint from sequence X. Applying proper sequencing to the constituent systems and system of systems domain, the sequence X (from the constituent systems domain) and the sequence Y (from the systems of systems domain) have a common referent in which there exists proper meaning for integration and interoperability for Z, notwithstanding W.

7.1.6.4 Proper Part, Proper Sequencing

A proper part is considered to be a trait of an object that is comprised of multiple objects, e.g., an artifact that has properties, traits, and attributes (Langford 2017). A proper sequence is considered to be ontologically dependent on the artifacts from the perspective of integration and interoperability. An object that is sequenced with another object with a different referent is an improper sequence, the combining of which has interactions with different emergents than the emergents from proper parts and proper sequences. The systems engineer will determine a different result if the sequencing is proper or improper. In summary, the mereological sum of proper parts in proper sequence is mandatory for bringing inputs from the domains of constituent systems and system of systems so that the resultant emergence reflects the fullest extent of integration and interoperability of constituent systems and the system of systems.

7.2 Objects, Functions, Behaviors, Processes, Cognition, Mechanisms, and Modeling

Objects are entities that are things or represent things physical. Objects are of obvious interest to every living thing, and as such, objects have been accommodated, considered, or discussed since life's beginning. Before Parmenides and Plato, to Einstein (Einstein 1950), Gödel (Gödel 1951 [1995]), and after (Korman 2011), objects continue to reveal different aspects of their existence and importance (Langford 2012). Objects have location in space (Everett III 1957) and are measurable in that position space (Lyre 1995), and are representable by ideas that are the agents of physical/intellectual things. Objects are both physical and intellectual. Physical objects are made up of parts that are also physical objects; therefore objects are divisible. An object is an object regardless of the number of objects that comprise the parts of an object. Physical and intellectual objects make up the world and all within it, the solar systems, and the universe. The cognitive representation of that object is an object (see Chapter 6).

Some objects are artifactual and are designed and made to have intended functions. Not all objects have intended functions. But, if an object has no function, then it is not interactive with any other object. The teleological discourse is broadly extended to artifactual and natural objects. Objects are absolutes (Martin 1988). Physical/intellectual objects, people, and ideas are representative of objects. As absolutes, objects are viewed independent of other objects. An object's independence results from its individual nature that is differentiable from other objects. Each object is bounded, i.e., separated from other objects.

7.2.1 Object Boundaries

All objects are bounded. Boundedness is perfected by three types of boundaries that are the limits of physical objects. These limits are expressed as the physical limits of the object, the results of the uses of the object (functional space), and the consequences of either having the object or the function or not having the object or the function (behavioral domains). The physical limits of an object have continuity of parts that comprise the limit of an object. That continuity is expressed as the molecular components of the objects, commonly referred to as the surface of the object. At the micro-scale, the object's surface will have unbounded molecular components that may be in proximity but not strongly bonded. The micro-scale propinquity is beyond the scope of this discussion. However, the meso-scale issues derive from the context and environmental situation in which the object's physical, functional, and behavioral boundaries are considered. If the temperature and pressure is such that the physical surface is reactive with its environment, then significant quantities of the object's molecular material may be associated with the object, but neither strongly bonded nor in close proximity to the bulk of the object's material. The meso-scale propinquity is also beyond the scope of this discussion. In the classical physical sense, the boundary at the macro-scale is as determined to be the surface, i.e., the commonly referred to surface of an object that can be seen and touched.

When objects interact with other objects, the effects are determined by the condition of connectivity. Physically touching the surface of an object by another object (macro-scale connectivity) results in an exchange of EMMI. The matter-to-matter contact between two objects enacts a transfer of energy; matter, e.g., mass; material wealth, e.g., money; or information. (Langford 2012). The physical boundaries of the two objects are made observable both by the lack of exchange of EMMI as well as the transfer of EMMI.

Definition: An object is a bounded set of material properties and traits whose structure is comprised of constituent objects and mechanisms. An object is or represents a tangible, material thing. Objects have influence. Objects have boundaries. Objects can change. Objects have properties. Objects have traits. Objects have attributes. Objects can be produced. Objects can interact with other objects. EMMI expresses the interactions between objects.

7.2.2 Object Boundary Conditions

An object is influenced by its sensitivities to conditions. For example, conditions that define the mediation of exchange of EMMI across a boundary are termed boundary conditions. A convenient means of describing objects is to characterize their intrinsic properties, situational traits, and attributes. A condition is the circumstances that encompass an object; the factors that affect the manner and ways in which the object interacts; the situation in which the object operates; or the terms under which an object behaves (Langford 2012, p. 42). Citing Norman Swartz, "A condition *A* is said to be necessary for a condition *B*, if (and only if) the falsity (/nonexistence/non-occurrence) [as the case may be] of *A* guarantees (or brings about) the falsity (/nonexistence/non-occurrence) of *B*" (Swartz 1997). Many conditions (e.g., circumstances) may apply. An object is influenced by its sensitivities to these conditions. Boundary conditions mediate the flow of EMMI across interfaces at boundaries.

7.2.3 Object Interaction

Objects interact. Objects interact with other objects through the exchange of EMMI. A sampling from the many authors who have discussed interactions based on various inputs and outputs for functions is presented here: energy and matter (Sage and Armstrong 2000); energy and information (Morris and Pinto 2004); energy, material, and information (Wieringa 1996; Oliver, Kelliher, and Keegan 1997; Kossiakoff 2003); and energy, matter, and information (Miller 1978; Bornemann and Wenzel 2006; White 2007; Edwards 2009; Tan, Hao, and Yang 2003; Wells and Sage 2009). However, the expression of EMMI captures both the physical domain of inputs and outputs as well as the intellectual domain of inputs and outputs, according to the general expression of objects as physical and intellectual. Therefore, EMMI is the means of expressing interaction for the purpose of integration (Langford 2012, 2013), and emergence (Langford 2017).

Interaction leads to constraints on the two objects (Langford 2013). Limitations (boundaries) are given by the domain of the interactions, whereas, constraints are a structural property of the interactions. Therefore, interactions cause a change in boundaries, which in turn changes the constraints on the objects, which results in emergence.

7.2.4 Object Constraints

A constraint exists due to the conditions that govern within the context of objects or processes. These conditions are specified by the amount of EMMI that is available for an object or process. The mechanism, which drives the physical manifestations of an object, acts according to the capacity, the needs of the mechanism(s), and the kinds and amounts of EMMI. The object's output EMMI is a transformation of the input EMMI and the effects of EMMI on the object's mechanism. In all cases of interaction, there is loss of EMMI. Similarly, the conditions under which a process is enabled and driven by input EMMI through its procedures and mechanistic behaviors and then carried out by objects are likewise constraining for the process. As well, the processes interact to constrain processes that result in objects being constrained if those objects also interact with the constrained process. An object or process under such conditions is said to be constrained. One constrained process within a network of constrained processes can constrain the whole of the network, depending on the architecture and design of that network.

Constraints change the boundaries and boundary conditions of objects. Changes in boundaries or boundary conditions precipitate emergence. The amount of change in the constraints and therefore the amount of change in the boundary conditions leads to non-linear changes that are observed as emergence. There are threshold effects that limit the amount of emergence that is observable. In other words, a small, imperceptible change in boundary conditions may not be observable, but the cumulative effects of small changes in boundary conditions may exceed a threshold after which the effects can cause subsequent interactions to be observable, but without traceability to the precipitating sub-threshold effects. Interacting entities are subject to limitations, conditions, and constraints (Langford 2012).

Boundary conditions are either satisfied by incident EMMI or not satisfied. If satisfied, then the EMMI becomes an input to the object. If not satisfied, the incident EMMI is either not accepted or in some fashion enters the physical boundary of the object. The object may be disrupted or destroyed if its boundary conditions are not satisfied. As with constraints, a limit imposes a boundary and changing a limit changes a boundary and perhaps a boundary condition.

7.2.5 Objects in Context

Premised on a state-context-property theory (Aerts and Gabora 2005), objects exhibit various traits based on context. While objects may be categorized according to any number of priorities or measures, a rule-based categorization process is inconsistent with the results of modern empirical and qualitative research (Rosch 1978, 1983, 1999). Observing objects within various contexts, from moving frames of reference (e.g., velocity) to changes due to processes, the result of contexts is often noticed as a change in traits of an object. Relating observable traits to an internal mechanism of an object introduces the concept of "state." The conception of "state" is a level of detail unnecessary for this chapter and is therefore beyond its scope. However, without adopting the concept of "state," as is done by Aerts and Gabora (Aerts and Gabora 2005), objects exist in context(s). It is that context that imbues objects with various traits that we observe and measure. As the context changes, the measurements may change. The characterization of objects as perhaps exhibiting different observables by changes in their traits is not meant to imply that context is only the mechanism that is correlated with traits. Changes that are internal to the object also may result in a host of different observables. This chapter posits that objects may exhibit various traits due to context, and context may change due to its interactions with the object. Similarly, the object may change its context. For example, moving an object with mass m may result in changes in mass as a function of velocity v, and in addition, space-time may be changed by its interactions with the object.

If certain traits are in some way related to or identifiable as properties in a given context, then those traits might in fact be emergent properties. This formulation of emergent behaviors is consistent with Lewes' definition and current interpretations as expressed in systems engineering literature (Polani 2006; Fromm 2005; Valerdi et al. 2011). Rather than ascribe emergent behaviors as an attribute which is intrinsic to objects, that context and objects are operands on each other through interaction offers additional insight into the context by which emergence arises. Systems integration creates contexts within which objects interact and it is that context coupled with the object from which emergence arises. In a general sense, integration may be merely an object with its context. The impact of context is observed "... as an element of complexity ...," "... in the case of adversary-created emergence, changes in enemy tactics serve to adapt or interact with our efforts to ..." interpret those (Cabana et al. 2006). In these interpretations of observed emergence: "... the principle that whole entities exhibit properties which are meaningful only when attributed to the whole, not to its parts ..." is discussed in (Hitchins 1992, p. 10); and (Weiss and Glanville 2002).

The discussion of objects in context relates to emergence through precipitant events arising out of objects as they change contexts. This change is shown to be a testable hypothesis with the result that there exists a threshold for emergence that can be defined and is observable (Langford 2013). Whether there is actually a gradient of change that is at some point versus one event that results in a "tipping point" (Gladwell 2002) is identified as a topic for further research, suggesting a transition from an element (as an object that is changed due to a process) based on a threshold(s) that represents some resistance to the transition to from one object to a new object.

7.2.6 Functions

The exchange of EMMI creates a function that did not exist before the interaction between two objects, notionally based on Gottlob Frege's ontology of properties (concepts as part of realism) and relations of objects, i.e., functions (Frege 1979, p. 177). It is the interaction that provides the use of one object by another. Without interaction, there is no access to another object and therefore no means to use that other object. The minimum requirement for an object to use another object is interaction sufficient to provide for that use.

A function is the action realized when objects interact (Langford 2012). A function is measurable by performances, referred to as functional performances. Functions arise due to the enactments of mechanisms of at least two objects that are interacting. Functions have inputs of EMMI and outputs of EMMI. More formally, Object O and Object P create a function f if and only if:

a. Object O interacts with Object P.
b. Function f is the set of all actions that are a consequence of Object O interacting with Object P.
c. An action is defined as the release or receipt of EMMI.
d. EMMI activates mechanisms.
e. Object O is changed by its mechanism(s) because of an interaction with Object P, or Object P is changed by its mechanism(s) because of an interaction with Object O, or both Object O and Object P are changed by their respective mechanism(s) because of an interaction.
f. Mechanism is set of rules and logic constrained by context and environment that govern the transformation of EMMI by objects.
g. Mechanisms are the means or methods by which objects change, i.e., the execution of activities and acts.
h. A change is defined as the difference between an object before interaction and the same object after interaction.

An example of interaction to produce a function with emergence is to use a carpenter's hammer to nail a 2 x 4 brace and brackets to provide restraints on an installed, gas-fired water heater. The purpose of the project is to provide bracing of the water heater to the structural wood-framed wall to secure the water heater in case of an earthquake. This situation is commonplace in a residential setting within an earthquake prone area. The function of "to secure (water heater)" is in response to the problem of preventing a gas-line attached to the water heater from becoming dislodged during an earthquake and causing a fire that results in loss of life or property due to inadequate response from overwhelmed emergency services. A hammer, nails, and bracing materials are needed to secure a water heater to a structural wall. Procedurally, the gas should be turned off before working in the area surrounding the water heater, then turned on once the task of securing the water heated is completed. Buffo, the homeowner's pet dog, stays with the tradesperson from the start to finish of these activities at the home. The interaction between the tradesperson and the hammer with a mass of 473 grams to pick up, swing, and strike nail, produces the function of "use (hammer)." The subfunction of "F: to strike (nail)" begins the interaction between the hammer and the nail. Since functions are measurable and testable, the functional performance specification for "FP: to use (hammer)" is an 850 +1100/-730 Newton force to perform the subfunction of "F: to strike (nail)." Buffo places his mouth over the gas shut-off valve and cracks it open slightly after the tradesperson leaves the work area to respond to a distressing text message. When the tradesperson returns, he finds Buffo waging his tail, quite content that he has helped. Thinking about the text message, the tradesperson slams the hammer onto a 16-penny nail (posed to secure a strap to the wall), clipping the edge of the nail, then ricocheting the nail point into the nearby metal corrugated connector pipe, puncturing the gas line. Gas begins to leak into the immediate area. Without realizing the damage and still reeling from the text message, the tradesperson jams another nail into the same nail-point site and winds up taking quick aim, then punishes the nail with a thrashing blow. A spark flies from the contact point at the physical boundary of the hammer and the nail. Neither the tradesperson nor Buffo are harmed by the explosion or during the writing of this example. The resultant function of "F: to nail" was enacted with less functional performance than required to complete the intended task. The functional performance was inadequate and "out of spec"; the boundary conditions were not satisfied; yet the function of "to use" was enacted. The interactions between the tradesperson and the hammer and then two combinations of the tradesperson, hammer, nail resulted in the additional functions of "F: to fly (nail)"; "F: to puncture (pipe)"; "F: to create spark"; "F: to ignite gas"; and "F: to knock bejezes out of the cognitive processing of stakeholders (tradesperson and Buffo)." The function of "F: to fly (nail)" results from interaction between tradesperson-hammer-nail, as measured at 2 meters/second; the function of "F: to puncture (pipe)" results from the interaction between nail-pipe, as measured at 150 Newton force to penetrate gas pipe; the function of "F: to create spark" results from interaction between tradesperson-hammer-nail, as measure at 0.05 Kilojoule; the function of "F: ignite gas" results from the interaction between spark and mixture of gas and air, as measured at 5–15% natural gas concentration in a volume of air with 0.33 Millijoules of spark energy; and the function of "to shock cognitive processing" as measured by blast overpressure of 700-1400 kilograms per square meter. All of the interactions in this example show emergence, some intended, some not. Regardless of intention, the results of interaction are the enabling of functions, i.e., uses that may or may not be wanted or used.

If two objects interact such that one or both of the objects change, then a function is derived as trait of the relation or behaviors that is evoked because of the interaction. A function is a set of actions that result when two objects interact. A function is measured by performance of the bi-object's properties or traits, either individually or pairwise. Functions arise due to the enactments of the mechanisms of at least two objects. Intended functions require proper EMMI.

7.2.7 Ontology of Causation: The Process Frame

A process is a partially ordered set of activities, use of tools, and practices set down in a structured order, that are carried out within limitations, subject to constraints to provide a result that is intended to satisfy a need (adapted from Humphrey [1989] and referring to the work of Lonchamp [1993]. Processes can be measured relative to other processes and improved relative to themselves (Goldberg et al. 1994). In the main, processes are sets of activities, of which the make-up and sequencing is often accommodated by the participants' skills, resource availability, and supporting infrastructure. There is no set of required acts, activities, or procedures; no absolute sequencing; and no fixed or unalterable juxtaposition of work.

A process is a systematic pattern, a coordinated set of procedures, tasks, activities, or acts that convert inputs of EMMI into outputs of EMMI. Process is an amalgamation of cogitation, acts, activities, procedures, and developing models or invoking representations of concepts. In other words, all things necessary to convert an input into an output are accomplish by process. The basic unit of a process is an act—a single factor that can be evaluated in isolation. An act is a single step in a series of steps that when combined are recognizable as an activity. Activities are orchestrated by procedures to form process.

Processes rely on inputs and incur losses to achieve outputs. Processes are measurable with objective measures such as cost, number of people, and the amount of loss to achieve certain results. Processes are comparable to other processes, but on a subjective basis, i.e., without a standard or reference. Consequently, it is quite difficult to say that one process is better than another.

Processes are of two kinds: those that are continuous and expressive within the postulates of four-dimensional space-time, and those that are discrete. Discrete processes exist within boundaries, but with piece-wise continuous expressions that concatenate interrupted segments of activities into the semblance of continuity. At the fundamental level, all processes are discrete. In this regard, the use of process in this chapter is generalized to include both continuous and discrete kinds.

Process is a metaphor—a core metaphor (Martin 1988, p. 38). A core metaphor is a set of metaphors linked by association that extends to all activities that make up or support the commonly shared metaphor. In this regard, process can be considered an enclosure (Martin 1988, p. 1), or as an environment (Martin 1988, p. 2). That processes have inputs and outputs corresponds with the concept of enclosure by which the wrapper that delineates both process and the contexts of process as the process environment.

There are noticeable differences between processes, in general, with some consuming more labor, some requiring greater resources, some costing more money. In isolation, the differences can be viewed as objective measures. Since one process differs from another by its make-up of activities and its environment, any comparison may have little meaning since their basis may be quite different.

An activity is the means by which a mechanism applies force to an object. Processes have inputs of EMMI and outputs of EMMI. Process provides for the enactment of a function, i.e., the capacity to do work. Every function has multiple measurable performances associated with the perspectives of stakeholders. To achieve various levels of functional performances, processes must be enabled by EMMI to enable mechanisms. Processes derive from mechanisms as the capacity to do work on objects. Note: excess capacity is a trade-space (Emerson 2011), by which the design and architecture for systems and system of systems is a primary concern. Tradeoff studies create an objective mechanism (Daniels, Werner, and Bahill 2001). Objective mechanisms are those that are evaluated either by whether or not formal rules have been followed (i.e., process) or by measurements relative to a standard unit of measure, e.g., rule of length, time, mass, energy, power, velocity, or temperature, (i.e., function).

7.2.8 Mechanism

A mechanism is defined as the set of processes (procedures, activities, and acts—whether a constituency or acting individually) that operates in the context of forces. More formally, Machamer, Darden, and Carver (2000), explain that "Mechanisms are entities and activities organised such that they are productive of regular changes from start or set-up to finish or termination conditions" (Machamer, Darden, and Carver 2000). Additionally, Machamer, Darden, and Carver state that two distinct kinds of building blocks make-up a mechanism—entities and activities. Mechanisms "force" objects or processes to be conducted. Mechanisms are carried out by objects that interact with objects and processes (Gabora 1995); and by objects, systems, and processes, all interacting in various combinations with each other (NASA 2004).

Mechanisms explain (Illari and Williamson 2011) how things change. Change is caused by the input of EMMI into an object or parts of an object. Without an input of EMMI, there is no change. Change means that an object is different because of its use of EMMI. That difference is observable from one instance in time to another instance in time. An event is the result of the enactment of a mechanism by input EMMI which is transformed into output EMMI. The output EMMI is termed functional performance, i.e., the performance that a user derives from applying or using the function. Functions occur at the physical interface of or between two objects. Behaviors occur due to the existence or non-existence of object(s) and due to function(s) or no function(s). Every enactment of a mechanism results in a change, the results of which are an event.

EMMI stimulate the enactments of mechanism(s) that exert(s) force(s) that influence(s) some or all of the structures that comprise the object or its parts. The structures (Reed 2008), e.g., mechanical, electrical, and chemical activities and procedures within an object, give way to the influences of the forces derived from EMMI. Structures may give way to these influences depending on their susceptibility to such influence. When structures give way, they can have an impact on other structures, influencing these other structures depending on the coupling and cohesion between the structures. Structures that give way in turn may influence other structures to give way. The properties of structures (those intrinsic characteristics that have mechanistic and process resilience, i.e., relaxation and restoring action), have a semblance of reliability if their variations in performance about a central nominal value stay within a range of upper- and lower-boundedness.

Objects have structures with mechanical characteristics that result in giving way to influences from EMMI. Mechanisms can be characterized by an enabling space, a transacting space, and an outcome space (Trockel 1999). The output EMMI is parsed for convenience into two components: one describable as performance (functional performance), the other indicated as losses of EMMI. Functional performance is measurable (with appropriate instruments with sufficient accuracy and precision). Output EMMI may be uni- or multi-variant. Therefore, there can be multi performances for a single function.

Each object transforms the input EMMI into an output EMMI by the process of changing the input through the actions of the object's mechanism(s). The mechanism is a form of force, whether mechanical, chemical, or electrical. Force is defined as the influences of EMMI on objects—no influence on an object, no force, and no change. The test for influence is determined by the net of power (i.e., work done) on an object as observed by the output EMMIs of that object; changes in the object's properties, traits, or attributes; or other changes in boundary, boundary conditions, physical issues, and functional or behavioral issues (Kocsis 2008).

An object sends EMMI (e.g., releases, sets free, or give ups) through the process(es) of an object's internal mechanisms. Similarly, an object receives EMMI (e.g., accepts, collects, or takes in) through the process(es) of an object's internal mechanisms. An interaction is defined by identifying the sending and receiving objects. James Lake argued that "functional congruence must exist between phenomena that underlie a specified symptom pattern and phenomena operationalized as the mechanism of action ..." (Lake 2007).

7.2.9 Force and Energy

Processes are akin to force (Feynman 1970). Processes enable mechanisms to do work. For an object to do work, processes must build the capacity to do work by enabling mechanisms within an object to accept input EMMI and to release EMMI. EMMI is the capacity to do work. Work may be observable by the behaviors of the objects that are responsive to interactions. The interaction (the input to the mechanism) facilitates work.

The broader definition of "force" captures the general notion of overcoming resistance to change, i.e., the enactment of processes to "P: take-in (EMMI)" and "P: to release (EMMI)."

That change in the object due to the interaction also results in a change in the object's energy. The energy represents the capacity of an object; and a change in capacity is indicated by an inputs and outputs of EMMI. Interaction results in a change in capacity with process being responsible for building or reducing capacity.

The empirical (in contrast to the theoretical) form of interaction takes place through processes (forces) that act to enable mechanisms. A metaphorical interpretation of interaction is the exchange of capacity that results in interactions of objects. It is likely that the local environments have received EMMI lost from the interactions, for example, the absorption of heat resultant from the physical heating of the atmosphere along the path of transmissions of an electromagnetic beam; or the reduction in the interpretable information due to a reduced signal-to-noise ratio caused by dissipation of energy over distance transverse and longitudinal to the primary motion. Therefore, the aggregate consequence of two-object interactions is not the simple sum of one action (from the sending object with that of the action of the receiving object). The objects may be different, including their internal mechanisms, their actions, and their losses, and the resulting changes they undergo are different than each other's changes.

Mechanisms are the means by which objects and processes change. Change means that an object or process is different at one instance from the previous or next instance. Change is precipitated by EMMI. If there is no change in an object, there is no energy loss.

7.2.10 Cognition

Cognition is the acts and activities carried out to think, reason, perceive, meditate, and learn. The result of these activities is to form the psychological make-up of being. Applying cognition to processes, the activities span the gamut from thinking (e.g., about thinking), formulating (e.g., the thought patterns to detect proximate events), determining (e.g., how to trade off approaches and methods), determining (e.g., how to identify the best course of action), laying out (e.g., how to carry out planning), reasoning (e.g., through the kinds of issues), meditating (e.g., to calm the soul); rationalizing (e.g., the efficacy of a set of procedures); considering (e.g., feasibility of a decision), to give some examples. The kinds and types of thinking are embodied in the process ontology that determines all that follows in the causal frame.

7.2.11 Modeling and Representing

To model objects (the process) is to imitate or represent the dynamic relations between objects, the interactions of objects, the consequences of interactions, and the behaviors of objects on a scale that is convenient for imitating causal properties, traits, and attributes of objects. To model processes (the process) is to imitate or represent the dynamic relations of objects regarding causal cognition, causal procedures, and causal mechanisms. The enactments of modeling or representing objects and process are by abstracting, granularizing, reduction, and partitioning.

7.3 Basis for Metamodel, Principles, and MULE

Heuristics are inherited from a structure of objects, processes, and principles that are formed into an explanation that applies in a variety of circumstances. This structure is neither fact nor evidence. It is also not a guide. It is often without standards or substance. Its aim is to influence practice. It is called theory. The role of theory is to inspire an association of its principles with the proper circumstances under which to accomplish the requisite work, to bring about critical thinking (Williams 2011; Kasser et al. 2010) to develop relevant and appropriate heuristics, and to expose cogent and useable information. Much of the discussion on "best practices" is more fittingly focused on the normative assumptions (Ulrich 2006) and worldview that are strong determinants of the outcome of research and their application in practice (Ulrich 1988). Theory has necessary and sufficient conditions that suggest how and why it should be used, and the boundaries and boundary conditions for its use. In essence, these conditions stipulate the minimum criteria for validity (Bailey 1987). Validity is established for a theory through (1) an explanation of phenomena that is both consistent with observation and representative of the needs of integration, i.e., that the successful systems integration is both recognized and achieved (Valerdi and Davidz (2009); (2) development of a framework by which it can be evaluated, i.e., the interplay of frames that fairly and accurately represent mereology (Schlager and Blomquist 1999); (3) a determination of the commensurability of what is said and what is known; and (4) broad agreement as to the principles that permeate the practice of systems and systems of system. The primary goal was to establish a meta-paradigm (Powers and Knapp 1990), a metamodel (MULE) that, following (Kuhn 1962), carries out six items:

1. It is formed on a set of basic assumptions that show relevance and utility in multiple disciplines, but are not derived solely from practice;
2. It presents a unifying perspective by defining terms to encompass the meanings that are generally in use in systems thinking, but not the nuances that may have evolved due to local norms or conventions;

3. It provides a wholeness of character through a common structure and set of processes, i.e., integration and interoperability;
4. It communicates the essence, nature, and purpose of engineering systems and systems of systems, i.e., integration and interoperability;
5. It refrains from developing or providing specific strategies for practice, but rather encourages the development of heuristics based on principles derived from theory, rather than from best industry practice; and
6. It recommends that best practices be brought under configuration management and vetted by their environment, their context, their broadness of applicability, their congruency, and their outcome(s).

Therefore, principles need to be categorized by forms of differentiation. Principles derived from various forms of categorizations may be abstractable to just the point where they remain meaningful in the context of the specifics of programs or projects. From these abstracted and principled notions, heuristics (or stratagems) can be devised to not be "... guaranteed to give right answers in all cases to which they apply."

Maier recognized five principles for systems of systems (Maier 1998) that fulfill the intent and promise of a widely-accepted body of literature that is representative of the primitives and advanced thinking about systems and systems of systems.

Principle 1: Operational independence of the constituent systems, when separate and distinct from the system of system.

Principle 2: Managerial independence of the systems when working with and when separate and distinct from the system of systems.

Principle 3: Evolutionary development occurs throughout the system of systems lifecycle with changing functions, functional performances, and performance quality as constituent systems join and depart from the system of systems. However, if a constituent system was to experience non-reciprocal emergence due to its interactions with other constituent systems and evolve due to non-reciprocity, it might endanger its autonomous behavior. Non-reciprocal behaviors are inherent in a system, but not in systems of systems. The principle of operational independence and the condition that joining, participating, or departing from a system of systems shall do no harm to constituent systems (Langford 2012) prohibit (in strongest terms) non-reciprocal behaviors in any constituent system. The outcomes may be surmised from the example of a hydroelectric dam, what Maier would refer to as a directed type of system of systems (Maier 1998). The recreational system is created by the reservoir (with water) and the electrical generation system is created by water flowing (in this case from the reservoir). In the extremum, a low water condition in the reservoir eliminates the recreational system and eliminates the electrical output of the electricity generation system. Both systems experience non-reciprocal emergence from which both lose all functional performance for their primary purpose and goal. Both systems cease to be systems and their contributions to their system of systems degrades their system of systems performances to naught. The lack of EMMI (water in this case) caused both systems to perish. They reside as protasystems, but without a system of systems with functional performances. Evolutionary development will only occur throughout the system of systems lifecycle without violating the principle of operational independence. The conditions for systems and systems of systems are discussed in Sections 7.14, 7.15, and 7.4 of this chapter.

This chapter adds to the body of knowledge with an ontologically-premised set of conditions that distinguish systems from notasystems and systems from systems of systems that builds on the five principles of Maier.

Principle 4: Emergent behavior is characteristic of systems of systems. It is typical that systems of systems carry out functions and purposes that are not present in any constituent system. These behaviors are not localized to any constituent system.

Principle 5: Geographic distribution of the constituent systems is commonly large. In the case of a hydroelectric dam, the system that generates electricity exchanges EMMI with the recreational system. The dam is an essential part of both systems, that is both co-located and juxtaposed. Arguably, the hydroelectric dam–recreational area can have a large or small geographic distribution depending on the considerations of the physical, functional, and behavioral boundaries of the two constituent systems. Changing any one or all of the physical, functional, or behavioral boundaries between parts affects the whole, as seen in both the resultants and emergents. A part before a boundary change can indicate a different part after that change. A change in the boundary of the whole also can indicate a different whole, each part potentially affected by the change in the whole. The amounts of EMMI that are exchanged would seem dependent on variables other than geographic distribution as indicated by Maier. That leaves open to further consideration what variables are significant for geographic distribution.

7.4 Essential Foundation for Systems of Systems and Systems

A metamodel based on MULE is proposed to prevent disjunction between the idiosyncratic nature of systems of systems in contrast to the non-reciprocal essence of systems through emergence. The systems of systems harbor systems, and as such can be thought of as a system by its whole. But a system of systems is in actuality not a system in the same manner that a system is a system. Systems must satisfy the condition of non-reciprocity—inability to reverse emergence to return to the previous state of part-whole relations.

In keeping with the five principles of systems of systems (Maier 1998) and in agreement with the four conditions that must be satisfied for an agglomeration of objects to be a system (Langford 2012), if a system of systems were to exhibit emergent behaviors (as a system), one or all of the constituent systems would have changed, the result of which is a constituent system that will have changed by its interaction with the system of systems.

The MULE metamodel (MULE) specifically structures the artifacts from 21 interactions that provide inputs and feedback to the systems engineering process model for building, delivering, operating, and sustaining systems of systems and their constituent systems so as to assure that constituent systems retain their systemic, autonomous behaviors and that no unintended emergence results by which any of the constituent systems lose its systemness. Figure 7.2 shows the MULE interactions by adjacent trapezoids, each of which interacts concurrently from the perspective of a vertical line drawn through the trapezoids.

This metamodel describes the development process as 18 recursive phases of iterations appropriately sequenced and interleaved according to common referents between the intended system of systems and each of its constituent systems (whether existing or new).

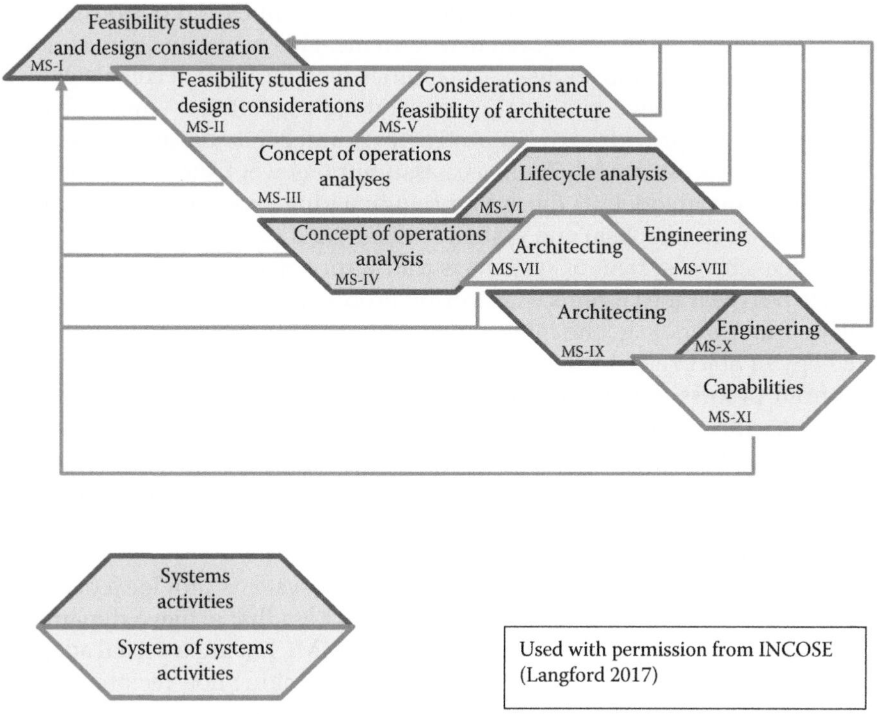

FIGURE 7.2
Stages and feedback in the multi-ubieties liaise expressions meta-model (MULE) for the systems of systems lifecycle.

MULE provides a meta flow of language and conjunctions of artifacts by which to use existing systems process models to develop and integrate physical infrastructures, computer hardware and computer software, and human/robotic processes (synchronous, asynchronous, and concurrent) for lifecycle capability. MULE accommodates software developments in the same fashion as the systems engineering process models or commercial process models take in inputs from different domains. During participation, the domains of the systems are interactive with the domain of the system of systems in a reflexive, self-referent manner. Whether for conceptualization, design, architecture, system or application coding, maintenance, or upgrade, MULE acts simply as an organizing structure, a model, whose procedures are inherent in the MULE domain relations.

7.5 Conclusions

In this chapter, the use of a metamodel (MULE) was introduced to improve the lifecycle outcomes of systems of systems using standard systems engineering process models. Building on the development and practical use of a metamodel that organizes data and information inputs from constituent systems within a system of systems, the systems engineering process model can be extended to deal with the complexities beyond its intended usefulness. This research followed from the realization that both systems and systems of systems can coexist to complement and augment functionalities and performances during a unique lifecycle, referred to as the system of systems. This view considers *m*ultiple states of existence and spatial localizations of capabilities (*u*bieties) that reconcile conflicts and differences (*l*iaise) to achieve non-local *e*xpressions of capabilities—multi-ubieties liaise expressions (MULE) to provide artifactual structures and behaviors as inputs into the systems engineering process model. Two basic assumptions were shown to underscore the validity of such a metamodel: (1) the systems engineering process model appositely captures the ramifications of holism and system realisms; and (2) fundamentally, a system is a system of systems but a system of systems is a structure of systems. The first assumption was shown to be true if the systems engineering process model (1) strictly adheres to a formal ontology framework (e.g., the interpretive integrative framework); (2) follows the formal mereology of object and process (e.g., Leśniewski, 1988); and (3) differentiates energy (function) from process (force) and distinguishes between interacting objects (function) and mechanisms (process). The second assumption was shown to be true through the conditional definitions by which a system is a system of systems and a system of systems is a system. Both assumptions are the paramountcy of all considerations for systems integration, interoperability, model-based systems thinking and engineering, and modeling and simulation for system of systems.

The conditional definitions of systems and systems of systems provide a decisive means to characterize a system as a group of objects and process that achieve dynamic stability by being metastable, having internal agility to move EMMI, having external adaptability to react appropriately to their environs and context, and exhibiting non-reciprocal emergence. Further, a system of systems acts much in the same manner as a system and satisfies nearly the same conditions as a system, with the caveat that constituent systems may join or leave.

With the purpose of reflecting design, architecture, integration, interoperability, performance, testing, verification, and validation, MULE incorporates emergence that occurs when artifacts from the various constituent systems domains interact with the domains of the system of systems. The joining of two or more constituent systems begins with the corporeal lifecycle of the system of systems, whereas when only one constituent system remains (and all others have therefore left), the end of the corporeal lifecycle is attained.

With the context of the interactions between domain artifacts from constituent systems and the system of systems, it is shown that intermingling artifacts from the constituent systems domain with those artifacts from the system of systems domain without proper sequencing obscures the meaning and reason for integration. The issue is context for the artifacts. Context with proper sequencing results in convergence to the appropriate meaning of the relationship between a domain artifact from the constituent system and a concomitant domain artifact from the system of systems. Sequencing is as crucial for aligning inputs and outputs as is the interactions between the artifacts from the constituent systems domains and the system of systems domain.

References

Aerts, D., Broekaert, J., and Gabora, L. (Eds.). 2003. *A case for applying an abstracted quantum formalism to cognition. Mind in Interaction.* Amsterdam, John Benjamins.

Aerts, D., and Gabora, L. (2005). "A Theory of Concepts and Their Combinations I," Kybernetes 34(1/2), pp. 167–191.

Bailey, K.D. 1987. *Methods of Social Research*. 3rd ed. New York, The Free Press.
Bornemann, F., and Wenzel, S. 2006. "Managing Compatibility Throughout the Product Life Cycle of Embedded Systems—Definition and Application of an Effective Process to Control Compatibility." INCOSE 2006—16th Annual International Symposium Proceedings: Systems Engineering: Shining Light on the Tough Issues, Toulouse, International Council on Systems Engineering (INCOSE).
Cabana, K.A., Boiney, L.G., Lesch, R.J., Berube, C.D., Loren, L.A., O'Brien, L.B., Bonaceto, C.A., Singh, H., Anapol, R.L. et al. 2006. "Enterprise Research and Development: Agile Functionality for Decision Superiority," Volume 9, Enterprise Systems Engineering Theory and Practice, The MITRE Corporation, Bedford, MA, Report Number MP05B0000043, February.
Daniels, J., Werner, P.W., and Bahill, A.T. 2001. "Quantitative Methods for Tradeoff Analyses," Systems Engineering 4(3):190–212.
Edwards, M. 2009. "The Impact of Technical Regulation on the Technical Integrity of Complex Engineered Systems," Proceedings of the INCOSE International Symposium. Singapore, International Council on Systems Engineering.
Einstein, A. 1950. *The Meaning of Relativity*. Princeton: Princeton University Press.
Emerson, R.F. 2011. "Infrastructure Systems Characteristics and Implications," CSER 2011, Redondo Beach, University of Southern California.
Everett III, H. 1957. "Relative State Formulation of Quantum Mechanics," Review of Modern Physics 29(3), pp. 454–462.
Feynman, R. 1970. *The Feynman Lectures on Physics*. 3 Volume Set, Addison-Wesley Longman.
Frege, G. 1979. "Posthumous Writings," Hermes, Kambartel, F., Kaulbach, F. (Eds.), translated by Long, P., and White, R., Blackwell, Oxford.
Fromm, J., "Types and Forms of Emergence," Complexity Digest, vol. 25, 2005.
Gabora, L. 1995. *Meme and Variations: A Computational Model of Cultural Evolution*, SFI Studies in the Sciences of Complexity, Lecture Volume VI, Addison-Wesley.
Gladwell, M. 2002. *The Tipping Point: How Little Things Can Make a Big Difference*. New York, Little, Brown and Company.
Gödel, K. 1951 (1995). Collected Works, Volume III. Oxford: Oxford University Press.
Goldberg, B.E., Everhart, K., Stevens, R., Babbitt III, N., Clemens, P., and Stout, L. 1994. System Engineering "Toolbox" for Design-Oriented Engineers. Marshall Space Flight Center, Alabama, National Aeronautics and Space Administration (NASA): 306.
Hitchins, D.K., 1992. *Putting Systems to Work*. Chichester, New York: John Wiley & Sons.
Holland, O. 2007. "Taxonomy of the Modeling and Simulation of Emergent Behavior Systems," Proceedings of the 2007 Spring Simulation Multiconference, vol. 2.
Humphrey, W.S. 1989. *Managing the Software Process*. Reading: Addison-Wesley.
Illari, P., and Williamson, J. 2011. "Mechanisms Are Real and Local. Causalities in the Sciences." Illari, P., Russo, F., and Williamson, J. (Eds.) Oxford University Press, Oxford.
Kahn, H., Brown, W., and Martel, L. 1967. *The Next 200 Years: A Scenario for America and the World*. New York: William Morrow.
Kasser, J.E., Frank, M., Zhao, Y.Y. et al. 2010. "Assessing the Competencies of Systems Engineers," 7th Bi-annual European Systems Engineering Conference (EuSEC), Stockholm, Sweden.
Kocsis, J.G. 2008. Determining Success for the Naval Systems Engineering Resource Center. Department of Systems Engineering. Monterey, CA, United States Naval Postgraduate School. M.S.: 101.
Korman, D.Z. 2011. "Ordinary Objects." Stanford Encyclopedia of Philosophy, published online by Stanford University, http://plato.stanford.edu/archives/spr2007/entries/ordinary-objects/ (accessed 13 August 2011).
Kossiakoff, A., and Sweet, W.N. 2003. *Systems Engineering Principles and Practice*. Hoboken, New Jersey: John Wiley & Sons.
Kuhn, T.S. (1962). *The Structure of Scientific Revolutions*. Chicago, University of Chicago Press.
Lake, J. 2007. *Textbook of Integrative Mental Health Care*. New York: Thieme Medical Publishers, Inc.

Langford, G. 2007. "Reducing Risk in Designing New Products Using Rapid Systems Engineering," Paper # 18, 23–24 March, Asia-Pacific Systems Engineering Conference.
Langford, G.O. 2012. *Engineering Systems Integration: Theory, Metrics, and Methods.* Boca Raton: CRC Press/Taylor & Francis.
Langford, G.O. 2013. Toward a general theory of systems integration: Research in the context of systems engineering. PhD diss., University of South Australia.
Langford, G.O. 2016. Maintenance Scheduling Using Systems Engineering Integration. 26th Annual INCOSE International Symposium (IS 2016), International Council on Systems Engineering, Edinburgh, Scotland, July 18–21.
Langford, G.O., and Langford, T.S.E. 2017. The making of a system of systems: Ontology reveals the true nature of emergence. IEEE SoSE Conference, Hawaii.
Langford, G.O. 2017. Verification of requirements: System of systems theory, framework, formalisms, validity. 27th Annual INCOSE International Symposium (IS 2017), International Council on Systems Engineering, Adelaide, Australia, July 15–20.
Langford, G.O. 2018. Phenomenological and ontological models for predicting emergence. In *Engineering Emergence: A Modeling and Simulation Approach*, Rainey, L. and Jamshidi, M. (Eds.). Boca Raton: CRC Press/Taylor & Francis.
Langford, G.O., and Ferris, T. 2014. A Systems Thinking Society: Integrative Framework for Technology. The International Journal of Technology, Knowledge and Society. Common Ground, ISSN # 1832-3669.
Lewes, G.H. 1875. *Problems of Life and Mind.* Boston: James R. Osgood and Company.
Lonchamp, J. 1993. A Structured Conceptual and Terminological Framework for Software Process Engineering. In Proceedings of the 2nd International Conference on the Software Process (ICSP 2), Berlin, Germany, IEEE Computer Society Press.
Lyre, H. 1995. "The Quantum Theory of Ur-Objects as a Theory of Information." International Journal of Theoretical Physics 34(8): 1541–1552.
Machamer, P., Darden, L., and Carver, C.F. 2000. "Thinking About Mechanisms," Philosophy of Science 67(1):1–25.
Maier, M.W. 1998. "Architecting Principles for Systems-of-Systems." Systems Engineering. 1(4): 267–284.
Martin, J.A. 1988. Computational Theory of Metaphor. Electrical Engineering Computer Science Department, Berkeley, University of California. Ph.D.: 224, page 48.
Miller, J.G. 1978. *Living Systems.* New York, McGraw-Hill Book Company, Inc.
Morris, P.W.G., and Pinto, J.K. (Eds.). 2004. *The Wiley Guide to Managing Projects.* New Jersey, John Wiley & Sons, Inc.
NASA. 2004. National Airspace System: System Engineering Manual, version 3.0, 30 September, Washington, D.C. (http://www.faa.gov/about/office_org/headquarters_offices/ato/service_units/operations/sys engsaf/seman/ (accessed 14 December 2008).
Oliver, D.W., Kelliher, T.P., and Keegan, J.G. (1997). *Engineering Complex Systems with Models and Objects.* New York, McGraw-Hill.
Polani, D. 2006. "Emergence, Intrinsic Structure of Information, and Agenthood," International Conference on Complex Systems (ICCS 2006): Boston, MA.
Powers, B.A., and Knapp, T.R. (1990). *A Dictionary of Nursing Theory and Research.* Newbury Park, Sage Publications.
Reed, I. 2008. "Justifying Sociological Knowledge: From Realism to Interpretation." Sociological Theory 26(2): 101–129.
Rosch, E. 1978. "Principles of Categorization," in *Cognition and Categorization*, pp. 133–159, Rosch, E., and Lloyd, B. (Eds.) Lawrence Erlbaum, Hillsdale, NJ.
Rosch, E. 1983. "Prototype classification and logical classification: The two systems," in *New Trends in Conceptual Representation: Challenges to Piaget's Theory?*, pp. 133–159, Scholnick, E.K. (Ed.) Lawrence Erlbaum, Hillsdale, NJ.
Rosch, E. 1999, "Reclaiming Concepts," Journal of Consciousness Studies, 6: 61–78.

Sage, A.P., and Armstrong, J.E. 2000. *Introduction to Systems Engineering.* New York: John Wiley & Sons, Inc.

Schlager, E., and Blomquist, W. 1999. "A Comparison of Three Emerging Theories of the Policy Process," Political Research Quarterly 49: 651–672.

Swartz, N. 1997. "Definitions, Dictionaries, and Meanings," Philosophia Vol. 3:2-3:167–178 April–July 1973, http://www.sfu.ca/~swartz/definition.htm (accessed 8 November 2016).

Tan, H.B.K, Hao, L., and Yang, Y. 2003. On formalizations of the whole-part relationship in the Unified Modeling Language. IEEE Transactions on Software Engineering, 29(11): 1054–1055.

Trockel, W. 1999. "Integrating the Nash Program into Mechanism Theory, Economics," Working Paper 787, Department of Economics. University of California, Los Angeles, and Bielefeld University.

Ulrich, W. (1988). "Systems Thinking, Systems Practice, and Practical Philosophy: A Program of Research," Systems Practice 1 (2), pp. 137–163.

Ulrich, W. (2006). "Rethinking critically reflective research practice: Beyond Popper's critical rationalism," Journal of Research Practice 2(2), Article P1., http://jrp.icaap.org/index.php/jrp/article/view/64/63 (retrieved 4 March 2012).

Valerdi, R., and Davidz, H.L. (2009). "Empirical Research in Systems Engineering: Challenges and Opportunities of a New Frontier," Systems Engineering 12(2), pp. 169–181.

Valerdi, R., Friedman, G., Marticello, G.D. et al. 2011. "Diseconomies of Scale in Systems Engineering," Proceedings Conference on Systems Engineering Research (CSER), Redondo Beach, CA, University of Southern California.

Weiss, F., and Glanville, D. 2002. "The Need for Descriptions of a System's Dynamic Behavior on Projects Involving the Integration of Large and Complex Systems," Systems Engineering, Test and Evaluation Conference, Sydney, Australia, October.

Wells, G.D., and Sage, A.P. 2009. *Engineering of a System of Systems.* John Wiley & Sons, Inc., Hoboken.

White, D.R. 2007. Innovation in the Context of Networks, Hierarchies, and Cohesion. Dordrecht, Springer Methodos Series.

Wieringa, R.J. 1996. *Requirements Engineering: Frameworks for Understanding.* John Wiley & Sons, Inc., Amsterdam.

Williams, J.D. (2011). "Are You in Touch with Your System Engineering "'Softer-Side?,'" Conference on Systems Engineering Research (CSER 2011), Los Angeles, University of Southern California, 15–16 April.

8

An Ontology of Emergence

John J. Johnson IV and Jose J. Padilla

CONTENTS

8.1 Introduction

The term emergence often brings to mind a familiar paradox: "… the whole is greater than the sum of its parts." The actual statement, "… the whole is something beside the parts," is from Aristotle's discussion on the nature of material things and the substances of which they are composed (Aristotle, 350 B.C.). The implication is that an ensemble of material things has something that their parts don't have. Engineered systems (i.e., systems produced by humans) are ensembles that are designed to produce effects (behaviors, properties, qualities, etc.) that fulfill a purpose that cannot be satisfied by individual parts alone (Ackoff, 1971; Checkland, 1999; Blanchard & Fabrycky, 2006). Aristotle might say that engineered systems are wholes that are something beside their parts (i.e., they have emergent capabilities). The additional capabilities of an ensemble of parts that has capabilities beyond those of its parts makes emergence a potential opportunity and a worthy engineering objective (Valerdi et al., 2008; Tolk and Rainey, 2015). However, another basic aspect of the emergence concept is the apparent absence of traceability between the system design (i.e., relationship between parts) and the system effects (Lewes, 1875). This is the aspect of the emergence concept that presents a potential problem for engineers and system stakeholders.

There is no single generally accepted definition of emergence or emergents that captures all of their important elements. Having such a definition would contribute to efforts to model, simulate, and ultimately engineer emergence. One approach is to build an otology. An ontology is a structured model of a domain of knowledge. It is a formal representation of the nature of a concept, its terms, properties, relationships, restrictions, and otherwise its essential elements (Gruber, 1993). The ontology contributes to modeling and simulation efforts for engineering emergence by providing a congruent and unambiguous representation of the concept. It serves as the basis for developing specifications for building a model of emergent phenomena.

The first step in building an ontology is a comprehensive study of the knowledge domain for the concept that captures its theories, definitions, descriptors, distinctions, and other important elements. One of the artifacts of the study is a reference model that captures what is known and assumed about emergence/emergents. The reference model informs the ontology by providing a comprehensive view of the domain for concepts, including its conflicting and inconsistent aspects.

8.2 The Emergence Concept

There is no shortage of theories of emergence. However, it is still not clear what the term "emergence" denotes or, more important, how emergence emerges (Corning, 2002). Corning is not alone in pointing out the "multifarious," "confusing," and "contradictory" claims about emergence:

- **Silberstein and McGeever (1999)** discuss the confusion of emergence as a concept where system properties are in no way determined by or derivable from their constituents, vs. properties that are actually determined by constituents but are very difficult in reality to derive.
- **Corning (2002)** list examples of "ambiguous" and "contradictory" claims from theorists and the science community, such as: the relevance vs. irrelevance of perception by an observer; whether or not emergents are irreducible or predicable; the requirement for interactions vs. a change in scale of observation.
- **Campbell (2015)** identifies issues in the concepts of emergence that need clarification, including: what it means to emerge; what it is that emerges (entities, properties, behaviors, etc.); what constitutes an emergent property as novel; and whether the spacio-temporal aspects of emergence are synchronic or diachronic, or both.
- **Sartenaer (2016)** describes the insufficiency or "emptiness" in emergence concepts in three categories: 1) positivity (defining what emergence is *not* rather than what it is); 2) consistency (a system simultaneously determined yet unexplainable by its constituents is contradictory); and 3) triviality (the unqualified definition that emergents are properties of wholes that are not properties of parts can be an obvious observation with insignificant consequences).

An initial review of the literature on emergence supports the assertions that there are gaps in the body of knowledge on emergence. A sample of the prevailing theories that are frequently cited in the literature is summarized in Table 8.1.

The initial survey in Table 8.1 highlights the variety of definitions and contributing factors that are found in the literature. On the surface, several conflicts seem apparent and raise questions as to what emergents are: 1) properties, patterns, or behaviors; 2) new or reoccurring/persistent; 3) originates from the bottom up or the top down; 4) provisional or permanent; 5) observer dependent or independent. An equally wide range of assertions about the factors that cause emergence is also observed, including: coalescence of components; lack of knowledge; evolution; "mathematical incompetence"; interactions; variations in the size; and inherent difficulty, just to name a few. While these and other available definitions of emergents provide some insight into the concept, ambiguity still persists.

TABLE 8.1

Prevailing Theories of Emergence

Source	Theory/Definition	Factors Causing Emergence
Lewes, 1875	Emergents are system effects caused by the coalescence of dissimilar components, and are not traceable/reducible to the steps of the coalescence process.	Coalescence of dissimilar components.
Broad, 1925	Emergents are behaviors theoretically unexplainable by their components due to lack of knowledge of component microscopic structures and "mathematical incompetence."	Lack of component knowledge; "mathematical incompetence."
Morgan, 1929	Emergence is the process of producing qualities or properties of material existents (i.e., wholes) that form over time and are only predicable/deducible with observation/experimentation.	Lack of experience or observation.
Ashby, 1956	Emergents are properties that are not shared by the system and its components due to large variations in the size of the parts relative to the size of the system (i.e., scale).	Large part to system size variations.
Crutchfield, 1994	Emergence is the dynamic interaction of subsystems and components that create new patterns that produce new system capabilities/functions.	Dynamic interaction of subsystems.
Bedau, 1997	Emergents are properties possessed by macro objects that are caused by interwoven non-linear relationships between their micro constituents, but cannot be posed by them and are only derivable by simulation (i.e., apparently underivable).	Interwoven non-linear relationships.
Holland, J.H., 1998	Emergents are recognizable, persistent, and reoccurring patterns that are not predicated due to the inherent difficulty of calculation and the size of potential state space.	Inherent difficulty.
Bar-Yam, 2004	Emergents are properties of the system that cannot be inferred from observations of components and are the result of system level constraints.	System level constraints.
Maier, 2015	Emergence is the production of properties possessed by an assemblage of things that are not possessed by its members; that vary in degree of derivability in models; and is the result of interactions between components and time scales.	Interactions between components.

A reference model can be instrumental in removing some of the ambiguity by providing a level of structure for a concept and capturing its essentials elements.

8.3 A Reference Model for Emergence

Reference models are sets of unstructured statements representing the claims, assumptions, and constraints for the subject of the model (Tolk et al., 2013). Claims are statements explicitly stated by the laws and axioms; assumptions are not explicitly stated but are logical conclusions based on combinations of claims; and constraints are statements that define boundaries. The intent is to capture a comprehensive view of the subject from relevant perspectives, including inconsistent or conflicting interpretations. There are easily hundreds of papers and books on emergence concepts. In our research on emergence, an attempt was made to capture original concepts and those frequently cited in the literature. The reference model developed from this research covers a wide variety of historical and contemporary concepts. The theories and concepts of emergence were studied and summarized into statements (claims, assumptions, and constraints). The set of statements forms the reference model. The full reference model and background material that was included in the study is discussed in Johnson (2016). Samples of the reference model statements are presented in Table 8.2.

The reference model statements are analyzed and the reoccurring themes or essential elements that define the emergence concept are identified. These conceptual elements become the inputs for constructing an ontology according to the Web Ontology Language standard (McGuinness et al., 2004) using the Protégé tool for developing and maintaining ontologies (Protégé, 2016).

8.4 Building an Ontology of Emergence

Using the reference model to build an ontology further explains the concept by providing a structured representation of its essential elements. The essential elements formally represented by the ontology are the main ideas that are used to explain what emergence is and how it occurs. Each idea has its own set of defining characteristics and relationships that provide the structure for the concept. In the ontology, the main ideas are called "classes" and their structural characteristics are "properties" (Noy & McGuinness, 2001). The ontology is built through an iterative process of: a) searching the reference model for its main ideas (classes); b) defining characteristics (properties); and c) mapping the relationships between classes and properties to specific theories and concepts of emergence. The tool is used to create logical links between the classes, subclasses, properties, and specific instances of theories and concepts of emergence. The ontology is actually a database of the essential elements of the concept and their logical relationships. It is depicted as a directed tree in Figure 8.1 and as a matrix in Figure 8.2.

While the reference model captures what is known and assumed about a concept without regard to consistency between its statements, the ontology puts the concept into a logical structure of congruent elements. The logical structure of the ontology provides new insights into the nature of emergence—insights with potential benefits for efforts to model emergence phenomena in systems.

8.5 Insights from the Ontology of Emergence

Building and studying the ontology of emergence led to the discovery of common classes and properties that can be used to define the concept and specify its essential elements. Figure 8.3 depicts the hierarchy of classes and subclasses for the emergence ontology.

TABLE 8.2

Emergence Reference Model

Claims, Assumptions, and Constraints	Source
Emergents are system effects that are not presently traceable to or deducible from the properties and interaction processes of its components, but may become so as knowledge improves.	Mill (1846) Lewes (1875)
Emergents are indicated by non-linear interactions and increase uncertainty about the effects produced by the interactions.	Mill (1846) Lewes (1875) Crutchfield (1994) Bedau (1997) J.H. Holland (1998)
The uncertainty of the system properties increases with the addition of new configuration.	Alexander (1920) Morgan (1929)
Emergents are qualities or properties that form a structure over time and are only predicable/deducible with observation/ experimentation.	Alexander (1920) Morgan (1929)
Physical emergents are latent qualities unique to component configurations and are only theoretically deducible after observation.	Broad (1925)
Emergents are actually predictable behaviors of the system but are apparently unpredictable due to limited visibility of system parts and their coupling relationships.	Ashby (1956)
Weaker emergents are properties/behaviors of groups of components (i.e., subsystems) that are difficult to deduce because of the density of information at lower scales.	Bar-Yam (2004)
Stronger emergents are properties of the system or subsystems that are not applicable to its members and vice versa and are only derivable by observing the system at large.	Bar-Yam (2004)
Novelty of emergents is intrinsic and observer independent because they provide new internal capabilities.	Crutchfield (1994) Kim (1999)
An increase in information processing (Shannon entropy) is an indication of emergence.	Crutchfield (1994)
Emergents are properties of systems that cannot be possessed by their constituents, but can be reductively explained through inherently difficult iterative aggregation processes (i.e., they are intrinsic).	Bedau (1997) Maier (2015)
Emergents are produced by information as well as material exchanges between independent systems and system components.	Maier (2015)
Only reductively explainable emergents are scientifically relevant and applicable to natural and human constructed things.	Bedau (1997) Maier (2015)
Emergents are reoccurring system patterns/behaviors that are apparently unexplainable in terms of system components.	J.H. Holland (1998)
The number of system states is so great that it is inherently difficult (yet possible) to explain, reduce, or predict emergents based on their constituents and governing rules.	J.H. Holland (1998)
Emergence is applicable to Physics (inanimate objects and events, i.e., phenomena), Physiology (feelings, sensations, consciousness, mind/ body relationships), Biology (living bodies); Chemistry (chemical reactions); Philosophy (certitude, truth).	Lewes (1875) Mill (1846) Alexander (1920) Morgan (1929) Ashby (1956) Bar-Yam (2004) Crutchfield (1994) J.H. Holland (1998)

The authors posit that the ontology at the level depicted in Figure 8.3 represents an unambiguous and unifying general definition of the emergence concept. Emergence is defined by two primary classes: characteristics of the emergent effects, and characteristics of systems where emergent effects take place. The primary classes are broken down into eight subclasses: type; logical relationship; perspective; indicators; temporality; structure;

FIGURE 8.1
Emergence ontology tree.

knowledge constraint; and application domain. Conflicts among emergence concepts are reconciled by grouping them into the eight subclasses. For example, the conflict between emergence as a diachronic vs. synchronic concept is resolved by grouping both concepts in the category of temporality. Whether emergence is time dependent (diachronic) or independent (synchronic), there is an aspect of time (temporality) that defines the emergence concept.

			Lewes (1875)	Mill (1846)	Alexander (1920)	Morgan (1929)	Broad (1925)-physical	Broad (1925)-mental	Ashby (1956)	Bar-Yam (2004)-weak	Bar-Yam (2004)-strong	Crutchfield (1994)	Kim (1999)	Bedau (1997)-weak	Bedau (1997)-strong	Maier (2015)-weak/strong	Maier (2015)-spooky	Holland, J.H. (1998)
Phenemona characterisitics																		
	Type																	
		Qualities/properties	X	X	X	X	X	X		X	X		X	X	X	X	X	
		Behaviors							X	X								X
		Patterns										X						X
		Structures										X						
	Logical reltionships																	
		Explainable/derivable										X		X		X		X
		Not explainable/derivable						X					X		X		X	
		Theoretically explainable/derivable	X	X	X	X	X		X	X	X							
	Perspective																	
		Intrinsic					X	X		X	X	X	X	X	X	X	X	X
		Extrinsic	X	X	X	X			X									
	Indicator																	
		New configurations			X	X												
		Complexity							X						X			X
		Nonlinearity	X	X								X						X
		Uncertainty	X	X	X	X							X					
		Information										X		X		X	X	
		Shannon entropy										X						
System characteristics																		
	Application domain																	
		Engineered (physical)	X	X	X	X	X		X	X	X	X		X		X		X
		Natural (physical)	X	X	X	X	X		X	X	X	X		X		X		X
		Natutal (metaphysical)	X					X					X		X		X	
	Structure																	
		Coupled/interconnected	X	X					X							X	X	X
		Dissimilar size/complexity of parts vs system							X									
		Dissimular laws		X														
		Dissimular parts	X															
		Governing rules << Component types																X
		Hierarchical			X	X	X	X		X	X	X	X					
		Learning/adapability (Δ based on feedback vs goals)																X
		Multiple component types and instances																X
		Multiple paths per system state																X
		Non-linear interactions										X		X	X			X
	Temporality																	
		Synchronic	X	X			X	X	X	X	X				X	X		
		Diachronic			X	X					X	X	X	X				X
	Knowledge constraint																	
		Experience/observations	X		X	X	X											
		Observation inaccessibility							X									
		Information density								X								
		Visibility of whole system									X							
		Visibility of coupling relationships									X							
		Modeling capability										X						
		Human comprehension						X					X		X		X	
		Inherent difficulty (iterative aggregations)												X		X		
		Inherent difficulty (size of system state space)																X

FIGURE 8.2
Emergence ontology matrix.

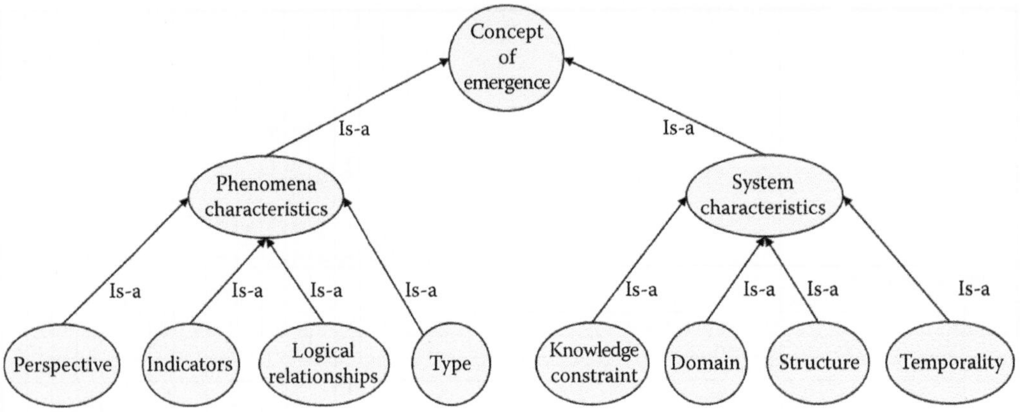

FIGURE 8.3
Ontology of emergence: Unifying level.

The concept is further defined by propositions for each subclass:

Phenomena Characteristics

- *Perspectives.* The existence of emergents depends on the point of view that the novelty of the effect is determined. If the significance is determined and varies according to the observer, then emergence is extrinsic (i.e., observer dependent). If the significance is inherent to the system and does not vary according to the observer, then emergence is intrinsic (i.e., observer independent).
- *Indicators.* The occurrence of emergent effects is marked by certain measurable facts, i.e., parameters. The parameters are not the causes of emergents; they are the quantitative signs that emergent effects have taken place. Parameters have magnitude (size or amount) and direction (increasing or decreasing). Examples include: non-linear results from components' interactions; uncertainty, randomness, or disorder of the future system states (i.e., entropy); variety of potential system states relative to the initial variety; the number of distinguishable states or the number of variables required to define a state (i.e., complexity); and data that describe or are actionable by the system (i.e., information).
- *Logical Relationships.* The relationship between system components and emergent effects is characterized by the ability to derive or explain an effect from system components and their interactions. Derive is used in the context of being able to start with an initial point of knowledge about the system's parts and interactions, and make logical progressions to arrive at the system level effects. Explain is used in a similar way but in the opposite direction. To explain is to start with the system level effect and logically trace its origin back to knowledge about the parts in the system. Relationships are either currently derivable/explainable; theoretically derivable/explainable in the future; or completely underivable/explainable.
- *Type.* Emergents are the consequence of the interaction of components in a system. They include behaviors (the particular way in which the system functions); qualities (characteristics of the system); patterns (a reoccurring sequence or identifiable form); and structures (particular configurations of the system parts).

System Characteristics

- *Domain.* Emergent concepts apply to different types of systems. There are three general domains of systems types to which emergence concepts apply: physical systems that are engineered by humans; physical systems that occur naturally in the environment; and metaphysical systems that occur naturally in the environment.
- *Knowledge Constraint.* The ability to derive/explain emergents' effects is limited by insufficient knowledge of component properties and their interrelationships in the system. The various type of constraints that inhibit deriving/explaining effects include but are not limited to: experience of the observer; density of information in the system; capability to view the system as a whole; and inherent difficulty due to iterative aggregations.
- *Structure.* Nature of the configurations of components that form the system are described by certain characteristics, which include but are not limited to: hierarchal order; assemblage of dissimilar parts; components with non-linear functions; and coupled/interconnected components.
- *Temporality.* The state of the system is the configuration of its parts and the values of their variables at a point in time. The state may or may not be time dependent. If the state that produces emergent effects develops over time, the system temporality is diachronic. If the emergent state is constant and always exists, the system temporality is synchronic.

It was determined that each instance of the emergence concept that was studied contained claims and/or assumptions that fit into the eight subclasses. Variation among the concepts was found below the level depicted in Figure 8.3. However, the consistency among concepts at the class and subclass level supports the ontology as a unifying definition of emergence.

The ontology covers physical and metaphysical domains. Given that the focus of this text is on engineered systems, operational definitions of emergents and emergence in the physical domain are synthesized from the ontology:

a. Emergents are system effects that are approximately underivable based on system components and their interrelationships.
b. Emergence is the action of producing system effects that are approximately underivable based on system components and their interrelationships.

Definitions for both emergent and emergence are offered to provide additional clarity, given that both words are frequently used throughout the literature. The operative words in the definitions are "underivable" and "approximately." Underivable describes the limitation on logically determining in advance the effects of component interactions based on knowledge of the components and their interrelationships. A corollary for the underivable limitation is unexplainable. It is assumed that a limit on the ability to derive and effect is also a limitation on the ability to explain it. "Approximately" is used to qualify the underivable limitation. It captures the concept that emergents in the physical domain are theoretically derivable, but that doing so is inherently difficult and may not have occurred. The qualifier of being "approximately" underivable is necessary to distinguish the concept of emergence in the physical domain from its application in the metaphysical domain.

The metaphysical domain refers to that which is unperceivable by the senses (i.e., the mind, consciousness, etc.). In the metaphysical domain, emergents are completely underivable vs. being theoretically derivable in the physical domain. Restricting the operational definition to the physical domain is appropriate, as this text is concerned with engineered systems.

8.6 Implications for Modeling and Simulation

We posit that the ontology for emergence provides an unambiguous definition and specifications of the concept; it can be used to inform and validate the design of the simulation models of emergence in engineered systems. As an unambiguous definition of the concept, the ontology can be used to objectively determine if a phenomena formally corresponds with the concept of emergence. As a specification, the ontology can be used to identify attributes of a phenomena that are related to emergence, and to validate that key aspects of emergence are represented in models of the phenomena. For example, consider the New York Stock Exchange (NYSE) System and the Flash Crash phenomena. We use the ontology of emergence to construct a conceptual model in the form of a system dynamics causal loop diagram (CLD). CLDs are a System Dynamic modeling concept that represents the feedback structures and causal relationships between system variables, and graphically depicts the behavior of the system (Sterman, 2000). They are an important step in building system dynamics simulation models.

8.6.1 Use Case: New York Stock Exchange System

On May 6, 2010, the market saw a complete evaporation of liquidity and individual stock prices swings from $.01 to $100K. The phenomenon is known as a "flash crash" and is characterized by: 1) evaporation of liquidity; 2) extreme price swings and a return to normalcy in a relatively short period of time; 3) irrational prices that are not based on economic information. The rules, properties, and behaviors of the system components do not explain or provide a means for predicting the flash crash phenomenon (Bowley, 2010; SEC, 2011; Serritella, 2010; Sommerville et al., 2012; Aldridge, 2014).

Even though there have been several flash crashes, the 2010 event was the first documented occurrence of this phenomenon (Aldridge, 2014). There are expectations that future flash crashes will occur. However, to date there have been no successful predictions of a flash crash. While the exact cause of the flash crash is unknown, the behavior of the NYSE system is completely determined by its components and their relationships. It was found that the primary contributing factors to the crash were the high frequency and speed of the trades by computer-generated algorithms, along with the high degree of coupling between the components of the system. In terms of the ontology of emergence, flash crashes are approximately underivable based on system components and their relationships. Flash crashes are emergent events.

8.6.2 Modeling the NYSE Flash Crash Based on the Ontology of Emergence

Flash crash phenomena can be defined in terms of the unifying level of the ontology of emergence that was depicted in Figure 8.3. Table 8.3 captures attributes of flash crashes in each of the classes/subclasses of the emergence ontology.

TABLE 8.3

Mapping the Flash Crash Phenomena to the Ontology of Emergence

Ontology Classes/ Subclasses	Attributes of Flash Crashes in the NYSE System
Indicators	*Market liquidity:* The ratio of the total capital required to meet seller's ask price on the supply side to total capital available for purchase (i.e, bid) on the demand side.
Perspectives	*Intrinsic:* a) Sellers are attracted to high liquidity (demand > supply); b) Buyers are attracted to low liquidity (supply > demand); c) Designated market makers (DMM) attempt to maintain liquidity.
Logical relationships	*Theoretically derivable/explainable in the future:* Flash crashes are approximately underivable based on system components and their interrelationship, but may be derivable in the future with new knowledge of systems components and their relationships.
Type	*Quality:* The system is designed to produce stable liquidity in the market (i.e., ability to execute contracts in a timely fashion between buyers and sellers without affecting price).
Domain	*Physical systems engineered by humans:* While some specific details have evolved, the components and general system structure were designed and constructed by humans.
Knowledge constraint	*Density of information in the system:* The volume and speed of buyer/seller transactions are so great that it is not currently possible to derive the impact on liquidity.
Structure	a. *Assemblage of dissimilar parts:* Human buyers/sellers are dissimilar to computers. b. *Components with non-linear functions:* Supply/demand relationship is nonlinear. c. *Highly coupled/interconnected components:* Automated trading algorithms link buyer/seller and market maker computers across the NYSE system.
Temporality	*Diachronic:* The state of the NYSE system changes over time as capital enters and vacates the market based on changes in supply and demand.

Mapping the flash crash phenomena to the ontology of emergence establishes concurrence between the phenomena and the emergence concept. Doing so also provides specifications for constructing a conceptual model in the form of a causal loop diagram (CLD). The causal loop diagram for flash crashes in the NYSE is depicted in Figure 8.4.

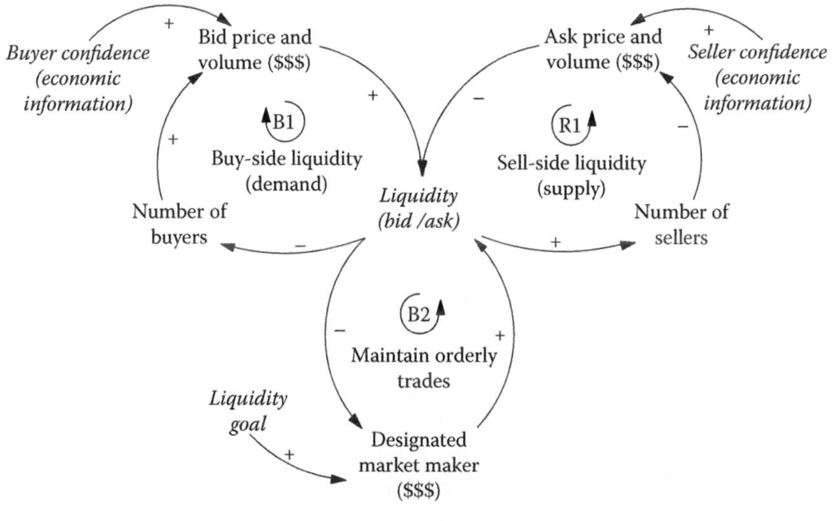

FIGURE 8.4
NYSE flash crash causal lop diagram.

Feedback processes in the system are identified by selecting any point in the diagram and tracing the arrows back to that point. All of the variables connected during the trace are interrelated through feedback mechanisms. The positive (+) signs in the loops of a CLD indicate that variables connected by the arrows change in the same directions, while the negative (–) signs indicates that they change in opposite directions. Note that any variable can have an increasing or decreasing change.

There are three propositions for the desired state of the NYSE system. Each proposition is represented by a causal loop. The direction and polarity of the arrows connecting the system components describes the propositions for how the desired states are achieved.

R1 Reinforcing Loop (Sell-side Liquidity). Sellers are attracted to high liquidity. Increasing liquidity indicates that there is an abundance of buyers (i.e., demand is high), so more sellers enter the market and increase the ask price for their stocks. As long as liquidity is increasing, sellers will continue to enter the market, and if economic information gives sellers confidence about their securities, they will increase their prices. However, the increasing price works to reduce liquidity, which will eventually reduce the number of sellers.

B1 Balancing Loop (Buy-side Liquidity). Buyers are attracted to low liquidity. Decreasing liquidity indicates that the competition among buyers is low (i.e., demand is low). This presents a gap that causes buyers to enter the market. If economic information causes buyers to be confident about the market, buyers will be willing to bid more for stocks to beat the competition which will increase liquidity.

B2 Balancing Loop (Maintain Orderly Trades). Designated market makers (DMM) attempt to maintain balance between sellers and buyers (i.e., supply and demand). The DMM achieves balance by responding to the discrepancy between the market liquidity and the liquidity goal for the system. Purchase contracts are executed to fill gaps between insufficient buyers relative to sellers as indicated by declining liquidity. Sell contracts are executed to fill gaps between insufficient sellers relative to buyers as indicated by increasing liquidity.

The flash crash occurs when one or more of the propositions fails. The flash crash is emergent when the proposition failures are not explainable or predicable in terms of the components and available information in the system. With these two conditions in mind, the NYSE CLD might conceptually be considered an example of emergence detection.

8.7 Conclusions

An ontology is a formal representation of the nature of a concept. Developing an ontology for emergence provides an unambiguous definition of the concept. The representation contributes to the clarity of the emergence concept and provides the formal basis for developing specifications to build models of emergence in engineered systems. Modeling emergence in engineered systems should address at least two basic questions: a) is the phenomena consistent with the concept of emergence? and b) does the model represents the essential aspects of the emergence concept? One way to answer these questions is to establish concurrence between the subject phenomena being modeled and the classes/subclasses of the emergence ontology. A next step to advance modeling emergence in engineered systems would be applying the ontology to real-world system phenomena. This would provide an opportunity to identify and close gaps in the ontology, and further develop it as a source of specification for modeling emergence.

References

Ackoff, R. L. (1971). Towards a system of systems concepts. *Management Science, 17*(11), 661–671.

Aldridge, I. (2014). High-frequency runs and flash-crash predictability. *The Journal of Portfolio Management, 40*(3), 113–123.

Alexander, S. (1920). Space, time, and deity: The Gifford lectures at Glasgow, 1916-1918 (Vol. 2). Macmillan.

Ashby, W. R. (1956). An introduction to cybernetics. Chapman & Hall, London.

Bar Yam, Y. (2004). A mathematical theory of strong emergence using multiscale variety. *Complexity, 9*(6), 15–24.

Bedau, M. A. (1997). Weak emergence. *Noûs, 31*(s11), 375–399.

Blanchard, B. S., and Fabrycky, W. J. (2006). Systems engineering and analysis / Benjamin S. Blanchard, Wolter J. Fabrycky. Upper Saddle River, N.J.: Pearson/Prentice Hall, c2006.

Bowley, G. (2010). Lone $4.1 billion sale led to "Flash Crash" in May. *The New York Times,* 1.

Broad, C. D. (1925). The mind and its place in nature (pp. 97–113). Paul, T. K. (Ed.). London: Routledge & Kegan Paul.

Campbell, R. (2015). The concept of emergence. In The Metaphysics of Emergence (pp. 192–231). Palgrave Macmillan UK.

Checkland, P. (1999). Systems thinking, systems practice: Includes a 30-year retrospective. John Wiley & Sons Ltd, 1999.

Corning, P. A. (2002). The re-emergence of "emergence": A venerable concept in search of a theory. *Complexity, 7*(6), 18–30.

Crutchfield, J. P. (1994). The calculi of emergence: Computation, dynamics and induction. *Physica D: Nonlinear Phenomenon, 75*(1), 11–54.

Gruber, T. R. (1993). A translation approach to portable ontology specifications. *Knowledge Acquisition, 5*(2), 199–220.

Holland, J. H. (1998). Emergence: From chaos to order. Reading, Mass.: Addison-Wesley.

Johnson IV, John J. (2016). "A general theory of emergence in engineered systems." PhD diss., Old Dominion University, 2016.

Kim, J. (1999). Making sense of emergence. *Philosophical Studies, 95*(1), 3–36.

Lewes, G. H. (1875). Problems of life and mind, First series, The foundations of a creed, Vol. II. Cambridge, MA. The Riverside Press.

Maier, M. (2015). Chapter 2, The role of modeling and simulation in systems-of-systems development. In Rainey, L. B., & Tolk, A. (Eds.). Modeling and Simulation Support for System of Systems Engineering Applications. John Wiley & Sons.

McGuinness, D. L., and Van Harmelen, F. (2004). OWL web ontology language overview. W3C recommendation 10, no. 10.

Mill, J. S. (1846). A system of logic, ratiocinative and inductive [electronic resource]: Being a connected view of the principles of evidence and the methods of scientific investigation. New York: Harper, 1846.

Morgan, C. Lloyd (1929). The case for emergent evolution. Philosophy, 4, pp. 23–38 doi:10.1017/S0031819100031077.

Noy, N. F., and McGuinness, D. L. (2001). Ontology development 101: A guide to creating your first ontology. Book/journal.

Protégé (2016). Stanford Center for Biomedical Informatics Research (BMIR) at the Stanford University School of Medicine. www.protege.stanford.edu.

Sartenaer, O. (2016). Sixteen years later: Making sense of emergence (again). *Journal for General Philosophy of Science, 47*(1), 79–103.

SEC, U. (2011). Recommendations regarding regulatory responses to the market events of May 6, 2010: Summary Report of the Joint CFTC-SEC Advisory Committee on Emerging Regulatory Issues.

Serritella, D. M. (2010). High speed trading begets high speed regulation: SEC response to Flash Crash, Rash. *University of Illinois Journal of Law, Technology & Policy*, 433.

Silberstein, M., and McGeever, J. (1999). The search for ontological emergence. *The Philosophical Quarterly, 49*(195), 201–214.

Sommerville, I., Cliff, D., Calinescu, R., Keen, J., Kelly, T., Kwiatkowska, M., McDermid, J., and Paige, R. (2012). Large-scale complex IT systems. *Communications of the ACM, 55*(7), 71–77.

Sterman, J. (2000). Business dynamics, systems thinking for a complex world. McGraw-Hill.

Tolk, A., Diallo, S. Y., Padilla, J. J., and Herencia-Zapana, H. (2013). Reference modelling in support of M&S—Foundations and applications. *Journal of Simulation, 7*(2), 69–82.

Tolk, A., and Rainey, L. B. (2015). Chapter 22, Towards a Research Agenda for M&S Support of System of Systems Engineering. In Rainey, L. B., and Tolk, A. (Eds.). Modeling and Simulation Support for System of Systems Engineering Applications. John Wiley & Sons.

Valerdi, R., Axelband, E., Baehren, T., Boehm, B., Dorenbos, D., Jackson, S., and Settles, S. (2008). A research agenda for systems of systems architecting. *International Journal of System of Systems Engineering, 1*(1–2), 171–188.

9

Modeling and Validation Challenges for Complex Systems

Mikel D. Petty

CONTENTS

9.1 Introduction

> Our best equations for the weather differ from our best computer models based on those equations, and both of those systems differ from the real thing … (Smith, 2007)
>
> Complexity lies somewhere between order and chaos. (Miller, 2007)

Complex systems are often modeled.* Among several potential reasons for this, two stand out. First, the systems of the greatest practical interest, and thus those most likely to be worth the effort and expense of being modeled, tend to be complex. Second, as a result of their special characteristics, complex systems generally resist closed form mathematical analysis, and so modeling is often the best or even the only way to study and experiment with them. This is true of both engineered and natural systems.

Complex systems have a number of special defining characteristics, including sensitivity to initial conditions, emergent behavior, and composition of components. Unfortunately

* In this chapter, "complex" is meant in the sense of complexity theory, as opposed to simply a synonym for "complicated." The term "complex system" will be defined in Section 9.2.

for those involved in modeling complex systems, these special characteristics of complex systems lead to additional challenges beyond those encountered with non-complex systems in both modeling them accurately and effectively and in reliably and completely validating the models.

This chapter, which is meant as an introductory tutorial and brief literature survey, has four main sections.* The first describes complex systems and lists their defining characteristics, and motivates the interest in validating models of complex systems by discussing validation risk. Then, each of the following sections discusses one of three selected defining characteristics of complex systems (sensitivity to initial conditions, emergent behavior, and composition of components), explaining why the characteristic in question makes modeling and validation more difficult and offering some approaches to dealing with and mitigating the difficulties.

9.2 Complex Systems

Complex systems were recognized as qualitatively distinct from non-complex systems at least as early as 1984, with the founding of the Santa Fe Institute, a research institute devoted to complexity theory (Waldrop, 1992). Since then, a body of specialized knowledge has been developed on the subject, driven by both theoretical and experimental investigations (Lewin, 1992).

9.2.1 Definitions of Complex System

A range of definitions of a *complex system* are available.† Although the definitions are far from as reassuringly consistent or precise as that of, say, an equivalence relation (e.g., see [Epp, 2011]), they are nevertheless informative.

> A system comprised of a (usually large) number of (usually strongly) interacting entities, processes, or agents, the understanding of which requires the development, or the use of, new scientific tools, nonlinear models, out-of equilibrium descriptions and computer simulations. (Schweitzer, 2003)
>
> A complex system is one whose evolution is very sensitive to initial conditions or to small perturbations, one in which the number of independent interacting components is large, or one in which there are multiple pathways by which the system can evolve. (Whitesides, 1999)
>
> Complex systems are neither ordered nor random, but combine elements of both kinds of behaviour in a very elusive but striking manner. (Stewart, 2002)

Both engineering and natural systems satisfy these definitions; systems that are widely considered to be complex are illustrated in Figure 9.1.

* Earlier versions of this chapter appeared in the *Spring 2012 Simulation Interoperability Workshop* (Petty, 2012) and the *M&S Journal* (Petty, 2014). This version has been updated and enhanced.

† The similar term *complex adaptive system* is also found in the literature, e.g., (Law, 2015); (Keller-McNulty, 2006). *Complex system* and *complex adaptive system* seem to be referring to the same class of systems. We will use the former term exclusively.

FIGURE 9.1
Examples of complex systems: (top) air traffic control, (center) weather, and (bottom) stock market.

Certain defining characteristics or properties are associated with complex systems. These characteristics are individually arguable, in that not every complex system necessarily exhibits every one of them; e.g., some definitions seem to require the system be composed of agents, whereas others do not. However, these characteristics are collectively definitive, in the sense that most complex systems will exhibit most of them. Taken together, they define the class of complex systems and serve to distinguish them from non-complex systems. A list of the defining characteristics with brief descriptions follows.* The first three are described in detail in later sections of this chapter, whereas the others are briefly described here:

1. *Sensitivity to initial conditions.* See Section 9.3.
2. *Emergent behavior.* See Section 9.4.
3. *Composition of components.* See Section 9.5.
4. *Uncertain boundaries.* Determining the boundary between a complex system and the environment in which it is situated and with which it interacts can be difficult.
5. *Nesting.* Components of a complex system may themselves be complex systems.
6. *State memory.* Future states of a complex system often depend on past states in ways that are difficult to understand or model.
7. *Non-linear relationships.* Relationships between components of a complex system may be non-linear, which means a small cause may have a large effect.
8. *Feedback loops.* Negative (damping) and positive (amplifying) feedback loops exist between elements of complex system.

9.2.2 Modeling and Validation Risk for Complex Systems

Important systems, complex systems, and modeled systems overlap to a great extent. Systems that are important to their users, for reasons of safety, economics, or ubiquity, are often complex; the reverse is also true. For example, financial markets are important to those who participate in them, whether voluntarily or involuntarily, because of their potential impact on the participants' quality of life and long-term security, and they exhibit all of the defining characteristics listed earlier. Similarly, systems that are important are also often modeled, because their importance makes them more likely to be worth the effort and expense of being modeled; and again the reverse is also true. The air traffic control system is an example. Finally, complex systems are often modeled, and once more the reverse is true. The weather is an example. Because of their inherent structure, complex systems are often difficult to study using closed form mathematical analysis (Miller, 2007). Consequently, modeling is often the best or even the only way to study or experiment with them.

In M&S, verification is "the process of determining that a model implementation and its associated data accurately represents the developer's conceptual description and specifications"; validation is "determining the degree to which a model or simulation and its associated data are an accurate representation of the real world from the perspective of the intended uses of the model"; and accreditation is "official certification [by a responsible authority] that a model or simulation is acceptable for use for a specific purpose"

* An overlapping but somewhat different list is given in (Williams, 1997); that list includes *adaptiveness* and *self-organization.*

	Model valid	Model not valid	Model not relevant
Model not used	Type I error Non-use of valid model Insufficient V and V Model builder's risk Less serious error	Correct	Correct
Model used	Correct	**Type II error** Use of invalid model Incorrect V and V Model user's risk More serious error	**Type III error** Use of irrelevant model Incorrect accreditation Model user's risk More serious error

FIGURE 9.2
Validation error types and risks (Petty, 2010) (adapted from [Balci, 1998]).

(DOD, 2009). Further elaboration of these definitions, and descriptions of methods to perform verification and validation, can be found in (Balci, 1998; Petty, 2010). Validation is our primary interest here.

Models are subject to validation risk. The general concept of validation risk is that improperly or incompletely validating a model can result in risk to the developers and/or the users of the model. This general notion has been refined into specific types of validation error and the type of validation risk that results from each. Figure 9.2 summarizes these error types. A Type I error is the failure to use a valid model in a situation where the model's use would be valuable; it is normally the result of insufficient verification and validation to convince the accreditation authority that the model can be used. A Type II error is the use of an invalid model; it is often the result of verification and validation done incorrectly but convincingly enough that the accreditation authority accredits the model for use.* A Type III error is the use of a model, even a valid one, for a purpose for which it is not relevant; this error is usually the result of an accreditation that is too broad, or a misunderstanding by the model user of the extent of the model's accreditation. Among M&S practitioners, Type II and Type III errors are generally considered to be more serious, in that they can lead to incorrect actions taken or decisions made on the basis of a model, but Type I errors can be very serious as well.†

Whenever a model is used validation risk exists, and for a model of an important system, that risk is a function of both the importance of the system and of the model's intended use. Obviously, a Type II validation error clearly has less potential consequences for a model of ant behavior being used for a video game than a model of metal fatigue being used to design the airframe of an airliner. Decisions made about important systems using models can have major impact. As an example, consider the 2008 financial crisis in the

* Type I and Type II errors in model validation are defined in a manner that directly corresponds to the like-named error types in statistical hypothesis testing. In statistics, a Type I error is rejecting the null hypothesis H_0 when H_0 is true. The conventional formulation of hypothesis tests in model validation uses H_0 to express the assumption that the model is valid. Thus a Type I statistics error (rejecting H_0 incorrectly) corresponds to a Type I validation error (not using a valid model). Similarly, in statistics a Type II error is failing to reject H_0 when H_0 is false, which corresponds to using an invalid model (Banks, 2010).

† A Type I validation error was involved in the tragic 2003 loss of the space shuttle *Columbia* and its crew. After the collision of the foam piece with the orbiter's heat shielding during launch was detected, a model used for ground-based analysis of the impact conducted while the orbiter was still in orbit correctly predicted that the heat shielding had been penetrated. However, the model had not been validated for debris volume up to the size involved, and consequently the model's results were discounted by the analysts (Gehman, 2003).

United States. Some financial analysts have argued that that crisis was in significant part triggered by a financial model, namely the famous (or infamous) Gaussian copula, which is a model of the prices of collateralized debt obligations (Salmon, 2009):

$$\Pr[T_A < 1, T_B < 1] = \phi_2\left(\phi^{-1}(F_A(1)), \phi^{-1}(F_B(1)), \gamma\right) \tag{9.1}$$

The model was based on the assumption that the price of a credit default swap was correlated with, and thus could be used to predict, the price of mortgage backed securities. Beyond that, the mathematical and notational details of this model need not concern us here. What is important to the matter at hand is that the bounds of validity of this widely-used model were not fully understood by its users. Because the model was easy to use and compute, it was soon employed by a large portion of mortgage issuers, rating agencies, and financial investors. In fact, the model was ultimately invalid and its use constituted a Type II error. The result of that error is all too well known:

> Then the model fell apart. ... Financial markets began behaving in ways that users of [the] formula hadn't expected. ... Ruptures in the financial system's foundation swallowed up trillions of dollars and put the survival of the global banking system in serious peril. (Salmon, 2009)

The significant overlap of important systems, complex systems, and modeled systems means that our models are often of systems that are both important and complex; their importance magnifies validation risk, and their complexity complicates validation. Given the validation risk associated with models of important and complex systems, it is prudent to expend validation effort proportional to the risk, and to adapt or develop validation methods suitable for complex systems.

9.3 Sensitivity to Initial Conditions

> "Like causes produce like effects" ... is only true when small variations in the initial circumstances produce only small variations in the final state of the system. In a great many physical phenomena this condition is satisfied; but there are other cases in which small initial variation may produce a very great change in the final state of the system ... (Maxwell, 1873)
>
> Non-linear processes with the complex system can potentially amplify microscopic heterogeneity hidden within it ... (Allen, 2005)
>
> Small differences can build upon themselves and create large differences, making precise prediction difficult. (Miller, 2007)

The first of the three defining characteristics of complex systems to be examined for its effect on modeling and validation is sensitivity to initial conditions. Here the phrase "initial conditions" refers, of course, to either the starting state of the system (e.g., a rocket motor at ignition), or, if the system has an effectively continuous existence (e.g., the weather), the state of the system at the beginning of the time period being studied or modeled. The state evolution of a complex system can be highly sensitive to its initial conditions, with the result that small differences in initial state can become magnified over time into large

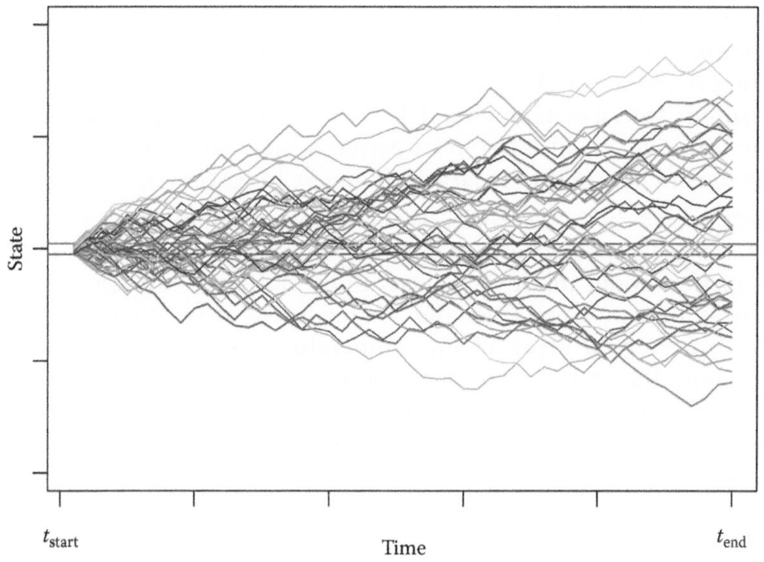

FIGURE 9.3
Sensitivity to initial conditions; possible system states diverge over time.

differences in future state (Smith, 2007). Figure 9.3 illustrates this with a set of notional system (or simulation) histories; in the figure, the horizontal axis represents time, advancing from left to right, and the vertical axis represents system state, simplified to a single variable. States that are only slightly different at some initial time t_{start} can evolve away from each other, becoming arbitrarily different at some future time t_{end}.

Models of complex systems, if they accurately represent the system's characteristics, can be similarly sensitive to initial conditions. From two model starting states that are quite similar, the execution of a model of a complex system can produce widely divergent end states.

9.3.1 Modeling

Sensitivity to initial conditions can introduce modeling challenges in these ways:

1. *Implementation side effects.* Technical aspects of the model that are purely implementation details and do not correspond to any aspect of the simuland* can have significant side effects that influence, or even overwhelm, the results. A well-known example is the effect of the numerical precision of the implementation language on numerical integration of differential equations in physical models (Colley, 2010).† In stochastic models that rely on random number generators, the seed and cycle length of the random number stream can, through the magnifying effect of sensitivity, significantly affect the model's results (Banks, 2010).

* A *simuland* is the system, phenomenon, or process that is the subject of a model, i.e., the modeled system (Petty, 2010).

† For example, a seemingly reasonable fourth-order Runge-Kutta integration with a fixed time step used to calculate an orbit in a two-body (sun and planet) gravitational system completely breaks down in the vicinity of the sun due to numerical precision issues, with the result that the simulated planet incorrectly "flies off completely into space" (Colley, 2010).

2. *Sensitivity consistency.* If a complex system is sensitive to initial conditions, the modeler may seek similarly sensitivity in a model of that system. However, given the nature of the sensitivity, it can be quite difficult to accurately match the model's sensitivity to that of its simuland. Even if both the complex system and a model of it are sensitive, small differences between the simuland's sensitivity and the model's sensitivity can lead to large differences in outcomes.
3. *Input imprecision.* Because sensitivity magnifies small differences in initial conditions, a small difference between the simuland's true initial state and the values of the model input data describing that state can again lead to large differences in outcomes. Consider, for example, a weather model that uses a three-dimensional array of air temperature, pressure, and humidity values to define the initial state of the atmosphere. Small errors in measuring those values can be magnified as the model executes into large discrepancies between the model's prediction and the actual weather. The input data precision needed by the model to accurately predict the simuland's future may exceed that obtainable due to limits in instrumentation accuracy or observation availability. This observational uncertainty is one reason that the useful predictive power of current weather models is currently limited to a few days, and the maximum achievable limit, even with perfect models, is considered to be "about two weeks" (Casti, 1990).

These methods can mitigate the modeling challenges associated with sensitivity to initial conditions:

1. *Selective abstraction.* During conceptual modeling, identify simuland features and state variables that are not required for the model to satisfy its intended purposes. Omit them in the implemented model, thereby eliminating them as possible sources of sensitivity.
2. *Ensemble forecasting.* The core idea of ensemble forecasting is to execute multiple runs of a model, each of which was initialized with slightly different initial states, and then develop a prediction based on the multiple results.* The differences in the inputs are intended to reflect the uncertainty in the knowledge of the initial state. The multiple results may be aggregated or averaged, and the variation and divergences between them analyzed; the details of aggregation and analysis depend on the application, but statistical methods are often employed. In some forms and contexts this is a familiar idea; modelers using a discrete event simulation to study a queueing system often conduct multiple trials, each beginning with a different random number seed (Law, 2015). In the case of weather models, different values for the initial conditions of the atmosphere may be used, with the differences generated based on the noise or uncertainty in the observations upon which the input is based (Smith, 2007). The uncertainty of the forecast may be estimated based the variation in the different forecasts generated.

* In addition to multiple runs of a single model, ensemble forecasting may also refer to an aggregating or merging of the results of multiple models. This approach is used to predict hurricane tracks.

9.3.2 Validation

Sensitivity to initial conditions can introduce validation challenges in these ways:

1. *Results distributions.* Broad distributions (i.e., large variance) in both simuland observations and model results caused by sensitivity to initial conditions can reduce the power of statistical comparisons of the two (Brase, 2015).
2. *Sensitivity analysis.* The potential for widely divergent outcomes from closely similar initial conditions can complicate conventional sensitivity analysis by requiring more closely spaced sampling of the response surface to capture the response variation.
3. *Input imprecision.* Measurement errors unavoidably introduce uncertainty into measurements of physical systems (Adams, 2012). Because of sensitivity to initial conditions, small uncertainties in model input values based on measurements of the simuland may be magnified, making comparisons of simuland observations and model results more difficult.

These methods can mitigate the validation challenges associated with sensitivity to initial conditions:

1. *Increased trials.* Increase the number of trials (i.e., executions of the model) when possible, thereby regaining some statistical power through larger sample sizes.
2. *Sensitivity analysis.* Use sensitivity analysis as a validation method by statistically comparing the magnitude and variability in the simuland observations to the magnitude and variability in the model results, in effect using sensitivity as a response variable for validation comparison (Balci, 1998; Liu, 2005).
3. *Precision awareness.* Understand the precision available in simuland observation data, and based on that precision, use an appropriate comparison threshold when comparing simuland observations and model results. For example, it is a mistake to expect the model to match the simuland within one unit when the observations are only accurate to within five units.
4. *Variance reduction.* Techniques known in discrete event simulation, including common random numbers, antithetic variates, control variates, indirect estimation, and conditioning can be applied to reduce variance in simulation output (Law, 2015), perhaps mitigating some of the effects of sensitivity to initial conditions.

9.4 Emergent Behavior

> We are dealing with a [complex] system when ... the entire system exhibits properties and behaviors that are different from those of the parts. (Jervis, 1998)
>
> A working definition of a complex system is of an entity which is coherent in some recognizable way but whose elements, interactions, and dynamics generate structures admitting surprise and novelty which cannot be defined a priori. (Batty, 2001)
>
> The philosophical core of complexity theory is the concept of emergence, in which a system may transcend its components ... (Stewart, 2002)

> Much of the focus of complex systems is how … interacting agents can lead to emergent phenomena. … Individual, localized behavior aggregates into global behavior that is, in some sense, disconnected from its origins. (Miller, 2007)
>
> Complex systems can have interactions that produce unexpected results in seemingly benign situations. (MITRE, 2017)

The second of the three defining characteristics of complex systems to be examined for its effect on modeling and validation is emergent behavior. Emergent behavior is behavior that is not explicitly encoded in the agents or components that make up the simuland or the model; rather, it emerges in reality or during a simulation from the interaction of agents or components with each other and their environment (Williams, 1997).

An important aspect of emergent behavior is that it is not directly predictable or anticipatable from the individual agents' or components' behaviors, even if they are known completely. Figure 9.4 illustrates a form of natural emergent behavior that exhibits this. Emergent behavior is, in some intuitive sense, unexpected; it produces "surprise" in the observer (Miller, 2007). There is the possibility of multiple levels of emergence, with mesoscale behavior that emerges from microscale interactions itself contributing to the emergence of even higher level macroscale behaviors (Miller, 2007). Emergent behavior can be present and observed in both simulands that are complex systems and in models of those simulands.

FIGURE 9.4
Emergent behavior in a natural system; flocking emerges from individual bird actions.

9.4.1 Modeling

Emergent behavior can introduce modeling challenges in these ways:

1. *Incomplete observations.* Because emergent behavior is potentially unpredictable, available observations of a simuland may not include all of the simuland's possible emergent behavior. Indeed, the modeler may not even be aware of some potential simuland emergent behaviors.
2. *Indirect representation.* Because emergent behavior is not, in general, predictable from the individual behavior of agents or components within the complex system, those aspects or characteristics of it that produce emergent behavior can be difficult to identify and include in the model.
3. *Overabstraction risk.* Because emergent behavior is produced indirectly from potentially non-obvious aspects of a simuland, a modeler may unintentionally abstract away those aspects, inadvertently eliminating the possibility of the model generating interesting or important emergent behavior.

These methods can mitigate the modeling challenges associated with emergent behavior:

1. *Additional observations.* Increase the number or duration of simuland observations, and broaden the range of conditions under which the simuland is observed, thereby increasing the likelihood of observing and detecting the full repertoire of emergent behaviors.
2. *Conceptual modeling focus.* When developing the conceptual model of a complex system, give explicit attention to the inclusion of emergent behaviors, or aspects of the complex system that may give rise to emergent behaviors (such as inter-agent interactions). Those aspects of may be identified in the relevant literature or by simuland subject matter experts.

9.4.2 Validation

Emergent behavior can introduce validation challenges in these ways:

1. *Incomplete observations.* Emergent behavior is, by its nature, difficult to predict, observe, measure in the simuland; this was already noted as a modeling challenge. It is also a validation challenge, as some emergent behavior observed in the model results may not have been observed in the simuland, thus leaving gaps in the data for use in validating the model's behavior.
2. *Incomplete results.* Conversely, emergent behavior observed in the simuland can be similarly difficult to generate in the model results. Of course, if the behavior is not in the model results, it cannot be validated beyond noting that it is absent.
3. *Face validation unreliability.* Face validation is a frequently used validation method. Essentially, in face validation subject matter experts observe a model's behavior or examine its results and compare them subjectively to their knowledge of the simuland (Petty, 2010). Because emergent behavior is unpredictable, face validation of

emergent behavior is less reliable. The experts may overestimate or underestimate the likelihood of occurrence of emergent behavior, or they may have little direct knowledge of it.*

4. *Test case uncertainty.* Because emergent behavior is not directly predictable, designing model validation test cases (trials) which will generate specific emergent behaviors for validation can be difficult.

These methods can mitigate the validation challenges associated with emergent behavior:

1. *Additional observations.* Increase the number or duration of simuland observations, and broaden the range of conditions under which the simuland is observed, thereby increasing the likelihood of acquiring the data needed to validate emergent behavior.
2. *Structured face validation.* To overcome deficiencies in the knowledge of any particular subject matter expert, use teams of experts and conduct organized face validation assessments. The latter may be based on pre-planned validation scenarios designed to cover the full range of simuland behaviors (Petty, 2010; Belfore, 2004) and employ Delphi methods, wherein panels of experts make forecasts and examine the model's results over multiple rounds, eventually converging on a consensus assessment of validity (Rowe, 2001).
3. *Scenario space search.* Generate validation test cases automatically via heuristic search in scenario space, i.e., generating new test cases based on previous trials that elicit some emergent behavior; this method requires metrics for emergent aspects of complex systems.
4. *Semi-automated model adaptation.* For each abstraction (i.e., simplification or estimation) within a model, exploit pre-defined alternative model abstractions embedded in the source code using an optimization-based adaptation process to generate emergent behavior under user specific conditions of interest (Gore, 2008). Essentially, those conditions that produce emergent behavior are predicted by the user, found by the adaptation process, and compared.
5. *Multi-methods.* Employ multiple validation methods which assess the model's results in different ways and compare them to different representations of the simuland. A combination of "breadth" methods (resource efficient but relatively low power, e.g., face validation) with "depth" methods (resource intensive but relatively high power, e.g., statistical hypothesis testing) is recommended in (Reynolds, 2010). Multi-dimensional multi-method approaches to validating models of complex systems are also recommended in (Batty, 2001; Bharathy, 2010); both emphasize the importance of domain knowledge in the validation process. The combination of validation methods can mitigate each method's weaknesses with another method's strengths.

* Experts often underestimate the probability of an unlikely event, implicitly assuming a normal probability distribution when a "fatter tailed" distribution would be more appropriate (Miller, 2007). Examples of such distributions and their asserted applications include power laws for city sizes (Simon, 1955), deaths in warfare (Richardson, 1960), and Lévy stable laws for stock market price changes (Casti, 1990).

9.5 Composition of Components

> Because of the prevalence of inter-connections, we cannot understand systems by summing the characteristics of the parts or the bilateral relations between pairs of them. (Jervis, 1998)
>
> We would, however, like to make a distinction between complicated worlds and complex ones. In a complicated world, the various elements that make up the system maintain a degree of independence from one another. ... Complexity arises when the dependencies among the elements become important. (Miller, 2007)
>
> [A]nalysis of complex systems, particularly in the netcentric operations and warfare domain, which has proven particularly challenging to the modeling, simulation, and analysis community ... (Topper, 2013)

The third of the three defining characteristics of complex systems to be examined for its effect on modeling and validation is composition of components. Complex systems are, by definition, composed of interacting components.* Similarly, models of complex systems are often composed of submodels, and those submodels are typically organized in a structure that reflects the structure of the complex system itself. For example, a spacecraft model may be composed of power system and thermal submodels, with the thermal submodel providing input to power system model to predict power loading. Figure 9.5 illustrates a notional model-submodel structure. In the figure, the overall model is composed of three submodels. The connecting arrows show data from model inputs, through submodel inputs and outputs, to model outputs.

9.5.1 Modeling

Composition of components can introduce modeling challenges in these ways:

1. *Interface compliance.* The existence of multiple submodels, and thus the need for interfaces between them, adds new opportunities for modeling errors, such as mismatches in data types, measurement units, and execution sequence.†
2. *Architecture selection.* The appropriate software architecture framework for organizing and connecting the component models (such as hierarchy, blackboard, or agent-based) may not be obvious, and it may have unintended effects on the model results (Shaw, 1996).
3. *Model correlation.* Different component models may have differences (such as underlying assumptions, representational granularity, or level of fidelity) that negatively affect the overall model's results (Spiegel, 2005).

These methods can mitigate the modeling challenges associated with composition of components:

1. *Interface analysis.* Specifically examine submodel-to-submodel interfaces to determine if interface structures are consistent and accurate (Balci, 1998).
2. *Known problem review.* Review available lists of known interoperability problems typically encountered to see if they apply (Gross, 2007).

* As discussed earlier, interactions between those components can lead to emergent behavior.

† Arguably, the entire subject of simulation interoperability, which is elaborated at length in (Tolk, 2012; Clarke, 1995), is embedded in this modeling challenge. This is no small matter.

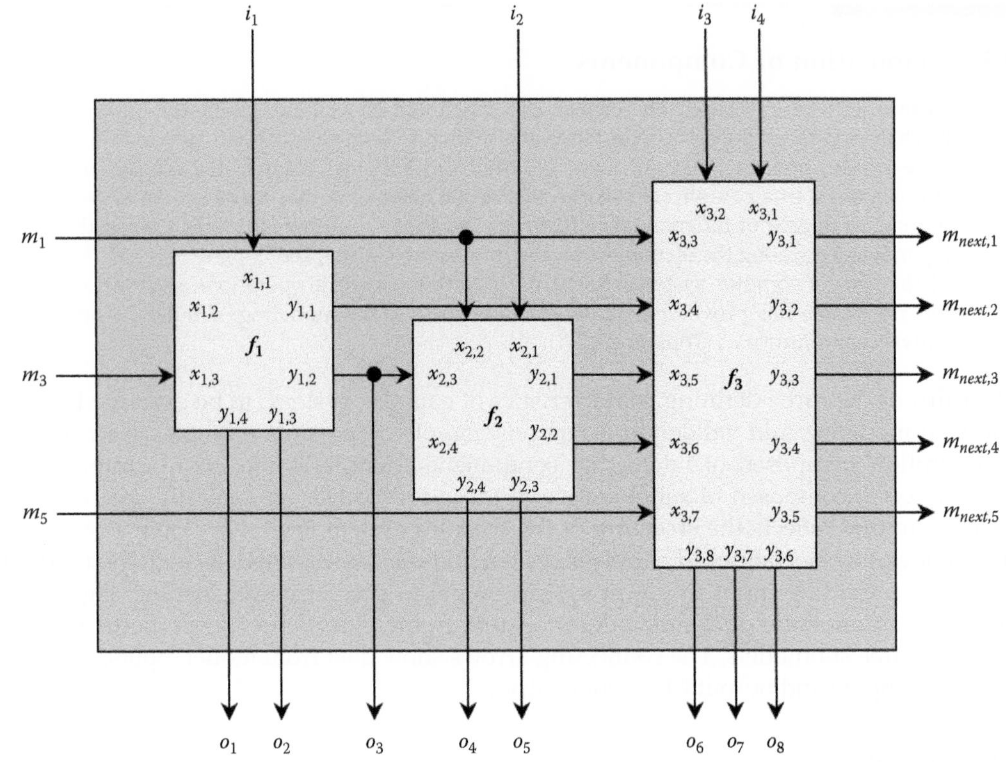

FIGURE 9.5
Composition of components; a model composed of three submodels (adapted from [Weisel, 2003]).

3. *Architecture reuse.* Reuse and revise known model architectures when appropriate, and exploit available architecture-based systems engineering processes (e.g., the Distributed Simulation Engineering and Execution Process [IEEE, 2010]).
4. *Conceptual model verification.* Compare component models' conceptual models to detect model correlation errors.
5. *Model-based systems engineering.* Use model-based systems engineering methods, particularly the Unified Modeling Language (UML) and the Systems Modeling Language (SysML), to develop conceptual models of the simuland. The detailed information and precise semantics of these tools support representations of composition of components when modeling in a system-of-systems domain (Topper, 2013).

9.5.2 Validation

Composition of components can introduce validation challenges in these ways:

1. *Weakest link validity.* The overall validity of a model assembled as a composition of component models may be limited by the lowest fidelity component model. For example, a high fidelity ground vehicle movement model composed with a low fidelity terrain model will likely not produce accurate movement speeds.

2. *Error location ambiguity.* Errors in model results detected during model validation may be difficult to associate with correct component model; indeed, they may result from an interface error, rather than one of the component models.
3. *Statistical method unsuitability.* The statistical methods used most often in validation typically compare single variables, e.g., the Student t test compares the means of two populations (Brase, 2015), or the Mann-Whitney U test determines whether two independent samples of observations are drawn from populations with the same distribution (Hollander, 2014). Models of complex systems have states represented by multiple non-linear variables related non-linearly, requiring the use of multivariate methods that accommodate non-linear effects (Balci, 1998).
4. *Noncomposability of validity.* In a model assembled as a composition of components, i.e., from submodels, the submodels are typically validated individually. Unfortunately, submodel validity does not ensure composite model validity; even if the submodels are separately valid, the composite models may not be (Davis, 2003). It has been mathematically proven that for non-trivial models separately valid component models cannot be assumed to be valid when composed (Weisel, 2003). Indeed, composability was recently described as "still our biggest simulation challenge" (Taylor, 2015).

These methods can mitigate the validation challenges associated with composition of components:

1. *Uncertainty estimation.* Determine or estimate the possible error range for key model results variables for each component model. Then propagate and accumulate those errors to find the overall error range for the same variables for the composite model (Oberkampf, 2000). If the overall error is too large, revise the model.
2. *Non-linear multivariate statistics.* Apply multivariate statistical methods to validation of non-complex systems models. For example, the Bonferroni correction should be used when calculating confidence intervals for multiple model response variables to be used for validation (Petty, 2013), and the Hotelling T^2-statistic, which is a generalization of Student's t statistic that is used in multivariate hypothesis testing, can be used for constructing ellipsoidal joint confidence intervals in validation (Balci, 1984).
3. *Composition validation.* During validation of a composite model, validate both the component models individually and overall composite model. This is directly analogous to conventional unit and system testing in software engineering practice.

9.6 Summary

Complex systems, which are increasingly often the subject of modeling efforts, have certain defining characteristics that make them more difficult to model and make models of them more difficult to validate. The specific modeling and validation challenges can be associated with the complex system characteristic that causes them. Although these challenges can be problematic, and in some cases are in principle impossible to overcome entirely, they can often be mitigated through informed application of appropriate methods.

Acknowledgments

Earlier versions of this chapter were published by the Simulation Interoperability Standards Organization and the Defense Modeling and Simulation Coordination Office. Both gave permission to reuse the content of those earlier versions. The late William F. Waite, formerly of AEgis Technologies, originally suggested the topic.

References

Adams, M. L., Higdon, D. M., Berger, J. O., Bingham, D., Chen, W., Ghanem, R., Ghattas, O., Meza, J., Michielssen, E., Nair, V. N., Nakhleh, C. W., Nychka, D., Pollock, S. M., Stone, H. A., Wilson, A. G., and Zika, M. R., *Assessing the Reliability of Complex Models: Mathematical and Statistical Foundations of Verification, Validation, and Uncertainty Quantification*, The National Academies Press, Washington DC, 2012.

Allen, P. M., and Torrens, P. M., "Knowledge and complexity," *Futures*, Vol. 37, Iss. 7, September 2005, pp. 581–584, doi:10.1016/j.futures.2004.11.004.

Balci, O., and Sargent, R., "Validation of simulation models via simultaneous confidence intervals," *American Journal of Mathematical and Management Science*, Vol. 4, No. 3–4, 1984, pp. 375–406.

Balci, O., "Verification, Validation, and Testing," in Banks J. (Ed.), *Handbook of Simulation: Principles, Methodology, Advances, Applications, and Practice*, John Wiley & Sons, New York, 1998, pp. 335–393.

Banks, J., Carson, J. S., and Nelson, B. L., *Discrete-Event System Simulation, Second Edition*, Prentice-Hall, Upper Saddle River NJ, 1996.

Banks, J., Carson, J. S., Nelson, B. L., and Nicol, D. M., *Discrete-Event System Simulation, Fifth Edition*, Prentice Hall, Upper Saddle River NJ, 2010.

Batty, M., and Torrens, P. M., "Modeling Complexity: The Limits to Prediction," *Cybergeo: European Journal of Geography*, Online at http://cybergeo.revues.org/1035, Accessed May 26, 2017, doi:10.4000/cybergeo.1035.

Belfore, L. A., Garcia, J. J., Lada, E. K., Petty, M. D., and Quinones, W. P., "Capabilities and Intended Uses of the Joint Operations Feasibility Tool," *Proceedings of the Spring 2004 Simulation Interoperability Workshop*, Arlington VA, April 18–23, 2004, pp. 596–604.

Bharathy, G. K., and Silverman, B. G., "Validating Agent Based Social Systems Models," *Proceedings of the 2010 Winter Simulation Conference*, Baltimore MD, December 5–8, 2010, pp. 441–453.

Brase, C. H., and Brase, C. P., *Understandable Statistics: Concepts and Methods, Eleventh Edition*, Houghton Mifflin, Boston MA, 2015.

Casti, J. L., *Searching for Certainty: What Scientists Can Know About the Future*, William Morrow and Company, New York, 1990.

Clarke T. L., (Ed.), *Distributed Interactive Simulation Systems for Simulation and Training in the Aerospace Environment*, SPIE Critical Reviews of Optical Science and Technology, Vol. CR58, SPIE Press, Bellingham WA, 1995.

Colley, W. N., "Modeling Continuous Systems," in Sokolowski, J. A., and Banks C. M., (Eds.), *Modeling and Simulation Fundamentals: Theoretical Underpinnings and Practical Domains*, John Wiley & Sons, Hoboken NJ, 2010, pp. 99–130.

Davis, P. K., and Anderson, R. H., *Improving the Composability of Department of Defense Models and Simulations*, RAND National Defense Research Institute, Santa Monica CA, 2003.

Gehman, H. W. et al., *Columbia Accident Investigation Board Report Volume I*, National Aeronautics and Space Administration, August 2003.

Gore, R., and Reynolds, P. F., "Applying Causal Inference to Understand Emergent Behavior," *Proceedings of the 2008 Winter Simulation Conference*, Miami FL, December 7–10, 2008, pp. 712–721.

Gross, D., and Tucker, W. V., "A Foundation for Semantic Interoperability," *Proceedings of the Fall 2007 Simulation Interoperability Workshop*, Orlando FL, September 16–21, 2007.

Hollander, M., Wolfe, D. A., and Chicken, E., *Nonparametric Statistical Methods, Third Edition*, John Wiley & Sons, Hoboken NJ, 2014.

Institute of Electrical and Electronics Engineers, *IEEE Recommended Practice for Distributed Simulation Engineering and Execution Process (DSEEP)*, IEEE Standard 1730™–2010.

Jervis, R., "From Complex Systems: The Role of Interactions," in T. Czerwinski (Ed.), *Coping with the Bounds: Speculations on Nonlinearity in Military Affairs*, National Defense University, Washington DC, 1998.

Keller-McNulty, S., Bellman, K. L., Carley, K. M., Davis, P. K., Ivanetich, R., Laskey, K. B., Loftin, R. B., Maddox, D. M., McBride, D. K., McGinnis, M., Pollock, S., Pratt, D. R., Robinson, S. M., von Winterfeldt, D., and Zyda, M., *Defense Modeling, Simulation, and Analysis: Meeting the Challenge*, National Academies Press, Washington DC, 2006.

Law, A. M., *Simulation Modeling and Analysis, Fifth Edition*, McGraw-Hill Education, New York, 2015.

Lewin, R., *Complexity: Life at the Edge of Chaos*, Macmillan, New York, 1992.

Liu, F., and Yang, M., "Validation of System Models," *Proceedings of the IEEE International Conference on Mechatronics and Automation*, Niagara Falls Canada, July 29–August 1, 2005, pp. 1721–1725.

Maxwell, J. C., *Matter and Motion*, Society for Promoting Christian Knowledge, London UK, 1873.

Miller, J. H., and Page, S. E., *Complex Adaptive Systems: An Introduction to Computational Models of Social Life*, Princeton University Press, Princeton NJ, 2007.

Oberkampf, W. L., DeLand, S. M., Rutherford, B. M., Diegart, K. V., and Alvin, K. F., *Estimation of Total Uncertainty in Modeling and Simulation*, Sandia National Laboratories, SAND2000-0824, April 2000.

Petty, M. D., "Verification, Validation, and Accreditation," in Sokolowski, J. A. and Banks (Eds.), C. M., *Modeling and Simulation Fundamentals: Theoretical Underpinnings and Practical Domains*, John Wiley & Sons, Hoboken NJ, 2010, pp. 325–372.

Petty, M. D., "Modeling and Validation Challenges for Complex Systems," *Proceedings of the Spring 2012 Simulation Interoperability Workshop*, Orlando FL, March 26–30, 2012, pp. 178–187.

Petty, M. D., "Advanced Topics in Calculating and Using Confidence Intervals for Model Validation," *Proceedings of the Spring 2013 Simulation Interoperability Workshop*, San Diego CA, April 8–12, 2013, pp. 194–204.

Petty, M. D., "Modeling and Validation Challenges for Complex Systems," *M&S Journal*, Vol. 9, Iss. 1, Spring 2014, pp. 25–35.

Reynolds, W. N., "Breadth-Depth Triangulation for Validation of Modeling and Simulation of Complex Systems," *Proceedings of the 2010 IEEE International Conference on Intelligence and Security Informatics*, Vancouver BC, May 23–26, 2010, pp. 190–105, doi:10.1109/ISI.2010.5484739.

Richardson, L., *Statistics of Deadly Quarrels*, Boxwood Press, Pittsburgh PA, 1960.

Rowe, G., and Wright, G., "Expert Opinions in Forecasting: Role of the Delphi Technique," in Armstrong (Ed.), J., *Principles of Forecasting: A Handbook for Researchers and Practitioners*, Kluwer, Boston MA, 2001.

Salmon, F., "Recipe for Disaster: The Formula That Killed Wall Street," *Wired*, Vol. 17, No. 3, March 2009.

Schweitzer, F., *Brownian Agents and Active Particles: Collective Dynamics in the Natural and Social Sciences*, Springer, Berlin Germany, 2003.

Shaw, M., and Garlan, D., *Software Architecture: Perspectives on an Emerging Discipline*, Prentice Hall, Upper Saddle River NJ, 1996.

Simon, H., "On a Class of Skew Distribution Functions," *Biometrika*, Vol. 42, No. 3–4, December 1955, pp. 425–440.

Smith, L., *Chaos: A Very Short Introduction*, Oxford University Press, Oxford England, 2007.

Spiegel, M., P. F. Reynolds, and D. C. Brogan, "A Case Study of Model Context for Simulation Composability and Reusability," *Proceedings of the 2005 Winter Simulation Conference*, Orlando FL, December 4–7, 2005, pp. 437–444.

Epp, S. S., *Discrete Mathematics with Applications, Fourth Edition*, Brooks/Cole, Boston MA, 2011.

Stewart, I., *Does God Play Dice?*, Blackwell Publishers, Malden MA, 2002.

Taylor, S. J. E., Khan, A., Morse, K. L., Tolk, A., Yilmaz, L., Zander, J., and Mosterman, P. J., "Grand challenges for modeling and simulation: Simulation everywhere—From cyberinfrastructure to clouds to citizens," *SIMULATION: Transactions of the Society for Modeling and Simulation International*, Vol. 91, No. 7, pp. 648–665.

The MITRE Corporation, "Verification and Validation of Simulation Models," Online at https://www.mitre.org/publications/systems-engineering-guide/se-lifecycle-building-blocks/other-se-lifecycle-building-blocks-articles/verification-and-validation-of-simulation-models, Accessed May 25, 2017.

Tolk A., (Ed.), *Engineering Principles of Combat Modeling and Distributed Simulation*, John Wiley & Sons, Hoboken NJ, 2012.

Topper, J. S., and Horner, N. C., "Model-Based Systems Engineering in Support of Complex Systems Development," *Johns Hopkins APL Technical Digest*, Vol. 32, No. 1, June 2013, pp. 419–432.

U.S. Department of Defense, *Instruction 5000.61*, M&S VV&A, 2009.

Waldrop, M. M., *Complexity: The Emerging Science at the Edge of Order and Chaos*, Simon & Schuster, New York, 1992.

Weisel, E. W., Mielke, R. R., and Petty, M. D. "Validity of Models and Classes of Models in Semantic Composability," *Proceedings of the Fall 2003 Simulation Interoperability Workshop*, Orlando FL, September 14–19, 2003, pp. 526–536.

Whitesides, G. M., and Ismagilov, R. F., "Complexity in Chemistry," *Science*, Vol. 284, No. 5411, April 2, 1999, pp. 89–92.

Williams, P., *Chaos Theory Tamed*, Joseph Henry Press, Washington DC, 1997.

10

Foundations for the Modeling and Simulation of Emergent Behavior Systems

Thomas Holland

CONTENTS

10.1 Defining Emergent Behavior Systems

Often describing some interesting and usually unanticipated pattern or behavior, the term *emergence* is often associated with time-evolutionary complex systems comprised of relatively large numbers of interacting yet simple entities. A key characteristic of these systems is that the measurements used to describe the overall system are fundamentally different from those used to describe the entities comprising it. That is to say, "The emergent is unlike its components in so far as these are incommensurable, and it cannot be reduced to their sum or their difference" (Lewes, 1874). This quality is readily apparent in naturally occurring systems. Consider, for example, a flock of birds; the qualities and descriptors for a flock are meaningless if applied to a single bird. Researchers have recognized the emergence phenomena in many man-made systems, such as collaborative robotics, supply chain analysis, social science, economics, and ecology. Often the desire is not only to recognize emergence, but also to control it; either to reduce the likelihood of an undesirable emergent event, or to attempt to capitalize on the non-linear increase in ability, efficiency, and robustness that results from the dynamic interactions of the entities in these systems.

Whereas obtaining empirical data for such systems is problematic at best, the computer simulation of these and other complex systems is increasingly receiving greater interest as a surrogate environment in which to study the phenomena of emergence. As improvements in computational technologies combined with new modeling paradigms allow the simulation of ever more dynamic and complex systems, the generation of data from simulations of these systems can provide a controlled environment in which to explore the phenomena of emergence. Darley states, "... emergent systems are those in which even perfect knowledge and understanding may give us no predictive information. In them the optimal means of prediction is simulation" (Darley, 1994). In this chapter, we are considering complex systems that can be simulated in terms of interacting entities defined by a set of behavioral rules, hence the term Emergent Behavior System (EBS).

Although there have been some significant efforts to describe emergence, there remains no universally accepted definition that results in axiomatic representation of the phenomena. As such, researchers adopt working definitions closely aligned with their domain of interest. Furthermore, the concept of emergence seems to elicit conflict between reductionist and holistic perspectives to modeling (Scholl, 2001), (Parunak et al., 1998). Nevertheless, both camps admit that there seems to be some phenomena whereby quantitative interactions between entities in a system can lead to qualitative changes in that system which are different from, and irreducible to, the entities comprising it. The late nineteenth century/early twentieth century psychologist C. Lloyd Morgan expresses this observation in his 1923 text "Emergent Evolution" (Morgan, 1923). Peter Coming summarizes the history of the concept of emergence in an excellent article in *Complexity* (Coming, 2002) and presents the many facets regarding definition (or the lack thereof) of the term.

Typically, the term emergence is used to describe some interesting and usually unanticipated state or sequence of states while observing a system as it evolves over time. This leaves the system modeler faced with the question, "What is emergent behavior and how should I deal with it?"

10.2 A Brief History of Emergence

Any discussion of emergence would be remiss without considering the contributions of John Holland. Known as the originator of genetic algorithms, Holland has done much to popularize the concepts and promise of emergence in his texts "Hidden Order: How Adaptation Builds Complexity" (Holland, 1995) and "Emergence: From Chaos to Order" (Holland, 1998). Holland describes emergence as "... much coming from little" and notes later in his text that "[i]t is unlikely that a topic as complicated as emergence will submit meekly to a concise definition ..." and as such he does not present a concise definition. Rather Holland's texts, although rich in content and perhaps the best of any yet written, is more of a call to further research into the nature and analytical representation of emergence, stating that "[i]n short, we will not understand life and living organisms until we understand emergence." It is within the descriptions of emergence that Holland sets forth that a study (and broad definition) of emergent behavior systems takes form specifically; such systems are "... composed of copies of a relatively small number of components that obey simple laws."

The casual reader might be inclined to attribute the notion of emergence (or the hypothesis that emergence may be a fundamental phenomenon) to Holland. However, the concept is rooted in philosophy, biology, social sciences, and mathematics reaching back hundreds of years. Not to diminish the significant contributions of Holland, in order to unearth a consistent intuition of emergence and to eventually scope a definition by which models of systems exhibiting emergent behaviors can be quantitatively and parametrically represented, it is good to understand the very beginnings of interest in emergence. Searching beyond Lorenz's beginnings of chaos theory (Lorenz, 1963), Forrester's System Dynamics (Forrester, 1961), Ashby's Cybernetics (Ashby, 1957), and the artificial intelligence concepts of Turing (Turing, 1950), this search leads to the efforts of Reuben Ablowitz, who in 1939 published a philosophical theory of emergence (Ablowitz, 1939), and (indirectly at least) identified many fundamental characteristics of emergence, such as the ideas of relationship and scale of observation. Citing references as early as 1843 that addressed philosophical issues such as causation, Ablowitz drew on the works of nineteenth and early twentieth century scientists and philosophers. In his work, we can see the beginnings of what will later become known as *complexity theory*.

Ablowitz set forth a fundamental definition of emergence that begins by making a distinction between "emergents" and "resultants." He states that "Emergent" is a "new quality of existence which results from the structural relation of its component parts." "Resultant" is a "property of the combination that can be foretold exhaustively from the individual elements." Ablowitz illustrates the distinction between the two by considering bricks used to build a house (see Figure 10.1):

> [T]he weight of the house is a resultant of the individual weight of the bricks, but the peculiar characteristic of being a new entity called a "house" is an emergent; you could not possibly tell from looking at a single brick what manner of object a house would be, unless you considered the structural relation the bricks were to assume.

FIGURE 10.1
"Emergent" compared to "Resultant".

He continues to point out that resultant properties are additive, whereas emergent properties are not. Continued research in the decades since Ablowitz has led to additional insight to the nature of emergent phenomena. Although researchers have yet to form a consensus as to the definition of emergence, today most researchers do agree that emergence is a non-linear effect.

10.3 Modern Thoughts on Emergence

More recent interests in emergence have included the perspectives of Artificial Life and Cognitive Science, synthesizing aspects from modern biology, chemistry, economics, and physics (Holland, 1998), (Bonabeau et al., 1995). Such examples suggest that emergent systems demonstrate the existence of levels such that lower level components give rise to new characteristics at higher levels with global coherence arising from local rules, and the necessity of an observer by which emergence can be recognized. Although the modern understanding of emergence indicates that it takes on different forms depending on the domain of study, a ubiquitous observation is that the phenomena of emergence does often arise from very simple entities interacting in various ways and suggests that there is indeed some underlying phenomena that remains to be understood.

Although the concept of emergence is old indeed, the advent of modern computing has fueled an explosion of research into complexity and the rediscovery of emergence. It is no wonder then that the complexity made possible with modern computers would lead to emergent behaviors, whether intentional or not. Given the rise of complexity in such systems, assuring the correct emergent behavior poses a challenging task. Lyons and Arkin (Lyons & Arkin, 2003) explored this very problem with regard to robot-environment interaction and developed an analytical approach that reduces the combinatorial scale between robot actions and the environment. As early as the mid 1980s, systems engineers have observed the problems associated with systems of increasing complexity—that they are typically more event-driven than transformational and continuously having to react to both external and internal stimuli. Whether the intent is to produce a simulation of an existing system, or to achieve an advanced behavior such as in autonomous vehicles, simulation is critical. Lyons and Arkin observed that the current design of such complex systems is analogous to civil engineering prior to the advent of physics and calculus when major structures such as bridges were built without any way to predict their performance. With varied success, projects became more ambitious wherein "… some spectacular failures ensued due to the absence of effective performance guarantees" (Lyons & Arkin, 2003). Emergence appears to be happening all around us. We can either build highly constrained and rigid systems in an attempt to avoid dealing with it, or we can attempt to understand it with the hope to exploit its resilience and efficiency.

10.4 The Role of Complexity

To a large extent, researchers in system dynamics, complexity, chaos, and computer science have made significant inroads to the mathematical definitions relative to emergent behavior systems. Unfortunately, because of little cross-disciplinary interchange (Scholl, 2001), a fair amount of effort must be made to recognize that a varied terminology is frequently used to express essentially similar ideas. The varied definitions for emergence related terms need to be viewed in light of their mathematical basis and as such stated with careful consideration to their current or eventual mathematical expression. It should be stressed however, that wherever possible, the foremost goal is the pretext that whatever is defined in words must be expressible mathematically.

It is easy to see that Ablowitz's philosophical treatment of emergence begins to lend itself to a mathematical treatment. For instance, we can make the observation that the properties of the Resultant are necessarily linear, e.g., the weight of all the bricks is a simple sum, whereas the structure imparted by interactions among components results in the inherently non-linear properties of the Emergent. Albowitz's theory of emergence foreshadowed the modern scientific interest in emergent behaviors and laid a philosophical foundation on which the applications of chaos and complexity theories intertwine with computer science in the modern discipline of modeling and simulation.

Ablowitz defined emergence as "the non-additive quality of existence which results from the structural relation of component parts." Although such a definition suggests the non-linear quality of emergence, continued research for the subsequent six decades has led to additional insight to the nature of emergent phenomena. Michel Baranger asserts that "... the enormous success of calculus is in large part responsible for the decidedly reductionist attitude of most twentieth century science, the belief in absolute control arising from detailed knowledge" (Baranger, 2000). In short, Chaos was developed because the determinism of calculus fell apart with the discovery of certain equations with hypersensitivity to initial conditions (Lorenz, 1963) with great impact in the area of dynamical systems where there are many examples of systems where uncertainties in initial conditions lead to exponential changes in system state. We now know, for instance, that chaos in the spatial domain manifests itself in fractals, whereas chaos in the time domain results in a complex system-dynamics (Baranger, 2000). Baranger connects spatial chaos (fractals) with temporal chaos and observes, "Every chaotic dynamical system is a fractal-manufacturing machine. Conversely, every fractal can be seen as the possible result of the prolonged action of time-chaos." The mathematical relationship among components, whether we study chaos or complexity, is inherently non-linear. Interestingly, the concepts of statistical self-similarity and chaotic dynamical systems align nicely with the ideas of scale (or resolution) in the study of emergent phenomena. The field of Complexity then can shed a great deal of light on the nature of emergent phenomena, and therefore the properties of complex systems can provide the basis for a deeper understanding of emergent behavior systems. Baranger identified the following properties that characterize complex systems:

1. They contain many constituents interacting non-linearly.
2. Their constituents are interdependent.
3. They possess a structure spanning several scales.
4. They involve interplay between chaos and non-chaos.
5. They involves interplay between cooperation and competition.

Although complex systems are often associated with what might be called emergent phenomena, Baranger's statements do not necessarily imply that systems exhibiting emergence must always be complex. Indeed, although the behaviors exhibited by EBS can be very complex, such complexity need not be the result of *complicated* components. For example, Rodney Brooks has shown that highly adaptive, seemingly intelligent behavior can arise from simple stimulus/response entities or "reactive agents" (Brooks, 1991).

10.5 Information—Energy Duality

While naturalists were considering the intricacies of the interactions of many entities, engineers were considering the practical aspects of statistical mechanics to electronic communications. In particular Shannon showed in 1948 that there was a duality between energy and information. This discovery has resulted in a measure of uncertainty for the correct transference of information between transmitters and receivers, now known as *Shannon Entropy*, or *Information Entropy* (Shannon, 1948). Information Entropy becomes particularly salient in regards to the study of systems demonstrating emergence as such systems are essentially communication systems. Owing to Ablowitz's assertion that emergence is about the interrelationships between system components, in most systems of interest, those interrelationships are manifested by the communication of some information. Within systems demonstrating emergent phenomena, communication can occur either directly between entities or indirectly by the entities modifying their local environment (Chira et al., 2007). Shannon's Theory does not depend upon a particular means of communication aside from the conceptual model of there being a transmitter, a receiver, noise source, and a channel by which communication can occur. The information shared among entities should be of great interest to the modeler charged with representing such systems, because what the entities in the system do in response to that information affects the overall system behavior.

Statistical mechanics asserts that the critical exponents describing the divergence of physically measurable properties such as specific heat, magnetization, etc., are universal at a phase transition. In other words, these phase transitions depend only on a few fundamental parameters such as the dimension of the space and the symmetry of the underlying order parameter. Similarly, the emergence of phenomena observed at the higher levels in complex systems is believed to be the result of fundamental principles acting at the local level. Shannon showed that there is information–energy duality that can be understood by means of statistical mechanics, i.e., thermodynamics (Shannon, 1948). Thermodynamics describes the relationship between ordered and disordered energy. In physics and chemistry disordered energy is heat. Statistical mechanics strives to derive the laws of thermodynamics using statistics. Entropy (the Second Law of Thermodynamics) is a measure of heat, or more specifically disorder in a system, and observes that closed systems progress to disorder over time. Fundamentally, entropy S is defined in terms of temperature T and heat Q by

$$\Delta S = \frac{\Delta Q}{T} \tag{10.1}$$

This is referred to as the *macro definition of entropy* and is commonly used in physics and chemistry. On the other hand, Statistical Mechanics takes a micro view of physics and relates the macro definition of entropy to the number of microscopically defined states Ω *accessible* to a system, such that

$$S = k \ln \Omega \tag{10.2}$$

where k is Boltzmann's constant (1.4×10^{-16} erg/deg).

The Second Law of Thermodynamics might seem counterintuitive to the observation of emergent phenomena that appear to become more organized, i.e., increasing order, as time progresses; indeed, we observe natural emergent behavior systems organizing with great efficiency. However, this is achieved without violation of the Second Law because of the coupling between the macro-levels of the system with the disorganizing process at the micro-level. This phenomenon has been described as an *entropy leak* that drains disorder away from the macro-level to the micro-level (Parunak, 1997). Shannon's formulation of the Second Law of Thermodynamics considers the rate at which information is produced and by taking a statistical mechanics approach considers a set of possible events $p_1, p_2, \ldots, p_n$ as the microscopically defined states that describe the system. The question Shannon asked is how much choice is involved in the selection of an event? Or rather, how uncertain is the outcome? Shannon showed that if there is such a measure $H(p_1, p_2, \ldots, p_n)$ then it must have the following properties:

1. H should be continuous in the p_i.
2. Uncertainty should increase with equally likely events as the number of events increase. That is, if all p_i are equal, then $p_i = 1/n$ and H should be a monotonically increasing function of n.
3. If H is decomposed then the result is a weighted sum of the decomposition of H.

Shannon showed that the only H that can satisfy all three properties is a function of the form

$$H(x) = -K \sum_{i=0}^{N-1} p_i \log p_i \tag{10.3}$$

where x is a chance variable and K is a positive constant which amounts to a selection of units of measure. From the similarity of his result with the measure of entropy in statistical mechanics, Shannon referred to this as the entropy of the set of probabilities, which we now call *information entropy*. An information source provides large entropy ($|H(x)|$ large) if its contents are of random nature; entropy is small if the source contains regular structures. This phenomenon can be observed in time varying systems of interacting entities when some spatial patterns appear more frequently than others. If we consider this formation of structure with regard to Ashby's Law of Requisite Variety (Ashby), a cause of decreasing disorder is increasing constraint; in a system this reduction of the uncertainty is indicative of increasing structure.

10.6 The Role of Variety and Constraint

Shortly after Shannon's formulation, Ashby explored the nature of controllability in a complex system and noted in his seminal text "Introduction to Cybernetics" that a complex system can only be effectively regulated if the variety of the system regulator is equal to or greater than the possible variety of the system being regulated. This means that within any system there is a possible variety of processes, but it is by the selective constraining of that variety by which useful results are obtained (Ashby, 1957). Ashby's *Law of Requisite Variety* relates the complexity of an overall system to the complexity of the entities comprising that system. Statistical mechanics relates states of individual entities to overall system states. The Law of Requisite Variety similarly provides a means to express the constraint required on the entities of a system to achieve system performance. The two together suggest underlying phenomena critical to system behavior that is manifested by observable, i.e., measurable, self-organization (Holland, 2007).

Ashby's Law of Requisite Variety further relates the variety of a system's components to the variety of the system itself, the proportion of this variety being identified as an *intensity of constraint*, i.e., the reduction in the number of possible arrangements that the controller allows. In general, this intensity of constraint I can be expressed as

$$I = \frac{V_p - V_s}{V_s} \tag{10.4}$$

where V_s is the variety found in the constrained system and V_p is the variety available to the system if it were unconstrained. Within models of EBS, the value of I is closely related to specific classes of EBS in which communication between entities comprising the system provides local constraining mechanisms with information flows tending to provide positive or negative feedback with regards to regulation of the system (Holland, 2007). These constraints in EBS tend to be self-reinforcing and evolutionary within the modeler's context of observation, manifesting as system states which can be recognized by such things as phase transitions, correlation lengths, mean free paths, and mean relaxation times.

10.6.1 Relating Global to Local

Holland showed that the intensity of constraint for a system comprised of N identical entities can be written as

$$I(t) = \frac{V_p}{V_s(t)} - 1 = \frac{N \log_2 m}{\log_2 M(t)} - 1 \tag{10.5}$$

where m is the number of unique states possible, $N \log_2 m$ is the variety available to the system if it was unconstrained, and $M(t)$ is the number of unique states exhibited by the constrained system at time t (Holland, 2007). The problem with (10.5) is that it assumes that the only influence on intensity of constraint is the number of states of the system. However, I could have just as easily been increased by increasing V_p: for simplicity, it has been assumed that V_p itself is constant. However, we must consider under what conditions V_p can change; if the components of the system are identical finite state automata then m does not change. However, as the components interact in response to influences of others and their environment they will at times combine to form structures within the system, i.e., new *meta-components* with influences affecting the system being some combination of the states of the components.

10.7 A Context-Free Definition of Emergence

Clearly the most obvious characteristic of systems exhibiting emergent behaviors is that at some scale of observation they consist of multiple entities that share information in such a way as to constrain their inherent variety and to dissipate their energy (or information) to achieve a (perceived useful) equilibrium. This perspective is couched in statistical mechanics and system dynamics and lends a sense of mathematical formalism for specification. However, the ideas explored by more recent researchers, benefiting from the parametric simulations made possible with modern computing hardware, would seem to provide a complementary view; nowhere in the discussions of systems exhibiting emergence can we find the requirement that the entities comprising such systems must be of a certain kind, or be of a certain degree of sophistication. With these observations in mind let us now assert a context-free definition of an Emergent Behavior System (EBS):

> An Emergent Behavior System is a natural or synthetic system that produces observable changes in state in the form of either spatial or temporal patterns as a result of interactions between the components comprising the system.

This high-level definition would seem to address all the ideas noted by earlier researchers and spans the breadth of systems from those comprised of very simple components to those where the individuals are varied and very complicated. The key word is "interactions" because that is what separates the resultant characteristics of a collection of entities from the emergent characteristics of a system of entities. This is called a "context-free definition" because it is very general and too broad for a working definition within a specific context. In fact, we cannot escape the dependence on context, and have seen in fact that there exist classes of EBS that lend themselves to different modeling approaches. Nevertheless, much remains to be done and the context-free definition supports the continued development of a modeling and simulation lexicon by which EBS can be described. Furthermore, we can begin with this definition to show that such systems can be abstracted to a useful degree by considering the complexity of the entities comprising the system, their interactions with each other, and the major paths of information feedback in the system. In this regard, the context-free definition does help us develop an intuition about modeling EBS.

10.8 The Importance of Modeling Emergence

It is common to assume that emergent behavior systems have their beginnings in the study of complexity, but it is rather a chicken-and-egg issue: clearly the idea of emergence (at least philosophically) predates the concept of complexity, yet it has only been the advances in recent history, primarily the advent of modern computing equipment, that has allowed the explosion of formal research into complexity that has arguably lead to the rediscovery of emergence. As our ability to simulate complex systems increases, the opportunity to experience emergence in the simulation also increases. These simulations allow a multitude of independent entities to interact, reacting individually to their environment and to each other. Although the entities and their interactions are within a simulation, this more closely approximates the real entities and interactions of the real world. In the simulation, just as in the real world, the entities react to the various influences on them, whether from interactions with each other or their environment. It allows us to consider both the analysis and design of large and complex reactive systems. Advances in computer simulation have made this possible, but we saw it coming. As Harel stated three decades ago:

> The literature on software and systems engineering is almost unanimous in recognizing the existence of a major problem in the specification and design of large and complex reactive systems. A reactive system, in contrast with a transformational system, is characterized by being, to a large extent, event-driven, continuously having to react to external and internal stimuli. (Harel, 1987)

Whether the intent is to produce a simulation of an existing system, or to achieve an advanced behavior, e.g., autonomous weapons, simulation is critical. In the latter case, the simulation is the system. The assertion of Lyons and Arkin describes our current state of engineering for emergence:

> There is an analogy with the history of civil engineering: bridges and other major structures were constructed for thousands of years before the necessary mathematical tools were developed to guarantee their performance. In the 19th and 20th centuries, as such projects became more ambitious, some spectacular failures ensued due to the absence of effective performance guarantees. (Lyons & Arkin, 2003)

Without the establishment of axiomatic formalisms to describe systems exhibiting emergent phenomena, advances in this area will not keep pace with needs, and perhaps more importantly, it will be very difficult to assure the behavior of such systems, which could lead to unintended consequences and in some cases devastating results.

10.8.1 Inconsistency of the Agent Metaphor

Clearly Ablowitz had no idea of the modern thinking on agents and considered even the simplest of entities as components of an EBS (e.g., bricks). A review of more recent literature would suggest that almost any component of a system can be represented by an agent metaphor; the more simple components being modeled as agents with profoundly limited sense and respond behaviors (reactionary agents, like fish in an aquarium) and the most sophisticated components modeled as cutting-edge artificially intelligent engines (cognitive agents). One must ask, "Why agents?" and along with that, "Is then everything some kind of agent?" The answer to both questions necessarily degenerates to a philosophical debate founded on inconsistent and poorly defined terms used in rather broad applications. Broad and inconsistent terms give system modelers intense headaches. Since so many like to create systems of agents and then marvel at the often unexpected yet interesting ensemble behavior, let us consider the components of a system in general and then the concept of agent specifically.

The current literature is replete with references to the term "agent" whether it is with regard to "agent-based" (Parunak et al., 1998), "agent-oriented" (Kim et al., 1999), "intelligent agent" (Franklin & Graesser, 1996), rational agent (Russell & Norvig, 1995), "multi-agent" (Ferber, 1999), etc. Unfortunately, one must struggle a little to identify the unique characteristics of these terms when reading current literature, and many working definitions seem to be tightly coupled to the specific problem domain that is the area of interest of the respective researchers. This has led to much inconsistency and confusion among the varied domain perspectives. For example, Franklin and Graesser presented a formal definition of an "autonomous agent" as a means to distinguish software agents from other programs. Researchers have made some progress toward a more general description of such behavior based entities, such as Ashlock's and Kim's examination of representation of agents by Cellular Automata (CA) which found that using the CA representation resulted in less cooperation in solving the prisoner's dilemma problem (Ashlock & Kim, 2005). It must be understood that in dealing with the concept of emergent behavior, especially where we are concerned with the discovery of emergence or with the design of systems with intentional emergent behavior, we readily observe that such behavior can result from either collectives of interacting simple entities such as finite state automata and CA or collectives of more sophisticated entities, i.e., agents with all the complexity that the modern term connotes. The fact is that in both naturally occurring and synthetic emergent systems we typically observe interactions among three readily recognizable components; those being the boundaries of the system or the environment, objects within that environment, and actors in the environment that tend to affect objects and other actors. Unfortunately, given the many available definitions of agent, any and all of these components are often implemented as agents, leading to the need for innumerable modifiers. The modeling lexicon presented here distinguishes between the "agents" (Actors and Objects) in a system and the "agent-based methods" of implementing them.

10.8.2 Multi-Agent Systems and EBS

Although not required by definition, emergent phenomena are often associated with MAS. Ferber observes that the term "agent" "is used in rather a vague way." He then presents what he says is a "minimal definition" as follows (key characteristics have been emphasized in bold):

An agent is a physical or virtual **entity**:

a. which is capable of acting in an **environment**,
b. which can **communicate directly** with other agents,
c. which is driven by a set of tendencies (in the form of **individual objectives** or of a **satisfaction/survival function** which it tries to optimize),
d. which possesses **resources** of its own,
e. which is capable of **perceiving** its environment (but to a limited extent),
f. which has only a partial **representation of this environment** (and perhaps none at all),
g. which possesses **skills** and **can offer services**,
h. which may be able to reproduce itself,
i. whose behavior tends towards satisfying its objectives, taking account of the resources and skills available to it and depending on its perception, its representations and the communications it receives. (Ferber, 1999)

The fact that this is a "minimal definition" means that there is essentially no limit on what can be considered an agent. But is the agent metaphor and the specification of MAS the best choice for the modeling of EBS?

Beginning with Ferber's definition of agent within MAS, perhaps we can synthesize a workable definition of the fundamental entities comprising a model of an EBS. First, we can observe that an agent must exist within some **environment**: that would imply it has no meaning outside of its environment (or at least not the same meaning in different contexts) and the agent is itself something apart from its environment; so, an agent is not the system itself, but rather a part of a greater system. Second, not only do agents **communicate** according to Ferber, but they also communicate directly with other agents. This implies that an agent must not only be able to transmit information but also be able to receive it. Notice that this does not preclude communication with something other than agents as well. So, what does it really mean for agents to communicate? At its lowest level, communication can be defined as the transference of data. Ferber suggests no restriction on what form that communication must take or by what medium it must be achieved, but he does indicate that it is direct. This implies that for something to be an agent it must be able to transfer data directly to other agents and not only by some consequential or secondary means; that is, agent-A sends a message intended to be received by agent-B.

However, do entities comprising and EBS need such awareness? This is not a trivial consideration since communication in natural systems can be both direct and indirect; that is, in some cases one entity purposefully communicates to another entity with the expectation of a response, and in other cases the communication is the passing of information such as the posting of traffic signs by road designers or the pheromones deposited by certain social insects. In both cases information is transmitted and received, but in the first case there is intent by the transmitting *agent* to communicate and that a receiving *agent* will respond. In the second case there is no such intended recipient or expectation, but data rather appears as part of the environment whether it is created or received by entities in that environment. A key characteristic of EBS is constraint that leads to organization, so in the EBS there is no requirement that an entity need know that the information it has just received is from another entity; it only needs to act on the information it receives. So this begs the question, what constitutes information in an EBS, especially in light of the previous discussion concerning information entropy? The EBS modeler might consider that information is anything that imposes constraint on the entities in a system such that there is a change of state in an entity—or to be more specific, *any actionable data that elicits a change of state in an entity*. Either form of this working definition of information is necessarily recipient-oriented; that is, the recipient's response (or lack of it) determines whether in fact received data is information.

Ferber's definition requires an agent to be **goal seeking** or seeking to optimize some utility function. To do so would require that it have some way of ascertaining the extent to which its actions allow it to attain its goal. Interrelated, **perception** relates to an agent's ability to sense (or measure) to some degree its environment or other entities and so leads to an internal representation within the agent of its world. Ferber points out that this perception is typically incomplete. This is in keeping with our previous observation that in an EBS responses are based on local information and entities have no understanding of the larger system. Having **skills**, and **offering services** implies that something an agent can do is of use to something else and that it can communicate (or it is understood) what it can offer to others. Reproduction is the only "may" meaning that it is not critical to the definition—just if it can still be an agent. Clearly, modern definitions of agent applied to EBS would rule out a great deal of systems covered by our context-free definition.

For example, particles, although very simple entities, can establish complex relationships and give rise to synergies of scale and threshold effects producing phenomena such as avalanches (Coming, 2002). Typically, one would never refer to a grain of sand or a pebble as an agent, but nevertheless agent methods can be (and increasingly are) utilized in the implementation of models of such systems.

10.8.3 Observer Significance

Although Ablowitz's definition of emergence suggests a non-linear quality stemming from entity interactions, continued research for the subsequent six decades has led to additional insight into the nature of emergent phenomena. Holland points out that a key characteristic of emergent systems is the persistence of patterns even though the components change (Holland, 1998). Bonabeau et al. observed that "there is no real agreement on what it should imply for a phenomenon to be emergent" (Bonabeau et al., 1995). They examined emergence from the perspective of Artificial Life and Cognitive Science, presenting examples from modern biology, chemistry, economics, and physics. Their examples identified additional characteristics of emergent systems, such as the existence of levels whereby lower level components give rise to new characteristics at a higher level, global coherence arising from local rules, and the necessity of an observer by which emergence can be recognized. The upshot of the modern understanding of emergence is that it takes on different forms depending on the domain of study, but can indeed arise from very simple entities. Bonabeau et al. reasoned the need for a framework for characterizing emergence, a result in keeping with Holland's assertion that "[t]he uncertainty of definition forces us to rely on partial descriptions, which in turn rely heavily on context."

Bonabeau's subsequent paper presents the concept of emergence through a framework built on levels of organization, levels of detection, and theories of complexity, which reveals some characteristics of emergent phenomena and asserts that any recognition of a phenomena to be emergent is in itself subject to the existence of an *observer* (Bonabeau et al., 1995). The authors include the local perception of actors and their ability to act locally, initial states of organization, evolution over time, etc., and note "... the emergent aspect of a phenomena is related to the point of view of an observer of this phenomena: it is not intrinsic to the phenomena, but related to the global system (phenomenon + observer)." This insight has major implications for the EBS modeler; specifically, what is (or what should be) the point of view of the observer, i.e., the necessary perspective and level of resolution required for the model?

Can we formalize the definition of an EBS that is consistent with the fundamentals discussed earlier and building on more recent progress in MAS? To do so will require us to further decompose the concept of an EBS and revisit our context-free definition.

10.8.4 Recognizing Context

First, to call something a *System* it must be distinguishable apart from other systems, i.e., it must be bounded or rather it must exist within some context. Spatially, we observe systems within a volume. Typically, we call such a context the Environment. If we were to ask a child (the observer) to identify the parts of a scene of fish schooling in the water, he might identify that there are a school of fish that seem to stay together as they move in the water. The water (or at least the child's perception of it) is the environment in which the school exists. The fish don't really do fishy things if taken out of the water, so they only have meaning in the water, i.e., in their environment. They certainly can't school on the sidewalk!

The child will also observe that there might be octopi and seahorses in the water, as well as jellyfish and seaweed. The child will tell you they are not the same as the school of fish, but the fish must swim around them. Those entities are not the fish, but they also occupy the environment. So there is the school of fish, and their environment composed of water along with the other things in the water. Although the fish swim in the water and respond to it, they are unable to substantially affect the water. However, they can bump into, avoid, communicate with, etc., the other animals in the water. But from their perspective these are just Entities in the Environment. So it seems that the concept of an EBS is meaningful only within the context of the environment perceived by the observer. We must conclude that a formal definition of an EBS must include a definition of Environment.

Second, by definition, a system is composed of discrete units, i.e., Entities. Entities occupy the Environment, but are there characteristics that clearly distinguish entities within and from the Environment; that is, what is an entity? Referring again to Ferber, an agent is obviously an entity, but should all entities be treated the same or referred to the same way in specification? Clearly the fish in the previous example are agents in the Ferber sense, and that can be easily extended to the octopi and seahorses. But what about the seaweed or a piece of flotsam; surely neither would qualify as an agent? Still, the flotsam has characteristics that govern how it behaves in the water, and the fish can influence it by pushing it, eating it, etc. The real question is not so much how something is represented, but rather, what role does it play? If we assume that entities are the atomic components of an EBS that are observable at the scale of observation and that entities through some means can interact with other entities in the environment, then how they interact and those means of interaction become crucial to the EBS modeler. Entities must be observable in the environment. Entities can influence each other. But consider what happens if one of the fish dies—that fish is now flotsam. It is still an entity within the system, but from the modeler's perspective it now belongs more to the Environment than to the set of agents in the school. Additionally, it is the fish that form the school, a system within the system, i.e., a structure with aggregate characteristics not readily deducible from the parts.

Finally, recall that we have made the distinction that the term "agent" refers more to a method of implementation than to a specific role in an EBS. Therefore, we adopt the term "actor" for a fish in the school. For the modeler, objects apart from those that are the primary *actors* in the system would seem to define the Environment of an EBS. It is the difference between a fish in a school and a dead fish; an entity that was once an agent has become an *object*. The flotsam, whether it is a dead fish or something else, is still contained in the Environment and if it were to fall to the ocean floor and become encrusted with barnacles it might then be considered part of, that is, an object making up, the Environment. With this trite example, we conclude that within any *context of observation* an Entity may be an Actor, or Object, within the Environment. Additionally, we observe that although these entities form structures that persist, as time progresses the roles of entities can and do often change as shown in Figure 10.2. Furthermore, Ferber's definition of "agent" no longer confuses the discussion since we now recognize "agent" as an implementation means of EBS Entities.

Of utmost concern to the EBS modeler is the Context of Observation, C, for it is the initial consideration that will define the resolution of interest. The specification of C defines what is regarded as *micro-*, *meso-*, or *macro-*scale in the EBS. Micro-scale concerns what describes the internal working of entities comprising the EBS; it typically measures in units unnecessary at the scale of the EBS modeler. This includes states internal to actors or properties of objects such as molecular structure. In artificial intelligence or in advanced

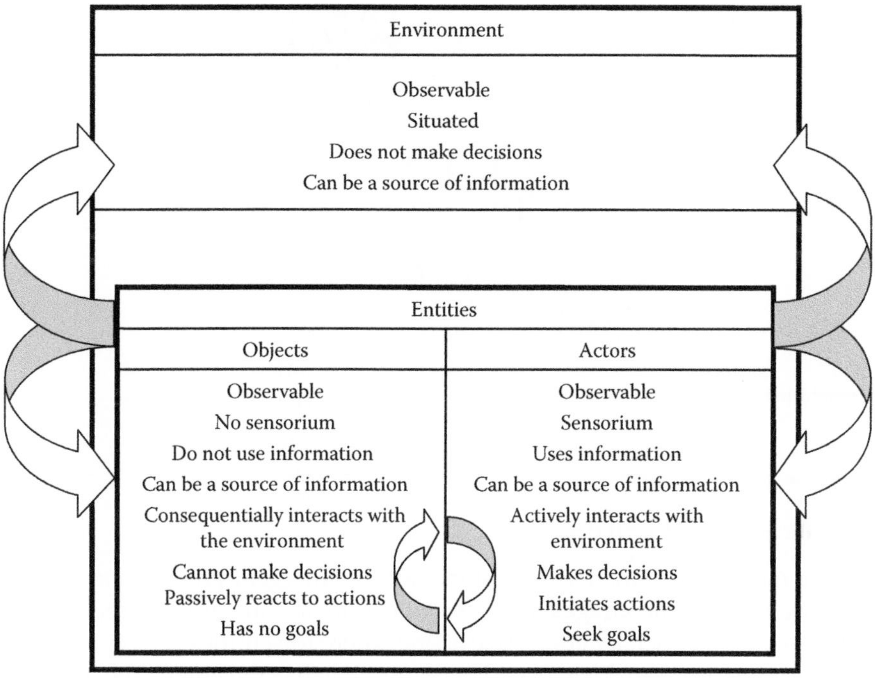

FIGURE 10.2
Actor—Object agents in the Context of Observation.

software agents, the micro-scale often refers to the result of many complex processes. Macro-scale describes the measures applicable to the overall EBS. An example of this is the wave front produced by the motion of many particles which is measured in units such as wavelength and amplitude. Meso-scale is the scale in between the micro and macro, and is most often the resolution of concern of the EBS modeler. This is the scale at which an entity interacts with other entities to influence the states of the system. Meso-scale units are apart from the micro-scale properties that produce them. Consider, for example, a traffic simulation with obvious micro-, meso-, and macro-scales; at the micro-scale level cars are measured in terms of horsepower, wheelbase, curb weight, etc., but the modeler is not typically concerned with that level of resolution when simulating traffic. Instead, the modeler is concerned with meso-scale measurements of velocity, acceleration, minimum braking distance, and such. What constitutes micro, meso, and macro defines the Context of Observation; for an EBS to be successfully modeled, the modeler must be able to discriminate these levels as shown in Figure 10.3.

Actors can be thought of as the traditional definition of an agent, whereas Objects are distinguished as being devoid of any goal-seeking behavior and merely react to external actions. Both are implementable by agent-based methods, and it is easy to see how an agent can be an Actor or an Object depending on its role; consider how an aquarium simulation may have Actors that are fish, but if a fish dies and becomes flotsam, then that Actor has become an Object within the modeling Context of Observation.

Structures, considered a key feature of EBS, form at the meso-scale as entities establish sustained interactions and form a collective. Structures are measurable apart from the entities forming them and define a new entity in the system.

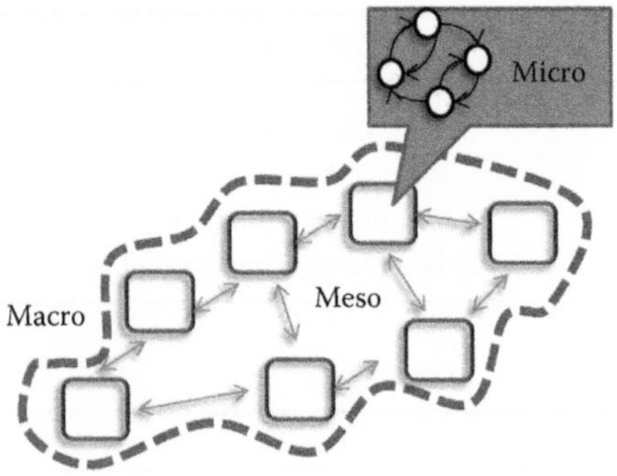

FIGURE 10.3
Structure resolution in EBS.

10.9 A Concise EBS Lexicon

To encourage more formal models of systems where emergent phenomena tend to dominate, a perspective that encourages the EBS modeler to take an entity-based perspective is useful (Holland, 2008). Such a perspective is one where the entities are actors, objects, and structures with clearly definable roles. This perspective has the advantage of reducing the vagaries arising from over-dependence on agent formalisms by considering agents not as the fundamental components of an EBS but rather as a paradigm for implementing the entities comprising the EBS. The following specifies a declarative model of an EBS, identifying entities, relationships, system states, and time within a context of observation, and views the EBS as a sequence of changes in state:

$$EBS \equiv \langle C, \{E\}, \{R\}, \{\Psi\}, T \rangle \tag{10.6}$$

where

C is the spatial and temporal *context of observation* defining the environment that contains all entities, i.e., $C(t) \equiv \{e : P_e(t)\}$ where $P_e(t)$ is the position of an entity e at time of observation t

E is the collection of all entities that are *actors* A and *objects* O or *structures* S within the context of observation, i.e.,

$$E \equiv \{e : e \in (A \cup O \cup S) \cap C\}$$

O is the subset of entities called *objects* where $\forall o \notin A : O \in E$
A is the subset of entities called *actors* where $\forall a \notin O : A \in E$
S is the set of *structures* formed by entities in relationship, i.e., $S \equiv \{e : e_i \mapsto E \times R\}$.

These structures can take on the roles of actors or objects within the context of observation.

R is the set of *relationships* forming the links between entities such that $R \equiv \{r : r_{i,j} \mapsto (B_i \times B_j)\}$ where B_i and B_j are the *behavior sets* governing the i_{th} and j_{th} entities forming a relationship such that B is the collection of entity *behaviors* that can potentially affect other entities within the context of observation.

$$B \equiv \{\forall e_i : e_i \in E, \exists\{b_i\} : \{b_i\} \subseteq B\}$$

Ψ is the set of *states attainable by the system* which emerges from the relationships among the entities.

$$\Psi \equiv \{\psi \Rightarrow (E \times R)\}$$

and

T is the time base at which state transitions are observed.

A natural lexicon follows where,

Environment is defined as the spatial and temporal context of observation defined by objects and containing all actors.

Actors are a special class of entity best described as autonomous stochastic dynamical systems that attempt to build and maintain a maximally predictive internal model of their local environment within the context of their sensorium, behavior sets, and effectors.

Objects are entities that contribute to the environment of the EBS but are not actors within the context of observation.

Structures are formed by the grouping of entities within the context of observation, are in relationship and persist for some time.

Behaviors are the set of rules inherent to an entity that tends to govern its relationships to other entities.

Information is any actionable data that elicits a change of state in an entity.

10.10 The Five Types of EBS

Holland showed that in modeling EBS, systems can be classified according five types (Holland, 2007). These five types support the declarative specification in (10.6) with regard to the context of observation and form the basis of an analysis framework for exploring entity relationships that give rise to emergent phenomena. Referring to Figures 10.4 through 10.7, the following describes each type.

Type 0 emergent behavior (T0 emergence) is no emergence at all but is necessary within the specification of an EBS in order to establish the fact that measures applicable to the entities comprising the system do not describe the greater complex system. Intensity of constraint for such a system is undefined.

Type 1 emergent behavior (T1 emergence) describes systems where information flow is local-to-global, i.e., feed-forward. Intensity of constraint in such systems is either extremely strong ($I \to \infty$) or none at all ($I = 0$) depending on whether there is a strict ordering of relationships between entities (T1a) or not (T1b). These types can also be considered intentional or unintentional respectively; the former describing ordinary machines, the latter particle systems.

Type 2 emergent behavior sees information flows from the entities of the system affecting the global system, which in turn provides information to the entities and therefore moderates the behavior of the system.

The net effect of this feedback from global-to-local tends to be either negative (T2a) resulting in stable systems or positive (T2b) with rapid, often exponential, changes in the system and general instability. Intensity of constraint varies as the system's variety changes over time in which case (10.4) can be written in time dependent form as,

$$I(t) \approx \frac{V_p}{V_s(t)} - 1 \tag{10.7}$$

If I is increasing, the system is tending toward T2a, if decreasing then T2b.

Type 3 emergent behavior describes a system in which both T2a and T2b characteristics are in play, often exhibiting rapid rise in pattern formation (T2b effect) followed by stabilization during a T2a episode. This form of EBS is typified by reaction-diffusion systems where short-term positive feedback is held in check by long-term negative feedback. The effect on (10.7) is a damped oscillation, tending toward increasing I within the context-of-observation.

Type 4 emergence, the fifth type, is a product of evolutionary changes within the entities comprising the system. In the previous types, the assumption is that the entities

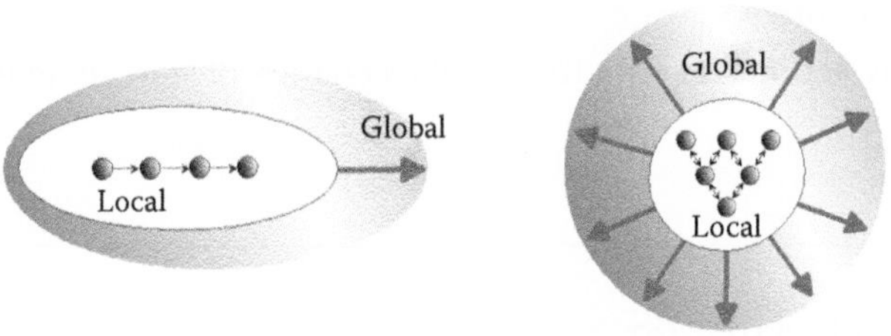

FIGURE 10.4
Type 1 (nominal emergence) receives no feedback from the global system.

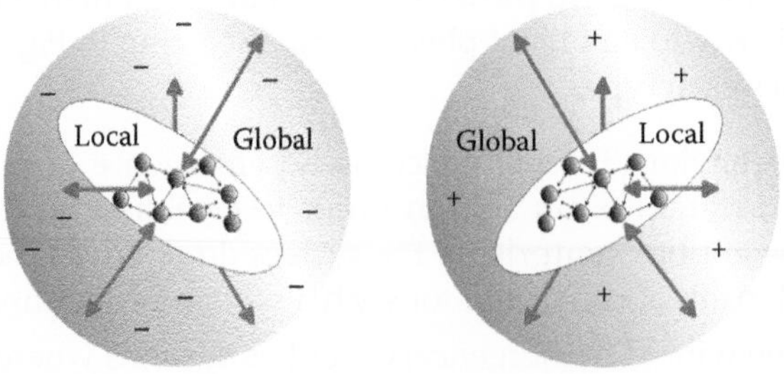

FIGURE 10.5
Type 2 (moderated emergence) receives feedback from the global system.

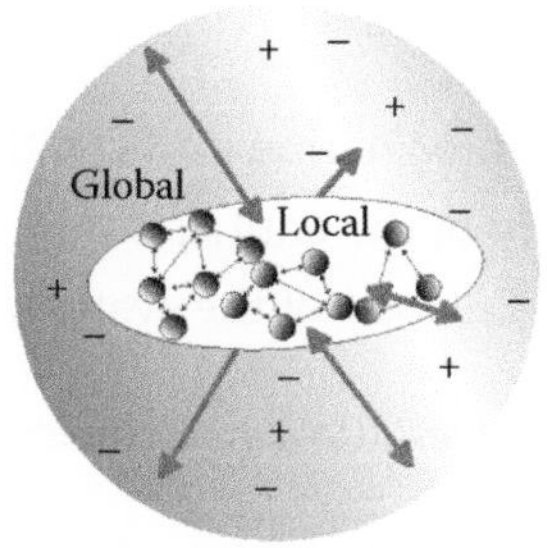

FIGURE 10.6
Type 3 (multiple emergence) receives both positive and negative feedback as the system evolves.

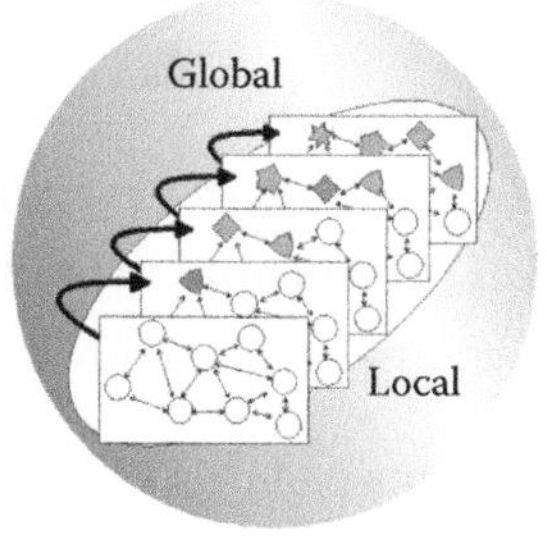

FIGURE 10.7
Type 4 (evolutionary emergence) components change at fundamental levels in response to global or local feedback.

can be represented by finite state automata; that is, regardless of the information flow, the entities maintain a finite rule-base from which they develop relationships to other entities. In T4 I is affected by changes in V_p. In such highly-sophisticated systems, the changing dynamics of the entities make aggregate observations of I intractable for all but the most limited contexts of observation.

10.11 Can We Measure Emergence?

So far, we have made a case that the emergence phenomena in simulations of EBS can be described by system properties related to the complexity associated with the role of the entities comprising a system, the information (or energy) flow between the entities, and the information uncertainty associated with the system. We can examine these properties by relating the entities in a simulation to a graph representation and defining metrics for:

1. The principal information (or energy) flow through the entities,
2. The complexity associated with the role of the entities comprising the system, and
3. The information uncertainty (a duality with energy) associated with the system in the form of information entropy.

We can denote these three measures as:

f the measure of the principle information flows in the system,
Ω the measure of the complexity of the entities comprising the system, and
S the measure of the Shannon Entropy of the system.

We will see that these three measures allow us to examine and distinguish at least Types 1, 2, and 3 EBS in simulation.

The regulation of variety, i.e., *constraint* manifests as phenomena that tend to reduce the variety at the meso-scale level and are related to the primary mediums of communication among the entities in the system. In natural systems exhibiting emergent phenomena such as social insects, we often observe that interactions between entities occur both directly and indirectly. These interactions are regulated by the following constraining modes:

1. Inherent Constraint (self-imposed limits),
2. Contextual Constraint (objects in the environment reduce variety),
3. Peer Constraint (obtrudent, i.e., direct entity to entity communication),
4. Stigmergic Constraint (entities communicate indirectly by affecting the environment).

These different constraint mechanisms can be interpreted as information flows within a system, both in regards to the taxonomic delineation of EBS types and the principal paths of information flow in the system.

Foundational to the modeling of EBS is the distinction between *data, information,* and *communication*. For our purposes, information is any actionable data, i.e., data that causes an entity to change state within the context of observation. Communication is the sharing of information; whether or not data produces an effect is the key. Relationship is established when communication occurs; structures form when relationships are sustained. Graph theory methods provide a means to measure the ability of data to flow through a structure, i.e., a network of entities in relationship. In particular, representing the EBS in graph terms supports the examination of information flows and feedback, i.e., paths that form loops in a graph.

10.11.1 Measuring Information Flow

The flow measure f is related to the sequence of vertices that can represent the principal actors in a structure and is determined by the characteristic path length, i.e., the mean over all pairs of vertices of the number of edges in the shortest path between two vertices.

Figure 10.8 shows some test graphs and the flow metric of characteristic path.

10.11.2 Measuring Complexity

The complexity measure Ω deals with the relative importance an actor maintains in the emergent structures in the system. Stated another way, we are interested in a measure of the significance of an entity to the structure of which it is a part, i.e., its "relatedness" of the structure. Again, graph theoretic measures readily apply.

Watts and Strogatz introduced the measure of the *clustering coefficient* as a measure of connectedness in networks that "rewire" and exhibit varying amounts of disorder (Watts & Strogatz, 1998). Referring to the examples in Figure 10.9, the local clustering coefficient

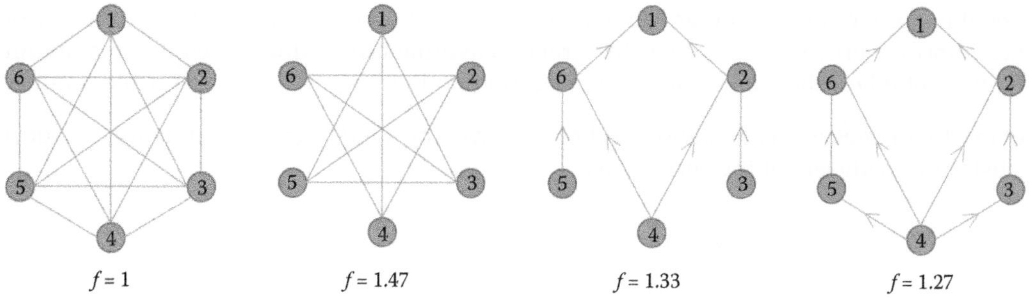

FIGURE 10.8
Characteristic path length as a measure of information flow.

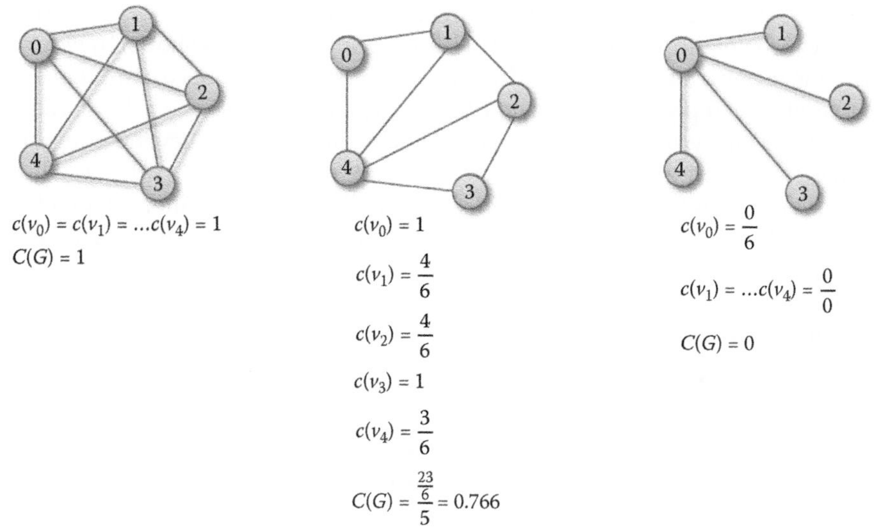

FIGURE 10.9
Complexity measure in EBS.

$c(v_i)$ for a vertex v_i is given by the proportion of links between the vertices within its neighborhood and the number of links that could possibly exist between them. For a directed graph, e_{ij} is distinct from e_{ji}, and therefore for each neighborhood N_i there are $k_i(k_i - 1)$ links that could exist among the vertices within the neighborhood (k_i is the total (in + out) degree of the vertex). Thus, the local clustering coefficient for directed graphs is given as

$$c(v_i) = \frac{|\{e_{jk}\}|}{k_i(k_i - 1)} : v_j, v_k \in N_i, e_{jk} \in E \tag{10.8}$$

We can observe that for an undirected graph e_{ij} and e_{ji} can be considered identical. Therefore, if a vertex v_i has k_i neighbors, edges could exist among the vertices within the neighborhood. We can then consider the idea of a *clustering coefficient* of a particular vertex v defined for the neighborhood of v denoted as $c(v)$ (Newman et al., 2001). We define $c(v)$ as the fraction of edges that actually exist in the neighborhood of v relative to the total possible number of edges that can exist in the neighborhood of v. If k_i is the number of neighboring vertices of v_i, then, if the graph is assumed to be undirected, the maximum number of edges that can exist in the neighborhood of v is $\frac{k_i(k_i-1)}{2}$ and if the number of existing edges in the neighborhood of v is given by $|\{e_v\}|$ then we can define the local clustering coefficient of the vertex v_i as

$$c(v_i) = \begin{cases} \dfrac{2|\{e_{jk}\}|}{k_i(k_i-1)} & ,k_i \geq 2 \\ 0 & ,k_i = 1 \end{cases} \quad : v_j, v_k \in N_i, e_{jk} \in E \tag{10.9}$$

The significance of $c(v)$ is that it indicates the extent to which the vertices in the neighborhood of v are also neighbors to each other. The three simple graphs shown in Figure 10.9 all have the same number of vertices but different edge sets. The first is a complete graph and can be thought of as representative of an EBS in which each entity has a direct relationship with every other entity. The one in the middle is incomplete with several subgraphs, and the one on the right is a tree, representative of an EBS where the structure, i.e., all relationships, is dependent on the single entity v_0. The smaller $c(v)$ is, the more dependent those in the neighborhood of v are on the existence of v. If $c(v)$ is 1, then the neighbors of v are still connected even if v is removed. If $c(v)$ is 0 then all the neighbors of v depend on it as illustrated by v_0 in the third graph of Figure 10.9.

We can further define a global clustering coefficient $C(G)$ of a graph G as the mean of the local clustering coefficients in G

$$C(G) = \frac{1}{|V|}\sum_{i=0}^{|V|-1} c(v_i) \tag{10.10}$$

where $|V|$ is the number of vertices in the graph with degree $d(v) \geq 2$.

A common measure of complexity associated with a vertex in a graph is its degree. Indeed, both degree and clustering coefficient are local graph measures of a vertex's connectedness. However, the clustering coefficient factors in not only the connectedness of a vertex but also the importance of a vertex to those to which it is connected. The global clustering coefficient then provides insight into both the connectedness and the strength of dependencies of entities in the system; this is a measure not only of the information paths associated with an entity, but of how reliant the emergent structure is on each entity.

10.11.3 Measuring Entropy

Recall that Shannon Entropy is a measure of uncertainty or disorganization of the system. In its essence, it provides a measure of system disorder as it relates to the entities comprising the system. Shannon showed that Equation (10.3) is the only form of a solution that can satisfy all three properties of the macro definition of entropy with respect to the number of microscopically defined states. As such, we should expect to see decreasing entropy as an EBS evolves to an orderly state. The challenge for the EBS modeler is to determine what defines the state of a system. Two approaches to the estimation of Shannon Entropy S can be considered. The first deals with representations of the system state in terms of the system's spatial geometry, i.e., the location of entities at discrete times. This approach is in agreement with that described by Parunak and Brueckner (Parunak & Brueckner, 2001) and is well suited to systems where system state is easily defined as some spatial relationship between actors. A more general approach is in keeping with the emphasis on relationship within the EBS; as opposed to a Euclidean-space perspective, this approach depends on what could be called the Relation space of the system, i.e., the state vector defining the system at any time is a measure of the number of structures in the system and normalized mean degree of the structures in the system. For systems comprised of multiple entities in real-world geometry, we may wish to consider the state of the system to be described by the positions of the entities in Euclidean space, but in other cases we should consider the Relation space. In either case the premise is the same.

Computationally implementing (3) to analyze emergent behavior systems requires observations of the states accessible to the system and the determination of the probability of finding the system in each of those states. This can be accomplished parametrically with a simulation of a system and using the Monte Carlo method to build an estimate of $H(x)$ by counting the occurrence of each observed state. Since many EBS manifest as spatial-temporal patterns, the states of the EBS can be expressed in terms of the locations of the entities comprising the system. This location-based approach and the sampling in time to produce the necessary observations suggest an approach that is discrete in both space and time. The following steps explain the location-based approach.

Step 1: Gridding

Parunak and Brueckner developed a gridding method to specify system states in multi-agent systems. Figure 10.10 depicts the method. Here an $m \times m$ context of observation (8×8 in Figure 10.10) is gridded into $n \times n$ cells (here 4×4) to produce a state vector of the form

$$\Psi(t) = \begin{matrix} c_{0,0} & \cdot & \cdot & c_{0,3} \\ \cdot & \cdot & & \\ \cdot & & \cdot & \\ c_{3,0} & & & c_{3,3} \end{matrix} = \begin{matrix} 0 & 0 & 0 & 0 \\ 0 & 2 & 2 & 0 \\ 0 & 2 & 2 & 0 \\ 0 & 1 & 1 & 0 \end{matrix} \qquad (10.11)$$

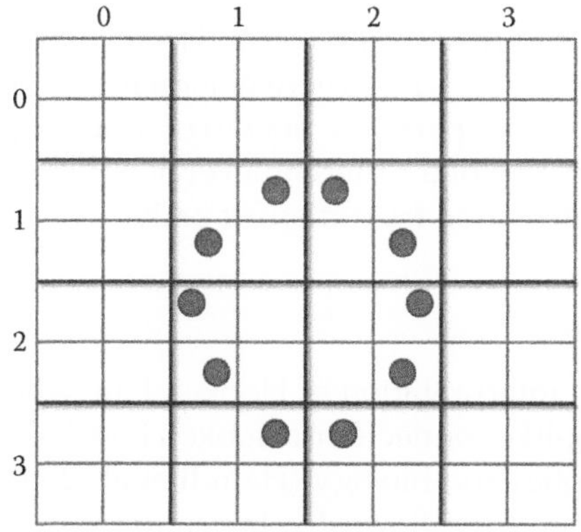

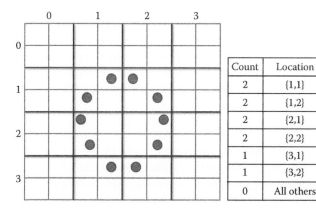

Count	Location
2	{1,1}
2	{1,2}
2	{2,1}
2	{2,2}
1	{3,1}
1	{3,2}
0	All others

FIGURE 10.10
Spatial gridding method for system state representation.

Each c in the vector is the count of entities in that cell at time t. For the 4×4 grid shown, each state Ψ is given by a vector of 16 elements. Each element is the entity count in that cell.

Step 2: Determine Probability of States

Multiple simulation runs beginning with random initial conditions can be used to generate the set of p_i needed for (10.3). At each time-step of the simulation, the state vector Ψ is produced. The count of each Ψ across the simulation runs at each time-step can be used to estimate the probability p_i at each time-step t in the manner,

$$p_i(t) \approx \frac{count\_of\_\Psi_i(t)}{R} \tag{10.12}$$

where R is the number of simulation runs.

Step 3: Compute Entropy

Once the states are identified and counted for each time step, (10.3) can be computed. In computing (10.3), recall that K is a constant, so for this analysis, K is assumed to be 1 so we can rewrite (10.3) as

$$S(t) = \frac{-\sum_{i=0}^{R-1} p_i(t)\log p_i(t)}{\log R} \tag{10.13}$$

Equation (10.13) is the Shannon Entropy estimate at time t for the system but normalized by dividing by $\log R$ so that the case $S = 1$ indicates maximum disorder.

10.12 Examples

Three EBS inspired by natural systems are presented here. These three systems represent three fundamentally different classes of systems where emergence phenomena are traditionally observed. These three systems are particles, flocking (or herding), and stigmergy. The actors in each EBS are representative of a different level of entity sophistication, and each system derives its principal information from distinct paths.

10.12.1 Particle System

The particle system simulation is an interpretation of Hooke's Law where unit mass particles form structures through spring-like connections. Hooke's Law has found wide application in domains as diverse as robotics and biology (Hamdi et al., 2005), (Oddershede et al., 2002). Particles interact directly and the flow of information is very physical, either through direct contact with each other or their environment, or by the formation and breaking of the spring connections. Without any global to local feedback mechanisms, the particle system simulation can be an example of Type 1b emergence. A damping factor ensures that the net feedback is negative and as such is an example of Type 2a emergence.

The actors in this simulation are not typically associated with the common usage of the term "agent." Indeed, this simulation was developed in part to demonstrate that the agent metaphor might perhaps be more properly considered a programming paradigm (much like object-oriented programming) as opposed to a specific modeling construct only useful for certain types of representations. Information flow in the particle system occurs when one particle interacts with another in such a way as to influence its behavior. The idea that information is "actionable data" is maintained; specifically, data is transferred from one particle to another by means of their relationship. For example, billiard balls transfer data when one strikes another. The data (velocity, momentum) is transferred on striking and the resulting reaction is the effect of the "actionable data." For the system of masses and springs here, the relationship is given by $F = -kx$, so any action taken by one particle is transferred to particles to which it is connected (in this example, all have the same k but x varies). As such, any time a particle strikes an object, information from that strike propagates throughout the system. In this way, the state of the particle system is affected whenever any particle of the system strikes another particle, the floor (ceiling) or wall. Figure 10.11 identifies the entities in the particle system simulation as well as how energy can propagate through the structures.

One of the parameters of interest in the particle system is the *shear-point*, i.e., the distance at which the connection between particles breaks. Increasing the value of *shear-point* has the effect of increasing the likelihood that one particle can sustain relation to another. In graph theory terms, the EBS in the minimally constrained case can be considered a *nearly-random* graph with the probability of a vertex adjacent to another being dependent on the value of the *shear-point*. The effect of increasing *shear-point* for a single random arrangement of particles in the particle system can be seen in Figure 10.12 where the value of *shear-point* is increased from 5 to 25.

Another parameter of interest in the particle system is the *rest-length*, i.e., the spring length when no forces are acting on it. When the *rest-length* is very small relative to the

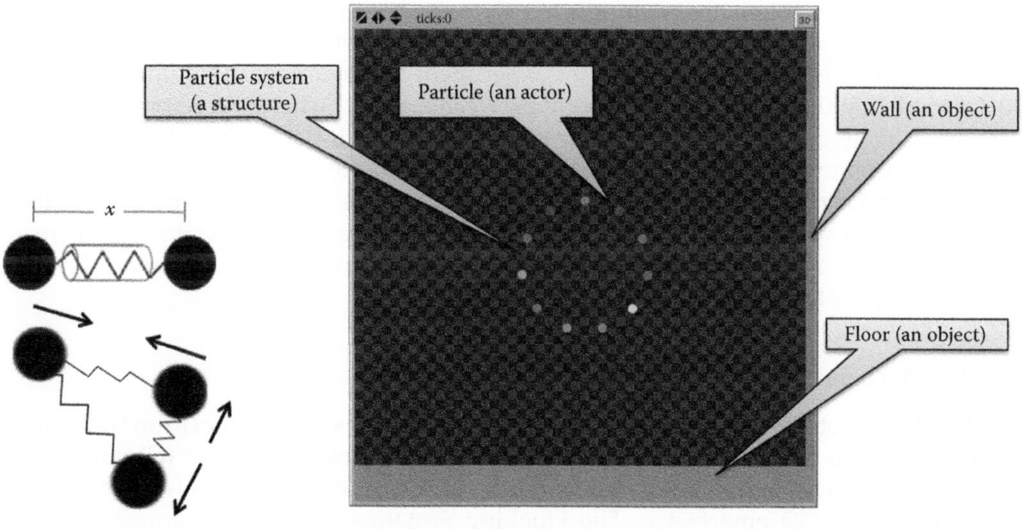

FIGURE 10.11
Particle system simulation is an interpretation of Hooke's Law.

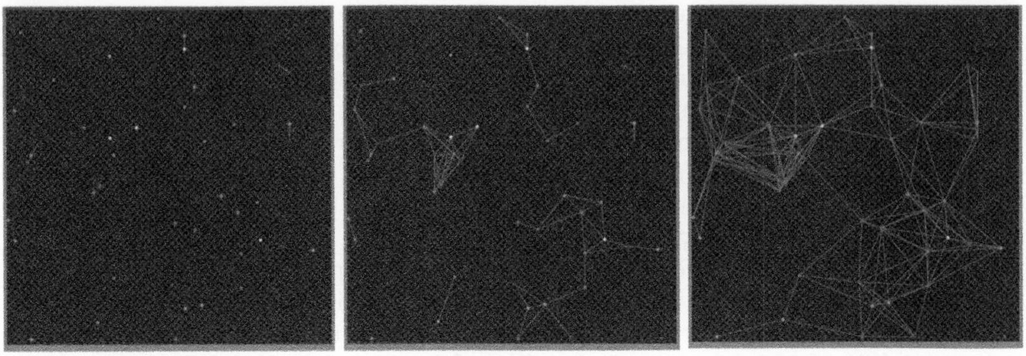

FIGURE 10.12
The effect of increasing *shear-point* in the particle system.

shear-point, the particles have more opportunity to establish relationships without breaking relation when the *shear-point* is exceeded. As the *rest-length* increases, there is less opportunity to stretch without exceeding the *shear-point*. The intensity of constraint then in this system is given by:

$$I_C = \frac{rest-length}{shear-point} \tag{10.14}$$

If *rest-length* is equal to *shear-point*, then particles are beyond the influence of other particles and the $I_C = 1$, the maximum constraint. As *rest-length* decreases, there is more variety available to the particles, i.e., the intensity of constraint is lower. Figure 10.13 shows the changes in the flow measure *f* as *shear-point* is increased. When we examine this data, which is measured when the system is in equilibrium, we see consistent formation of structure with the system always producing the same result for strong constraints.

Similar results are recorded for the complexity metric of global clustering coefficient (GCC). Figure 10.14 shows the GCC for the experiment as I_C increases in columns 1 through 10. Observe that there is a transition point between 9 and 10 where the system's complexity would likely be similar to the nearly-random scenario if data were taken at 8 < *rest-length* < 9. (Column 11 is the data for the nearly-random scenario.) The data analysis for the Shannon Entropy (SE) is shown on the right side. Notice how this metric reveals a condition in the system where there is a sudden drop in entropy after a period of increasing entropy. The SE for the nearly-random scenario is in column 11 of the plot, showing relatively high entropy, as expected.

10.12.2 Flocking System

The Flocking System simulation is based on Craig Reynolds' Boids algorithm (Reynolds, 1999), which has found broad application such as to crowd behavior (Chiang et al., 2008). Three simple behaviors based on each actor's observation of other actors near it suggest a Type 2a emergence. The Flocking System simulation demonstrates self-organization of a group of entities like a flock or herd. Based on only a few simple rules of behavior, namely Reynolds' three rules of "alignment," "separation," and

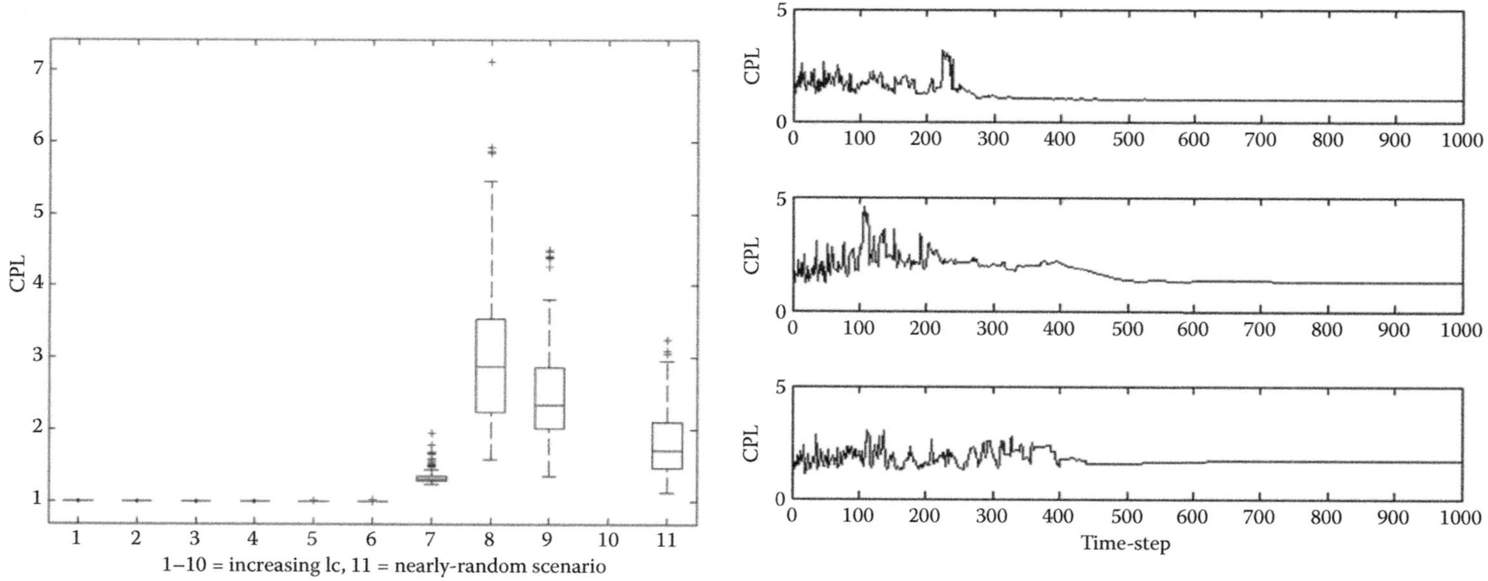

FIGURE 10.13
Characteristic path length (*f*) with increasing *shear-point* in the particle system.

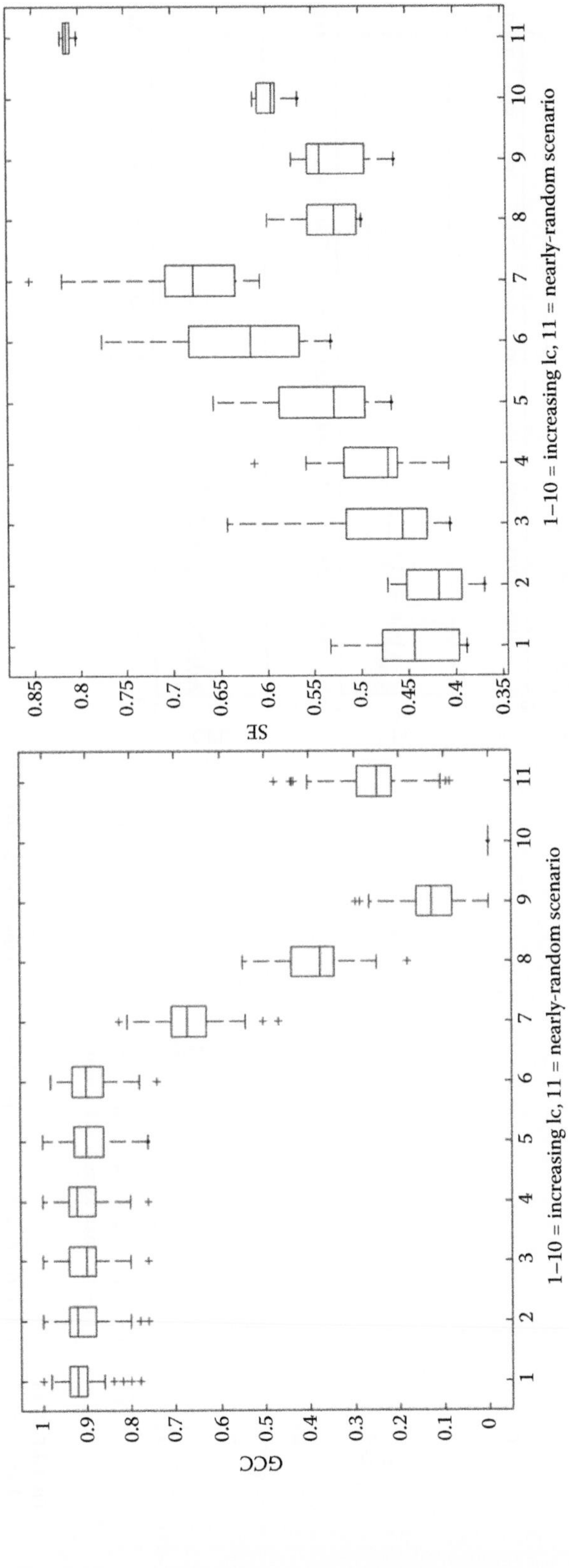

FIGURE 10.14
Complexity (left) and entropy (right) in the particle system.

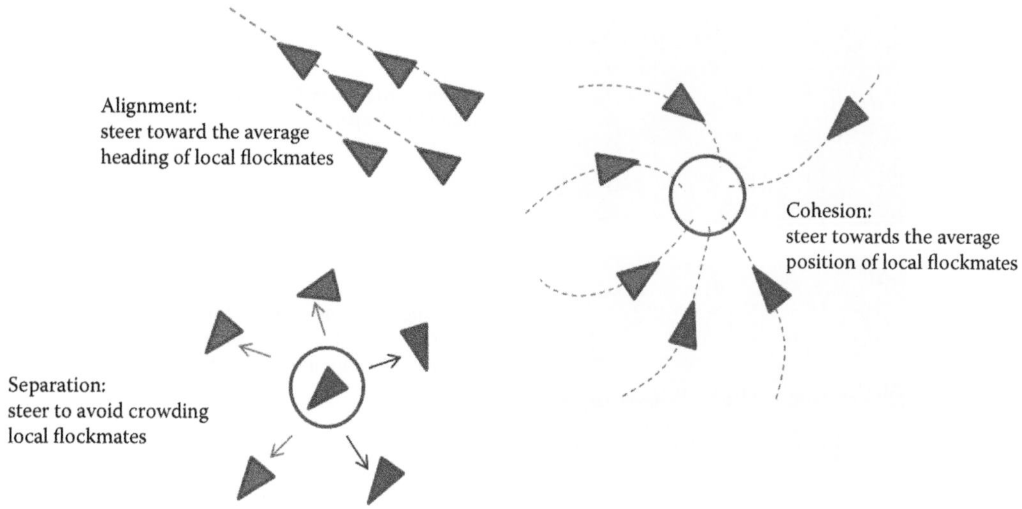

FIGURE 10.15
Flocking system simulation is an interpretation of Reynolds' "Boids".

"cohesion," the flocking simulation illustrates a system where entities make decisions based on their perception of other entities. Figure 10.15 illustrates the three behaviors. "Alignment" means that a bird tends to turn so that it is moving in the same direction that nearby birds are moving. "Separation" means that a bird will turn to avoid another bird that gets too close. "Cohesion" means that a bird will move towards other nearby birds (unless another bird is too close). When two birds are too close, the "separation" rule overrides the other two, which are deactivated until the minimum separation is achieved.

A parameter of interest is the maximum turn a bird can make to either steer toward or away from other birds (Figure 10.16). We call this parameter in the simulation *max-align-turn*. The intensity of constraint decreases as the value of *max-align-turn* increases, i.e., more variety is available to a bird with increasing ability to turn to align itself. For convenience, we state the intensity of constraint as

$$I_C = 1 - \frac{m}{\max(m)} \tag{10.15}$$

where m = *max-align-turn*.

The box plot on the left of Figure 10.17 is the flow metric, CPL, for the flocking system as I_C increases. (CPL for the nearly-random scenario is in column 14.) Here it can be seen that the system becomes more random-like in a transition between columns 10 and 11, corresponding to I_C of 0.75 and 0.83. The plot on the right shows the value of the flow metric over 1000 time-steps at three different values of *max-align-turn*. Here the random-like behavior of the highly-constrained case (bottom plot) is evident.

Similar results are recorded for the complexity metric of global clustering coefficient (GCC). The left graph in Figure 10.18 shows the progression of GCC for the flocking system as I_C increases. Similarly, to the flow metric, we observe random-like behavior at higher constraint.

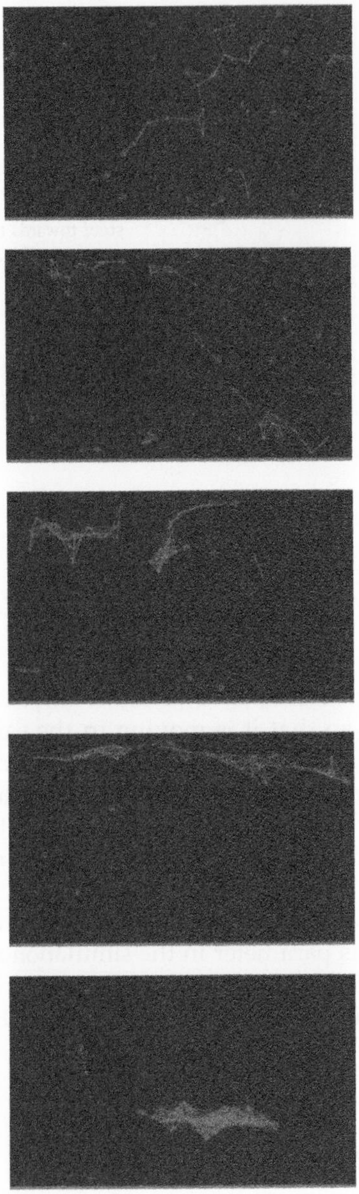
With a *max-align-turn* = 0, many flocks form and break apart.
As *max-align-turn* increases, in this case to 2 degrees, flocks become larger and fewer. Still, flocks are not very stable.
At a *max-align-turn* = *4* degrees, we see still fewer flocks with more birds.
With a *max-align-turn* = *6* degrees, we see the trend toward a single large flock.
In the extreme case of *max-align-turn* = *180*, a tight flock forms quickly as birds rapidly adjust to maintain alignment with their flockmates.

FIGURE 10.16
Effects of increasing *max-align-turn* in the flocking system.

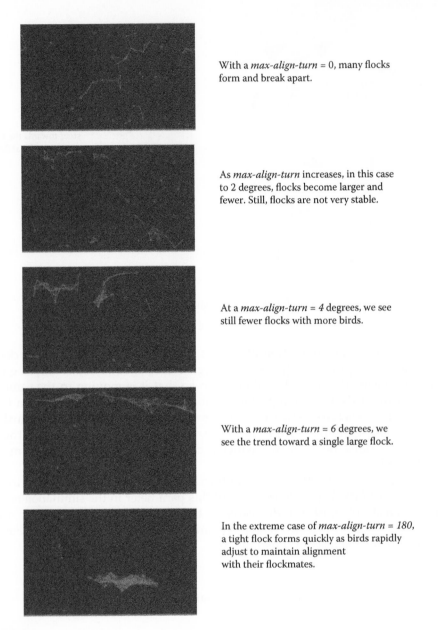

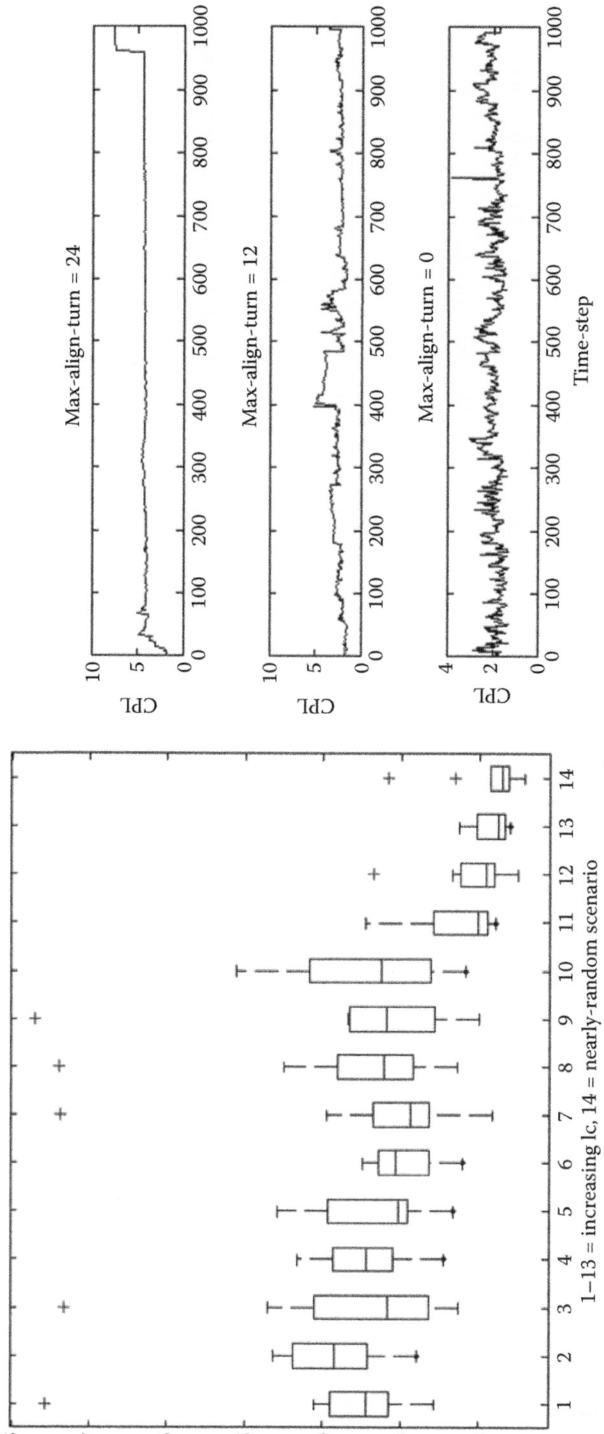

FIGURE 10.17
Comparing cpl for varying constraint in the flocking system.

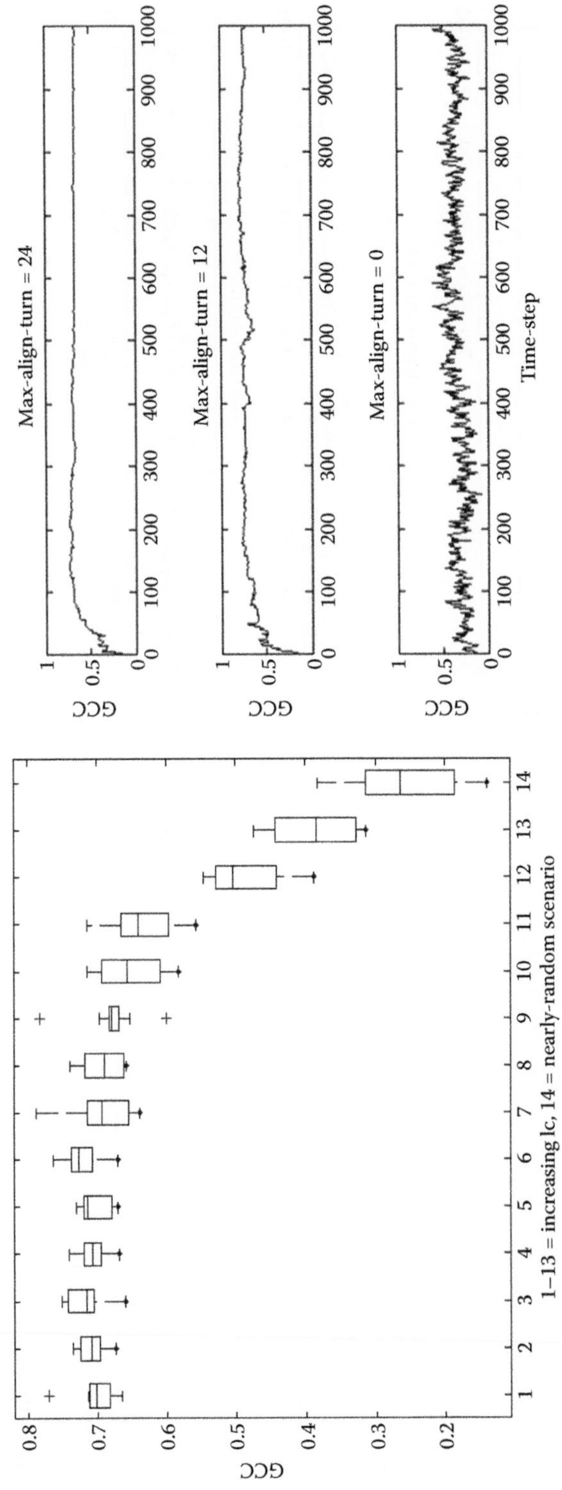

FIGURE 10.18
Comparing complexity for varying constraint in the flocking system.

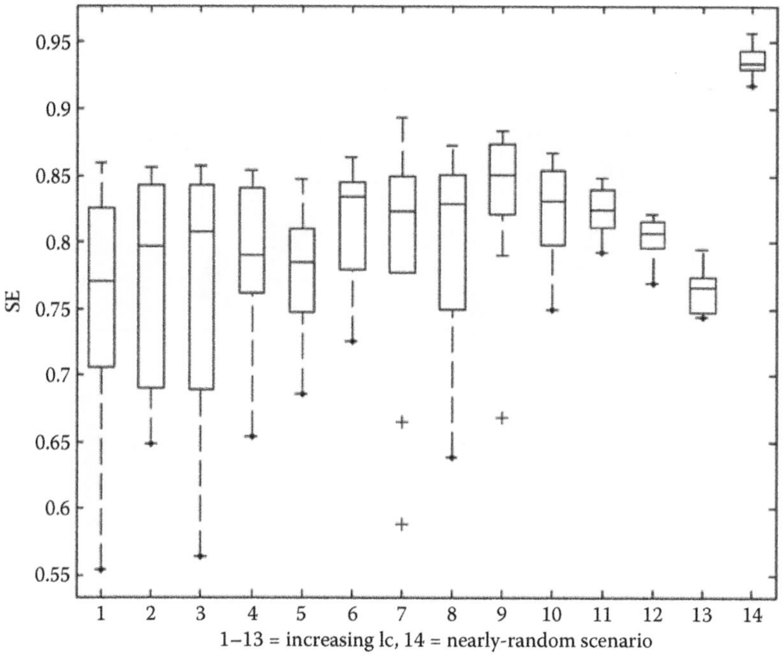

FIGURE 10.19
Comparing entropy (*s*) of the flocking system as *max-align-turn* increases.

Figure 10.19 shows the change in the measure of entropy in the flocking system as constraint increases. Column 14 in the box plot is the maximally constrained system and, as can be seen, this system tends to self-organization until severely constrained.

10.12.3 Stigmergy System

The Stigmergy System simulation illustrates information flow between actors through the environment. It is also an example of collective intelligence, in this case where the ant-hill "knows" where the food is located, but the individual ants do not. Ants search randomly for food until they come upon a chemical trail deposited by ants that were carrying food. A form of stability is achieved when the number of ants is sufficient to sustain a path to the food. This is the popular pheromone following behavior that is central to ant-inspired self-organizing and path following algorithms (Dorigo et al., 1999), coordination of mobile robots (Beckers et al., 1994), and control of unmanned military robotic vehicles (Parunak et al., 2001). Figure 10.20 illustrates the conceptual representation of ants bringing food back to their anthill. The flow of information in this stigmergic system (indicated by the arrows) is from an ant, to an object in the environment (a patch), then to another ant when it traverses that location in the environment. In the approach taken here, ants wander about in search of food. When they find food they pick some up and begin dropping pheromone while they make their way back to the nest. This pheromone tends to spread out (diffuse) and to decrease in strength over time (evaporate). Wandering ants move about randomly searching for food or a trail to follow. When a wandering ant encounters a trail, it attempts to determine the direction that the pheromone is strongest (sniffing). The implication is that there will be an increase in activity and therefore more pheromone deposits closest

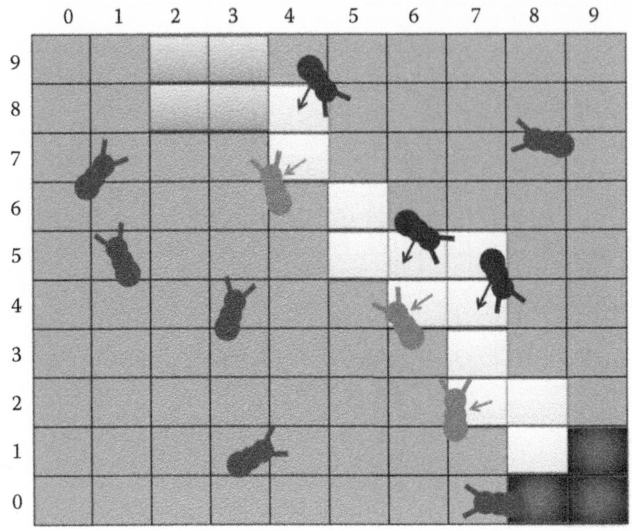

FIGURE 10.20
Simulating a stigmergy system: pheromone following ants.

to the found food supply. Ants that find a trail follow it in the direction of increasing intensity (uphill), which will usually be toward the food. The ants hold a limited supply of pheromone that is replenished each time they pick up food. They can exhaust their supply if the nest is too far from the food. A nest-scent helps the ants find their direction to the nest once they pick up food. Information flow is from a carrying ant to a patch in the environment, then from the environment to a sniffing ant. There is also information flow from patch to patch as the pheromone diffuses.

Information flow is characteristically bidirectional in the particle simulation and directional in the flocking simulation, but in the stigmergy system we see directional information flow associated with one mechanism, namely between the ants and the patches upon which they place pheromone. Bidirectional flow is then associated with the patches. There is no direct communication between the ants in this model (unlike real ants that will communicate through touching of antennae), but instead communication is through the intermediation of the patches with pheromone. Figure 10.21 shows the EBS entity types for this simulation.

The "boids" of the flocking simulation and the ants in the Stigmergic simulation are examples of more sophisticated traditional agents; however, the resulting graph of the stigmergy system is fundamentally different from that of the previous systems. In both of those cases, relationships are primarily between like actors and the structures that result are comprised of those primary actors. Information flow is characteristically bidirectional in the particle simulation and directional in the flocking simulation. In the stigmergy example here, we see directional information flow associated with one mechanism, namely between the ants and the patches, and bidirectional flow associated with the patches. Notice that there is no direct communication between the ants in this model, but instead communication is through the intermediation of the patches with pheromone as shown in Figure 10.22.

Observe that if in our simulation of this system we were to identify each patch as an actor, very large adjacency matrices can result. For example, a context of 100 × 100 patches

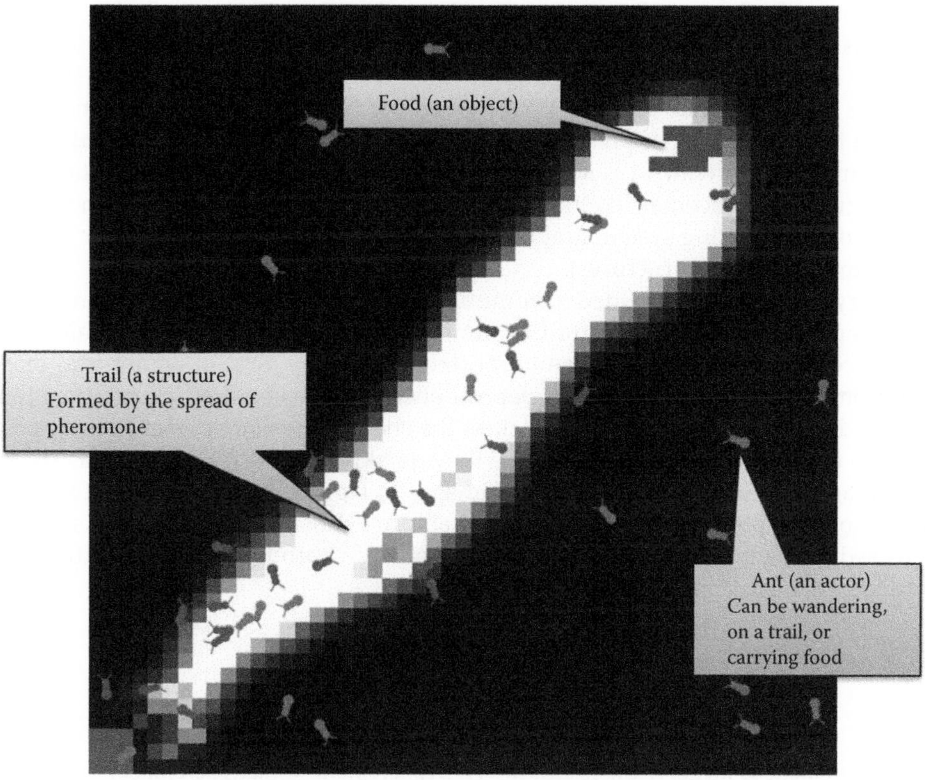

FIGURE 10.21
The stigmergic system in terms of the EBS specification.

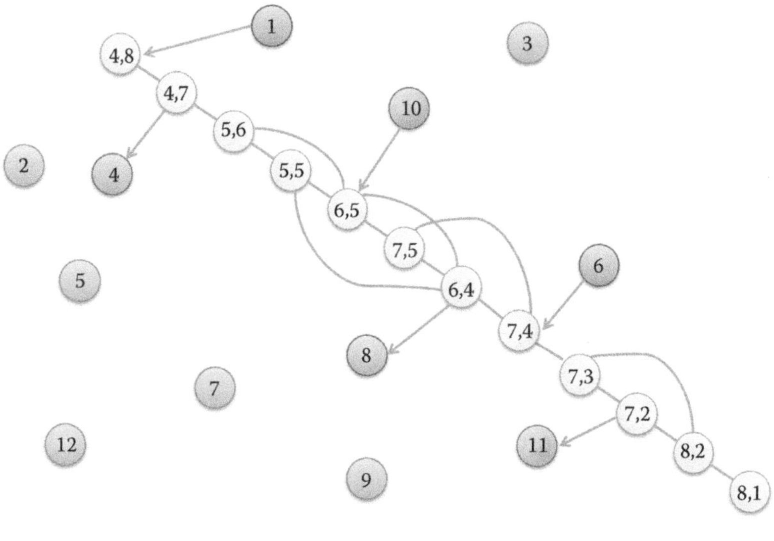

FIGURE 10.22
Information flow in the stigmergic system.

with 100 ants produces an adjacency matrix that is 10,100 × 10,100; computing graph metrics such as those used in this study on matrices of this order is challenging to available computing resources. However, the unique characteristic of the stigmergy system can reduce this computational demand if we remember to apply our EBS specification and so observe that the spread of pheromone in the patches produces structures that we can aggregate into a single entity, i.e., a vertex, or *super-node*. In the graph representation of the system, Figure 10.22 becomes like that of Figure 10.23. This greatly reduces the computational size of the resulting system graph to the number of ants and the number of contiguous pheromone patches (structures).

Stigmergy can be loosely defined as communication through the environment. The simulation examined here is that of trail forming ants and demonstrates the traits of Type 3 emergence. In this case, a spatiotemporal structure in the form of an ant trail emerges rapidly from positive feedback as ants deposit pheromone when they are carrying food. It then is moderated with negative feedback as the chemical trail disperses and evaporates. About 10% of ants are carrying food since they quickly deposit it at the nest as shown in the top image of Figure 10.24. Subsequent images in Figure 10.24 show the effect of varying the amount of pheromone an ant has available to deposit on the patches. As *pheromone* is increased, ants have a greater likelihood of finding a path to the food. The intensity of constraint, I_C for this experiment is simply,

$$I_C = 1 - \frac{pheromone}{\max(pheromone)} \tag{10.16}$$

When *pheromone* is large, the I_C is small. When *pheromone* = *max*(*pheromone*) in the experiment, the I_C is minimized and indicated as 0. The unique nature of this stigmergy system has significant implications for the analysis metrics. Since the information flow is from

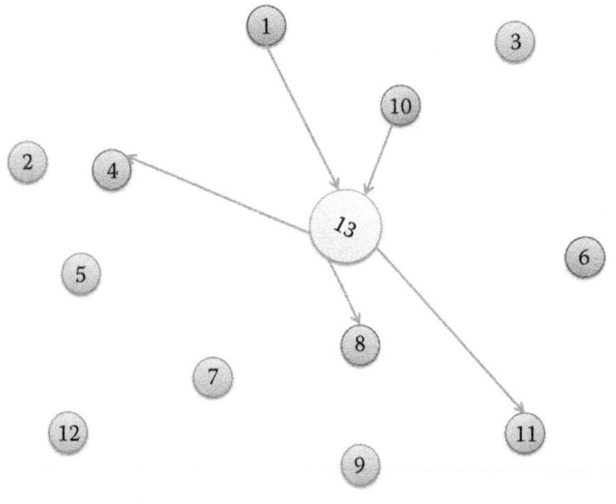

FIGURE 10.23
Super-node representation of the stigmergic system.

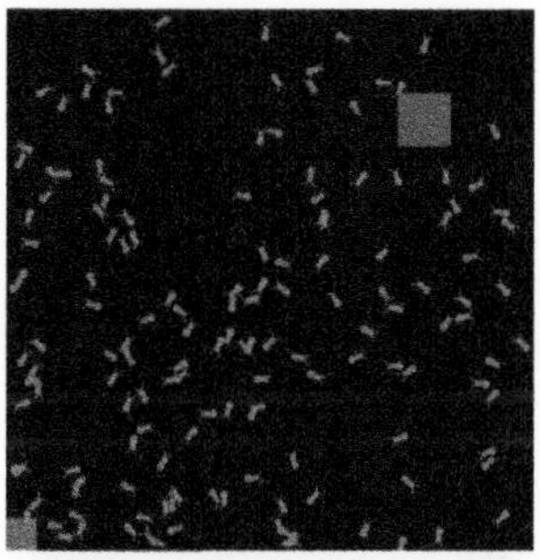

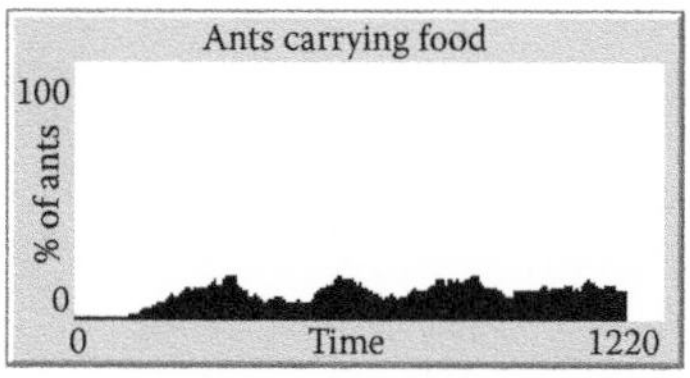

With *use-nest-scent?* on, ants return food to the nest more efficiently, but finding food is still random. Approximately 10% of ants are carrying food.

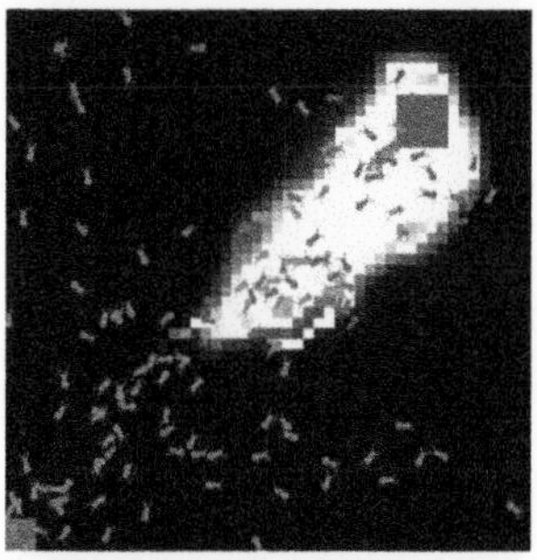

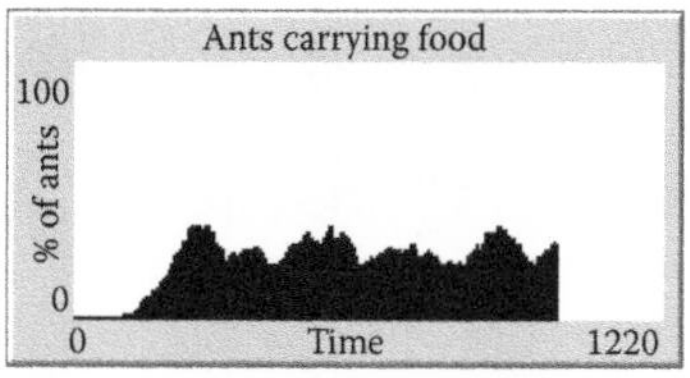

With *pheromone* = 50, approximately 25% of ants are carrying food.

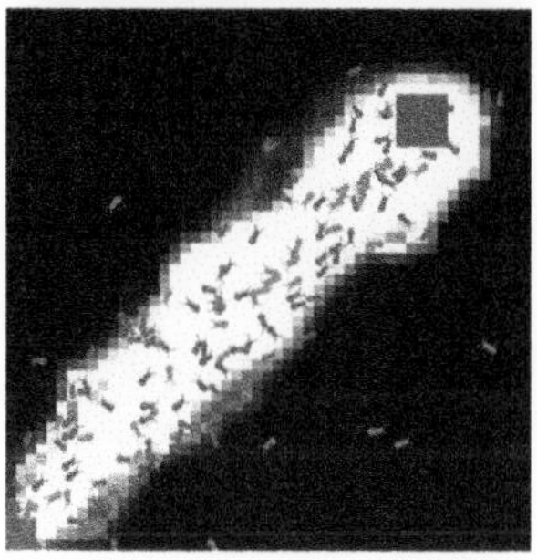

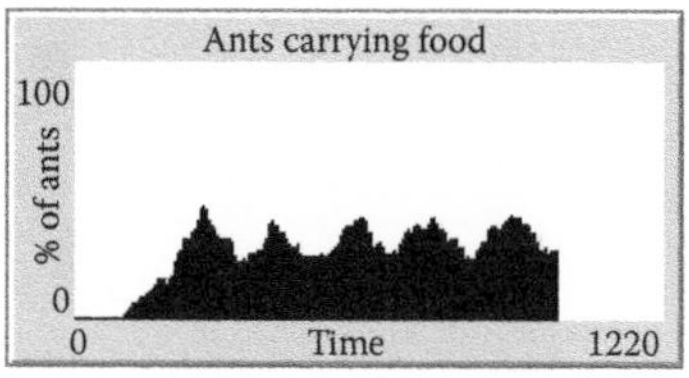

At *pheromone* = 100, approximately 30% of ants are carrying food.

FIGURE 10.24
Effects of increasing pheromone level in the stigmergy system.

the ants carrying food, i.e., those depositing pheromone to the super-node, then from the super-node to those ants who are following the trail, the path lengths between ants will always be either 0 or 2. The characteristic path length for the system will then always be between 0 and 2. Similarly, since all information flow is through the structure represented by the super-node, the global clustering coefficient will always be 0.

The left plot in Figure 10.25 shows how the CPL changes as *pheromone* increases. The first approximate 100 time-steps correspond to the time it takes the ants to move from the nest to the food. The increasing value of *pheromone* supports increasingly larger patch structures. With smaller *pheromone* values, a frequency of increasing and decreasing CPL can be seen as in the top plot on the right. Here, ants rapidly deplete their pheromone before new ants arrive and relationships are quickly established but then broken. Similarly, the effects of increasing constraint are seen in the plot for entropy of the stigmergy system in Figure 10.26.

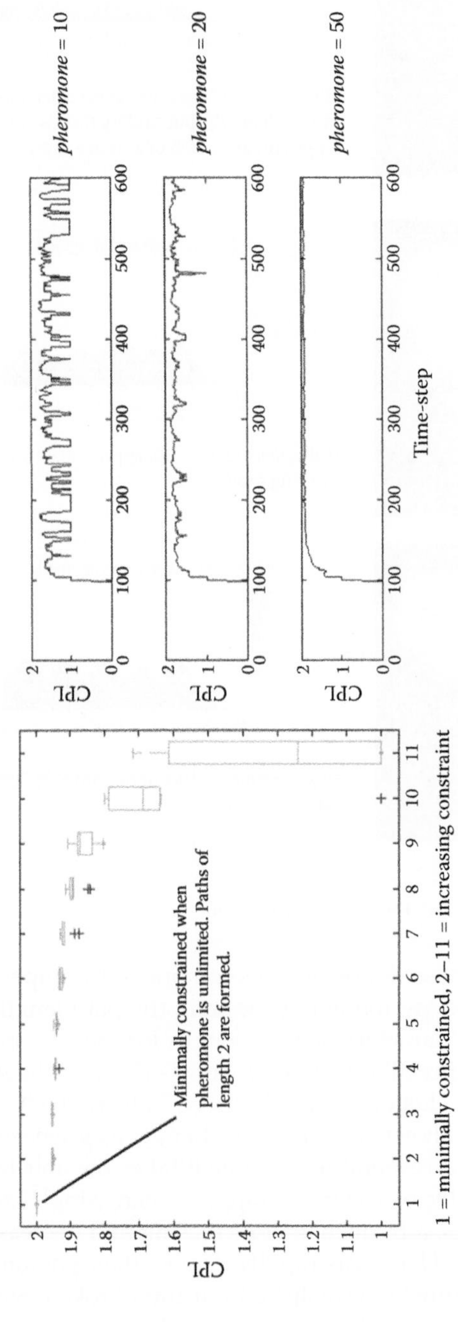

FIGURE 10.25
Flow in the stigmergy system as *Ic* increases.

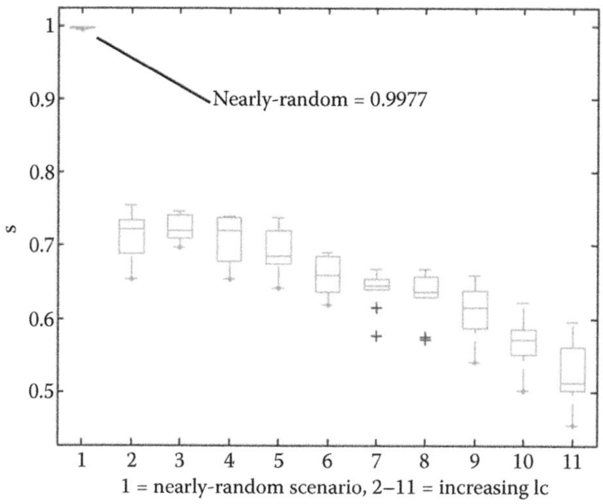

FIGURE 10.26
Entropy of the stigmergy system as *Ic* increases.

10.13 Conclusion

We see some interesting qualities of the systems when we look at the results of the three metrics across them when they are in equilibrium and examine the metrics using a one-way analysis of variance (ANOVA) where the null hypothesis is that all samples are drawn from populations with the same mean and thus indistinguishable. For example in Figure 10.27, we see that there is a good amount of separation between the systems in almost all three of the measures and in the cases where two of the measures are potentially ambiguous, e.g., where S is close between the flocking and stigmergy systems, Ω clearly distinguishes them.

In this chapter, we have considered the long history of emergence and how modern computing methods allow us to explore it. One might say that M&S is the Petri-dish where we can culture emergent phenomena and study its qualities. We have presented some concepts that support a Lexicon for EBS as a modeling specification. We have seen that agent-based methods are useful for simulating EBS and when combined with graph methods can provide a means to detect and possibly predict the onset of emergence. The three systems inspired by nature illustrate fundamental concepts related to the presented Taxonomy of EBS. We believe that these concepts are particularly suitable to developing a means to engineer EBS. Although we yet to have developed a "calculus of emergence," the simulation-based approach does encourage further experimentation in the emergence Petri-dish and hints that we are making strong progress toward the ability to exploit emergence in our engineering of systems.

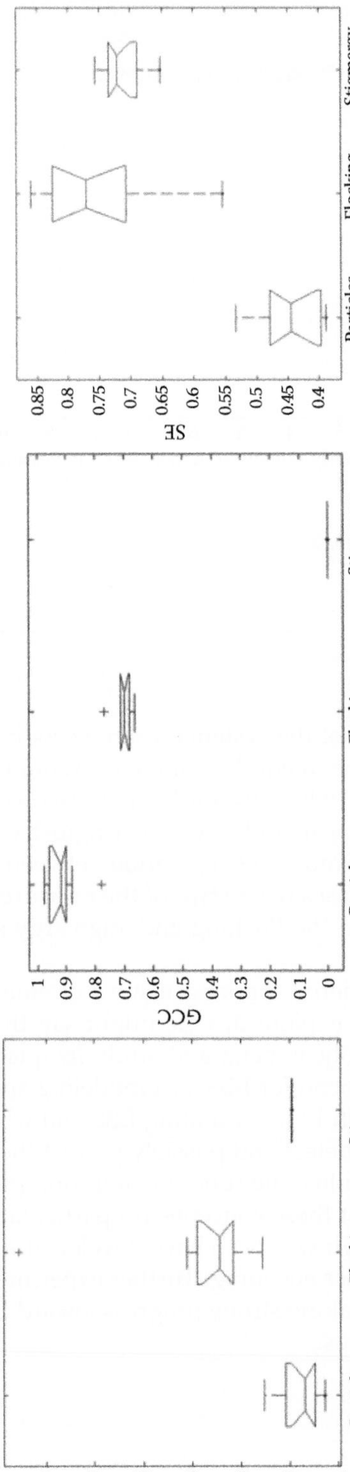

FIGURE 10.27
Comparing the three systems.

References

Ablowitz, R., 1939. The Theory of Emergence. *Philosophy of Science*, Issue 6, pp. 1–16.

Ashby, W. R., 1956. *An Introduction to Cybernetics*. s.l.:Chapman & Hall.

Ashlock, D. & Kim, E.-Y., 2005. *The Impact of Cellular Representations on Finite State Agents for Prisoner's Dilemma*. Washington, s.n.

Baranger, M., 2000. *Chaos, Complexity, and Entropy*. [Online] Available at: http://necsi.org/projects/baranger/cce.html.

Beckers, R., Holland, O. E. & Deneubourg, J. L., 1994. From Local Actions to Global Tasks: Stigmergy and Collective Robotics. *Artificial Life IV*, pp. 181–189.

Bonabeau, E., Dessalles, J.-L. & Grumbach, A., 1995. Characterizing emergent phenomena (2): A conceptual framework. *Revue Internationale de Systemique*, 9(3), pp. 347–371.

Bonabeau, E., Grumbach, A. & Dessalles, J.-L., 1995. Characterizing emergent phenomena (1): A critical review. *Revue Internationale de Systemique*, 9(3), pp. 327–346.

Brooks, R. A., 1991. *Intelligence Without Reason*. Sydney, s.n.

Chiang, C., Hoffman, C. & Mittal, S., 2008. Emergent Crowd Behavior. *Computer-Aided Design and Applications Vol. 6 No. 6.*, pp. 865–875.

Chira, C., Pintea, C. & Dumitrescu, D., 2007. Stigmergic Collaborative Agents. *Journal of Universal Computer Science*, 13(7), pp. 922–932.

Coming, P. A., 2002. The Re-emergence of "Emergence": A Venerable Concept in Search of a Theory. *Complexity*, 7(6), pp. 18–30.

Darley, V., 1994. *Emergent Phenomena and Complexity*. Cambridge, MIT Press, pp. 411–416.

Dorigo, M., Di Caro, G. & Gambardella, L., 1999. Ant Algorithms for Discrete Optimization. *Artificial Life, Vol. 5, No. 2.*

Ferber, J., 1999. *Multi-Agent Systems, An Introduction to Distributed Artificial Intelligence*. s.l.:Addison-Wesley Longman.

Forrester, J., 1961. *Industrial Dynamics*. Cambridge: MIT Press.

Franklin, S. & Graesser, A., 1996. *Is It an Agent or Just a Program?*. s.l., s.n.

Hamdi, M., Sharma, G., Ferreira, A. & Mavroidis, C., 2005. *Prototyping Bio-Nanorobots Using Molecular Dynamic Simulation*. Paris, s.n.

Harel, D., 1987. Statecharts: A Visual Formalism for Complex Systems. *Science of Computer Programming*, pp. 231–274.

Holland, J. H., 1995. *Hidden Order: How Adaptation Builds Complexity*. s.l.:Addison-Wesley Publishing Company.

Holland, J. H., 1998. *From Chaos to Order*. s.l.:Addison-Wesley Publishing Company.

Holland, O. T., 2007. *Taxonomy for the Modeling and Simulation of Emergent Behavior Systems*. s.l., s.n.

Holland, O. T., 2008. *Towards a Lexicon for the Modeling and Simulation of Emergent Behavior Systems*. s.l., s.n., pp. 28–35.

Kim, M. et al., 1999. *Agent-Oriented Software Modeling*. s.l., s.n., pp. 318–325.

Lewes, G. H., 1874. *Problems of Life and Mind*. London: Trubner & Co., Ludgate Hill.

Lorenz, E. N., 1963. Deterministic Nonperiodic Flow. *Journal of Atmospheric Science*, Issue 20, pp. 130–141.

Lyons, D. M. & Arkin, R. C., 2003. *Towards Performance Guarantees for Emergent Behavior*. s.l., s.n.

Morgan, C. L., 1923. *Emergent Evolution*. London: Williams & Norgate.

Newman, M., Strogatz, S. & Watts, J., 2001. Ranom graphs with arbitrary degree distributions and their applications. *Physical Review E 64*.

Oddershede, L. et al., 2002. The Motion of a Single Molecule, the λ-Recepter, in the Bacterial Outer Membrane. *Biophysical Journal, Vol. 83*, pp. 3152–3161.

Parunak, H., Brueckner, S., Sauter, J. & Posdamer, J., 2001. *Mechanisms and Military Applications for Synthetic Pheromones*. Montreal, ERIM, pp. 58–67.

Parunak, H. V. & Brueckner, S., 2001. *Entropy and Self-Organization in Multi-Agent Systems*. Montreal, ACM, pp. 124–130.

Parunak, H. V. D. & Brueckner, S., 2001. *Entropy and Self-Organization in Multi-Agent Systems.* s.l., s.n.
Parunak, H. V. D., Savit, R. & Riolo, R. L., 1998. Agent-based modeling vs. equation-based modeling: A case study and users' guide. In: *Multi-agent systems and agent-based simulation.* Heidelberg: Springer Berlin, pp. 10–25.
Parunak, V. D. H., 1997. "go to the ant": Engineering principles from natural agent systems. *Annals of Operations Research, 75,* pp. 69–101.
Reynolds, C., 1987. *Flocks, Herds, and Schools: A Distributed Behavior Model..* s.l., s.n., pp. 25–34.
Reynolds, C., 1999. *Steering Behaviors for Autonomous Characters.* s.l., s.n., pp. 763–782.
Russell, S. & Norvig, P., 1995. *Artificial Intelligence: A Modern Approach.* s.l.:Prentice-Hall.
Scholl, H. J., 2001. *Agent Based and System Dynamics Modeling: A Call for Cross Study and Joint Research.* Maui, University of Hawaii.
Shannon, C. E., 1948. A Mathematical Theory of Communication. *The Bell System Technical Journal, Vol. 27,* pp. 379–423.
Turing, A. M., 1950. Computing machinery and intelligence. *Mind,* pp. 433–460.
Watts, D. & Strogatz, S., 1998. Collective dynamics of "small-world" networks. *Nature,* pp. 409–10.

11

Characterizing Emergent Behavior in Systems of Systems

Donna H. Rhodes

CONTENTS

11.1 Introduction

The often repeated adage "the whole is more than the sum of the parts" captures the essential nature of emergence. In the case of a system such as an aircraft, the individual component parts are relatively stable over time and their interrelationships are reasonably understood. As a result, engineers have some ability, relative to the complexity of the system, to understand, predict, and design for emergent behavior. Characterizing emergence behavior for system of systems is more challenging.

A system of systems is comprised of constituent systems, and each of the constituents has its own unique emergent behavior at the system level. The system of systems emergent behavior is more than the sum of the constituent system emergent behaviors. As stated by Maier (2005), system of systems emergent behaviors "are the emergent properties of the entire SoS and not the behavior of any component system."

Silva et al. (2014) describe systems of systems as encompassing an emergent behavior "corresponding to functionalities yielded by the cooperation between the constituent systems, which must have the ability of connecting to each other, i.e., connectivity." Over the lifespan of a system, positive emergent behavior becomes increasingly understood and managed, as long as its architecture and connectivity are fairly stable. Potential for negative emergent behavior is often uncovered and mitigated before causing significant harm. The system of systems differs from a traditional system in that there is periodic re-architecting at various points throughout its operational life, leading to emergent behaviors triggered by these architectural changes.

In characterizing emergent behavior in systems of systems, it is first useful to consider the nature of complexity and how this may trigger emergent behavior. Considering emergence from five viewpoints (structural, operational, contextual, temporal, and perceptual) helps to shed light on the various aspects of emergent behavior in systems of systems. System of systems re-architecting is highlighted as a unique situation that accentuates emergence in systems of systems.

11.2 Nature of Complexity

Myriad aspects of complexity in a system of systems give rise to emergent behaviors, both positive and negative, including:

- nature of the constituent interconnections (e.g., degree of coupling)
- interdependencies of constituent capabilities and functions
- contention between constituent's local and global mission objectives
- social complexities arising from the multiplicity of stakeholder interactions
- external forces impacting actual and perceived constituent value delivery
- external agent influence or attempted influence on the system of systems
- internal agent influence or attempted influence from within the system of systems

The degree of coupling between the constituent interconnections has impact on emergence. For example, loosely coupled constituents in the system of systems enable more degrees for freedom. According to Meentemeyer et al. (2009), emergence in a system of systems "conveys a behaviour or capability coming forth in the system of systems that may or may not have been foreseen." They assert that emergence is the result of the system of systems having the ability to make independent choices, given that the system of systems has autonomous constituent systems that are permitted to make decisions.

Similarly, emergence may result from the interdependencies of required capabilities and functions in the system of systems. Interdependencies enable system of systems performance, but also are a potential source of cascading failures. If an external agent exerts negative influence on a required function of a particular constituent system, there may be cascading failures across the constituents leading to unanticipated negative emergence at the system of systems level. Emergence is often linked to the interaction of multiple autonomous agents: "an emergent phenomenon or behavior is defined as a novel phenomenon that was not explicitly encoded in the system but emerges as a result of concurrent interactions between multiple agents within the system" (Mittal, Doyle, & Portrey, 2015).

A system of systems, by virtue of being comprised of constituents with operational and managerial independence, often deals with issues of contention. The local objectives of a constituent can conflict with satisfying global mission objectives. A negative emergent behavior such as degraded performance at the system of systems level could be triggered by a constituent system's optimizing its independent operational mission. For example, in a collaborative transportation system of systems, a constituent may take an action to improve its own performance (bus company reduces number of stops on a bus route to achieve faster route time) that has a negative emergent impact at the global level (increasing taxi use back and forth along bus routes reduces taxi availability at the airport leading to stakeholder dissatisfaction with regional transportation services).

The nature of complexity in the system of systems is that these various aspects not only exist, but are entangled. This is precisely why understanding emergent behavior can exceed the cognitive limits of the human decision maker.

11.3 Viewpoints for Understanding Emergent Behavior

Simplifying the consideration of emergent behavior through a reductionist approach is not an effective strategy. A more holistic approach is necessary, such as investigating emergent behaviors through viewpoints. Five useful viewpoints for characterizing emergent behavior are structural, operational, contextual, temporal, and perceptual, each briefly discussed below.

11.3.1 Structural Viewpoint

The structural aspect is often used in discussion of system of systems emergent behavior. From this perspective, some emergent behavior results from the set of constituent systems that comprise the systems of systems, and the means by which they are interconnected. Systems of systems are usually comprised of a combination of legacy constituent systems and new constituent systems, interconnected in some way such as a network (which itself can be viewed as a constituent). The examination of structure includes both physical and logical components, and their interrelationships.

Saunders et al. (2005) refer to the "connecting the parts" perspective of systems of systems capability development, which recognizes that the system of systems is "built from a collection of independently acquired and operating systems that must now be connected together," and there must be specific mechanisms that allow the component systems to come together. According to these authors, interoperability standards are the primary mechanism to enable effective connection of the parts, enabled by an environment that fosters standardization, asserting consistent protocol behavior is the property that enables interoperability.

While the practical aspects of taking a structural perspective are useful, it should be noted that ever more complex systems of systems call for re-examining the fundamentals. Sauser et al. (2010) assert that a distinguishing feature of a system of systems is its internal connectivity, which is "is not presciently designed but emerges as a property of present interactions among holons." They propose connectivity "has to do with a lot more than just topologies and protocols and interoperability standards, although it does address these practical matters, and is more concerned with the agility of structures for essential connectivity…." These authors make a cogent case for holism over reductionism.

11.3.2 Operational Viewpoint

The operational viewpoint considers designed and emergent behavior resulting from the interconnected constituents, somewhat independent of the specific physical constituents. This is sometimes referred to as a mission-based approach.

Silva et al. (2015) describe a mission-based approach as involving "the definition of the missions, the capabilities of the constituent systems required to accomplish such missions, and the interactions among these systems." The authors believe that given the fundamental role played by missions, a mission-based approach could support describing a system of systems even when information related to implementation is not available.

According to Silva et al. (2014), the representation of emergent behavior is not trivial because it relates to the interactions among the constituent systems, and a "natural solution would be to represent system parts and to describe their interactions and emergent behaviors," noting this is better applied to foreseen emergent behaviors. The authors argue the solution would be to represent the emergent behavior through interaction patterns, which can be understood as a potential contract between constituent systems. The desired emergent behavior, then, could be implemented with the agreement of such a contract. The authors assert it would then be possible to "separate the mission description from system models, thus allowing the description of emergent behaviors with no further knowledge of the constituent systems."

11.3.3 Contextual Viewpoint

Shah et al. (2007) discuss the contextual perspective, stating the context of a system of systems is "a complex entity that results from the interaction between the system of systems constituents and may not be readily understood by looking at those constituents." From an analytic standpoint, context is defined as the external entities and conditions that need to be taken into account in order to understand system behavior. From the architectural standpoint, context is the mold to which the architecture must conform, as a result of the objectives and constraints that the architect is attempting to satisfy. The authors give the example of system context for a communication satellite as consisting of both physical laws (e.g., space environment and the realities of signal transmission physics) and requirements imposed by stakeholders (e.g., requirements on communication capability and regulatory constraints on spectrum usage). Additionally, the acquisition environment (schedule, budget, resources) may come into play. Within the systems of systems, there are constituent system contexts and their interactions. The system of systems level context has the unique requirement for system of systems behavior, and likely other aspects resulting from context interactions and overlaps.

System of systems emergent behavior can arise from constituent systems that operate under different contexts, for example, constituents in a defense system that are operating in land-based, sea-based, and air-based contexts. Accomplishing the system of systems mission requires constituent systems to share contextual information with each other in a timely manner, such that components have the same system awareness at any given time. In order for that to happen, three things must occur: (1) the important differences in context must be apparent; (2) stakeholders must be willing to share this information (not always the case given policies and situations); and (3) mechanisms must exist for this information to be shared in a timely manner. For these reasons, constituents within systems of systems that operate under different contexts may be operating under incorrect or incomplete information about the overall system itself (Mekdeci et al., 2015).

The system of systems operational environment changes over time, and as a result may need to operate in contexts for which it was not originally intended to operate. Changing contexts may trigger emergent behavior in various ways, including in technical performance of the system constituents and agents, ability to perform the intended mission, and stakeholder perceived value of the system of systems.

11.3.4 Temporal Viewpoint

Most systems of systems have a long lifespan, and continuously evolve over long periods. As such, a temporal perspective is very important in considering emergent behavior. Rhodes and Ross (2010) point out that a system exists in a dynamic world, with potentially many shifts in context and needs during the system lifetime that can be decoupled from the acquisition phases. They assert that a temporal viewpoint is necessary to "characterize changes over time, as well as time-based properties such as survivability or adaptability of the system over its lifespan." Systems of systems "ilities" such as evolvability, resilience, and survivability are positive emergent behaviors that necessitate a temporal perspective. Survivability, for example, concerns continued value delivery to stakeholders over time in the face of disruptions and disturbances.

In special cases, a system of systems might be rapidly constructed from constituents for a special purpose. For example, a natural disaster may result in drawing together existing independent systems into a collaborative system of systems, intended to exist for a sole purpose and defined period of time. The positive emergent behavior ceases to exist when the system of systems is dissolved, either when disaster recovery is completed or due to another factor such as loss of funding.

11.3.5 Perceptual Viewpoint

Weinberg (1975) describes emergence as "a relationship between system and observer," where properties emerge for a particular observer who could not or did not predict it. He states, "[W]e can always find cases where a property will be 'emergent' for a particular observer and 'predictable' for another."

The system/observer relationship drives stakeholder perception of the system of systems value delivery, and accordingly stakeholder behavior. For example, during peacetime a border protection system of systems may be perceived by a stakeholder as intrusive to her personal privacy. In an increased threat environment, the stakeholder's perception is likely to shift to a higher level of perceived value of surveillance. Accordingly, stakeholder public pressure could result in reduction of a level of surveillance in peacetime, potentially introducing vulnerabilities that may lead to negative emergent behavior. While local law enforcement might find this somewhat predictable, the general public might not have anticipated this emergence given their limited understanding of the dynamics.

The perceptual viewpoint also relates to the changing preferences of systems of systems stakeholders in response to context shifts over time as these stakeholders interact with the system in its environment (Rhodes & Ross, 2010). Accordingly, cognitive limitations, biases, and preferences of the stakeholders are considerations in understanding emergent behavior in systems of systems.

11.4 Re-Architecting Impact

Traditional engineering of a system involves systems architecting early in the lifecycle. A distinguishing trait of systems of systems is that they undergo periodic re-architecting throughout the operational lifespan. Ricci et al. (2013) state that "evolving an SoS from a current architecture to a future one (to meet emerging and anticipated needs, often reflecting changes in operational contexts) requires coordination and agreement among the constituents." The constituent system decision makers need to be empowered to make local decisions without constantly consulting all other constituents, and while having the best intentions of avoiding negative impacts on others, sometimes negative impacts can occur.

The architect's intent in a re-architecting effort is to influence the behavior of the system of systems as a whole in accordance with a particular desire. While not exhaustive, six types of architect's intent include: (1) add a new constituent system to the system of systems; (2) add more of an existing type of constituent system to the systems of systems; (3) replace or upgrade capabilities of an existing constituent system for systems of systems level benefit; (4) add more of an existing function to the system of systems; (5) change the way a system of systems function is performed; and (6) relocate resources or capabilities within the system of systems. Given that this re-architecting results in system of systems level impact, there can be a variety of emergent behaviors that result, both the desired positive emergence and unintended negative consequences (Ricci et al. 2014).

While expected positive emergence may result from an architectural change, the inherent complexities of the system of systems can potentially induce unexpected (strong) emergence, especially given limitations to fully test or simulate the change in advance. A further complication is that there may be numerous changes ongoing within the system of systems at any one point in time, which are not always coordinated. This situation is more likely in collaborative and acknowledged systems of systems. Yet, even in the case of the directed system of systems, changes within a single constituent system aimed at optimizing local performance or benefits may result in unforeseen impacts to the global system of systems value delivery.

11.5 Summary

Emergent behavior is a necessary property for an assemblage of systems to be considered a system of systems (Boardman & Sauser, 2006; Keating & Katina, 2011; Maier, 2015). Given that emergent behavior is not expected to diminish over the lifespan of the system of systems, there is a need for continuous monitoring of potential opportunities for fostering positive emergence and for mitigating situations that may give rise to negative emergence. Additionally, periodic re-architecting activity provides the opportunity for investigating potential emergent behavior in the context of changes to be made. Given the multi-faceted challenges of understanding system of systems emergent behavior, modeling and simulation offers an effective means for enabling responsive and proactive decision making through analytic capability.

References

Boardman, J., & Sauser, B. System of Systems—The meaning of "of." In Proceedings of the 2006 IEEE/SMC International Conference on System of Systems Engineering Los Angeles, CA, 2006, pp. 118–123.

Keating, C.B., & Katina, P.F. (2011). Systems of Systems Engineering: Prospects and Challenges for the Emerging Field. International Journal of System of Systems Engineering, 2(2), pp. 234–256.

Maier, M.W., "Research Challenges for Systems-of-Systems," 2005 IEEE International Conference on Systems, Man and Cybernetics, 2005, pp. 3149–3154.

Maier, M.W. (2015). "The Role of Modeling and Simulation in System of Systems Development," in Rainey, L. and Tolk, A. (Eds.), Modeling and Simulation Support for System of Systems Engineering Applications, Hoboken, NJ: John Wiley & Sons, pp. 11–41.

Meentemeyer, S.M., Sauser, B., & Boardman, J. (2009). Analysing a System of Systems Characterisation to Define System of Systems Engineering Practices. International Journal of System of Systems Engineering, 1(3), pp. 329–346.

Mekdeci, B., Ross, A.M., Rhodes, D.H., & Hastings, D.E., "Pliability and Viable Systems: Maintaining Value Under Changing Conditions," IEEE Systems Journal, Volume 9, Issue 4, December 2015, pp. 1173–1184.

Mittal, S., Doyle, M.J., & Portrey, A.M. (2015). Human in the Loop in System of Systems (SoS) Modeling and Simulation: Applications to Live, Virtual and Constructive (LVC) Distributed Mission Operations (DMO) Training. In Modeling and Simulation Support for System of Systems Engineering Applications. John Wiley & Sons, pp. 415–451.

Ricci, N., Fitzgerald, M.E., Ross, A.M., & Rhodes, D.H., "Architecting Systems of Systems with Ilities: An Overview of the SAI Method," 12th Conference on Systems Engineering Research, Redondo Beach, CA, March 2014.

Ricci, N., Ross, A.M., Rhodes, D.H., & Fitzgerald, M.E., "Considering Alternative Strategies for Value Sustainment in Systems-of-Systems," 7th Annual IEEE Systems Conference, Orlando, FL, April 2013.

Rhodes, D.H., & Ross, A.M., "Five Aspects of Engineering Complex Systems: Emerging Constructs and Methods," 4th Annual IEEE Systems Conference, San Diego, CA, April 2010.

Saunders, T. et al., System-of-Systems Engineering for Air Force Capability Development: Executive Summary and Annotated Brief, US Air Force Scientific Advisory Board, SAB-TR-05-04, 2005.

Sauser, B., Boardman, J., & Verma, D., Systomics: Toward a Biology of System of Systems, IEEE Transactions on Systems, Man, and Cybernetics—Part A: Systems and Humans, Volume 40, Issue 4, July 2010, pp. 803–814.

Shah, N.B., Hastings, D.E., & Rhodes, D.H., "Systems of Systems and Emergent System Context," 5th Conference on Systems Engineering Research 2007, Hoboken, NJ, March 2007.

Silva, E., Batista, T., & Cavalcante, E., "A Mission-Oriented Tool for System-Of-Systems Modeling," Proceedings of the Third International Workshop on Software Engineering for Systems-of-Systems, San Antonio, TX, 2015, pp. 31–36.

Silva, E., Cavalcante, E., Batista, T., Oquendo, F., Delicato, F., & Pires, P., "On the Characterization of Missions of Systems-of-Systems," Proceedings of the 2014 European Conference on Software Architecture Workshops, Article No. 26, Vienna, Austria, August 25–29, 2014.

Weinberg, G.M. (1975). An Introduction to General Systems Theory, New York: John Wiley & Sons, p. 60.

12

Engineered to Be Secure

Carol C. Woody

CONTENTS

12.1 Software in a System of Systems

Maier identifies five characteristics which set systems of systems apart from complex monolithic systems as follows (Maier 1998): (1) operational independence of their components, (2) managerial independence of the components, (3) their evolutionary nature, (4) emergent behaviors, and (5) a geographic extent that limits the interaction of their components to information exchange. Software has taken over an increasing portion of system functionality, as described in the Critical Code Report of the National Academies of Science (CSTB 2010). In addition, systems built to function independently are increasingly interconnected through software to share data and automate complex organizational and multi-organizational needs (also known as business processes or operational missions). Maier's system of systems characteristics were not initially defined to work with software-intensive systems, so there are differences that need to be taken into account, and these differences impact how systems need to be built and supported to operate securely in a systems of systems environment.

Software no longer lives within a component of a system. It is now part of each component, controlling the communication among the components within a system, handling many of the execution functions previously performed by infrastructure hardware, and controlling the linkages with other systems in the systems of systems. Each of these many types of software requires specialists to write and support. Because of the languages and tools now used to write software, only highly skilled developers, of which there are few, can produce the quality needed to avoid security vulnerabilities, so most of the software acquired and implemented today is average quality, containing many vulnerabilities.

Instead of engineering systems for a specific purpose, organizations now acquire software and link it together to address desired operational capabilities. Systems of systems are increasingly composed of multi-vendor sourced data-driven software products with known defects contributing to unexpected operational behaviors. These products are used by an acquirer but the vendor continues to control the product updates and enhancements. Table 12.1 summarizes the relationship of Maier's five characteristics to primarily software-driven systems and the challenges that those acquiring these software dominated products needed to address in order for the system to be engineered to be secure.

Because each system is *operationally independent*, engineering for security now requires consideration of each system within the system of systems context to determine how it contributes to a mission and how an attack could impact mission success. Attacks may come from any connected software and vulnerabilities that would endanger the mission; even if they are in a seemingly unimportant system, they can become high priority. A Target store point-of-sales system was attacked through a connection from the software providing remote control of the heating and cooling (Fatima 2014).

Because each software component is *managed independently*, and in most cases managed by external sources, acquirers must be aware of just how widely the software they rely on is deployed. Wider use provides a greater pool of potential attack targets. The Shellshock vulnerability in the widely used UNIX Bash shell was exploited within hours of discovery (Greenberg 2014).

Software is implemented, on average, with 600 to 1,000 defects per million lines of code (MLOC), and many of these could represent vulnerabilities. A vulnerability is a weakness that allows an attacker to bypass security controls, and over 90,000 have been identified and documented on the Common Vulnerabilities and Exposures since its initiation in the 1990s. *Evolutionary development* in the software context describes how the attackers are continually expanding their capabilities through the use of systems engineering tools and access to software and updates to exploit an ever growing volume of defects.

The challenge of *geographic distribution* augments the effort that acquirers must expend in determining where software is actually located so that vulnerabilities can be addressed.

TABLE 12.1

Maier's System of Systems (SoS) Characteristics for Software Engineered to Be Secure

SoS Characteristic (Maier 1998)	Engineering Software to Be Secure
Operational independence	Acquirers must identify and mitigate vulnerabilities in so much software that they have to prioritize their focus on software performing mission-critical functions.
Managerial independence	Acquirers select market dominants (build costs are more widely distributed, more resources for support, more functionality growth) so attacks on discovered vulnerabilities provide many targets.
Evolutionary development	Acquirers must patch critical software quickly to reduce their attack potential; attackers can reverse engineer patch code to quickly field exploits.
Emergent behavior	Acquirers must impose and monitor quality and security related requirements in their vendor contracts and ensure vendors manage their software supply chains effectively (increased costs and increased oversight).
Geographic distribution	Acquirer must impose secure-by-default requirements.

Expanding capabilities in distributed environments allows vulnerable software to creep into back up locations, cache, and other intermediate storage areas that delay the removal of vulnerabilities and provide additional sources of attack.

The security of software depends on the control applied to vulnerabilities and potential connectivity that would allow an attacker to reach and exploit a vulnerability. To be secure, engineering software requires a recognition that the security of a system is really the result of the *emergent behavior* of the system within the system of systems. Software controls the information exchange among the operationally independent systems, and software also handles the majority of the functionality within each system. To be secure, all of this software must be engineered to minimize the ability of an attacker to exploit potential vulnerabilities anywhere within the system of systems.

As software usage increases, coordinated attacks that leverage the interactions of software-driven systems at the system of systems level are becoming more common:

- Airline failures have increased in recent months as IT systems expand to accomplish far more than ever before, from reservations to merchandising to flight operations to customer service. There is much finger pointing as to whether the issues are design related, volume related, or security related; it is likely all of these issues are involved. The San Francisco crash of a Boeing 777 in March 2014 was traced to a design issue in the air-throttle—Boeing was notified in 2010 of a similar difficulty in a test flight and chose to put a note in the pilot manual instead of correcting the problem. All of these systems and components communicate with each other, and the more complex the interactions the more prone to failure it becomes (Wald 2014). Polish airline LOT says that the computer problems that grounded its flights over a weekend in 2015 were due to a distributed denial-of-service (DDoS) attack that prevented flight plans from being delivered to another system on time (Martin 2015).
- Swift, a financial messaging system used by banks to handle international transfers of money, has been used to steal over $81 million from the central bank of Bangladesh. The thieves were able to change the destination of the money from the Federal Reserve Bank of New York to accounts in the Philippines (Corkery 2016).
- Equifax failed to apply a patch for a bug that was documented in an open source product, Apache Strut. Two months later Equifax announced its failure to protect the personal information of 143,000 Americans sent to them by banks and organizations accepting credit card payments, which was stored in its organizational repositories accessible through the Apache Strut (Swamynathan 2017).

12.2 Transitioning Security Perspective from Secure Isolated Systems to Security of Highly Networked Software-Intensive Systems of Systems

Today software is rarely specifically written for the use to which it is applied. Software components in the system of systems are likely to be composed from commercial products, open source, and "glue code" that ties these various pre-existing pieces together to provide the desired functionality. Each of these components can also be composed of multiple pieces drawn from code libraries, automated code frameworks, and reused functions,

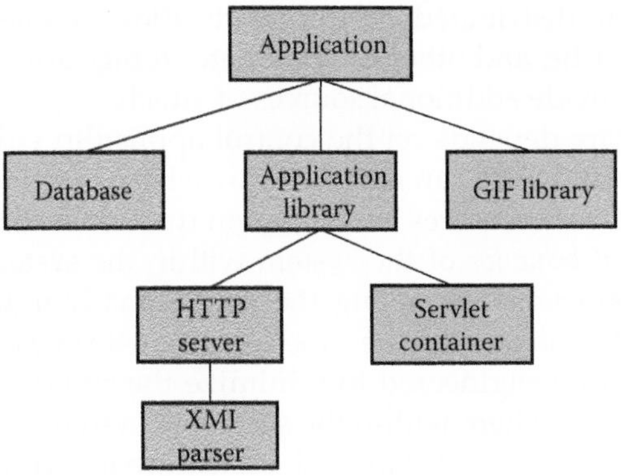

FIGURE 12.1
Notional software application component structure.

as shown in Figure 12.1. The supply chains can be long with many diverse sources. As software is assembled and integrated into a final product, each integrator inherits all of the complexity inserted by the developers at every sub-level, which can include additional custom, commercial, open source, and legacy software.

The software supply chain comes with an additional complication. Depending on the specific business agreement, an acquirer may have a very limited ability to verify the quality or integrity of the supplied software. A commercial software provider allows use of its product through a licensing arrangement, but does not give away the source code. Open source software is frequently also controlled through licensing but with a very different relationship structure—access to the source code is provided and copies of the code can become different products using the original baseline.

In a typical development effort, each separate system is engineered to deliver a specified level of functionality. Systems are subsequently interconnected to share data and support complex organizational and multi-organizational needs. The growing need for interconnected and often global operations in today's world means that an operational mission or critical work process is usually supported by multiple, independently developed systems that may also be globally distributed.

Unlike in the predominantly hardware driven systems built in the 1980s and 90s, the software driven components can be quickly assembled and quickly changed to meet rapidly changing and expanding needs. Software is also taking over the execution of functionality in the computing infrastructure as follows:

- Servers: Virtual CPUs
- Storage: SANs
- Switches: Soft switches
- Networks: Software defined networks
- Communications: Software defined radios
- Chips: Firmware

However, each piece of this software is riddled with weaknesses that represent potentially unacceptable security risks (McGraw 2006, p. 3). Even if the product developer had access to all of the code, there is insufficient time and resources available to deliver a vulnerability free product. As a result of the security of these interconnected systems, which were not designed to meet the same needs across all connected parts, the system of systems provides unexpected access to data as well as operational instability.

As noted in a recent report issued by Gartner (Driver 2017):

> By 2020, 50% of organizations will have suffered damage caused by failing to manage trust in their, or their partners', software development lifecycles (SDLC)—causing revenue loss of more than 15%. Application leaders responsible for modernizing application development should re-evaluate the SDLC in the form of a trusted software supply chain, with varied levels of trust.

The security problems we see reported daily are the direct result of how organizations acquire, integrate, and manage the use of software within their organizations, and of the trust relationships they establish, many of which are controlled by software, and which in many cases they fail to monitor internally and with partner organizations. In the rush to capitalize on the lower cost and faster delivery capabilities provided by software, high risk is accepted/ignored either through ignorance or failure to commit needed resources to identify and establish needed mitigations.

Systems have been successfully designed to operate in isolation with effective security for decades. Systems engineers incorporated security controls based on known needs for confidentiality, integrity, and availability. These were primarily centered on authentication and authorization (Saltzer and Schrader 1974). In 1974, when this approach to security was developed, the base for system execution was a System 360, a centralized mainframe with limited options for connectivity. Then much of the functionality was handled by hardware which provided a physical separation of the various components control barriers.

Increased options for connectivity among systems and data sharing, along with increased capabilities for use of software which was quickly developed and deployed instead of hardware which was time consuming and costly to build, provided a means of shifting increased amounts of functionality into systems that could capture, process, and share information.

These interdependent systems were not typically designed to address the ways in which they are expected to function today. The requirements were assembled independently, and each was built and fielded independently. Standard communication protocols such as File Transfer Protocol (FTP) and Transmission Control Protocol/Internet Protocol (TCP/IP) were adopted to move data from one system to another. Data encryption was an option in these communications but not a requirement. Variations in choices for security controls and differences in implementation options have provided opportunities for various type of security failures. The likelihood of failure increases when assumptions and decisions within one system are inconsistent with those of another. Good security for each component system does not guarantee good security for the composition of a system of systems.

Establishing effective security for a system of systems requires an evaluation of each system within the context of the composition:

- establishing the role of each system in system of systems success
- analyzing the impact for each potential system failure on the system of systems
- identifying critical system of systems dependencies and trust relationships
- considering threats to success for the system of systems and the role of each system in addressing them, such as operating conditions or actions triggered by threat agents

A key element to effectiveness is consideration of the actual operational usage and the opportunities for attackers to bypass security controls or exploit trust relationships. These highly interconnected operational environments allow linked systems to be used as staging platforms to access other systems. As an example, consider the following steps used to perform the data theft at Target in December 2013 (Fatima 2014):

- Heating and cooling service (HVAC) vendor's technology infrastructure is compromised through social engineering with a malicious email.
- Target store network access is achieved through HVAC remote access trusted connection.
- Malware injects itself into the running point-of-sale processes to identify credit card data in the computer memory and copy it prior to encryption, taking 11 gigabytes (GB) of data including credit card numbers and other customer data.
- Stolen data is transmitted to a File Transfer Protocol (FTP) server belonging to a hijacked website.
- Criminals then access and download the data files from the FTP server.

This type of attack was repeated at many other retail and entertainment operations to steal data from the point-of-sales systems. All used similar implementations of trust relationships and point-of-sales software with the same vulnerabilities.

Software is a key enabler providing connections among systems in multiple organizations (e.g., vendor providing remote heating and cooling monitoring at a retail store) and across organizational functions (e.g., heating and customer purchasing), increasing the emerging system of systems security challenges. All software, even the best in class, contains defects which can be potential security vulnerabilities. The average code developed in the United States is estimated at 6,000 defects per million lines of code (MLOC) for a high-level language (Jones and Bonsignour 2011). Very good levels would average 600 to 1,000 defects per MLOC, and exceptional levels would be below 600 defects per MLOC. Organizations should expect that 1–5% of these defects represent security vulnerabilities (Woody 2014). These will provide an attacker opportunities to trigger unexpected behavior in the software and potentially allow the attacker to bypass existing security controls and compromise the confidentiality, integrity, and availability of critical system assets.

As long as software was confined to small isolated areas of a system and these had limited interaction, the security vulnerabilities were of minimal consequence and largely ignored. However, as software has quickly taken over major portions of functionality at all layers of the system architecture—infrastructure, communications, and functional execution—the impact of these latent vulnerabilities across the system of systems is massive. As an example, we can trace the growth of software in military systems as documented in the Critical Code: Software Producibility for Defense Report (CSTB 2010). In 1960, software handled just 8% of the F-4 Phantom fighter's functionality. By 1982, it had expanded to 45% of the F-16 Fighting Falcon's functionality. By 2000, it handled 80% of the F-22 Raptor's functionality (CSTB 2010, p. 19).

So what does this mean in practice today? The Boeing 787 Dreamliner has around 14 MLOC, and if we assume all of it is exceptional code, which is highly unlikely, then 8,400 defects remain in the code, including potentially 420 vulnerabilities. But more likely the code is average to very good, which could have up to 84,000 defects and 4,200 vulnerabilities. The F-35 Lightning II is being fielded with 24 MLOC, which would be 14,400–144,000 defects and 720–7,200 vulnerabilities (Woody 2016).

There has also been a massive expansion in software reuse. Software for these complex systems is typically assembled from existing code such as code libraries, code that performs needed functionality that was written for other systems, commercial off the shelf (COTS) products, and open source code. Code developers are no longer working within the organization fielding a system; they are scattered worldwide and connected to each system through a complex and frequently convoluted supply chain. A single application may be composed of components that are commercial products, open source, and "glue code" that ties these various pre-existing pieces together to provide the desired functionality. Each of these components can also be composed of multiple pieces, and the supply chains can be long with many diverse sources. This greater complexity in both content and structure makes security harder to accomplish. Today's systems are composed of hundreds of these components and applications, and the resulting system functionality is an emergent result of this composition.

In the past, we could rely on access barriers such as network isolation and firewall restrictions to limit access to the software, but the growing reliance on connectivity among all platforms and applications to provide flexibility, constant communication, and situational awareness has greatly reduced the viability of access-control focused security techniques. The growth of remote server farms and cloud computing services to reduce operational costs and the expansion of access to mission capabilities from portable, and even sometimes autonomous, devices further limits our ability to use access barriers for protection. This increase in ubiquity and complexity carries with it an increase in access to latent security vulnerabilities; any element of our cyber infrastructures may be compromised and manipulated (Rice 2007).

12.3 Security of System of Systems Requires Consideration of Software and Its Multi-Layer Composition

In order to handle its role within a system of systems, the design of each system must have sufficient capability to support critical mission security requirements. This is a growing need, as the capabilities of the attacker expand leveraging the use of pervasive highly networked operational context.

Systems must be engineered to handle this increasingly complex technology environment securely. We need to apply a systematic lifecycle approach, just as is done for any other expected operational capability, for defining the security risk requirements that take into account the missions these systems will need to address. This will start with identifying and analyzing complex security risks in software-reliant systems and systems of systems across the lifecycle and supply chain. This can be done through the development of potential attack scenarios using a shared view of the mission attack surface constructed from information about the mission that is linked with the technology and data flows of the systems supporting the mission (system context). This complex, multi-system environment has implications for how security is analyzed and managed. This perspective also illustrates the complex nature of security attacks and how they typically include many systems that are managed by multiple, independent organizational entities.

Consider the following example of a complex risk scenario. In this scenario, an attacker intends to steal personally identifiable information about an organization's customer base.

The attacker's goal is to steal the identities of customers for financial gain. To carry out this risk scenario successfully, the attacker performs the following actions (Alberts 2014):

- The attacker performs reconnaissance on the organization's systems and networks.
- The attacker also performs reconnaissance on partners and collaborators that work with the organization and have trusted access to the organization's systems and networks.
- Reconnaissance indicates that the organization has strong perimeter security controls in place. As a result, the attacker targets a third-party collaborator that (1) has legitimate, trusted access to the organization's internal network and (2) has relatively weak perimeter security controls in place.
- The attacker gains access to the third-party collaborator's internal network by exploiting several common vulnerabilities.*
- The attacker uses the collaborator's trusted access to the organization's internal network to bypass the organization's perimeter security controls and gain access to its network.
- Additional reconnaissance indicates that the organization does not encrypt customer data as they are transmitted between an order entry system and an inventory system (to ensure quick processing of the data). In addition, the organization does not employ rigorous monitoring in its systems and networks. The organization's strategy is to focus primarily on its perimeter security. The attacker decides to exploit these vulnerabilities and installs malware (i.e., a sniffer) that is designed to
    - steal unencrypted customer data as it is being transmitted between systems on the internal network, and
    - send the stolen data to staging points at multiple external locations.
- Once installed, the malware collects unencrypted data and sends the data to the staging points. This data exchange is timed to occur during peak business hours to mask the attack.

As a result of this scenario, the organization could suffer significant financial, legal, and reputational consequences. This risk scenario example represents the inherent complexity of modern security attacks, where multiple actors (people and software) exploit multiple vulnerabilities in multiple inter-connected systems as part of a complex chain of events across a system of systems. Traditional risk analysis is focused on a single system and unable to model complex security attacks effectively.

Scenarios are widely used in system design, but there is a major difference in this approach. The focal point of this scenario is the threat actor. Systems must be designed to function within the context of the system of systems where they will be deployed, and success in any operational environment with the levels of connectivity that are becoming standard must take the attacker into consideration. A common goal of many threat actors is to inflict harm or loss on a mission's stakeholders. To accomplish that goal, a threat actor leverages the use of data as the connecting device among systems and targets data sources

* Vulnerabilities in the scenario include: lack of monitoring to detect the actor's reconnaissance activities; allowing trusted access to the organization's internal network by a third-party collaborator that employs poor security practices; the organization's lack of rigorous monitoring of its systems and networks; and lack of data encryption between the order entry and inventory systems.

that are shared among systems. A threat actor will navigate through the complex interactions of people, processes, and technologies, looking for weaknesses in organizational security practices (people) and vulnerabilities in software-reliant systems to exploit (software and firmware). Getting to the critical mission data is often not simple. A threat actor must connect from one computer to another when attempting to realize the goal of the attack. In many cases, an actor will target computers that are owned and maintained by trusted partners and third-party collaborators when conducting a cyber-attack, leveraging the trust relationships in the system of systems.

The threat actor is ultimately looking to violate the security attributes of mission data, with the hope of causing a range of direct and indirect, negative consequences for the mission. Data have three basic security attributes: confidentiality, integrity, and availability.* For a given risk, a threat actor generally is trying to produce one or more of the following outcomes:

- disclosure of data (violation of the confidentiality attribute)
- modification of data (violation of the integrity attribute)
- insertion of false data (violation of the integrity attribute)
- destruction of data (violation of the availability attribute)
- interruption of access to data (violation of the availability attribute)

Each outcome maps to a security attribute of the data. The example described above illustrates a scenario in which a threat actor steals unencrypted customer data as they are being transmitted between systems on an internal network and uses another system as an entry point.

In this example, the threat actor is targeting an organization's order entry and inventory processing to steal credit card data. As shown in Figure 12.2, the threat actor uses the emergent capabilities of the system of systems to reach critical data. The result is an impact on the organization which could range from a minor delay to a catastrophic outcome.

12.4 System of Systems Development and Sustainment Relies on a Global Multi-Organizational Software Supply Chain

The increasingly global nature of software development has also raised concerns that global supply chains could be compromised, allowing malicious code to be inserted into a delivered software product during development or enabling a compromised product to be substituted during delivery or installation. We have experienced the wide reaching impact of vulnerabilities in the software supply chain through open source software modules such as the Heartbleed security flaw (Wikipedia 2014). Since humans began trading goods

* *Confidentiality* is defined as keeping proprietary, sensitive, or personal information private and inaccessible to anyone who is not authorized to see it. *Integrity* is defined as the authenticity, accuracy, and completeness of data. *Availability* is defined as the extent to which, or frequency with which, data must be present or ready for use. These definitions are adapted from (Alberts 2002).

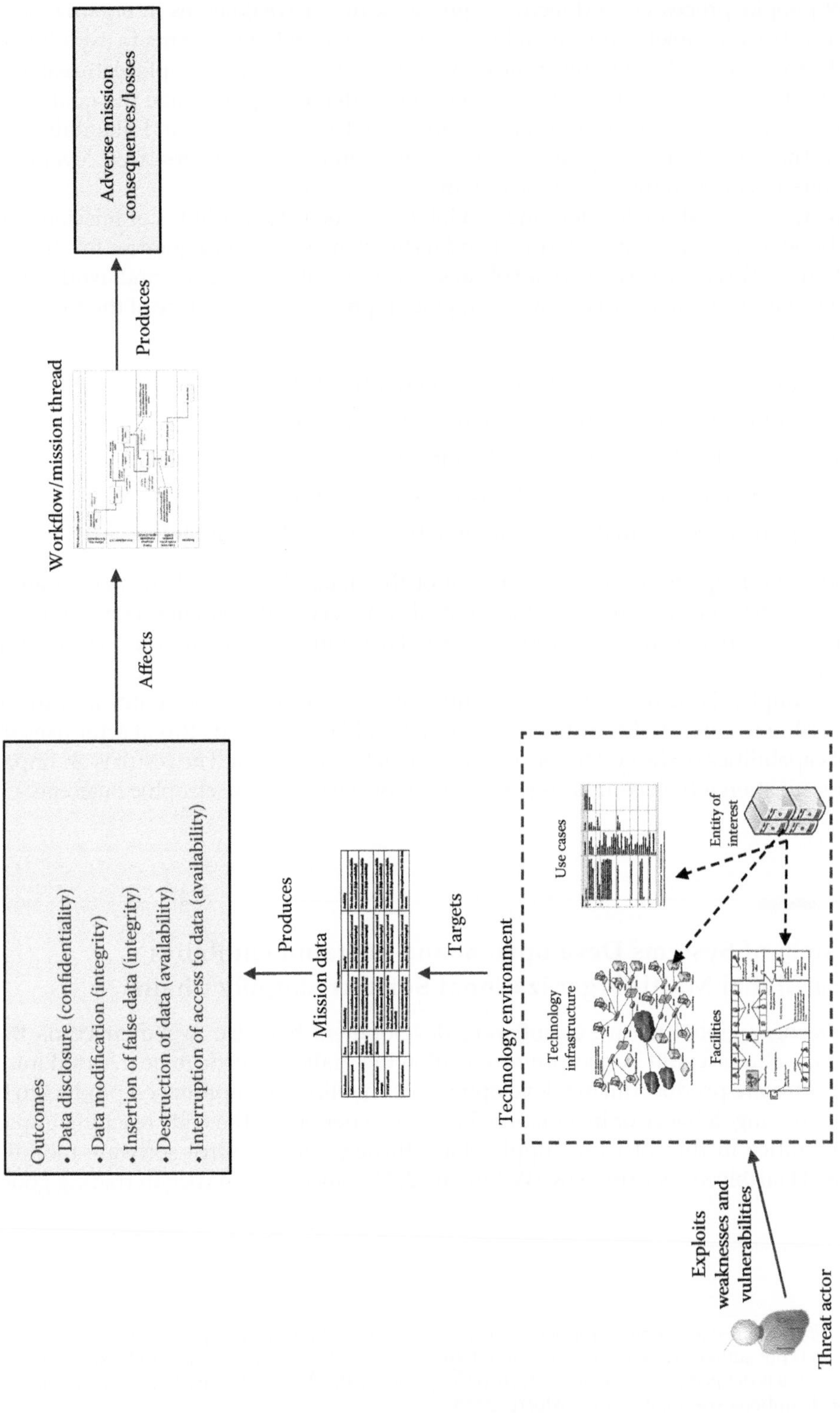

FIGURE 12.2
Steps of a mission thread attack. (Adapted from Albert, 2014.)

and services, concerns about counterfeit goods and bogus services have been part of the language of interaction. But we are now faced with a situation where the capabilities of today's software technology environment, the need to outsource to meet cost and schedule demands, and the interaction between off the shelf and open source software products have far outpaced our ability to effectively monitor and manage supply chain risk using traditional methods.

Like any supply chain, the ownership and responsibility for software may be scattered across several organizations, each of which may affect the security and integrity of the final system. For example, software is often assembled from existing code (as illustrated in Figure 12.1). An application may be composed from components that are commercial products, open source, and "glue code" that ties these various pre-existing pieces together to provide the desired functionality. Each of these components can also be composed of multiple pieces, and the supply chains can be long with many diverse sources.

As software is assembled and integrated into a final product, each integrator inherits all of the defects and vulnerabilities inserted by the developers at every sub-level, which can include additional custom, commercial, open source, and legacy software. Some behavior of this software composition can be validated through testing, some through modeling and simulation, and the remainder emerges as the software operates in a range of system contexts over time.

Special tools (static and dynamic analysis tools) have been developed to search for the most common types of vulnerabilities. The tools, however, can be complex to use, expensive to license, limited in the scope of vulnerabilities identified, and the findings may contain many false positives.

The most attractive means of an attack continues to be the intentional exploitation of software vulnerabilities inadvertently introduced during development (Verizon 2017). These weaknesses allow attackers to gain system access and bypass security controls. Each step in a supply chain can be a source of such vulnerabilities, and increased assurance for the final product requires the consistent use throughout the supply chain of development techniques demonstrated to reduce the likelihood of vulnerabilities.

As noted earlier, code quality plays an important role in the likelihood of vulnerabilities. We can estimate the level of remaining defects by estimating the injection rates during development and the effectiveness of removal activities. The estimates can be based on historic data and measures. SEI has data from many projects where developers using Team Software Process (TSP) have accurately predicted their injection rates and removal yields and measured the effects of process changes on outcomes (Davis 2003). The histogram in Figure 12.3 displays the measured defect removal in design, code, and unit tests for 114 projects from the TSP database. Software achieving levels below 1,000 defects per MLOC in system tests are considered to have good quality. It is important to note that testing is used to confirm defect removal and not to identify and fix defects.

We have identified projects that delivered effective operational security and safety results by addressing both defects and vulnerabilities. Five of the 114 projects in the TSP database also demonstrated exceptional results for security and safety. Figure 12.4 shows the relationship of the safety and security defect results to each project's quality results. Four of these projects reported no security or safety defects post-implementation, and one other had a very low reporting of 20 defects. In addition to tracking and removing defects early, these projects added detail safety and security inspections for requirements, designs, code, and test cases.

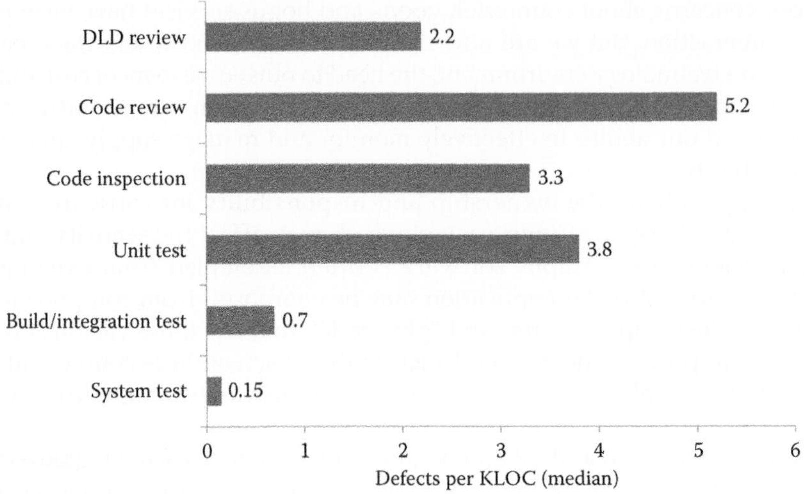

FIGURE 12.3
Defect removal density during software development (Woody, 2014, p. 10).

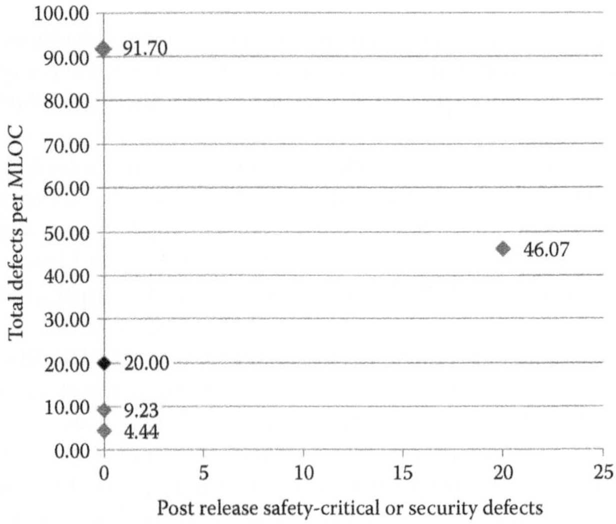

FIGURE 12.4
Post release defects compared to total defects for five projects (Woody, 2014, p. 8).

12.5 Systems Must Be Engineered for System of Systems Security

Many simplifying assumptions have been made in the past about software when it handled a limited portion of the system functionality. For many years, software failure rates were considered negligible compared to hardware failures, and this pattern has persisted inappropriately long beyond its usefulness (Goodenough 2010). Now that software is a major component of system functionality, system requirements for reliability, survivability, and operational effectiveness must expand to reflect the level of risk tolerance the system must support within the software-driven environment of the system of systems. Security of the information flowing into and out of the system is a major concern.

Security requirements are not absolutes and must reflect the risk tolerance critical to the system's operation. Those with a high risk tolerance will embrace change and rapidly adapt a system within its changing environment to meet their needs. These are typically early adopters who look for opportunities that technology can enhance. For this group, as much flexibility as possible is desired and tight security controls will be considered problematic. However, for those with a more limited risk tolerance, careful change and minimal disruption to the operational environment is expected in an environment where results can be well planned. Most systems are somewhere in between, requiring careful consideration of what risk can be tolerated and where investment will be needed to determine how to mitigate unacceptable risk (Brownsword 2010).

There is always uncertainty about a software system's behavior. Rather than performing exactly the same steps repeatedly, most software components function within a highly complex networked and interconnected system of systems that changes constantly. A measure of the design and implementation is the confidence we have that the delivered system will behave as specified. Once a level of risk tolerance has been established for security requirements, we can collect security measurement information indicating that security has been appropriately addressed from requirements to design, construction, and test to establish confidence that security is sufficient.

For hardware reliability, we can use statistical measures, such as the mean time between failures (MTBF), since hardware failures are often associated with wear and other errors that are frequently eliminated over time. The lack of hardware failures increases our confidence in a device's reliability. The software defect exists when the software is deployed, and the failure is the result of the occurrence of an unexpected operating condition. Too little reliability engineering was given as a key reason for the reliability failures by the DoD RAM guide (DoD 2005). This lack of reliability engineering was exhibited by

- failure to design-in reliability early in the development process
- reliance on predictions (use of reliability defect models) instead of conducting engineering design analysis

The same reasoning applies to software assurance. We need to engineer software assurance into the design of a software system. We must define the specific software goal for the system. From the goal we can identify ways in which the engineering and acquisition will ensure, through policy, practices, verification, and validation, that the goal is addressed.

We can use the Software Assurance Framework (SAF), a baseline of good software assurance practice for system engineers assembled by the SEI, to confirm completeness and identify gaps in current practices (Alberts 2017). The SAF defines a set of cybersecurity practices that programs should apply across the acquisition lifecycle and supply chain. The SAF can be used to assess a program's current cybersecurity practices and chart a course for improvement. To establish that the requirements will be met, a range of evidence can be collected by the SAF practices across the lifecycle for review during the following: milestone reviews, engineering design reviews, architecture evaluations, component acquisition reviews, code inspections, code analysis and testing, and certification and accreditation. By improving a program's cybersecurity practices, the SAF helps to (1) establish confidence in the program's ability to acquire software-reliant systems that are secure, and (2) reduce the cybersecurity risk of deployed software-reliant systems.

12.6 Modeling Mission Impact of Security Risks to Engineer System of Systems Security—Electrical Grid Example

Systems engineers can use mission threads to describe a group of behaviors that traverse the system of systems context. From this model, potential attacks can be described and analyzed to determine how each participation system will need to respond. As missions are increasingly supported by systems of systems, the security requirements for each participating system must take into consideration the risks from this perspective. The remainder of this section provides an example of how utilities are using mission thread models to improve their system of systems security.

Utilities build and support information technology, operational, communications, and cyber-physical systems (CPS) that support delivery of electricity, water, and gas to the general public, businesses, and the U.S. government. Combined, these systems support the modern way of life enjoyed by people across the United States and in other countries. These systems also comprise the lifeline sectors within the national critical infrastructure. Furthermore, the development of smart networking necessitates these systems to evolve and proliferate since modern communications and networks are required for the smart capabilities of such networks. As software takes over more and more controlling functions in all of these smart networked environments, the potential impact of security vulnerabilities and subsequent attacks, such as Stuxnet and Havex, demonstrate a growing need to improve capabilities to address cybersecurity risk (Wald 2013, Kovacs 2014).

Utilities have applied extensive resources to consider cybersecurity and implement risk management capabilities when delivering their services (e.g., water, electricity, and gas). (ESCSWG 2011). However, much remains to be improved due to the great diversity of networks, architectures, solutions, and resources available to the numerous U.S. utilities. In many utilities, security practices applied to IT systems are ahead of security practices applied to operational technology (OT) systems in sophistication and discipline due to longer term exposure of IT systems to the threats originating from the Internet. While similar practices can be applied to OT systems, they need to be applied differently to accommodate the cyber-physical structures of these systems; however, this difference is not uniformly well understood. Additionally, overall, cybersecurity still requires more attention and resources in many utilities. Situations such as the Northeast power failure of 2003 demonstrate the challenges of the growing interplay between OT and IT, and point to the need to strengthen both aspects of the organization (Wald 2013). Problems that contributed to the Northeast power outage were traced to multiple technology and human failures resulting from dependencies on technologies that were not well understood, as well as response decisions that focused only on individual systems and did not consider the interactions among systems.

The 2017 Data Breach Investigations Report indicated utilities are as vulnerable as other sectors and can expect the attacks to continue. Utilities have not yet experienced the recent major increase in attacks seen in many other sectors (Verizon 2017). Many utilities' systems are still disconnected from the Internet, but as their connectivity trends increases, the risk of more attacks can be expected to grow. Utility executives use mission threads to begin to understand the ways in which key operational needs depend on information and communication technology (ICT) and the security risks this dependency promotes. A risk analysis using a system of systems context helps utility executives integrate cybersecurity into existing utility enterprise risk management processes and identify the ways in which different functions and capabilities within utilities must work together to deliver organizational

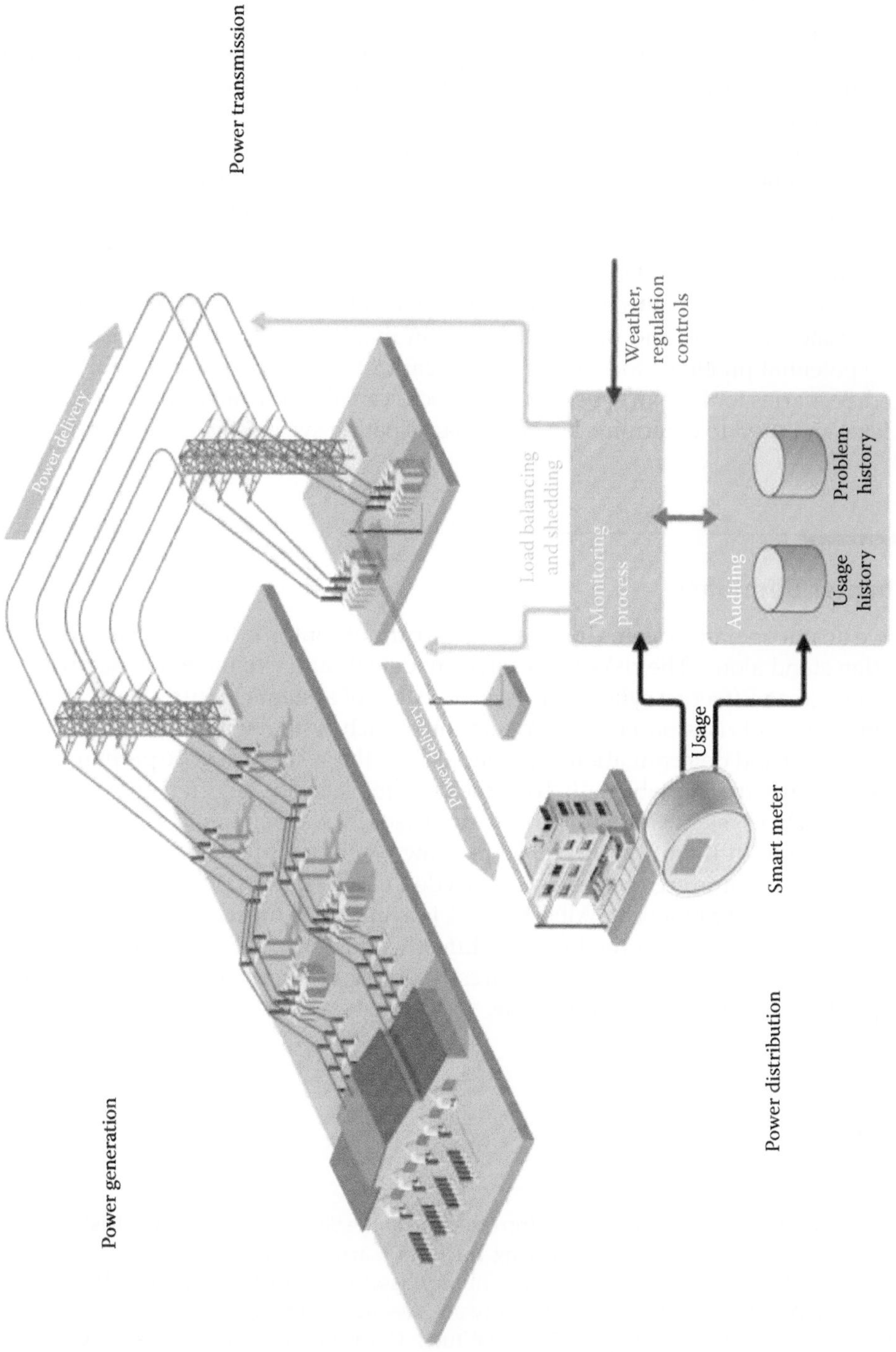

FIGURE 12.5
Example mission thread for U.S. utilities.

capability and address potential risks. The analysis begins with the construction of a mission thread that provides an operational context view. This view may be composed of a series of steps needed to deliver a capability. Figure 12.5 depicts an example system of systems view and shows the possible organizational elements associated with the delivery of power to a home and the subsequent bill the homeowner pays for the power used. A mission thread that shows the end-to-end steps needed from power generation through consumption can be overlaid on this diagram. Specific systems are then mapped to each step to define their role for operational success. There may be many variations in how an organizational capability can be assembled; it is not necessary to analyze all of them to understand how the various organizational and technology parts need to work together in a trusted relationship that needs to operate securely. Changes such as load balancing and load shedding, unexpected distribution demands, distributed generation, changes in usage as new technologies such as electric vehicles impact that technology base, and attackers could disrupt or overload the system, or create confusion and propagate errors in many ways. Scenarios are developed from these potential problems to evaluate how organizational and system capabilities will be affected. Scenarios for known cybersecurity attacks can also be evaluated at the system of systems level and used to determine how each participating system should respond.

12.7 Chapter Summary

Because we do not operate systems in isolation, we cannot continue to build them as if they will function stand alone. The risks from software defects and expanded connectivity are becoming too great. Effective cybersecurity in a system of systems requires keen attention to the identification of risks posed by the operation of each system within the larger system of systems context and determination of potential risks the system must be prepared to tolerate. Risks that are unacceptable will drive the need for security requirements to mitigate the ways in which a system would contribute to these risks. Operational scenarios can be modelled, as shown in the utility example, to show where a system fits within each scenario to assist in determining risk levels and needed requirements. The SAF, a framework of best software security practices (Alberts 2017), is available to meet engineering software security needs at each point in the lifecycle. Effective implementation and monitoring of these practices will provide a means to address security requirements within all aspects of the acquisition and development lifecycle.

References

Alberts, C., Woody, C., and Dorofee, A. 2014. Introduction to the Security Engineering Risk Analysis (SERA) Framework. Software Engineering Institute, Carnegie Mellon University, CMU/SEI-2014-TN-025. http://resources.sei.cmu.edu/library/asset-view.cfm?AssetID=427321.

Alberts, C., and Woody, C. 2017. Prototype Software Assurance Framework (SAF): Introduction and Overview. Software Engineering Institute, Carnegie Mellon University, CMU/SEI-2017-TN-001. http://resources.sei.cmu.edu/library/asset-view.cfm?AssetID=496134.

Brownsword, L., Woody, C., Alberts, C., and Moore, A. 2010. *A Framework for Modeling the Software Assurance Ecosystem: Insights from the Software Assurance Landscape Project* (CMU/SEI-2010-TR-028).

Retrieved March 09, 2018, from the Software Engineering Institute, Carnegie Mellon University website: http://resources.sei.cmu.edu/library/asset-view.cfm?AssetID=9617.

Computer Science and Telecommunications Board (CSTB). 2010. Critical Code: Software Producibility for Defense. Washington, DC (US): National Academies Press. http://www.nap.edu/openbook.php?record_id=12979&page=R1.

Corkery, M. 2016. Once Again, Thieves Enter Swift Financial Network and Steal. *New York Times* (May 12, 2016). https://www.nytimes.com/2016/05/13/business/dealbook/swift-global-bank-network-attack.html?_r=0.

Davis, N., and Mullaney, J. 2003. The Team Software Process (TSP) in Practice: A Summary of Recent Results. Software Engineering Institute, Carnegie Mellon University, CMU/SEI-2003-TR-014. http://resources.sei.cmu.edu/library/asset-view.cfm?assetid=6675.

Department of Defense (DoD). 2005. DoD Guide for Achieving Reliability, Availability, and Maintainability. Department of Defense (May 2005). http://www.acqnotes.com/Attachments/DoD%20Reliability%20Availability%20and%20Maintainability%20(RAM)%20Guide.pdf.

Driver, M., Gaehtgens, F., and O'Neill, M. 2017. Managing Digital Trust in the Software Development Life Cycle. *Gartner* (May 2017).

Energy Sector Control Systems Working Group (ESCSWG). 2011. Roadmap to Achieve Energy Delivery Systems Cybersecurity. http://energy.gov/sites/prod/files/Energy%20Delivery%20Systems%20Cybersecurity%20Roadmap_finalweb.pdf.

Fatima, R. 2014. How Criminals Attacked Target: Analysis. *Security Week* (January 20, 2014). http://www.securityweek.com/how-cybercriminals-attacked-target-analysis.

Goodenough, J. 2010. Evaluating Software's Impact on System and System of Systems Reliability. Software Engineering Institute, Carnegie Mellon University. http://www.sei.cmu.edu/library/assets/SW%20impact%20on%20system%20reliability.pdf.

Greenberg, A. 2014 Hackers Are Already Using the Shellshock Bug to Launch Botnet Attacks. *Wired* (September 24, 2014). https://www.wired.com/2014/09/hackers-already-using-shellshock-bug-create-botnets-ddos-attacks.

Jones, C., and Bonsignour, O. 2011. *The Economics of Software Quality*. Addison-Wesley Professional.

Kovacs, E. 2014. Attackers Using Havex RAT Against Industrial Control Systems. *Security Week* (June 24, 2014). http://www.securityweek.com/attackers-using-havex-rat-against-industrial-control-systems.

Maier, M. 1998. Architecting principles for systems-of-systems. The Information Architects Cooperative (TIAC) whitepaper, http://onlinelibrary.wiley.com/doi/10.1002/(SICI)1520-6858(1998)1:4%3C267::AID-SYS3%3E3.0.CO;2-D/epdf.

McGraw, G. 2006. *Software Security: Building Security In*. Upper Saddle River, NJ (US): Addison-Wesley.

Rice, D. 2007. *Geekonomics: The Real Cost of Insecure Software*. Addison-Wesley Professional.

Saltzer, J., and Schrader, M. 1974. The Protection of Information in Computer Systems. *Communications of the ACM* 17, 7. http://www.cs.virginia.edu/~evans/cs551/saltzer.

Swamynathan, Y. 2017. Equifax reveals hack that likely exposed data of 143 million customers. *Business News* (September 7, 2017). http://www.reuters.com/article/us-equifax-cyber/equifax-reveals-hack-that-likely-exposed-data-of-143-million-customers-idUSKCN1BI2VK.

Verizon. 2017. 2017 Data Breach Investigations Report (DBIR). Verizon (May 2017). http://www.verizonenterprise.com/verizon-insights-lab/dbir.

Wald, M. 2013. The Blackout That Exposed the Flaws in the Grid. *New York Times* (November 11, 2013). http://www.nytimes.com/2013/11/11/booming/the-blackout-that-exposed-the-flaws-in-the-grid.html?pagewanted=all&_r=1&.

Wald, M. 2014. Airline Blames Bad Software in San Francisco Crash. *New York Times* (March 31, 2017) https://www.nytimes.com/2014/04/01/us/asiana-airlines-says-secondary-cause-of-san-francisco-crash-was-bad-software.html.

Wikipedia. 2014. Heartbleed April 2014. https://en.wikipedia.org/wiki/Heartbleed.

Woody, C., Ellison, R., and Nichols, W. 2014. Predicting Software Assurance Using Quality and Reliability Measures. Software Engineering Institute, Carnegie Mellon University: CMU/SEI-2014-TN-026. http://resources.sei.cmu.edu/library/asset-view.cfm?AssetID=428589.

13

Cyber Insecurity Is Growing

Carol C. Woody

CONTENTS

13.1 Software in a System of Systems

Maier notes five characteristics which set systems of systems apart from complex monolithic systems as follows (Maier 1998): (1) operational independence of their components, (2) managerial independence of the components, (3) their evolutionary nature, (4) emergent behaviors, and (5) a geographic extent that limits the interaction of their components to information exchange. Each of these characteristics contribute through today's systems of systems to the growth of cyber insecurity. Organizations no longer build systems but instead primarily acquire the desired functionality. Systems of systems are increasingly composed of multi-vendor sourced data-driven software products with known defects contributing to unexpected operational behaviors. Developers and integrators in vendor and acquiring organizations assemble software components from many sources to build systems that motivate acquirers to implement the technology to address operational needs. It would take organizations too long and would cost too much money to develop everything themselves. While some of the acquired software may be embedded directly into a system, most components are linked through application programming interfaces (APIs) and structured messaging protocols (e.g., TCP/IP) that leave each component *operationally independent*. The software suppliers are driven by market realities to release products as quickly as possible to capture sales and dominate markets. Products are released with known defects, and patches for fixing errors and security vulnerabilities are released later as updates are demanded by the acquirers. Each of these software sources maintains ownership and control over its content (*managerial independence*), and fixes or new versions are provided on the vendor's schedule as it deems fit. These software owners determine when they will release a new product, and they may choose to no longer support a mature product thus leaving the acquirers with known problems in the software they have implemented (*evolutionary development*). It is incumbent on those selecting the software to keep the versions they are using up to date with new features and patches. The markets are world-wide, requiring delivery mechanisms that support *geographic distribution* (e.g., open source, websites) to reach the largest market as quickly and cheaply as possible. In many cases, vendors do not know who has their product so they broadcast information about problems and fixes. Most products are structured with default implementation mechanisms that can include embedded passwords and backdoors that must be changed/ removed by the acquirer, if the provider has given them a means.

Emergent behavior in system of systems, in this chapter, is an operational environment filled with security vulnerabilities which attackers are increasingly using to their advantage. It arises from efforts by acquirers and providers to reduce technology costs and increase flexibility by shifting functionality and information flow to software. The growing cyber insecurity is a negative emergent result of the increased accessibility attackers have through the growing systems of systems communication channels that transport data to and from the vulnerable software. Software communication relies on standardized interfaces (e.g., TCP/IP, UDP) to allow almost any product to be linked to any other product using a minimum amount of connecting software. Acquirers must accept licensing agreements that absolve the provider from any and all liabilities before they can touch a product, since it is known that the software has weaknesses. Attackers make use of the push by producers to get products to market quickly using low cost methods that do not address security weaknesses to compromise products at the source, in transit, and at the implementation. If the product creators and maintainers do not implement effective configuration controls over their software and update handling, their software can be corrupted by attackers to deliver malicious payloads to acquirers without their knowledge. Websites can be corrupted to deliver malicious payloads when someone opens the page to get an update. Attackers become acquirers (frequently at no cost) to analyze products and discover unreported vulnerabilities that provide attack vectors for unsuspecting acquirers. As our reliance on software increases, the risk of organizations experiencing the impact of an attack increases. Table 13.1 summarizes the relationship of Maier's five characteristics to primarily software-driven systems and the growth of insecurity this brings with it.

TABLE 13.1

Maier's System of Systems (SoS) Characteristics for Software Growing Insecurity

SoS Characteristic (Maier 1998)	Software Growing Insecurity
Operational independence	Acquirers/integrators/vendors assemble software from many sources to seamlessly deliver end-to-end mission capability.
Managerial independence	Vendors build and sell software for specialized niche markets (e.g., point-of-sales, printing, Cloud computing).
Evolutionary development	Vendors release new functionality to capture market share and drop support of older versions.
Emergent behavior	Vendors drive down costs through standardized interfaces (e.g., TCP/ IP) and reuse. Vendor demand licenses that absolve them of liabilities. Acquirer's focus on least cost and speed of delivery with extensive connectivity results in widespread vulnerability.
Geographic distribution	Vendors deliver insecure-by-default software (faster and easier).

13.2 Transition from Standalone Systems to Highly Networked Software-Intensive Systems of Systems

Vendors provide systems that are assembled from various products, and their components and subcomponents are created and maintained by many different technology providers, often globally distributed. These vendors make use of libraries and reusable components, which can be downloaded from many Internet accessible sources, and specialized manufacturing to create products that address some or all of the acquiring organization's desired functionality at a much lower cost and more quickly than each acquiring organization could do on their own. The acquirer uses data sharing capabilities to connect systems and components from vendors into a system of systems that performs what is needed.

Each vendor and sub-contractor will enhance its component to create a new version with new capabilities and repairs to problems found in operation in order to motivate the acquirer to buy the upgrade and swap out what they purchased for the "newer model." The portion of functionality handled by software in each vendor's products has steadily increased over time. Software is more easily changed and has few physical limitations for distribution, reducing logistical challenges for vendors globally distributed. Air Force studies indicate that well over 90% of the operations executed by a system in 2000 were performed and controlled by software (CSTB 2010) and there are no indicators that this expansion has diminished. Connectivity among all of the various pieces of software is controlled through data exchange. Structured application program interfaces (APIs) built into the software that support a standard set of data exchange capabilities allow acquirers to interconnect all of the various systems and components into a system of systems. Through the APIs, structured and semi-structured data can be shared among software written by different vendors.

This software is not defect free. Even the best quality software still contains many defects (Jones 2013). At least 5% of these defects must be considered vulnerabilities (Woody 2014), which, if accessed by an attacker, can allow it to gain unauthorized access to data that should be protected (confidentiality failure), change or destroy protected data (integrity failure), or block authorized users from accessing and using the system or network (availability failure such as denial of service).

Vendors are motivated to be the first to release a software-driven capability. The value of a product depends on how many other users adopt it. Initial development costs are high for initial creation and drop very low for any additional acquirers. For each acquirer there are large start-up costs to integrate the vendor product into their system of systems, creating a disincentive to switch to another vendor who might later produce a slightly better product (Anderson 2001).

Potential attackers as well as acquirers can have access to the software, providing increased opportunity for vulnerabilities to be found. As more software components with desired functionality are available through open source, acquirers are attracted to the cost (free), and the distribution model allows anyone with connectivity to access and scrutinize the behavior and potential vulnerability of the software component.

An ever-increasing system susceptibility to cyber-attack is an emergent characteristic resulting from this growth of software-controlled functionality with many inherent defects and the increased interconnection of systems and components built by various vendors for, in most cases, different uses. As the connected vendor components expand to include consumer devices such as tablets and cell phones to serve as input devices and component connection capabilities are augmented to use data input to drive autonomous behaviors, the potential for unexpected and unwanted behaviors within and among system of systems increases. Since an integrator has limited knowledge of each component and is primarily focused on delivering the planned functionality, little time is spent determining if unexpected behaviors exist and avoiding undesirable results. Products are shipped with known problems and are insecure by default. Problems are fixed in a later version if enough of the acquirers complain, and the acquirer must then expend time and resources to integrate the upgrade into its environment (Anderson 2001), which may break something else.

Software development is a resource intensive process performed by highly skilled individuals who can translate functional requirements into a structure that a machine can repeatedly perform to produce a desired outcome. These resources once worked directly for each organization, creating unique solutions that were implemented only within that organization. However, since many organizations have similar needs (e.g., financial management, inventory control, or cost analysis), the development of a shared software solution proved more cost effective. Many devices also share similar needs (e.g., flight control for all types of airplanes and cruise control for all types of automobiles). In these cases, the development resources work for an independent organization building the shareable solution, and each acquirer wishing to use the software buys a license (aka commercial off the shelf or COTS products). Over time developers built libraries of small pieces that could be reused and published these through various mechanisms, including language specific libraries, open source, and application stores. In addition, hardware production shifted from specific problem solutions to lower-cost generic equipment (e.g., servers, laptops, cell phones, and tablets) that could be manufactured in volume and, with the use of software, tailored to a variety of needs. The pre-existence of all of this working software allows today's acquiring organizations to compose their tailored solutions from commercial products, open source, commercial hardware, and "glue code" that ties these various pre-existing pieces together to provide the desired functionality. Each of these available components can also be composed of multiple pieces drawn from code libraries, automated code frameworks, and reused functions. In most cases, these solutions are assembled by individuals unfamiliar with the actual inner workings of the components they are assembling or of the capabilities of the developers who created them. To further reduce the need for acquirers to learn how to write software, coding frameworks are available, which allow the integrator to select components from a range of sources for assembly. Integrated Desktop Environments (IDEs) are available to support the developer/integrator in assembling and testing compositions to verify that the results satisfy the desired functionality. Unlike the predominantly hardware driven systems built in the 1980s and 90s, the software driven components can be quickly assembled and quickly changed to meet rapidly changing and expanding needs.

Software does not consume precious physical space in a size-challenged structure and provides a means for change and tailoring of capabilities after the chip is manufactured.

Commercial providers have leveraged the flexibility and low cost of software communications to add connectivity to anything and everything. A vast array of consumer products such as video cameras, microphones, sensors, and monitoring devices allow remote access and control of these devices. Parents can monitor their children at a distance; home owners can monitor the security of their homes from afar; businesses can monitor the heating and cooling of buildings remotely; elevators and access doors can be monitored remotely. Shared data can be housed with data storage services (e.g., Google Cloud and Amazon Cloud) that any of these access points can share and quickly expands as needed for a growing demand. All of these capabilities can be connected into a system of systems to collect data, minimize costs for operations, and improve physical access capabilities.

13.3 Cyber Security Risk Is Growing

As software takes over more system (and hardware) functionality and systems merge into systems of systems, the opportunities for security compromise have expanded. As an example, consider the growth of software in aircraft. The F-22 Raptor* was designed to provide the United States with air superiority, and 1.7 million lines of code were required to support its functionality. It was first flown in 1997 and went into full operational use in 2005. If we assume all of the code was of superior quality, we can expect that 1,020 defects remained in the operational code and a projected 51 vulnerabilities (Jones 2013). The F-35 that took its first flight in 2006 required 6.8 million lines of code to address operational plane functionality. Assuming all of this code is of superior quality, the defects that remain in the F-35 software now number around 4,080 with an expected 204 vulnerabilities. However, it is more likely that the code in these planes is average quality, which increases the F-35 expected defect level to 6,800 and the predicted vulnerabilities to 340. In 2011, the Air Force released a report (US AF SAB 2011) describing that in addition to the operational code on the plane, the F-35 requires an additional 24 million lines of code in support systems which provide critical functionality such as maintenance monitoring so that aircraft issues can be identified and ground crews ready to perform needed maintenance the minute the plane lands to keep flight capability at its peak. The plane is part of a system of systems connecting all F-35 aircraft to specialized support systems that improve its flight capabilities. However, this added code adds at least 14,400 defects if it is high quality and more likely 24,000 defects (average quality), which adds 720–1,200 vulnerabilities. Also, the added code is connected to other systems that support the needs of the aircraft, such as parts inventory and reorder systems and logistics systems to ensure the needed parts are in the right place at the right time. Each F-35 is now at risk for security problems in all of these interconnected systems.

As organizations expand the use of software and connectivity, risk considerations for cost and schedule are extensively evaluated, but other considerations such as security, safety, and privacy are typically not incorporated into the standard way of doing business (Geer 2010). The 2011 Air Force report (US AF SAB 2011) projected that the cost of software in 1997 was 45% of the total system cost. By 2010 this cost estimate had increased to 66% of the total system cost, and the expected cost by 2024 is expected to exceed 88% of the total system cost. All of this software carries vulnerabilities. Organizations buy technology to meet performance parameters but infrequently consider security risk as part of their selection process, and seldom consider the ways that a new acquisition will function within its intended deployment situation, which, with growing frequency, is a system of systems. The potential for unintended consequences is increasing as the use of software expands and connectivity increases (Ellison et al. 2010).

Over the past ten years, about one billion data records have been accessed inappropriately through compromises on systems that handle critical data (PRC 2014). Unfortunately, this count is estimated to be low since it only considers cybersecurity breaches that were

* https://en.wikipedia.org/wiki/Lockheed_Martin_F-22_Raptor.

reported. This number should be significantly higher but organizations are reluctant to disclose security failures since cyberattacks often lead to revenue losses, reputational damage, regulatory fines, litigation, increased costs of cybersecurity protection, the exposure of vulnerabilities to competitors, or potential loss of life in the case of critical national infrastructure breaches (Cherdantseva et al. 2016; Ogut et al. 2011).

Motivated by financial gain or extortion, the nature of cyberattacks today has evolved to more targeted and sophisticated types of attacks (Tankard 2011; Friedberg 2015). Advanced persistent threats (APTs) is a new class of threats with the goal of gaining access to targeted critical assets and to maintain a presence on the targeted system for long-term data collection and control. APTs can be extremely sophisticated attacks and are much harder to defend against (Tankard 2011). These types of attacks generally combine multiple techniques and are often customized to the targeted organization. Customized attacks that were unique to a single organization were found in over 70% of malware security breaches reported in the Verizon survey (Verizon 2015).

When Target was attacked in 2014, 11 gigabytes (GB) of data—110,000,000 records worth of payments, transactions, and other personally identifiable data—were stolen from over 40,000 customers who paid with credit and debit cards (Perlroth 2014). Secret service estimated that over 1,000 retail companies had been compromised in a similar manner based on information from seven companies that sell and manage in-store cash register systems. Only a few major retailers (e.g., Target, UPS, and Supervalu) notified customers of the attacks (DHS 2014). The initial access for an attacker comes through a compromised remote access point, then the system of systems connectivity is leveraged to reach the cash register system. All of these major retailers are using enterprise resource planning (ERP*), an integrated management of core business processes, often in real-time and mediated by software and technology for real-time tracking of inventory, cash, and profit to efficiently manage a retail operation and reap a profit from a highly competitive high volume business with tight margins. Because the retailers did not consider the possibility of data theft in the assembly of the system of systems, their internal monitoring did not include ensuring that the right kinds of data were flowing appropriately. The integration of business systems with manufacturing plant floor control systems (including Supervisory Control and Data Acquisition Systems (SCADA)) allowed attackers to gain control of a steel furnace at a German steelworks and cause massive damage to the plant. The attackers gained access through vulnerabilities in the business systems (FOIS 2016). In the 2016 Data Breach Investigation Report (Verizon 2016), participants reported 2,260 data breaches involving incidents in 82 countries.

Credit card processing capabilities have been added to gas pumps, taxi cabs, and parking garage access systems. These systems of systems connect card reader capabilities with credit confirmation systems and physical devices to support the convenience of immediate payment and use. In many cases, the systems used in individual locations are actually shared by many locations across the country. For example, a Chicago based parking garage company operates facilities in many cities, and data from all of their locations was compromised through vulnerabilities on the central data servers (DiGangi 2014).

In May 2017, hospitals, companies, and government offices around the world were hit with a new type of mass-attack which seized control of affected computers until the victims paid a ransom (aka ransomware) (Smith-Spark 2017). Major organizations such as the National Health Service in the United Kingdom and global firms such as FedEx reported hits. Variants of this ransomware attack continued to be launched almost hourly afterward. The ransomware is spread by taking advantage of a Windows vulnerability that

* ERP descriptions available at https://en.wikipedia.org/wiki/Enterprise_resource_planning.

Microsoft released a security patch for in March 2017. But computers and networks that didn't update their systems remained at risk. Once the core software capability for any attack becomes available, other less skilled developers use that software component as a base, make minor adjustments, and launch a variant.

For organizations, the impact can be devastating in terms of financial loss and irreparable harm to their reputation. The ransomware had real consequences at 16 organizations that are part of the National Health Service on that Friday, causing some surgical procedures to be canceled and ambulances diverted. The FBI reported that, on average, 4,000 ransomware attacks occurred per day in 2016 (Symantec 2016).

Many organizations do not know how much software they depend on and how it is interconnected. The broad connectivity and communication interchange facilitates denial of service attacks which fill the communication channels to prevent normal use. In addition, the connectivity provides access opportunities for brute-force attacks from malicious connected devices which have been used to breach systems. Upon successful access, attack capabilities such as the Bricker-bot performed a series of Linux commands that will lead to corrupted storage, followed by commands to disrupt Internet connectivity, device performance, and the wiping of all files on the device. These attackers corrupt the Boot capabilities of devices and the BIOS firmware, essentially destroying the technology (Guruburan 2017). NIST SP 800-193 was released in May 2017 to raise awareness of the risk to platforms across the system of systems (Regenscheid 2017).

13.4 Security Controls Can Add More Software and More Vulnerabilities

As an example of the broad consequences a vulnerability in a widely used software component can have, consider the recent Heartbleed vulnerability in OpenSSL and its impact. OpenSSL is an open-source implementation of the secure socket layer (SSL) and transport layer security (TLS) protocols used for securing web communications. The Heartbleed vulnerability occurred in the OpenSSL "assert" function, which is the initiator of a heartbeat protocol to verify that the OpenSSL server is live. The assert function allowed a violation of input parameter assumptions so the request can trigger a data leak. The security risk is that the additional data retrieved from the server's memory could contain passwords, user identification information, and other confidential information (Woody et al. 2014).

The defect appears to have been accidentally introduced by an update in December 2011. OpenSSL is a widely used and free tool. At the disclosure of Heartbleed, approximately 500,000 of the Internet's secure web servers certified by trusted authorities were believed to be vulnerable to the attack. The new OpenSSL version repaired this vulnerability by including a bounds check to ensure that the payload length specified is no longer than the data that is actually sent. Unfortunately, that check is only the start of an implemented correction because elimination of the vulnerability requires the 500,000 users of this software to upgrade to the new version. In addition, because this problem is related to security certificates, protecting systems from attacks that exploit the Heartbleed vulnerability requires that companies revoke old SSL certificates, generate new keys, and issue new certificates (Peters 2015). At least eight operating versions of OpenSSL were shipped containing the bug. While the OpenSSL program is complex, the cause of the vulnerability is simple. The software never verified the design assumption that the length of the content to be returned to the caller was less than or equal to the length of the payload sent.

Open source is extremely widely used. In a 2015 survey of 1,300 senior security professionals, 78% reported their organizations actively operate using open source—double from the previous survey performed in 2010 (Santinelli 2015).

Given this extensive use, it is disheartening to note that only 15.8 percent of open source projects actively fix vulnerabilities, and even then the mean time to remediation was 233 days (Sonatype 2017). Analysts documented that 1.8 billion vulnerable open source components were downloaded in 2015; 26% of the most common open source components have high risk vulnerabilities; on average, applications have 22.5 open source vulnerabilities (Sonatype 2016).

RSA, a security company that issues certificates that are used to electronically verify an information source, shared information describing how valid certificates for many of their customers were stolen. A single individual's machine was taken over though the use of an emailed spreadsheet that contained a zero-day exploit that installs a backdoor through an Adobe Flash vulnerability. Using this new access capability, the attacker installed a remote administration tool that allowed it to control the machine where the infected email was opened. From this staging point, system of systems connectivity is leveraged to identify RSA data sources of interest and copy them to a compromised hosting service (RSA 2011).

The shift to data hosting in Cloud (storage capabilities accessible via remote communication) services is promoted as ensuring, through a contract, that security issues are addressed. U.S. government acquisition is pushing to shift non-classified data handling to a Cloud environment, and the Federal Risk and Authorization Management Program (FedRAMP) now has 79 authorized Cloud services with another 64 in process.* The Cloud service now operates as a system of systems focused on efficiencies in sharing operating capacity with all of the customers connected to and using the service. A compromise on the Cloud service will provide access to data and processing capability for all of the organizations contracted to use these services. A new concern in this new type of infrastructure is virtual machine (VM) jumping, or hyper jumping, which can allow malicious users to gain access to several machines or hosts in an infrastructure and exploit vulnerabilities in the virtual machines. Virtualized platforms, heavily used in shared data services, offer powerful efficiency and significant savings on cooling, power, and floor space with the ability to host nearly unlimited applications, but the downside is that packing so many machines into a single infrastructure brings with it added susceptibility to cyber-attack (Rudy 2014).

New defects and vulnerabilities continue to be discovered. In environments characterized by these conditions, the effects of complex interrelationships and dependencies are not well understood (Brownsword et al. 2010). Once a vulnerability is successfully exploited, that knowledge is widely shared for reuse across the product acquiring communities. However, when patches are created and released, this content informs attackers who reverse engineer the software using broadly available tools (many in open source) to craft attacks on those who remain unpatched. Many libraries of information about vulnerabilities and appropriate mitigations, such as the National Vulnerability Database (https://nvd.nist.gov/) housed at the National Institute of Standards & Technology (NIST), have been collected and shared to further support broad accessibility to the information about vulnerabilities. Many forms of source code are also widely available to support opportunities for vulnerability discovery. Analysis tools that look for various types of vulnerabilities are widely available, many as free open source tools, and these are equally as accessible to the software owners as to the software exploiters.

* Retrieved May 2017 at https://marketplace.fedramp.gov/index.html#/products?status=Compliant&sort=productName.

13.5 Modeling the System of Systems Ecosystem That Supports Cyber Insecurity

New technologies can expand the susceptibility of system of systems to cyber-attacks, bringing in more vulnerabilities and expanding connectivity that increase access. Using standard communication tools, two-way communication has been added to a whole host of everyday things from children's toys and watches to devices such as Microsoft's Cortana and Amazon's Alexa, which will perform Internet searches based on verbal instructions and broadcast that information to anything within range that is listening. Applications are available to manage a home security system from a smart phone and potentially share that monitoring information with unintended recipients. These capabilities are being connected broadly into wide organizational use. The devices are all run by software which can be manipulated. Some organizations have seen security cameras converted to spy apparatus and heating sensors expanded to be microphones (Griffin 2014).

Just like all other software developers, attackers create components as building blocks that are supplied to other attackers through sources such as malware distribution websites. Attackers create an attack mechanism using the technical vulnerabilities within software products. Once enabled, the attack is disseminated through the software-enabled system of systems, potentially accessing key data or gaining control of the technology. Figure 13.1 provides a model of the growing insecurity environment as viewed from the exploiters' perspective. The layers are aligned from left to right as indicated by the numerals at the bottom of the diagram. Layers 1 through 3 focus on the supply side; layers 4 through 6 focus on the demand side. The resources or services at a given layer are used by the layers to its right—thus providing composite capabilities. Roles are abstract entities

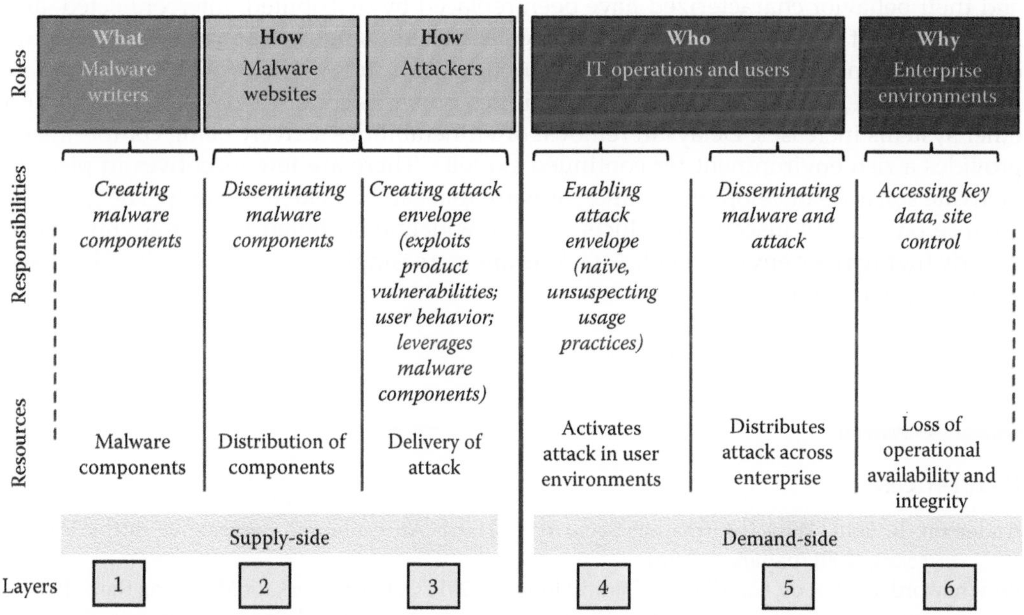

FIGURE 13.1
Anatomy of malicious attacks.

and not intended to represent specific people or job categories. This type of model is useful to improve the understanding of how such great impacts are possible and could, if applied to specific attacks, provide additional insight into deterrents.

13.6 Summary

Everything we do these days depends on software and yet we have no consistent and structured way of telling if what we are relying on is working securely. Even if we have the skills and knowledge to create all of our own code, we do not want to invest the time to do so—it's cheaper (often free) and faster to take what someone else has built and use it even if it is defective and riddled with vulnerabilities—so everyone is an acquirer in one form or another. Even when secure code is critical for important needs such as banking, transportation, critical infrastructure, or weapons systems, few are in a position to verify the quality and security of the code—it may be vendor intellectual property (IP) that the acquirer cannot access or provided to an integrator who incorporates it into a device, or it is legacy and no one wrote the acquisition contract to get access to the source code in the first place. Even with access, special tools are required to do an analysis, and identification of the vulnerabilities is only the first step. Only 38% of the organizations evaluated in version 7 of the Build Security in Maturity (BSIMM) do anything about software security in vendor contracts, mostly in the financial arena (McGraw et al. 2016).

Technologies make it easier to assemble systems and connect devices and systems into systems of systems, but the increasing reliance on software from vendors incentivized to distribute products with known defects has increased the susceptibility of these systems of systems to cyber-attack. Fairly simple computing architectures that could be understood and their behavior characterized have been replaced by distributed, interconnected, and interdependent networks linking systems constructed using a wide variety of software elements assembled from many sources. Vulnerabilities are widely available and increasingly accessible, and access to the vulnerabilities grows as connectivity from devices and other systems increases. Delays in removal of vulnerabilities actively being compromised provides a rich environment for continued exploits. There are few incentives in place to motivate software developers to build software that can be fielded securely. The growing reliance on free but insecure products that are widely distributed further aggravates an already highly risky environment. Attackers are increasingly organized to take advantage of these opportunities.

References

Anderson, R. 2001. Why Information Security Is Hard: An economic perspective. https://www.acsac.org/2001/papers/110.pdf.

Brownsword, L., Woody, C., Alberts, C., and Moore, A. 2010. A Framework for Modeling the Software Assurance Ecosystem: Insights from the Software Assurance Landscape Project. Software Engineering Institute, Carnegie Mellon University (CMU/SEI-2010-TR-028). http://resources.sei.cmu.edu/library/asset-view.cfm?AssetID=9617.

Cherdantseva, Y., Burnap, P., Blyth, A., Eden, P., Jones, K., Soulsby, H., and Stoddart, K. 2016. Review of cyber security risk assessment methods for SCADA systems. *Computers & Security*, No. 56 (February): 1–27.

Computer Science and Telecommunications Board (CSTB). 2010. Critical Code: Software Producibility for Defense. National Academies Press. http://www.nap.edu/openbook.php?record_id=12979&page=R1.

DHS. 2014. Cyber-attacks Have Affected More than 1000 US Businesses. Industry News & Analysis. http://www.cspdailynews.com/industry-news-analysis/technology/articles/cyber-attacks-have-affected-more-1000-us-businesses.

DiGangi, C. 2014. Parking Lots: The Next Threat to Your Credit Card. http://blog.credit.com/2014/12/parking-lots-the-next-threat-to-your-credit-card-103273.

Ellison, R., Goodenough, J., Weinstock, C., and Woody, C. 2010. Evaluating and Mitigating Software Supply Chain Security Risks. Software Engineering Institute, Carnegie Mellon University (CMU/SEI-2010-TN-016). http://resources.sei.cmu.edu/library/asset-view.cfm?AssetID=9337.

Federal Office of Information Security (FOIS). 2016. The State of IT Security in Germany 2016. https://www.bsi.bund.de/SharedDocs/Downloads/EN/BSI/Publications/Securitysituation/IT-Security-Situation-in-Germany-2016.pdf?__blob=publicationFile&v=3.

Friedberg, I., Skopik, F., Settanni, G., and Fiedler, R. 2015. Combating Advanced Persistent Threats. *Computers & Security*, no. 48:35–57.

Geer, D. 2010. Cybersecurity and National Policy. *Harvard National Security Journal*. No. 1: 207–219. http://harvardnsj.org/wp-content/uploads/2015/01/Volume-1_Geer_Final-Version.pdf.

Griffin, A. 2014. The only way to be absolutely sure that hackers aren't in your webcams is to turn them off. But there's a lot you can do to keep your cameras safe on the Internet. http://www.independent.co.uk/life-style/gadgets-and-tech/news/hundreds-of-baby-monitors-and-cctvs-hacked-how-to-keep-your-cameras-safe-9873777.html.

Guruburan. 2017. Permanent Denial-of-Service attack with IOT devices-BrickerBot. https://gbhackers.com/permanent-denial-service-iot-devices-phlashing/.

Jones, C. 2013. Software Quality in 2013: A Survey of the State of the Art. Namcook Analytics LLC. https://www.pnsqc.org/software-quality-in-2013-survey-of-the-state-of-the-art/.

Maier, M.W. 1998. Architecting principles for systems-of-systems. *Systems Engineering*, No. 1,4:251–313.

McGraw, G., Migues, S., and West, J. 2016. BSIMM 7. http://www.bsimm.com.

Ogut, H., Raghunathan, S., and Memon, N. 2011. Cyber Insurance and IT Security Investment: Impact of Interdependent Risk. http://infosecon.net/workshop/pdf/56.pdf.

Perlroth, N. 2014. Cyberattack that hit Target a widespread threat to consumers (August). *New York Times*. https://www.bostonglobe.com/business/2014/08/22/cyberattack-that-hit-target-affecting-businesses/AmsccErTlI4vLhQpUfSorL/story.html.

Peters, S. 2015. 3 of 4 Global 2000 Companies Still Vulnerable to Heartbleed (April). http://www.darkreading.com/vulnerabilities—threats/3-of-4-global2000companies-still-vulnerable-to-heartbleed/d/d-id/1319768.

PRC. 2014. Chronology of Data Breaches: Security Breaches 2005—Present. *Privacy Rights Clearinghouse*. http://www.privacyrights.org/data-breach.

Regenscheid, A. 2017. Platform Firmware Resiliency Guidelines (May). NIST SP 800-193. *National Institute of Standards and Technology*. http://csrc.nist.gov/publications/drafts/800-193/sp800-193-draft.pdf.

RSA Fraud Action Research Labs. 2011. Anatomy of an Attack Retrieved (April). https://blogs.rsa.com/anatomy-of-an-attack/.

Rudy, M. 2014. Virtualization Security: Tips to Prevent VM Hyper Jumping (October). *Techopedia*. https://www.techopedia.com/2/30822/trends/virtualization/virtualization-security-tips-toprevent-vm-hyper-jumping.

Santinelli P. 2015. 78% of Companies Say They Run Operations on Open Source (May). https://opensource.com/business/15/5/report-future-open-source-survey.

Smith-Spark, L. 2017. Global Ransomware Attack: 5 things to know (May). *CNN*. http://www.cnn.com/2017/05/13/world/ransomware-attack-things-to-know/index.html.

Sonatype. 2016. Open Source Development and Application Security Survey. Sonatype. https://www.sonatype.com/open-source-breaches.

Sonatype. 2017. Sonatype Open Source Development and Application Security Survey: 2017 State of the Software Supply Chain. https://www.sonatype.com/ssc2017.

Symantec. 2016. Ransomware and Businesses 2016. ISTR Special Report. http://www.symantec.com/content/en/us/enterprise/media/security_response/whitepapers/ISTR2016_Ransomware_and_Businesses.pdf.

Tankard, C. 2011. Advanced Persistent Threats and How to Monitor and Deter Them. *Network Security*, no. 8:16–19.

Verizon Risk Team. 2015. Data Breach Investigations Report. *Verizon*. https://msisac.cisecurity.org/whitepaper/documents/1.pdf.

Verizon Risk Team. 2016. Data Breach Investigations Report. *Verizon*. http://www.verizonenterprise.com/verizon-insights-lab/dbir/.

Woody, C., Ellison, R., and Nichols, W. 2014. Predicting Software Assurance Using Quality and Reliability Measures (CMU/SEI-2014-TN-026). Software Engineering Institute, Carnegie Mellon University. http://resources.sei.cmu.edu/library/asset-view.cfm?AssetID=428589.

U.S. Air Force Scientific Advisory Board (US AF SAB). 2011. Sustaining Air Force Aging Aircraft into the 21st Century (SAB-TR-11-01). U.S. Air Force.

14

The Challenge of Performing Research Which Will Contribute Helpful Engineering Knowledge Concerning Emergence

Timothy L.J. Ferris

CONTENTS

14.1 Introduction: Knowledge in Engineering

"Engineering is not a science, but a practice of the necessary to achieve a given goal" (Cripps 1991, p. 16). This quotation summarizing the nature of engineering is important in framing the nature and role of knowledge, and therefore of research, in engineering. In turn, the pragmatic nature of engineering specifically leads to a particular kind of knowledge of emergence being necessary in order to be useful. The issue of kinds of knowledge will be discussed throughout this chapter.

The task of engineering is to bring into being products and systems which provide, in a desirable manner, effective solutions to needs or means to produce a required effect. The purpose of science is to develop knowledge which is tested to provide confidence in the truth of the knowledge, which is commonly formulated as the testing of hypotheses. In this process, a scientist makes general observations of an area of interest which enable the generation of a hypothesis that becomes the subject of a test, and through that testing knowledge of the phenomenon relevant to the hypothesis is produced. During the 1950s, there was a general trend by engineers to assert the scientific credentials of engineering

(Ferris 2007a; Ferris 2007b; Ferris 2008). This followed the significant engineering advances made during World War II that were clearly associated with the application of scientific knowledge in the development of technologies which contributed significantly to the conduct and outcome of the war. The associated factor was that "science" was the recently emerged criterion for admission of a field of study into universities, so it was felt necessary for engineers to demonstrate the scientific credentials of engineering. At the extreme of this trend Hickey wrongly proceeded to assert that an engineer has a hypothesis which can only be tested through implementation in a product (Hickey 1960, p. 74). However, the engineer's goal, qua engineer, is not to test a hypothesis related to the possibility of developing a solution, but rather to provide the most appropriate solution for the need consistent with Cripps (1991).

The popular hendiadys "science and engineering" conflates and confounds engineering and science, treating them as part of the same enterprise, whereas they are distinct. Science seeks to generate knowledge about the world and things and phenomena within it. To this end science has developed methods of discovery which instantiate application of the view of knowledge and learning provided in the traditional western epistemology. The classical epistemological account of knowledge is expressed in the trifold conditional statement, or a variation.

S knows that p if and only if:

1. p is true;
2. S believes that p; and
3. S is justified in believing that p" (Gettier 1967).

Gettier's paper is very controversial because he challenged this account of the requisites of knowledge and several variants but, in turn, many epistemologists have objected to his view.

Engineering seeks to provide appropriate and useful outcomes. Thus, engineers use the best available knowledge about phenomena which they intend to exploit, including, and often predominantly, the knowledge developed through science, to predict results expected from design configurations which are posited as candidate solutions. In the acronym STEM (Science Technology Engineering and Mathematics) the science and engineering hendiadys is augmented by "technology" and "mathematics." Mathematics is used in engineering analysis and in engineering research, often to the extent that engineering appears like a branch of applied mathematics when, really, the power of mathematics in engineering follows because of the homomorphism of mathematics and the phenomena of real things, and successful engineers always keep in mind the limitations of their analytical models and seek empirical evidence in their work. Engineering, in the use of mathematics to predict expected outcomes of design configurations, tests the extent to which the homomorphism assumption is reasonable.

Although engineering research is mathematically intensive, engineering research is fundamentally different than research in mathematics. Research in mathematics is focused on the exploration of the implications of the axioms of a mathematical system, whereas physical science research is about the phenomena and the role of mathematics is description of observations. In engineering research the focus is on finding appropriate interventions in the world and the role of mathematics is as a tool to enable prediction of outcomes. A further purpose of mathematics in engineering research is to develop design tools to determine the parameters of design configurations that exploit known phenomena to achieve specified measures of performance.

In summary, knowledge in engineering concerns the subject matter of things that are useful in the engineering of solutions to needs and the methods required to apply the scientific knowledge of phenomena to the achievement of solutions to needs.

14.2 Definition of Emergence

Engineering is the endeavor in which things are combined according to a design with the purpose of achieving specific effects which are different than the effects or properties of the elements which are arranged through the design and building work. The goal of achieving effects through combining things that do not, by themselves, produce those effects is consistent with the common definition of emergence (Hitchins 1992). In the common definition emergence is the phenomenon that results in the effect of a combination of elements that form a system being different than the effect observed from any of the elements of the system taken alone. This juxtaposing of a description of the objectives of engineering and the definition of emergence makes it appear that all engineering work would be recognized as involving emergence, but this is not so universally accepted in the systems engineering community, which is the major engineering community in which emergence is discussed.

Discussion of emergence in the systems engineering community may produce a recitation of simplistic statements, such as "the whole is greater than the sum of the parts," or an argument that emergence refers to events that take one by surprise, or even just nasty surprises, sometimes euphemistically referred to as "undocumented features" (BKCASE Editorial Board 2016). None of these approaches is helpful in enabling substantial engagement with the idea of emergence.

To assert that "the whole is greater than the sum of the parts" does not help anyone design a system to provide desired services, performance, or some other characteristic, and simultaneously not produce either undesired or undesirable outcomes. The major problem with the simplistic statement is that it does not guide the achievement of the desired results.

The second approach is to regard emergence as concerning the things that take one by surprise, that is, the unexpected outcomes of assembling a system. The problem with this view is that it does not have a fixed boundary between what is, in contrast to that which is not, classified as emergent. Thus, a good designer, who knows a lot, may be able to predict what happens in certain circumstances and not be surprised, whereas a less knowledgeable engineer may be surprised by the same properties of a system under development. Alternatively, as knowledge develops, effects which were previously unobserved and unexpected have been observed on first occurrence and therefore transition from emergent to known, and presumably, expected phenomena. A definition of anything that results in such a fluid scope of that to which the term applies is not helpful. The only place in engineering where such a concept of unexpected and surprising is useful as a separate classification is in legal defense in cases where the argument is whether an appropriately competent engineer should have been aware of, and expected, systemic behaviors that caused a problem.

The third view is that emergence concerns undesirable effects found after assembly of the system. To define emergence as concerned with unexpected undesirable effects trivializes emergence in two ways. First, it associates emergence with the subclass of unpredicted or unpredictable outcomes which are also undesirable. This concern is addressed above. Second, it associates emergence with that which is considered undesirable. This is problematic because in the assembly of a system a particular property may be considered desirable or undesirable as a result of a stakeholder's viewpoint, with different stakeholders judging particular effects in opposite, or at least significantly different, ways.

Each of the three approaches to defining emergence is unsatisfactory because they yield different views concerning what is emergent, differing depending on viewpoint, time, or knowledge. Such a definition can do no better than allow, and even encourage, imprecise thought.

A useful definition must:

1. Provide terminology that has constant breadth of inclusion of cases.
2. Name something independently of connotations of value concerning the goodness, or otherwise, of the subject matter.

In this chapter we follow Checkland, as quoted in SEBoK (BKCASE Editorial Board 2016), who defined emergence as "the principle that entities exhibit properties which are meaningful only when attributed to the whole, not to its parts."

This definition has the desirable characteristics of:

1. Not ascribing a value judgment, good or bad, on anything described as emergent.
2. Providing a time, knowledge, and perspective invariant boundary between emergence and not-emergence.
3. Accommodating the primary goal of engineering, to make systems that satisfy needs through assembling components to produce effects that are not achieved by the components taken alone.

Another philosophical issue arises: if one knows certain facts, does one also know their logical consequences (Hintikka 1970)? This issue is important because emergence is the consequence of multiple elements, each with its own properties, being assembled into an interconnected construct. The question is whether, in adducing certain facts about the elements of a system, it can be said that one is aware of their consequences and interactions. If one were aware of the interactions, one would not be surprised by the effects of the combination of known entities.

Experience shows that surprises happen. We conclude that one may be aware of entities and their characteristics but not have immediate awareness of the characteristics of a system constructed from the entities. We also note that design relies on the consistency of matter, so that previous assemblies of entities enable prediction, based on precedence, of the properties that will be observed in future assemblies (Goodman 1973).

14.3 Framework to Organize Research Objectives

Research is commonly understood as the process by which new knowledge is created. Each academic field has developed traditions about both the nature of knowledge which is desired and the research methods to develop it. In most disciplines there are one or a few methods which have become accepted as the possible means of developing knowledge, and these have become so embedded in the culture of the discipline that they are part of the social and intellectual construct of the discipline. The famous workers in the discipline achieved their status through the use of those methods and those aspiring to recognition are effectively bound to pursue the same methods, on pain of ostracism. Those who have succeeded are rarely motivated to challenge the status quo. The consequence is that most disciplines, after an initial establishment phase, move to a continual refinement phase until, perhaps, something happens that results in a paradigm shift (Ackoff 1979a; Ackoff 1979b; Ackoff 2001; Kline 1995).

As we consider research as the activity that develops knowledge, we look to several forms of knowledge which have been identified. The first formulation is "know that," following Ryle's distinction between "knowing that" and "knowing how" (Ryle 1948). This distinction is significant in engineering, where both kinds of knowledge are required (Bucciarelli 2003). "Knowing that" is of the form of declarative knowledge, that is, knowledge of a kind that can be articulated to represent ideas. Knowledge of this type is about things "out there," objects which can be observed, but primarily in an object relation to the knower and speaker. This kind of knowledge is of a form which is relatively easy to teach because it is possible to reduce the teaching to statement of the representation. This is expanded in education in the development of kinds of ability to interact with the content as described in the kinds of learning described in the cognitive domain of Bloom's taxonomy (Bloom et al. 1979). "Know that" is associated with the conceptualization of knowledge as concerning the representations of facts rather than the assimilation of the significance of the facts into an action guiding construct.

The second formulation is "know how," Ryle's formulation to describe capacity to perform a function (Ryle 1948). The capacity to act appropriately is distinct from capacity to describe knowledge related to the function or the relevant theory. Knowing how does not preclude ability to articulate what is known, but emphasizes the ability to perform an act. The distinctive feature of knowing how is that the subject's orientation is with doing rather than describing something. "Knowing how" may be expressed in ability to use that which is the subject of "know that" but may also include abilities to perform action. Ability to perform action can be expanded using the psychomotor domain, for example as expanded in engineering (Ferris 2010). The ability to judge what is appropriate to do also involves development of the individual in the affective domain, which is concerned with developing value systems that lead to appropriate action (Krathwohl et al. 1964).

A third kind of knowledge is named "knowing" by Nissen (Nissen 2006) and "procedural knowledge," or "skill" by Biggs (Biggs 1999). The emphasis in "knowing" is on the ability to choose and perform some action in an appropriate and effective manner. Thus, Nissen says that knowing how to ride a bicycle is demonstrated by mounting and actually riding a bicycle. "Knowing" is tested or demonstrated through a practical test in which the candidate must perform the action in a fluent manner. Ability to articulate anything about the matter, its situation, or a theory about the action or its situation is irrelevant to "knowing." "Knowing" contrasts with "know how." In "know how" the emphasis is on performing a function as a capacity, but "knowing" is usually associated with ability to make sound judgements about the action and whether to act. Thus, knowing is a competency that arises from appropriate and fluent application of that capacity to act.

The three kinds of knowledge relate to abstraction, description, and theorization, through applied knowledge which enables performance of action, to ability to perform fluently. The three named kinds of knowledge are identifiable points in characterizing knowledge, but any particular instance of knowledge embodies a particular combination of abstract theorizing, ability to act, and fluency of action, thus positioning the knowledge at a particular place in a continuous knowledge space.

Identifying the three kinds of knowledge is useful for theorizing about knowledge and, by extension, research. Research creates new knowledge of some kind. The new knowledge developed through research must be new to everyone, not just new to the investigator. The increment of advance of knowledge may be of any size, possibly very small, and situationally constrained. This contrasts with science which seeks to generate context independent knowledge. But another very important factor is introduced through the recognition of the three kinds of knowledge. This is that research may be performed with a variety of objectives with respect to the kind of knowledge that is to be developed. There are two possible approaches to this challenge of finding an organization of objectives for research. One is to identify the kinds of knowledge produced by activities currently recognized as research. This approach has the problem that it limits the task to description of what has been done already.

This author followed this approach, with collaborators, but found it frustrating, as did the collaborators (Ferris et al. 2005; Ferris et al. 2007). The alternative is to look for a complete framework of kinds of knowledge, which could lead to generation of possible objectives of research. We choose the latter approach which has the advantage of holding potential for a greater completeness and seems to overcome the frustration described above.

Augustine of Hippo, in *De Civitate Dei* XIX:1, refers to Marcus Varro's now lost work, *On Philosophy*, which presented an analysis of the characteristics of all the actual and possible schools of philosophy (Augustine of Hippo 1984). Varro had organized philosophies according to their position, in a set of two or three categories, in each of six dimensions, leading to a taxonomy with 288 fundamentally different classes of philosophy. The relevance of a categorization of philosophies to the organization of objectives for research is that the philosophies represent views on the possible character and purpose of the knowledge that results from the research, so the research is an activity performed to discover knowledge, potentially of any of the possible kinds.

Varro's classification of possible philosophies has been transformed into a classification of research objectives, shown in Table 14.1. The "Life Goal" dimension which expresses

TABLE 14.1

Taxonomy of Research Objectives Built from Varro's Classification of Philosophies

Dimension	Code	Possible Categories	
Desired Outcome	*D1*	Theoretical development of the field	
	D2	Practical development in the field	
	D3	Development of theory and practice	
Knowledge Goal	*K1*	Goal is knowledge	
	K2	Goal is the field of application	
	K3	Goal is both knowledge and application	
Beneficiary	*B1*	Investigator	
	B2	Others in the community	
Certainty	*C1*	Knowledge is definite and certain	
	C2	Knowledge is contingent	
Tradition	*T1*	Build within the tradition of the field	
	T2	Challenge, replace, or reject the tradition of the field	
Life Goal	*L1*	Enjoy knowing	Leisure
	L2	Enjoy practicing	Work
	L3	Enjoy knowing and practicing	Work and leisure

a view on the relation between the knower and knowledge may, at first, be criticized as redundant, since it seems similar to the "Knowledge Goal" dimension with the effect that the factors {*K1,K2,K3*} and {*L1,L2,L3*} appear not orthogonal. The distinction of these dimensions is in the weltanschauung of the researcher. The "Knowledge Goal" dimension concerns the rational description of the research project purpose, while the "Life Goal" dimension concerns the existential purpose of knowledge and research in general. The other dimensions are distinct and distinguish characteristics of the possible purposes of research. Thus, in the research traditions of a field the "Knowledge Goal" dimension leads to the articulated description of what is considered research and how it should be done. The "Life Goal" dimension leads to the mores of the social community of the field, in which the research is done.

The dimensions in the taxonomy of Table 14.1 are:

Desired Outcome, {*D1,D2,D3*}, concerns the researcher's belief about the objective of the research project. The objective of a project may be to improve theory about the subject, practice related to the subject, or both. Development of theory emphasizes intellectual constructs describing the framework of ideas in the field. Development of practice concerns improving ability to act appropriately. For a particular subject matter, research conducted from each of these perspectives will involve different methods and have different indicators of success.

Knowledge Goal, {*K1,K2,K3*}, concerns whether the knowledge is valued for its intrinsic or its instrumental value. This contrast divides between disciplines oriented towards creation of knowledge, such as the sciences, and those oriented towards performance of some activity or achievement of particular outcomes, such as engineering.

Beneficiary, {*B1,B2*}, is concerned with the user of the research product. The researcher performs work intended to benefit a target audience. The target audience may be the researcher, for research to satisfy curiosity or for self-development, or others, where the intention is dissemination. This has implications, inter alia, for the extent to which generalizability is a focus of the work.

Certainty, {*C1,C2*}, concerns the researcher's epistemological stance. Knowledge may be viewed as being true in an absolute sense, that is always true, true everywhere, and the same for all knowers. Alternatively, knowledge may be viewed as relative or contingent, depending on the perspective of the observer. This dimension concerns whether it is possible to know things objectively or only experientially.

Tradition, {*T1,T2*}, concerns whether the tradition of the field is viewed as sufficient, so that the research is done to enhance knowledge of some aspect of the tradition within the overarching framework of the field. Alternatively, the iconoclastic position sets out to challenge or reject the tradition. Challenging the tradition may involve interpreting observations significantly differently than the current tradition, while rejecting the tradition may be done to be different or to radically reformulate the tradition.

Life Goal, {*L1,L2,L3*}, concerns the researcher's existential engagement with the knowledge under development. The Objective in Life may be the pleasure obtained through knowing, or through practice and achievement, or a combination. The Desiderata dimension concerns what is regarded as innately desirable, and Relation to Knowledge concerns a rational view of the nature and purpose of knowledge while the Objective in Life concerns the researcher's relation to the knowledge.

14.4 Discussion of Methods to Investigate Emergent Effects

Six approaches to research that may be used to investigate questions or issues related to emergence are discussed in this section.

14.4.1 Case Studies

"Case studies" is used here to refer to researcher initiated investigations motivated by the researcher's purpose of discovering the facts and relationships relevant to a case. A case study investigation is conducted by finding available materials and possibly obtaining primary data collection from participants or measurement of the case and other similar sources. The case study method is commonly used in some fields, among which are investigation of engineering projects, and engineering accidents and disasters. In these investigations "emergence" is often a phenomenon that makes the particular case interesting to investigate, for example because there is something instructive about how a project progressed, or an accident occurred because of an unexpected effect of the confluence of factors, matching one of the common views of "emergence" discussed in Section 14.2.

The investigator driven case study is limited because the study is constrained by the source materials that the participants with direct awareness of the situation choose to make available or which have otherwise been placed in the public domain. This limitation of available sources prevents assurance that the materials available present a complete and fair view of the situation, and that there is nothing that would lead to different conclusions in materials not divulged. This impacts the confidence that can be placed in any conclusions. A second major limitation of the case study method is that its subject matter is a particular case. This raises the question of whether what is found is just the description of a particular case or is, in some way, generalizable and the bounds of the generalizability. The impact of this limitation is whether the case should be read as a description of a past case or can be used to make recommendations for analogous cases.

Turning from the limitations of the case study method, we consider the strengths of the approach. The case study approach is well suited to addressing questions of how or why, and therefore for providing insight into a state of affairs (Yin 2014). If one understands the how and why relations that have been present in particular instances of a class of things, then the reasonable generalization is insight into the class of things of which the instance is an exemplar. The challenge is to recognize and define the attributes of the class appropriately so that the belief that one has gained insight is appropriately applied.

Depending on the purpose of the case study work, the researcher may use a relativist, interpretivist, or realist epistemology (Yin 2014). The choice of epistemological position to be used reflects a conclusion about a combination of the case, the information available about the case, and the researcher's purpose in performing the investigation. We now consider the case study method in relation to Table 14.1, to determine the most likely hex-tuple descriptors of the objectives of research.

$$\langle D3,K2,B2,C2,T1,L1\rangle \text{ for a relativist epistemology} \quad (14.1)$$

$$\langle D2,K2,B2,C2,T1,L1\rangle \text{ for an interpretivist epistemology} \quad (14.2)$$

$$\langle D1,K2,B2,C2,T1,L2\rangle \text{ for a realist epistemology} \quad (14.3)$$

Case studies are usually motivated by one of two scenarios. The first is understanding of an accident or disaster, usually on a newsworthy story scale, and in most cases the event that made for the problem was an unintended and disadvantageous emergent event. That is, case studies are usually investigations of emergence, the presence of which was discovered through the observation of unexpected or unintended effects and frequently these effects are disadvantageous. The usual purpose of the case study is to identify a crucial contributing factor in order to make a recommendation which would prevent similar problems.

14.4.2 Forensic Case Studies

We now turn our attention to a distinct kind of case study, the forensic case study. A forensic investigation is an investigation of a particular case, and always performed with a view to discovery of the cause of a problem, which may be motivated by the need to attribute blame or to formulate recommendations to improve future outcomes. The purpose of forensic case investigations may be legal or organizational. In a legal investigation the immediate purpose of the investigation is a formal investigation such as a court case or coroner's inquest. Organizational purposes for this type of investigation are driven by a need for rigorously founded recommendations for future practice, although the rigor required may not need to be of a standard that would be required for legal proceedings.

The forensic case study situation empowers the investigator with discovery powers much stronger than those available to the "normal" case study researcher. The legal or organizational principal directing the investigation gives the investigator strong powers to demand relevant information, in contrast to the curious investigator's need to rely on the willingness of information holders to divulge. This power is particularly important in situations where information holders may be afraid of either legal action or embarrassment associated with the matter. A forensic case study, like other investigations of the past, can only obtain primary evidence, either information or tangible, created while the events under investigation were unfolding. The other evidence available to the investigator is secondary, including testimony of participants, potentially taken under oath, and remnants of the things that remain at the time the investigation was initiated.

An investigation to identify responsibility for a situation will lead to an account of events in the case. The account does not, itself, provide recommendations for the future. In contrast, a coronial inquest, commission of inquiry, or less formal intra-organization equivalents are intended to yield recommendations to improve future scenarios. The hextuple descriptors of the research objectives of these two approaches to the forensic case study are:

$$\langle D2,K1,B2,C1,T1,L2\rangle \text{ to discover responsibility for past events} \quad (14.4)$$

$$\langle D2,K2,B2,C1,T1,L2\rangle \text{ to make recommendations to improve the future} \quad (14.5)$$

Like the general case study, the forensic case study is performed to understand emergent effects found through events which led to a bad outcome. The forensic case study is motivated by a bad outcome, whereas the general case study may be motivated by either a bad or an interesting outcome.

14.4.3 *Post Hoc* Empirical Studies

In some situations an observer is limited to observation of things and scenarios that already exist without the opportunity to make interventions but the observer desires to form conclusions about classes of cases. The limitations on the investigator intervening may be practical, such as cost or time required, or ethical, such as investigating an effect hypothesized to cause harm. Constraints such as these limit the investigator to observing cases that exist and seeking to form conclusions about them. Such investigations are *post hoc* empirical studies. The principal feature of *post hoc* studies is that the investigator cannot control the conditions of observations and may only have data collected by others available, with differences in observation methods for data from different sources.

In *post hoc* empirical investigations the researcher observes a sample of cases selected from a broader population with respect to a set of parameters that represent the researcher's model of the phenomenon of interest, with the result that conclusions can be formulated within the construct of the model but it is not possible to discern if the model is a valid method to view the phenomenon (Ferris 1997). This epistemological problem represents a serious limitation of this type of research. The difficulty caused by the epistemological problem is that one cannot know if correlated observations have a direct causal linkage or if the correlation could be explained by some other causal mechanism. A full account of the possible causal relations of two correlated sets of observations is given by Skrabanek and McCormick (1990).

Four difficulties with *post hoc* empirical studies are discussed here:

1. The cases studied pre-exist the study. Thus, the samples are chosen on the basis of presence of an outcome manifestation resulting in all observations of other factors being made after the action has occurred. This prevents investigation of whether the observed effect results from the matter of interest or from other known or unknown factors.
2. The cases receiving each of the treatments considered in the observations cannot be randomized because the treatments were applied by someone who believed that the treatment applied would increase the probability of achieving their desired outcome or the allocation of treatments to cases has no explanation. Observations cannot be used for determining any statistical relationship of application of a treatment to cases exhibiting particular characteristics because the characteristics that identify classes of case are not the basis for assignment of treatments. In turn, this prevents use of the results to predict the likely outcome if particular treatments are applied to a random member of the relevant class.
3. The pre-existing nature of the observed cases introduces the risk that determination of the surmised "background" factors may be biased by knowledge of the outcome class of the sample. Performing such a data collection process without observer awareness of the outcome classification of the sample is either difficult or impossible because in the *post hoc* observation situation the outcome exists before the initial condition is questioned.
4. Despite all attempts to the contrary by a diligent investigator, the sample actually studied may reflect a bias towards one sub-group in the population. For example, a medical investigation of patients presenting to a clinic is biased by the self-selection factor that the people presenting had something that made them believe

they needed to seek assistance, and therefore they are not representative of the broader population. This factor can bias a doctor's perception of frequency of a particular illness (Singer et al. 1997; Skrabanek and McCormick 1990).

The *post hoc* study may struggle to yield insight about the emergent effects which may have motivated it. The challenge may arise at several levels: the difficulty in obtaining suitable data to make a judgment about correlation of manifestations, and absence of control for extraneous factors makes conclusions about causality difficult.

The hex-tuple descriptor of the research objectives of *post hoc* empirical studies is:

$$\langle D1,K1,B2,C1,T1,L1\rangle \text{ for } post\ hoc \text{ empirical studies} \tag{14.6}$$

The *post hoc* empirical study is a method to discover emergent effects through observation of facts on the ground but is challenged to provide causal explanations for correlations of factors and therefore presents emergence as a phenomenological matter.

14.4.4 Experimental Empirical Investigation

Experimental investigation is a research method which is commonly regarded as the best method for discovery of the properties of things. While in *post hoc* studies it is only possible to observe outcomes and then investigate antecedent conditions which enables discovery of predictive correlations, an experimental study enables, with appropriate design, the investigation of causality. In experimental studies there are well-established protocols for randomly assigning cases to treatments and, to the greatest extent possible, blinding the investigator and other participants to the correlation of cases and treatments, which is done to reduce observation bias, in which observers may report outcomes they expect rather than what is objectively present.

Emergence presents a challenge in the context of experiments. Usually an experiment is performed to find the relationship of an independent variable applied to a class of entity to one or more dependent variables observed in the experiment. This requires an *a priori* theory of what the investigator believes may explain the situation represented by the experiment. In most cases an experiment is designed to investigate the theory posited through testing hypotheses about what one would expect to observe if the theory were true. Note, the hypothesis is usually reconstructed as the null hypothesis, because disproving the null hypothesis is as close as one can approach demonstrating truth of the positive hypothesis. A population of the entity of interest is sampled with the purpose of avoiding bias of which treatment is applied to samples, which in turn may bias conclusions about relationships of the variables. Enough samples are subjected to the experimental conditions to enable statistical testing of the observations using accepted inferential statistical techniques. This process demands that the experimental conditions exclude the possible effect of anything on the sample other than the independent variable under test. This enables discovery of one class of emergent effect: that which regularly occurs as a result of an underlying mechanism. This is the kind of emergent effect which is embedded in the normal practice of engineering as the scientific principles which are applied in design and action.

The aspects of emergence associated with the idea of surprising or unexpected effects are unlikely to be discovered through experiments because the method of performing experiments necessarily aims to separate the relationship under observation from the influence of any extraneous effect. This intentional isolation of the experiment from other effects results in physically excluding conditions in which unexpected things can be seen.

If something happens in an experiment which is of the form of "surprising emergence," it is likely that the result will be sufficiently different than the results obtained for other cases studied in the experiment that the observation will be classified as an unexplained outlier, and ignored.

Experimental investigations are capable of producing knowledge which can be used in the "forward path" of engineering design to propose, and through analysis, demonstrate, design ideas which should produce desired emergent effects. The problem is that the design of the experimental method will not enable discovery of emergent effects other than the relationship specifically investigated in the experiment. In addition, the presuppositions about the nature of knowledge embedded in the experimental method, such as repeatability and broad applicability, make it unlikely that experiments will identify unusual emergent effects because such effects, if not part of the phenomenon that the experiment is designed to investigate, will not be demonstrable through inferential statistical methods.

Experiments are motivated by one of three goals. The first is the pure scientific desire for knowledge about a phenomenon because the researcher finds the phenomenon interesting but without any particular intended application of the knowledge. Second is the applied knowledge motivation in which experiments are performed to investigate a particular phenomenon for the purpose of potential application. The intended application may cause the researcher to be concerned with only a certain subset of the potential range of independent variables, or to impose other constraints on what is investigated. A third motivation is the research and development scenario in which the subject matter of the experiments is things proposed as potential designs and the goal is to determine performance under specified conditions. These three motivations map differently to the objectives of research described in Table 14.1. The hex-tuple mapping is shown in Equations 14.7 to 14.9.

$$\langle D1, K1, B2, C1, T1, L1\rangle \text{ for a pure scientific motivation} \tag{14.7}$$

$$\langle D3, K3, B2, C1, T1, L3\rangle \text{ for an applied scientific motivation} \tag{14.8}$$

$$\langle D2, K2, B2, C1, T1, L2\rangle \text{ for a research and development purpose} \tag{14.9}$$

14.4.5 Research to Enable Engineering

The purpose of engineering work is to perform action which enables desired outcomes to be achieved. This work uses the knowledge of the various sciences with the purpose of informing action of design and building of things that will perform as intended in a range of situations relevant to the particular design. Success in this task demands willingness to use knowledge from any field of science and any experience, codified in the best currently available expression of the knowledge in order to inform proposed design ideas, to analyze and evaluate those ideas, and to review and criticize the design ideas for unintended emergent effects. The first three verbs in the previous sentence, "propose," "analyze," and "evaluate," work with the emergent effects that are known phenomena in the relevant branches of science. In this part of the engineering task the engineer works with known emergent effects to develop desired outcomes. The latter two verbs in the same sentence, "review" and "criticize," encounter the problem of unintended emergent effects. Some of this work can be done by finding out if there is existing scientific knowledge of effects which appear in the design proposal because of the juxtaposition of elements and for which there is an existing body of relevant knowledge.

For example, electronic design at the simpler level often taught in undergraduate degrees focuses on the circuit theory properties of the arrangement of components. A practical realization of the circuit will assemble those components onto a substrate and put it in a housing, and through that physical construction will introduce mechanical and thermodynamic effects which will interact with the environment of intended deployment. These factors can be analyzed by introducing knowledge from the relevant disciplines in the review process. If the engineering team challenges the original design proposal with the right review questions, the emergent effects of these kinds of issues can be identified. The difficulty is knowing the right review questions to properly challenge the design concepts. Once the project passes the tests applied to it by the engineering team analysis of proposals, the project progresses to instantiation in which various empirical tests can be performed on parts, or the whole, of the product, or experimental test-beds which represent aspects of the product. Tests may be performed to determine if design concepts are appropriate, or whether particular designed items are suitable for incorporation into the design.

In most cases the test purpose is to subject the designed entity to a set of conditions and observe parameters. If the measures are consistent with design predictions, in most cases, the entity is declared to have passed the test and the process progresses to the next test. In this situation it is assumed the models used sufficiently describe the situation as to provide assurance that the model is accurate enough to proceed with its use, and to trust its conclusions. This scenario is consistent with the use of Measures of Performance which are usually used to demonstrate compliance of the design to specification. Tests designed to determine Measures of Effectiveness, where the entity is tested under more realistic operational situations, are more likely to discover unintended emergent effects, and the limited resources which can be expended on such tests necessarily leave many conditions within the declared operational envelope of the entity which have not been tested. Whether an unexpected emergent effect is discovered in effectiveness tests or in use, the discovery is late enough in the lifecycle as to present a significant difficulty to the success of the project. This is the problem that research into emergence should aim to pre-empt.

Engineering projects conducted in the normal way produce both the deliverable products and the learning about what is achieved by particular designs. However, this learning is in the form of case specific outcomes, and does not provide a systematic foundation for future action. As each project is performed discoveries about the interaction of elements within the project, both the deliverables and the characteristics of the performing organization, will be made or at least enabled. These will form a scatter of points in a space with no systematization to assist interpretation that will enable recommendations about future action that can be put forward with assurance.

However, useful engineering knowledge about the design of systems, and therefore the emergence that occurs in the development of a system as related to both the things delivered by the project and the methods of performing the work, can be learned through appropriately designed research projects. Complex, or large scale, engineering projects of the kind that deliver systems, and therefore have the potential to inform the engineering of complex systems, are not amenable to experimental investigation for a range of reasons including resources and the impossibility of controlling to study the effect of any particular independent variable without confounding through a variety of other factors. This demands consideration of other methods which are capable of generating useful knowledge about the impact of design or method choices in such projects. These methods will differ from those used in other types of investigation.

Engineering-relevant research to discover the emergent effects in projects related to systems can approach the investigation from several perspectives. Projects may be performed to understand how to deliver in reasonably known areas of practice, or for the provision of radically new kinds of deliverables, or for investigation of the method a particular engineer should use to achieve desired results. These three purposes map differently to the objectives of research described in Table 14.1. The hex-tuple mapping is shown in Equations 14.10 to 14.13.

$$\left\langle \{D2, D3\}, K2, B2, C1, T1, L2 \right\rangle \text{ for knowledge about deliverables in a reasonably well-known field} \quad (14.10)$$

$$\left\langle \{D2, D3\}, K2, B2, C1, T2, L2 \right\rangle \text{ for a knowledge about deliverables in a radically new field} \quad (14.11)$$

$$\left\langle \{D2, D3\}, K2, B2, C2, T1, L2 \right\rangle \text{ to discover methods to suit particular engineers, personalized practice} \quad (14.12)$$

$$\left\langle \{D2, D3\}, K2, B1, C2, T1, L2 \right\rangle \text{ to discover methods to improve the researcher's own practice} \quad (14.13)$$

14.4.6 Research into Fundamental Theory

Fundamental theory research, as distinct from research in theoretical science, operates in an abstract space, generally of mathematical representations of things that either exist, or are postulated to exist. The entities under study are abstractions rather than tangible things. The method of such research is normally mathematical, and its practical impact relies on the existence of homomorphism of the mathematical constructs used in the investigation and the entities which they are used to represent. The mathematical abstraction of the investigation can identify results that would be observed if one can find a way to observe an effect, which, in turn, if observed would result in determination that the effect posited is real and, by extension, that other outcomes of the model are likely to be observed as instantiations of emergent effects.

Historically, fundamental research methods have been closely associated with theoretical physics and similar fields. However, it is reasonable to consider their use in engineering, in which the abstract space may enable the discovery of the presence of emergent effects, and judicious development of experiments to test critical issues could lead to the discovery of grounds to believe in the existence of emergent effects which may be rare, or at least, not yet actually observed.

Fundamental theory research can be mapped to the objectives of research described in Table 14.1 as described in Equations 14.14 and 14.15.

$$\langle D1, K1, B2, C1, T1, L1 \rangle \text{ for fundamental extensions to current theory} \quad (14.14)$$

$$\langle D1, K1, B2, C1, T2, L1 \rangle \text{ for fundamental revision to current theory} \quad (14.15)$$

	D1	*D2*	*D3*	*K1*	*K2*	*K3*	*B1*	*B2*	*C1*	*C2*	*T1*	*T2*	*L1*	*L2*	*L3*
Case Studies															
Relativist epistemology			■		▒			■		▒	■		▒		
Interpretivist epistemology		■			▒			■		▒	■		▒		
Realist epistemology	■				▒			■		▒	■			▒	
Forensic Case Studies															
Responsibility for past		■		▒				■	▒		■			▒	
Recommendation for future		■			▒			■	▒		■			▒	
***Post hoc* Empirical Studies**	■			▒				■	▒		■		▒		
Experimental Empirical Studies															
Pure scientific motivation	■			▒				■	▒		■		▒		
Applied scientific motivation			■			▒		■	▒		■				▒
Research and development		■			▒			■	▒		■			▒	
Research to Enable Engineering															
Deliverables in well-known area		■	■		▒			■	▒		■			▒	
Deliveries in radically new area		■	■		▒			■	▒			■		▒	
Methods for particular engineers		■	■		▒			■		▒	■			▒	
Methods for researcher's practice		■	■		▒		■			▒	■			▒	
Fundamental Theory															
Extensions to current theory	■			▒				■	▒		■		▒		
Revision to current theory	■			▒				■	▒			■	▒		

FIGURE 14.1
The hex-tuples describing the research objectives for various types of research that will inform understanding of emergence in a way that informs engineering practice.

14.4.7 Summarization of Research Objectives

We now collect together the hex-tuples describing the research objectives for investigations in each of the six types discussed above as relevant to, or potentially relevant to, the investigation of emergent effects in engineering and present them in Figure 4.1. In Figure 4.1 we observe that each kind of research, and each variation within each kind, involves a different combination of factors to describe the objective of the work. This shows that the objective in each case is to produce knowledge for a different purpose, and that the characteristics of the purpose vary in case specific ways, with the result that knowledge with different characteristics will be required. The effect is that different research methods are required to support the diversity of purposes. As discussed earlier in this section of the chapter, each of these research objectives and scenarios develops knowledge that in some way informs our understanding of emergence. It is not surprising that emergence can be elucidated by so many different methods of and approaches to research because emergence itself has many facets, as discussed Section 14.2.

14.5 Implications for Research in Engineering: What Is Considered Publishable

In Section 14.4 we outlined six research approaches often used in engineering research and investigations. The approaches are used to differing extents, and in different settings the driving motivations for research result in different emphasis on each of the types of research. In the academic community research achievement is one of the major drivers of careers, with the consequence that most academics seek to succeed in producing research publications in venues considered appropriate by their universities. A consequence of this drive to be published in the recognized venues is that the entire research activity is constructed in order to lead to recognized publications. This involves a series of steps rather like the list below:

1. Determine the intended publication venue, such as Journal X. The guidance to authors for the target journal describes the kind of papers published, which includes both the subject matter scope and, more importantly, the kind of research output that is considered suitable.
2. The researcher designs a research project, motivated by an interest in a topic within the scope of the journal.
3. The specific research questions are framed so that the method required to find an answer will produce results of a form that fit the normal published content of the journal. The act of framing the research questions to fit the usual published content of the journal may distort the original motivation for the research. The effect is that the results reported, while actually answering the formal research question do not form an answer to the issues which motivated the research.

In general, in the research journals there is a strong weighting towards the publication of reports of testing of a hypothesis. The reason for this is that the concept of knowledge in classical epistemology, see Section 14.1, demands justified true belief. The interpretation of "justified" in science and engineering has taken the position that justification is demonstrated through a process of positing an idea, the hypothesis, and performing an activity which tests that hypothesis. The practical effect of this is that, while other methods are acknowledged by many, statistical testing of observations conducted specifically with a view to testing a particular hypothesis are the most common form in research journals.

There are several causal factors:

1. A statistical hypothesis test study has clear measures of something which can be tested using established analysis methods to produce clear conclusions about support for the hypothesis. Also the process of analysis can be checked using the information included in a standard presentation of results. This characteristic provides confidence that the knowledge presented satisfies the justified true belief test (Haufe 2013). This characteristic leads to relative clarity in the review process and reduces the matters over which reviewers can contend, thus reducing the risk of delay or rejection.
2. If an investigation is designed to produce a statistical test of a hypothesis, the investigators have a basis for a higher level of confidence from the outset that if the work is performed following standard competence recognized in the field that the manuscripts reporting the work are likely to be publishable in their target journal.
3. Since publication is understood as the indicator of successful research, grant funding agencies, which have a remit to support good research leading to scientific knowledge, are more likely to fund projects designed to test a hypothesis because it is clearer how the result is likely to be accepted as a contribution (Haufe 2013). This satisfies the grant funder's "value for money" measure. Thus, proposals for funding are most likely to be designed to fit the form that is more likely to gain funding.

4. Even if research is not supported by a grant agency, the researchers will seek to publish, which biases their methodology towards the hypothesis test formulation, and their motivation for publication is further driven by personal track record of research outputs closely aligned with future funding proposals, further biasing them to perform projects that conform to the recognized format.
5. There is a current trend to "objectively" measure academic staff performance, with research achievement being the primary dimension for "success." The measures applied are often of things that are countable, for two reasons. Counts of things are "objective," which removes all the complications and effort associated with any attempt to evaluate "quality," other than using a countable proxy for "quality." Initially, output measure counts of papers published were used. In response many more venues for publication, such as new journals and a multiplication of conferences, appeared. This led to attempts to rank venues according to whole of venue esteem measures. Then it became recognized that many papers were either never cited, or perhaps cited only in self-citation or by co-workers, so citation counts, first of total citations and later independent citation counts, and further metrics building on citation counts were developed, but still the "quality" measure was a countable proxy for "quality," not a direct measure of quality. In addition, input measures for research, primarily amount of funding received for projects that the individual proposed and "won," are also applied. These measures may be direct money amounts, or amounts weighted by the kind of funding entity, with a ranking that may favor "pure science" funding over funding closer to application and product development. Measures of success as an academic, and primarily that means as a researcher, are used in career decisions including continued employment and promotion, leading many to distort their research to fit the evaluation constructs rather than to focus on investigation that directly addresses the questions that motivated their work.

This brief account of factors driving the kind of research performed shows challenges for the performance of research into emergence. We consider the six approaches to performing research discussed in Section 14.4 of this chapter. Both types of case study and the *post hoc* empirical approaches could be approached from the perspective of collecting and analyzing data in various ways until something interesting is found. This perspective is widely disparaged as a "fishing expedition" in which any discovery of an apparently interesting relationship is the result of stumbling into the result. Such an approach is viewed poorly by many researchers (Haufe 2013). The two case study approaches could be approached through positing a hypothesis and performing an investigation seeking to find support or refutation of the hypothesis. A hybrid of the two may be used if the researcher has no *a priori* basis for a hypothesis but the part of the work which will be better structured in reports and gain higher respect is the latter portion that works with a formulated hypothesis. Reports of the work may even be written, after the event, to be structured as if what was done was a hypothesis test.

The *post hoc* empirical investigation may be used to test a hypothesis. The challenge is that the lack of control of the situation limits the hypotheses to questions such as whether observed correlations evidence an underlying non-random factor, which requires considerable care to translate into a study that enables conclusions about causality.

The experimental approach is directly linked to hypothesis testing since most experiments are designed to enable testing a particular hypothesis. Setting up an experiment consumes resources and therefore, for the reasons described above, results in design that will enable testing of a hypothesis because that yields the highest probability of publication.

Research to enable engineering is concerned with enabling engineers to perform better the various kinds of work engineers perform. This includes all the contributions of engineers to any stage of the product lifecycle. The goal of this work is delivery of better engineered outputs, that is, things and processes in the real world. The goal is not perfecting knowledge about the effects and phenomena involved in the products and processes but, rather directly the quality and effectiveness of the things having impact in the world. Doing the work that ensures good results depends on the engineer being able to predict the effect of any design idea posited. This prediction demands reliable knowledge of the immediate effects of phenomena and also of the interactions of the elements with each other. Understanding the effects of interactions is difficult. Examples of the difficulty are the challenges around electromagnetic compatibility or the impossibility of complete testability of complex devices such as microprocessors because of the unmanageably large number of combinations of states in these scenarios. Engineering useful knowledge of emergence demands discovery of effective solutions to these problems and other problems with similar complexity.

In fundamental theoretical research the subject matter determines the place that any type of emergent effect has in the research. To the extent that the fundamental research investigates relationships, emergence is the subject matter.

14.6 What Research into Emergence Will Help Engineering

The existence of the phenomenon of emergence has been observed since ancient times and has been exploited in the practice of engineering throughout history. The history of the development of engineering knowledge until the rapid development of engineering and the parallel development of the sciences from the industrial revolution era was largely one of building according to known good practice with the occasional more ambitious project. From time to time an ambitious project would prove to have been too ambitious, with an ensuing disaster. The disaster often led to investigation and discovery of means to pre-empt a disaster of that class. Classically this pattern is recorded in the development of medieval cathedrals, but even in the "scientific" era of engineering of the past century there are many similar examples. A few, randomly selected examples include the Tacoma Narrows bridge failure, the collapse of the original box girder design of the Westgate Bridge in Melbourne, the loss of the first ancestor of the B-17 which proved too much for one pilot to handle, and the loss of several aircraft attempting to break the sound barrier. In each case additional research was done and means to successfully address the causal issues were found.

Events like those listed above are what have been noted as the outcomes of what is often called emergence, the sub-class of emergence associated with unexpected and deleterious outcomes. Specific instances such as these prompt research efforts to discover the causes of the unexpected outcomes, motivated by the specific case, but seeking understanding that is generic enough to apply to a class of cases of which the specific case is representative. The work that investigates the particular emergent effect may follow any of the approaches described in Section 14.4. This approach of investigating emergence through failures has a scope limitation because the emergence investigated is specific to the particular problem space and is performed after problems have been encountered.

Another approach to investigation of emergence is to investigate emergence as a general phenomenon. This approach describes emergence in general, but at that level of generality would be significantly challenged in the possibility of developing knowledge which would be usable by engineers in performing any of the various aspects of engineering.

What is needed is research which will enable the discovery of emergent effects before the product or system is constructed. It would appear that a potential path forward, with sufficient strength to be worth pursuing, would be to investigate the integration of computer modelling used in the design analysis process. Current modelling methods largely focus on analysis tools which operate on a model of the proposed product described from the viewpoint of a traditional engineering discipline. A design idea that appears to work in one domain can then be exported to the modelling tools for other domains of engagement to determine the properties of the design idea in those other domains. The work of exporting to other modelling tools is often considerable, reducing the amount of cross domain modelling views which are performed and the cycle time at which such analyses are performed. This allows for unfortunate design mistakes to be propagated until remediation is likely to become very difficult or expensive, resulting in problems that are likely to become evident too late in the system lifecycle.

The remedy would be research that enables the integration of the technical discipline views of things under development from the perspective of all technical disciplines concerned with the development project lifecycle within a shared modelling environment. For instance, the modelling environment for a product, for example a mobile phone, would have a common database representing the circuit at the levels of circuit theory, electromagnetic radiation, electromagnetic compatibility, thermal properties, mechanical properties of physical construction, vibration, shock and impact, lifecycle cost, reliability and maintenance characteristics, human factors, etc., all modelled within a single modelling environment so that where data is available for predicting effects in any of the domains, those effects can be found and used in design decisions. Where the necessary data is unavailable the absence of the data would be flagged so that a decision could be made whether to investigate a specific effect which would create the required data, or to proceed in the knowledge of the risk the uncertainty presents. Either way there would be a record of the conclusion and rationale, which would be useable in the event of an investigation.

An alternative approach, possibly as a stepping stone to the full modelling approach, may be the collection of information on all known emergent relationships in a construct that enables rapid identification of possibly relevant relationships in a structure like the TRIZ construct used to guide design conceptualization. A construct of this type would provide engineers with a manual tool that could enable review of all known emergence causing relationships. Such a construct would also assist in the design of the modelling analysis concept suggested above.

14.7 Conclusions

In this chapter we have discussed the nature of emergence in relation to the interests and concerns of engineers. Research that will assist engineers in the tasks of engineering must be focused on addressing issues arising from emergence. We then took an excursus through consideration of a classification of objectives of research because this highlights that there are a wide variety of activities and purposes which could frame research, and in turn the diversity of research objectives lead to a variety of research methods. These objectives were used to review six approaches used in research in engineering through which specific classes of research objectives were identified and the diversity of methods associated with the various objectives was discussed. Then we took a second excursus, discussing a current cultural pressure that is imposed on the nature of research performed in the broad set of fields associated with science and engineering. This excursus showed a significant tension between what is generally accepted as research and the kind of learning that is needed by engineers to support their work. All these threads were tied together in pointing the way forward with two goals for research which will provide engineers with practical benefit in relation to knowledge about emergence.

The simpler of these is research to develop a TRIZ-like construct that will guide engineers through the many possibilities of interaction of effects which could lead to emergent effects which need to be understood in the decision to proceed with a design idea. The other, more complex, goal is the development of a multi-engineering discipline approach to modelling in a seamless, or even automated approach so that all design decisions can be informed by analysis of the many kinds of interaction effect, which is the physical origin of emergence. A system that performs this task must be capable of using existing data concern in relationships which exist in the system and of identifying data that is missing, which in turn can be used to develop research investigation to generate the necessary data to use the relationship.

Through discussion of the underlying nature of emergence and research it has been possible to propose a research goal which can profoundly assist engineers in the process of dealing with the matters arising from emergence.

References

Ackoff, R.L., 1979a. The future of operational research is past. *The journal of the operational research society*. 30, no. 2, 93–104.

Ackoff, R.L., 1979b. Resurrecting the future of operational research. *The journal of the operational research society*. 30, no. 3, 189–199.

Ackoff, R.L., 2001. OR: After the post mortem. *Systems dynamics review*. 17 no. 4, 341–346.

Augustine of Hippo, 1984. *Concerning the city of God against the pagans*. Harmondsworth, Middlesex: Penguin Books.

Biggs, J.B., 1999. *Teaching for quality learning at university: What the student does*. Buckingham: Society for Research into Higher Education, Open University Press.

BKCASE Editorial Board, 2016. *The Guide to the Systems Engineering Body of Knowledge (SEBoK)*; v. 1.7 [online]. R.D. Adcock (EIC). Hoboken; NJ: The Trustees of the Stevens Institute of Technology. BKCASE is managed and maintained by the Stevens Institute of Technology Systems Engineering Research Center; the International Council on Systems Engineering; and the IEEE. Available from: www.sebokwiki.org. [Accessed 20 Mar 2017.]

Bloom, B.S., Engelhart, M.D., Furst, E.J., Hill, W.H., and Krathwohl, D.R., 1979. *Taxonomy of educational objectives: The classification of educational goals handbook I: Cognitive domain*. London: Longman Group Ltd.

Bucciarelli, L.L., 2003. *Engineering philosophy*. Delft, Netherlands: DUP Satellite.

Cripps, S.C., 1991. Old fashioned remedies for GaAsFET power amplifier designers. *IEEE society on microwave theory and techniques newsletter*. no. 129, 13–17.

Ferris, T.L.J., 1997. The concept of leap in measurement interpretation. *Measurement*. 21, no. 4, 137–146.

Ferris, T.L.J., 2007a. Some early history of systems engineering—1950's in IRE publications (part 1): the problem. In: *17th International Symposium Systems Engineering: Key to intelligent enterprises.* San Diego: International Council on Systems Engineering.

Ferris, T.L.J., 2007b. Some early history of systems engineering—1950's in IRE publications (part 2): the solution. In: *17th International Symposium Systems Engineering: Key to intelligent enterprises.* San Diego: International Council on Systems Engineering.

Ferris, T.L.J., 2008. Early History of Systems Engineering (Part 3)—1950's in Various Engineering Sources. In: *INCOSE International Symposium 2008.* Utrecht: International Council on Systems Engineering.

Ferris, T.L.J., 2010. Bloom's Taxonomy of Educational Objectives: A Psychomotor Skills Extension for Engineering and Science Education. *International journal of engineering education.* 26, no. 3, 699–707.

Ferris, T.L.J., Cook, S.C., and Honour, E.C., 2005. Towards a structure for systems engineering research. In: *15th Annual International Symposium INCOSE 2005 Systems Engineering: Bridging Industry, Government and Academia.* Rochester, New York: International Council on Systems Engineering: Paper 6.1.1.

Ferris, T.L.J., Cook, S.C., and Sitnikova, E., 2007. Research methods for systems engineering. In: *13th ANZSYS conference Systemic development local solutions in a global environment.* Auckland, New Zealand.

Gettier, E.L., 1967. Is justified true belief knowledge? In: A.P. Griffiths, ed. *Knowledge and belief.* London: Oxford University Press, 144–146.

Goodman, N., 1973. Fact, fiction, and forecast. Indianapolis, Indiana: The Bobbs-Merrill Company, Inc.

Haufe, C., 2013. Why do funding agencies favor hypothesis testing? *Studies in history and philosophy of science.* 44: 363–374.

Hickey Jr., A.E., 1960. The systems approach: Can engineers use the scientific method? *IRE transactions on engineering management.* 7, no. 2, 72–80.

Hintikka, J., 1970. Knowledge, belief, and logical consequence. *Ajatus.* 32: 32–47.

Hitchins, D.K., 1992. *Putting systems to work.* Chichester, New York: John Wiley & Sons.

Kline, S.J., 1995. *Conceptual foundations for multidisciplinary thinking.* California: Stanford University Press.

Krathwohl, D.R., Bloom, B.S., and Masia, B.B., 1964. *Taxonomy of educational objectives: The classification of educational goals handbook II: Affective domain.* New York: David McKay Company, Inc.

Nissen, M.E., 2006. *Harnessing knowledge dynamics: Principled organizational knowing and learning.* Hershey, Pennsylvania: IRM Press.

Ryle, G., 1948. Knowing how and knowing that. *Proceedings of the Aristotelian society.* 46: 1–16.

Singer, A.J., Homan, C.S., Stark, M.J., Werblud, M.C., Thode, Henry C.J., and Hollander, J.E., 1997. Comparison of types of research articles in emergency medicine and non-emergency medicine journals. *Academic emergency medicine.* 4, no. 12, 1153–1158.

Skrabanek, P. and McCormick, J., 1990. *Follies and fallacies in medicine.* Buffalo, New York: Prometheus Books.

Yin, R.K., 2014. *Case study research design and methods.* Los Angeles: Sage Publications.

Section III

Theoretical Perspectives with Practical Applications

15

Examination of Emergent Behavior in the Ballistic Missile Defense System: A Modeling and Simulation Approach

Saikou Y. Diallo and Christopher J. Lynch

CONTENTS

15.1 Introduction

Tolk et al. (2013) observe that the discipline of Modeling and Simulation is closely related but distinct from software engineering and systems engineering. For problems that are messy and prone to emergence, they propose the use of a modeling and simulation system development framework (MS-SDF) that helps simulationists accurately capture the system and the assumptions and constraints under which it was observed. They applied the methodology to the problem of sea level rise (Diallo et al. 2014) and showed that a consistent application of the framework is helpful in revealing emergence in complex systems. The MS-SDF culminates in the development of a simulation model that serves as a proxy for the real system. The idea is to use the model as the basis for experimentation to discover emergent conditions in the system. Diallo, Gore et al. (2016) propose a framework to verify and validate simulation models using statistical debugging with a specific adaptation for agent-based models provided in (Gore, Lynch, and Kavak 2017). The method allows for the experimental discovery of emergent conditions based on the ability of the model to achieve a set of expected outcomes defined by a subject matter expert. By doing so, the method provides a flexible way to describe and explore multiples forms of emergence and allows competing objectives to be compared under the same set of assumptions.

In this chapter, we apply these two methods to the Ballistic Missile Defense System (BMDS) and explore how they can shape the discussion around what it means to claim that a system is emergent. We aim to describe a practical experience of developing, implementing, and testing a model of the BMDS. The BMDS is a perfect candidate for an exploration of emergence because it fits all the requirements stipulated in Maier (1996, 2015) for a system of systems which makes it capable of exhibiting emergent behavior under certain conditions. Lynch, Diallo, and Tolk (2013) provide a basic explanation of the BMDS model that we describe in this chapter and Diallo, Lynch, Gore et al. (2016) present the design of experiments that led to an initial investigation into emergent conditions under which we observe emergent behavior. In this chapter, we focus on the activities centered around the journey we took to develop, test, and report on the model. The balance of the chapter is as follows: Section 15.2 briefly describes the goals and outcomes of the project; Section 15.3 describes how the model is configured; and Section 15.4 provides a discussion of emergent conditions before we conclude the chapter in Section 15.5.

15.2 Developing the Model

The goal of our project was to create a model comprised of both the agent-based modeling (ABM) and Discrete Event Simulation (DES) modeling paradigms in order to more accurately represent the command and control (C2) network and information flow of the BMDS for the Missile Defense Agency (MDA). The deliverables for the project include the model and its corresponding documentation. The model is a prototype that serves as a proof of concept for representing the engagement decision process of the BMDS.

The BMDS is designed for defense against ballistic missiles of all types with variable ranges and speeds. A combination of space-based, sea-based, and ground-based elements comprise the BMDS architecture (Garrett et al. 2011). These elements consist of both sensors and weapons systems which are designed to deal with ballistic missiles.

Why agent-based modeling and discrete event simulation? The BMDS is a layered and integrated system consisting of space-based, sea-based, and ground-based elements. It also includes elements and supporting efforts that make up the command, control, battle management, and communications (C2BMC) network which provides operation commanders a link between the sensors and interceptor missiles that are contained in their areas of responsibilities (AORs) (Garrett et al. 2011, Tolk 2012). The ability to communicate, make decisions, and for elements to be situated throughout an environment formed a natural fit with ABM (Sokolowski 2012). The C2BMC, space-based elements, sea-based elements, ground-based elements, interceptors, and threats are all represented as agents within the model.

The BMDS is often modeled using DES and therefore we had to explore the differences between ABM and DES as a motivation for using the former. Since ABM and DES share many characteristics (Brailsford 2014), we expected few additional challenges in building, running, verifying, or validating a simulation comprised of ABM and DES together (Lynch and Diallo 2015, Diallo, Lynch, Padilla et al. 2016). With respect to using the MS-SDF, the use of multiple paradigms requires that extra additional assumptions and constraints be captured to fully establish the reference and conceptual models (Lynch et al. 2014, Eldabi et al. 2016, Mustafee et al. 2015). The need to represent communication sequences and message saturation of communication elements formed a natural fit with DES.

Overall, we anticipated a low risk, high reward situation with respect to low additional burden in building, testing, and experimenting with the simulation versus the gained ability to identify and examine emergent conditions. While emergence is not mentioned in the figure above, we were able to demonstrate that ABM allows for the BMDS elements

TABLE 15.1

Comparison between ABM and DES in Modeling the BMDS

Application Areas of ABM to the Missile Defense Agency	ABM	DES
Represent *static* system processes—fixed sequence of events that are followed once a trigger occurs, such as firing a weapon system, sending a communication, probability-based events, etc.	X	X
Process data—C2BMC receiving reports, radar systems receiving updates, weapon systems adjusting aim, etc.	X	X
Represent *dynamic* system processes—non-ordered event paths that are followed once a trigger occurs, including making decisions, representing policy, capturing behavior, engaging targets, etc.	X	
Represent the integrated and layered architecture of the BMDS—the communication between sensor networks, interceptors, and the C2BMC	X	
Represent the behavior of decision makers or personnel—representing tactics, techniques, and procedures (TTPs), planning a mission, determining objectives, representing intent, force execution, etc.	X	
Represent characteristics of systems or elements—radar ranges, communication abilities, missile speeds, tracking abilities, etc.	X	
Represent the environment and capture environmental awareness—locations of elements, effects of terrain, weather, area boundaries, etc.	X	
Allow for dynamic placement of elements—movement of sea-based cruisers, location of C2 nodes, positioning of ground-based weapon systems, etc.	X	

and C2BMC to exchange information, have states and behaviors, sense and perceive their environment, and plan and conduct operations, when applicable. Table 15.1 shows a comparison between ABM and DES with respect to the BMDS.

We then proceeded to architect and implement the model using publicly released documents from the MDA and by gathering requirements and information from subject matter experts (SMEs). For instance, we took a trip to Colorado Springs in May, 2012 to meet with the SMEs in order to gather additional requirements for the model. These requirements were converted into a set of activity, sequence, use case, and class diagrams that were then checked and approved by the SMEs before beginning the actual model construction. Figure 15.1 shows the primary use case for the model.

The design of the model allows for different configurations of the C2 elements, sensors, interceptors, and communication satellites, and allows for a variable number of sensors and weapon systems (ground, sea, and space based) with different attributes to be initialized. The numbers and types of interceptors, the engagement area, the numbers and types of threats, the number of communication satellites, and the number of C2 elements can also be configured for each scenario. The model can provide insight into engagement sequences where communication gaps exist between the operational commanders of the weapons systems and the sensors or the C2BMC. The outcomes of having a single threat or a salvo of threats to defend against can also be examined in relation to different BMDS defensive configurations.

The model was constructed using AnyLogic Professional 6.8.1. We validated the model by (1) showing that the model passed the test cases presented by the SMEs and (2) demonstrating that the model is bounded correctly, i.e. things that always happen in the real

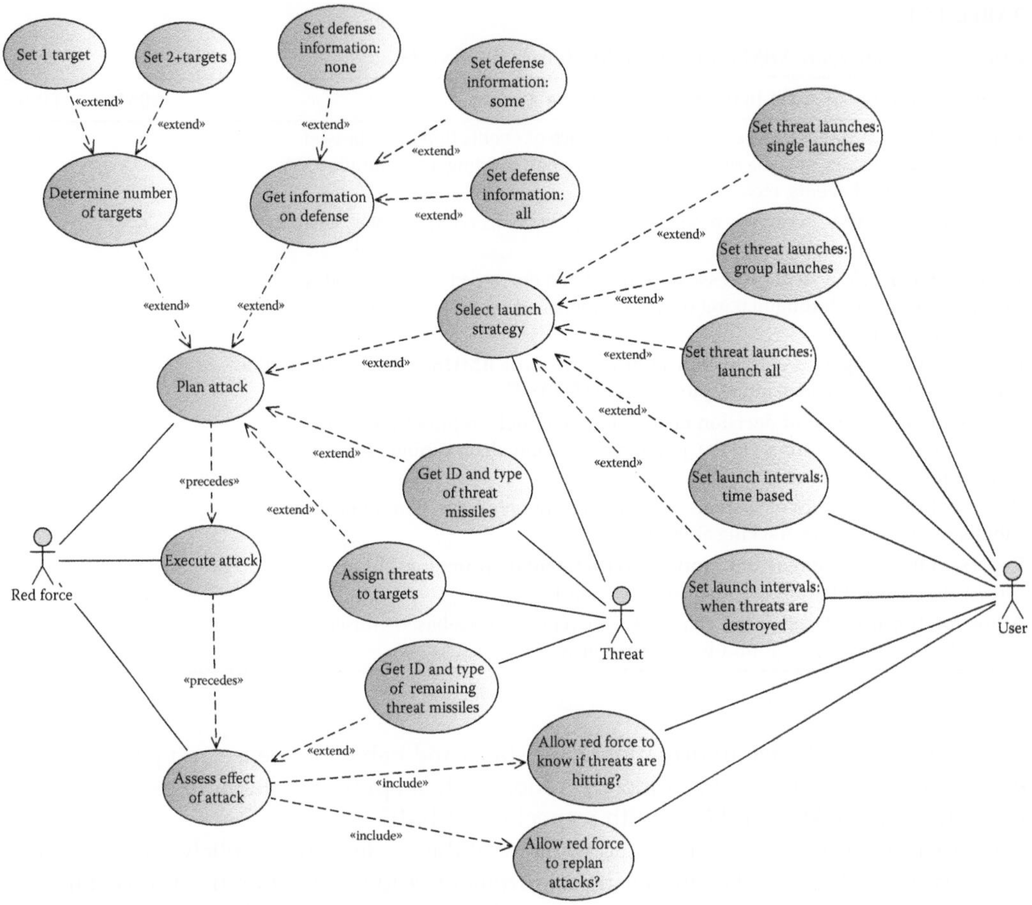

FIGURE 15.1
Primary Use Case Description.

system happen in the model and things that never happen in the system never happen in the model, by using statistical debugging (Diallo, Lynch, Gore et al. 2016).

15.3 Configuring the Simulation

This section outlines the various components associated with configuring and starting the simulation. The model configuration is divided into six categories: the environment, space-based elements, sea-based elements, ground-based elements, the C2BMC, and the enemy force. A Microsoft Excel template has been created for information to be imported and exported from the simulation in a consistent manner.

A home screen provides an intuitive starting point for configuring a scenario for the BMDS Simulation and provides the main menu from which all six categories can be configured. In addition, the home screen also provides the capability for users to select existing setup files to use for the simulation runs, both for input and output purposes. The first step in configuring the simulation is to set up the environment. Three main components are met through the configuration of the environment:

1. Environmental Dimensions: This allows for the physical dimensions of the map to be specified for the current scenario. This allows the user to specify the length, width, and height of the scenario in kilometers. This determines the distance between the red and blue forces which affects the types of ballistic missiles needed for the simulation. This also affects the total flight time of the missiles as well as the total amount of time available for defending against threats.
2. Communication Satellites: This allows for the specification of the communication satellites available in the simulation. This includes setting the total number of communication satellites available as well as the maximum number of transmissions that each satellite can send simultaneously. Specifying the maximum number of transmissions that can be sent is a simplification made for representing the bandwidth of the satellites.
3. Select Scenario: This contains a list of the different scenarios that can be selected for running the simulation. Each scenario has different locations for missile launch locations and different placements of defensive elements.

Next is the configuration of the space-based elements. Space-based elements are sensors that are in Earth's atmosphere with the purpose of detecting, tracking, classifying, and discriminating ballistic missiles that are in any stage of flight from takeoff to impact. The primary goal of these sensors is to detect any incoming objects and notify the C2BMC of a possible incoming threat. The sensors then track, classify, and discriminate any incoming threats to provide a better operational view of the situation to the C2BMC. The classification of the threat determines whether the incoming threats are Intercontinental Ballistic Missiles (ICBMs), Intermediate-Range Ballistic Missiles (IRBMs), Medium-Range Ballistic Missiles (MRBMs), or Short-Range Ballistic Missiles (SRBMs) and lets the C2BMC know what type of interceptor will be needed for the engagement. The discrimination of the incoming missile lets the C2BMC know whether the incoming missile is a threat or not. The goal is to provide as much information as possible to the C2BMC so that the best engagement plan can be determined to deal with the threat.

The configuration of the sea-based and ground-based elements for the scenario share a number of common characteristics, including setting the physical capabilities of their sensors and weapon systems, as well as any communication capabilities. Sea-based elements are any sensors or weapon systems that are located on and travel across water, while ground-based elements are located and travel across land. Both share the ultimate purpose of detecting, tracking, classifying, discriminating, or engaging ballistic missiles. These elements can be combination of sensors and/or weapon systems. The weapon systems contain missiles used to engage and intercept incoming ballistic missiles. Sensors have maximum and minimum ranges for detecting, tracking, discriminating, and classifying targets while weapon systems have maximum ranges for intercepting targets. Sensors and weapon systems can be assigned movement speeds dependent upon the type of real or experimental system that they are intended to represent.

Next, the C2BMC element is configured. The C2BMC is the operational system that allows for commanders at the strategic, regional, and operational levels to plan ballistic missile defense operations and manage the network of sensors and weapon systems to effectively engage incoming ballistic missiles. The C2BMC links, integrates, and synchronizes the individual sensors and weapon systems within its network to create a single integrated picture of the BMDS.

Finally, the red (enemy) force is configurable. The purpose of configuring the enemy force is to set up the plan of attack for incoming threats as well as specify the number and type of threats at their disposal. This includes setting the time interval between launches, the number of missiles to launch per launch, and the number of targets to target per launch. The physical properties connected to each of the missile types are assigned here as well, such as speed and range.

15.4 Exploring Emergence in the BDMS Using the Model

To explore emergence, we had to first decide on a working definition. After several discussions with our SMEs, we coalesced on defining emergence as "the apparition of the unexpected." This working definition has two facets:

- Subjectivity and Multi-perspective: This definition of emergence depends on the viewpoint of the SME. What is unexpected by one SME with one worldview can be expected by another with a different worldview. For instance, a person on the sea-based asset will have a different definition (a local view and focused on that asset perhaps) than a person sitting in the C2BMC (a global view focused on communication, command and control perhaps). Similarly, emergence can be centered around a technical worldview or centered around a human, social worldview. This process can be further challenged based on disagreements on how to address the problem or on which problem to solve (Lynch et al. 2014, Vennix 1999).
- Measurable and Observable: Even if the SMEs can define and describe what they mean by "unexpected outcome," the model must be able to show that this outcome is possible under the assumptions and constraints we used to build it. Even if that were the case, we realized that an unexpected outcome can be hard to observe in any simulation run because (1) it only appears under some unknown combination of input parameters; or (2) it is not persistent over time or is transient; or (3) a combination of (1) and (2) which would make it almost impossible to guess the existence of the unexpected event. Additionally, a compounding challenge exists with respect to examining the combinations and sequences of events leading to various system level outcomes within the model (Gore, Lynch, and Kavak 2017).

In addition to these observations, we have the challenge of deciding what is truly unexpected, since on one hand if the SME can describe the event one can claim that it is no longer "unexpected," but on the other hand if the SME cannot describe the "unexpected" we can only observe the model and hope for an "aha" moment. Ultimately, the compromise was two-fold. SMEs can describe what they mean by unexpected in terms of what they cannot explain. We asked them about outcomes from the real system that they cannot explain even without considering the variables of the model. From the discussions, it emerged that SMEs have a good understanding of how the BMDS works in terms of physical performance, i.e. time over target, accuracy, lethality, etc. However, when asked what constitutes success during an engagement, there was surprisingly no agreement. This pushed us to investigate success (or failure) with the following questions: (1) what is success and how do we measure it; and (2) what conditions are the most responsible for a successful engagement. The SMEs indicated that even if they do not agree on a single metric for success, they felt that they had a good understanding of its underlying causes. However, they had never considered the notion of a "tie" in which both sides of an engagement can claim success. It was then decided that the "tie" which no one could predict or explain (with or without the model) is one of the conditions that should it appear from a run of the model would be considered "emergent."

As we mentioned before, even with millions of runs of the model, the chances of observing a tie are not very high, especially since we do not have any indication on how to bias the parameters for exploration. This observation meant that we had to take an experimental approach. The first approach is to posit the "tie" and ask the model to self-configure inputs to produce the closest possible outcome to the posited tie. In addition, we ask it to bracket its prediction within a 95% confidence interval. This method produces a single solution for a tie that might or might not be realistic. It solves the problem of making emergence appear but does not help us explain it in terms of (1) the frequency at which we should expect the "tie" and (2) the leading combinations of inputs that led to the tie. Sensitivity analysis was considered, but the input variables are co-dependent which eliminates the traditional design of experiments. After further investigation, we decided to consider the software engineering technique of statistical debugging and treat the emergent condition as a rare condition that the simulation model can exhibit under certain conditions, much like a bug in software. Initial applications of predicate-level statistical debugging in the field of software engineering are explored in (Gore, Reynolds, and Kamensky 2011, Gore and Reynolds 2012). This approach was later adapted to simulation specific applications in (Diallo, Gore et al. 2016, Gore et al. 2015).

However, this second method requires an exhaustive sweep of the parameter space that requires an unreasonable amount of time to conduct (over 100 years). As a result, we used Latin Hypercube Sampling (Jansen 1999, Stein 1987, Collins et al. 2013) to generate a random sample of the input space. For each sample, we run N unique replications (N=30) and save the results of output variables. The resulting dataset should have records of wins and losses but also instances of "ties." If it does not, we resample the input space with more resolution and repeat the experiments. Our goal here is to generate a reasonable combination of "ties" to investigate using statistical debugging.

The results of our experiments are published in (Diallo, Lynch, Gore et al. 2016). Using the model, we could show that (1) there is no way to guarantee a win, (2) there are many more ways to lose than to win, and (3) there are several ways to end up in a tie.

15.5 Conclusion

We describe a journey to architect, develop, and build a prototype model that we used to investigate the nature of emergence in the BMDS. The journey points to the socio-technical nature of model development and its use. We spent more time in conversations and constructive arguments than we did in actual development. Arriving at a practical definition of emergence was particularly challenging. The use of a "tie" as the emergent behavior was helpful in devising a method to discover it and investigate its root causes, but it is an outcome that is not particularly interesting to the SME. The discovery that the model could not find a way to guarantee a win sparked interesting debates on whether this is due to the model being under-specified or the actual systems not being capable of such a guarantee.

From this exercise, we learned several lessons when searching for emergent behaviors, including considerations on relevance, being explainable, accounting for complexity, and accounting for verification and validation. An identified emergent behavior must be relevant to the SMEs examining the system. Behaviors not relevant to the question at hand are not useful for determining if the model produces emergent behaviors, as these behaviors may occur due to conditions outside of the constraints or assumptions underlying model implementation. An identified emergent behavior needs to be explainable. Identifying a behavior is not helpful if it cannot be traced back to its root cause within the model or within the real system. This process can help to facilitate interesting debates over whether the model is under-specified, over-constrained, or due to the inability of the real system to produce emergent behaviors when examining model outcomes.

Models of system of systems provide potentially many opportunities for increased complexities in design and analysis due to interactions between the parts of the system. This increases the quantity of outputs that can be tracked throughout execution, and it increases the combinations of factors potentially eligible for producing the emergent behavior. Methods such as our approach in using statistical debugging are very helpful in working backward from an identified behavior to identify the conditions that caused the behavior to emerge. These causes can then be examined by both the SMEs and the modelers to determine if the behavior is truly emergent or if it is caused by an error in the model. This leads to our final lesson learned with respect to verification and validation. Verification and validation processes are critical, as it is important to be able to identify whether emergent behaviors are really just errors manifesting themselves during execution. On the other hand, eliminating the possibility that a behavior is occurring due to an error helps to increase confidence that the behavior is in fact emergent. We found that statistical debugging greatly assisted in this endeavor by providing an avenue for breaking the model runs down into more easily digested pieces that the SMEs could compare against their specifications.

We are grateful for the opportunity to work alongside very knowledgeable and open-minded SMEs, and the collaboration led to an expansion of our understanding of emergence.

References

Brailsford, Sally C. 2014. "Modeling human behavior—an (id) entity crisis?" Proceedings of the 2014 Winter Simulation Conference, Savannah, GA, December 7–10 2014.

Collins, Andrew J, Michael J Seiler, Marshall Gangel, and Menion Croll. 2013. "Applying Latin Hypercube Sampling to Agent-Based Models: Understanding Foreclosure Contagion Effects." *International Journal of Housing Markets and Analysis* 6 (4):422–437. doi: 10.1108/IJHMA-Jul-2012-0027.

Diallo, Saikou Y, Ross Gore, Christopher J Lynch, and Jose J Padilla. 2016. "Formal Methods, Statistical Debugging and Exploratory Analysis in Support of System Development: Towards a Verification and Validation Calculator Tool." *International Journal of Modeling, Simulation, and Scientific Computing* 7 (01):1–22. doi: 10.1142/S1793962316410014.

Diallo, Saikou Y, Christopher J Lynch, Ross Gore, and Jose J Padilla. 2016. "Emergent Behavior Identification within an Agent-Based Model of the Ballistic Missile Defense System Using Statistical Debugging." *The Journal of Defense Modeling and Simulation: Applications, Methodology, Technology* 13 (3):275–289. doi: 10.1177/1548512915621973.

Diallo, Saikou Y, Christopher J Lynch, Jose J Padilla, and Ross Gore. 2016. "The Impact of Modeling Paradigms on the Outcome of Simulation Studies: An Experimental Case Study." Proceedings of the 2016 Winter Simulation Conference, Washington, DC, December 11–14 2016.

Diallo, Saikou Y, Andreas Tolk, Ross Gore, and Jose J Padilla. 2014. "Modeling and Simulation Framework for Systems Engineering." In *Modeling and Simulation-Based Systems Engineering Handbook*, edited by Daniele Gianni, Andrea D'Ambrogio and Andreas Tolk, 377-400. Boca Raton, FL: CRC Press.

Eldabi, Tillal, Mariusz Balaban, Sally Brailsford, Navonil Mustafee, Richard E Nance, Bhakti Stephan Onggo, and Robert G Sargent. 2016. "Hybrid Simulation: Historical Lessons, Present Challenges and Futures." Proceedings of the 2016 Winter Simulation Conference, Washington, DC.

Garrett, Robert K, Steve Anderson, Neil T Baron, and James D Moreland. 2011. "Managing the Interstitials, a System of Systems Framework Suited for the Ballistic Missile Defense System." *Systems Engineering* 14 (1):87–109. doi: 10.1002/sys.20173.

Gore, Ross, Christopher J Lynch, and Hamdi Kavak. 2017. "Applying Statistical Debugging for Enhanced Trace Validation of Agent-Based Models." *Simulation: Transactions of the Society for Modeling and Simulation International—Special Issue on Modeling and Simulation in the Era of Big Data and Cloud Computing: Theory, Framework, and Tools* 93 (4):273–284. doi: 10.1177/0037549716659707.

Gore, Ross, Paul F Reynolds Jr, David Kamensky, Saikou Y Diallo, and Jose Padilla. 2015. "Statistical Debugging for Simulations." *ACM Transactions on Modeling and Computer Simulation (TOMACS)* 25 (3):1–26. doi: 10.1145/2699722.

Gore, Ross, and Paul F Reynolds. 2012. "Reducing Confounding Bias in Predicate-level Statistical Debugging Metrics." Software Engineering (ICSE), 2012 34th International Conference on, Zurich, Switzerland, June 2–9 2012.

Gore, Ross, Paul F Reynolds, and David Kamensky. 2011. "Statistical Debugging with Elastic Predicates." Proceedings of the 26th IEEE/ACM International Conference on Automated Software Engineering, Lawrence, KS, November 6–10, 2011.

Jansen, Michiel JW. 1999. "Analysis of Variance Designs for Model Output." *Computer Physics Communications* 117:35-43. doi: 10.1016/S0010-4655(98)00154-4.

Lynch, Christopher J, and Saikou Y Diallo. 2015. "A Taxonomy for Classifying Terminologies that Describe Simulations with Multiple Models." Proceedings of the 2015 Winter Simulation Conference, Huntington Beach, CA, December 6-9 2015.

Lynch, Christopher J, Saikou Y Diallo, and Andreas Tolk. 2013. "Representing the Ballistic Missile Defense System Using Agent-Based Modeling." Proceedings of the 2013 Spring Simulation Multi-Conference—Military Modeling & Simulation Symposium, San Diego, CA, April 7–10.

Lynch, Christopher J, Jose Padilla, Saikou Y Diallo, John Sokolowski, and Catherine Banks. 2014. "A Multi-Paradigm Modeling Framework for Modeling and Simulating Problem Situations." Proceedings of the 2014 Winter Simulation Conference, Savannah, GA.

Maier, Mark W. 1996. "Architecting Principles for Systems-of-Systems." *INCOSE International Symposium* 6 (1):565–573. doi: 10.1002/j.2334-5837.1996.tb02054.x.

Maier, Mark W. 2015. "The Role of Modeling and Simulation in System of Systems Development." In *Modeling and Simulation Support for System of Systems Engineering Applications*, edited by Larry B. Rainey and Andreas Tolk, 11-41. Hoboken, NJ: Wiley.

Mustafee, Navonil, John Powell, Sally C Brailsford, Saikou Diallo, Jose Padilla, and Andreas Tolk. 2015. "Hybrid Simulation Studies and Hybrid Simulation Systems: Definitions, Challenges, and Benefits." Proceedings of the 2015 Winter Simulation Conference, Huntington Beach, CA, December 6–9 2015.

Sokolowski, J. A. 2012. "Human Behavior Modeling: A Real-World Application." In *Handbook of Real-World Applications in Modeling and Simulation*, edited by John A. Sokolowski and Catherine M. Banks, 26–92. Hoboken, NJ: John Wiley & Sons, Inc.

Stein, Michael. 1987. "Large Sample Properties of Simulations Using Latin Hypercube Sampling." *Technometrics* 29 (2):143–151. doi: 10.1080/00401706.1987.10488205.

Tolk, Andreas. 2012. "Scenario Elements." In *Engineering Principles of Combat Modeling and Distributed Simulation*, edited by Andreas Tolk, 79-92. Hoboken, NJ: John Wiley & Sons, Inc.

Tolk, Andreas, Saikou Y Diallo, Jose J Padilla, and Heber Herencia-Zapana. 2013. "Reference modelling in support of M&S—foundations and applications." *Journal of Simulation* 7 (2):69–82. doi: https://doi.org/10.1057/jos.2013.3.

Vennix, Jac AM. 1999. "Group Model-Building: Tackling Messy Problems." *System Dynamics Review* 15 (4):379–401. doi: 10.1002/(SICI)1099-1727(199924)15:4<379::AID-SDR179>3.0.CO;2-E.

16

Simulating Variable System Structures for Engineering Emergence

Umut Durak, Thorsten Pawletta, and Tuncer Ören

CONTENTS

16.1 Introduction: Background and Motivation

Complexity is becoming the future landscape of technical systems. Emergence is defined as the coherent and novel macro-level patterns, properties, behavior, or structures that arise from the micro-level interactions among the elements of the complex systems. It is a self-organized order. Self-organization can be pronounced as a designed system characteristic that yields an adaptive process to acquire and maintain a structure without external control. This process executes as micro-level interactions among the system elements and leads to a macro-effect, namely emergence. Organic Computing (OC) aims at utilization of self-organization and emergence as observed in nature for the design of future technical systems (Schmeck 2005). It promotes providing technical systems with some degrees of freedom in low-level behavior for allowing them to self-organize in order to adopt while controlling them with high-level objectives and goals. The observer/controller architecture was proposed as a generic architectural concept for OC (Branke et al. 2006). It offers system governance through an observer that monitors the underlying system in order to report a quantified situation and a controller that evaluates the situation against the user objectives and takes the control actions. So it is not fully autonomous; the users affect the system by changing the objectives. The key characteristic of the architecture is allowing the underlying technical system to be dynamic in a way that it can change its structure and thereby change its behavior.

Simulation has been evolving along with the advancement of systems themselves. Simulating systems with variable structures has long been studied in the modeling and simulation community (Zeigler and Praehofer 1989; Pawletta 1995; Pawletta 1996; Barros 1997; Uhrmacher 2001; Hu et al. 2005). Now it is again receiving attention with the simulation requirements for engineering the upcoming self-x systems with emergent behavior. This chapter will present a contemporary approach based on conceptualization of self-x systems in OC and observer/controller architecture.

After introducing the background about OC as a methodology for engineering emergence, a digest on the evolution of simulation will be given. Afterwards, the variable structure system simulation will be introduced as an upcoming challenge. The System Entity Structure (SES) and Model Base (MB) framework will be proposed as an approach to tackle this challenge. The comprehensive introduction of the SES/MB based approach is then exemplified with a sample case and supported with a prototype implementation within MATLAB/Simulink.

16.2 Organic Computing and Simulation

Organic computing is defined as a collection of methodologies and concepts for developing complex distributed systems that retain self-x properties such as self-organization or self-configuration to cope with dynamically changing environments and dependability to guarantee trustworthy response to control actions and systems objectives (Schmeck et al. 2010). OC stresses a controlled self-organization and advocates a paradigm shift in the design of future technical systems. Schmeck (2005) states: "It is not the question whether self-organised and adaptive systems will arise but how they will be designed and controlled."

Emergence in terms of patterns in space or time may occur with the self-organization property of technical systems which may be observed with the decreasing complexity. German Research Foundation (DFG) founded a priority program between 2005 and 2011 for research into emergent behavior in technical systems and developing technologies for OC. The observer/controller architecture was developed in this program as a generic architectural concept for design and analysis of OC (Branke et al. 2006). Observer/controller is composed of system under observation and control (SuOC), observer, and controller, which all together form an OC. SuOC is the system which allows a higher level of governance from an observer and a controller.

Observers are meant to collect data from the system and characterize the system state and dynamics, including the future status of the system (Richter et al. 2006). While an observer utilizes computational clustering techniques or some mathematical and statistical models to come up with a system state, it employs a prediction process, which is essentially simulation to forecast future system behavior for reducing the controller reaction time, preventing unwanted behavior, and assessing success of controller intervention.

The topology of this generic architecture can be customized depending on the realization scenarios. Typical topological classification is proposed as central, decentral, and multi-level. Central topology defines an observer/controller for the whole technical system, while decentral topology specifies an observer/controller for each system element. In the multi-level topology, while there is an observer/controller for each system element, there is also one for the whole technical system. These possible topologies are depicted using SysML in Figure 16.1.

The idea of OC is built upon controlled self-organization. In the central topology, the behavior of the system is top-down controlled. In contrast, decentral topology relies on

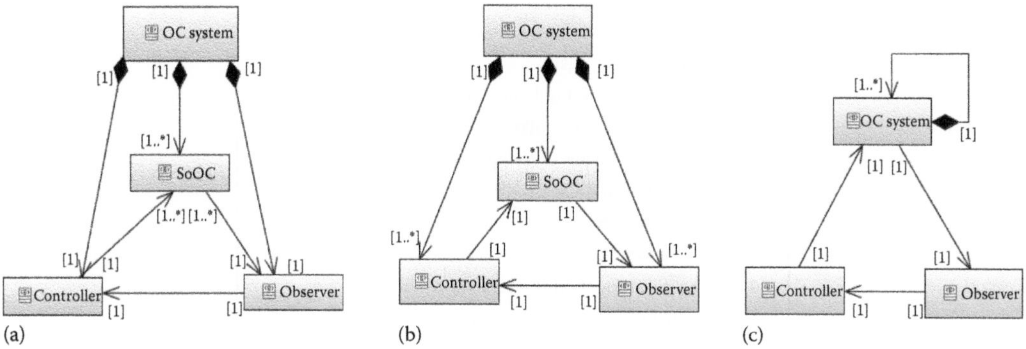

FIGURE 16.1
OC Topologies. (a) central, (b) decentralized, and (c) multi-level.

bottom-up generation of desired behavior with the effects of emergence. Multi-level topology resides in the middle and enables bottom-up emergence while providing a top-down control authority. The controller is about intervention based on the goal defined by the user and the situation parameters. The controller action is then categorized in three ways:

- *Affecting the local decision rules of SuOC elements.* This affects the local behavior of system elements.
- *Affecting the system structure of SuOC.* The system can be defined as a network of its components. Modifying this network will alter both the local behavior of the system elements and the overall system behavior. This modification can be changing the system structure by moving the elements in the network as well as removing or adding new elements to it.
- *Affecting the environment.* The controllers can change the environment via actuators. Such a change in the environment is grasped by the systems via their sensors and reacted to accordingly.

The controller basically gets the observed data from the observer, gets the objectives from the user, and controls the SuOC via interactions and reconfigurations. The controller uses the observed data to select the best action for the current system state. It may utilize mechanisms such as neural networks or evolutionary algorithms for machine learning. It can further utilize simulation for adaptation.

Simulations are not only essential for the development of emergent systems, but also are building blocks of them. As in various other autonomy architectures, in the observer/controller, simulations are crucial to make the information about the dynamics of the system available (Tolk and Durak 2017).

16.3 Evolution of Simulation

Simulation is used for two categories of problems related to experimentation and experience. From the experimentation point of view, simulation is goal-directed experimentation

1. Non-computational simulation
2. Computerized simulation
   2.1 Computerization of behavior generation
   2.2 Simulation tools, toolkits, and environments
   2.3 Influence of types of computers to the advancement of simulation
3. Theory-based simulation
4. Intelligent simulation
   4.1 Contribution of AI to simulation
   4.2 Contribution of software agents to simulation
5. Soft computing simulation
6. Simulation systems engineering
7. Simulation as a service
8. Simulation as infrastructure
9. Simulation for emergence

FIGURE 16.2
Aspects of evolution of simulation. (Adopted from Ören and Durak 2017.)

with dynamic models, i.e., goal-directed experimentation with models for which the behavior and/or the structure can change with time. From the experience point of view, simulation provides experience either for training (or enhancing three types of skills) or for entertainment. Some references about comprehensive views of simulation were developed by Ören (2009, 2010, 2011), as shown in Figure 16.2.

The term "simulation tool" evokes two concepts: (1) there are several tools to formulate and execute simulation studies, as well as analyze and display results of simulation studies; (2) simulation provides infrastructure for hundreds of application areas. Hence, simulation can be considered as a tool for these application areas.

Simulation has been advancing and maturing at an accelerating pace (Ören 2005). Hence, to properly appreciate power of simulation, it would be beneficial to consider its many aspects. Figure 16.2 outlines the nine aspects of the evolution of simulation.

Aspect 1: Non-computational simulation. This has two phases. The first phase, "thought experiments," consists of pure thinking. Even though thought experimenting was part of pre-Socratic thinking, it is still very useful (Brown and Fehige 2014). The second phase consists of use of physical aids for simulation, such as sand boxes and scale models.

Aspect 2: Computerized simulation. At the beginning, computers were used only to generate model behavior. Afterwards, tools, toolkits, and software environments were developed to support increasingly different functionalities for modeling, model processing, and processing of model behavior. Advances in the types of computers also influence advancement of simulation.

Aspect 3: Theory-based simulation, such as DEVS (Discrete Event System Specification), has been an important milestone in the maturity of simulation (Ören and Zeigler 2012).

Aspect 4: Intelligence in simulation. This aspect started with the contributions of Artificial Intelligence (or computational intelligence, or machine intelligence) to simulation. Currently, contribution of software agents to simulation (or ADS) is already a well-established field in simulation.

Aspect 5: Soft computing simulation. Soft computing (Zadeh 1994) induced neural network simulation, fuzzy simulation, evolutionary simulation, and swarm simulation (Thornton 2016).

Aspect 6: Simulation systems engineering. The advent of the simulation studies of complex systems necessitated the adoption of systems engineering in modeling and simulation (Gianni, D'Ambrogio, and Tolk 2014). Hence, simulation systems engineering and agent-directed simulation systems engineering are very powerful possibilities that simulation can offer.

Aspect 7: Simulation as a service. This is a relatively recent and very promising as well as important aspect of simulation (NATO MSG-136; Siegfried 2017).

Aspect 8: Simulation as infrastructure. Many disciplines already benefit from simulation to tackle more complex problems. These disciplines include: engineering and architecture (engineering, systems engineering, cyber-physical systems, materials engineering, complex adaptive systems, and architecture) (Samui 2015); simulation-based natural sciences (science, space exploration, cosmology/astronomy) (Winsberg 2015); health sciences (health care, pharmacology), social sciences and management (social sciences, econometric methods, enterprise management); education and training (education, training (civilian/military)) (Jansen 2015). Hence, studies of emergence in these disciplines can be enhanced by using simulation (Baker 2010).

Aspect 9: Simulation for emergence. This aspect is evolving with the increasing tendency of having simulation of physical systems within the systems. Simulation is becoming an essential part of emergent technical systems to make the information about the system dynamics readily available for observation and control segments (Richter et al. 2006; Tolk and Durak 2017). The term digital twin is used for simulations that consume data from system sensors and simulates the current and future state of the systems (Grieves and Vickers 2017; Marr 2017).

16.4 Simulating Emergence

16.4.1 The Challenge in Simulating Emergence

Simulation has long been a key methodology to study complex systems. As summarized above in the evolution section, it is becoming more than a method and in fact an enabling infrastructure for many domains. Emergence is one of them. Additionally, in accordance with the *Simulation as a digital twin* aspect, simulations are becoming a building block of emerging system architectures with self-x capabilities. Hence, it is arguable to state that simulation is essential for engineering emergence.

The chapter first introduced OC as a framework for understanding emergent technical systems. Then observer/controller was presented as reference architecture for OC systems. The main challenge in simulating emergent systems that the current simulation tools for technical systems fail to tackle naturally is the variation of the system structure during operation time.

Modeling and simulating variable structure and intelligent systems has long been researched (Zeigler and Praehofer 1989; Zeigler et al. 1991; Pawletta 1995; Pawletta 1996; Barros 1997; Uhrmacher 2001; Hu et al. 2005). Zeigler defines variable structure modeling as a methodology in which a model is able to transform from one structure state to a successor one (Zeigler and Praehofer 1989). He ranks these transformations into three levels. The easiest one is characterized with a single component that changes from one structure to another. That corresponds to an element of SuOC changing its structure in our reference OC architecture observer/controller. A good example of that is the parachute problem when the model structure changes from free fall to retarded fall with the opening of the parachute. A more complex transformation is characterized with a complex structure change of overall system, such as addition, removal, or modification of a component which is initiated within the system. Referring to the observer/controller architecture, SuOC changes its structure depending on the processes within the SuOC. The most complex case is when such a change is initiated with an outside model. In observer/controller architecture, this corresponds to the intended operation in which the controller is intervening to change the SuOC structure.

Variable system structures, as an essential characteristic of emergent technical systems, are required to be addressed by the simulation approaches that will be used for engineering the emergence. Representation schemas are required to describe the structure of emergent technical systems. These schemas need systems theoretic formal bases that allow structure synthesis for run-time structure state transitions. The challenge is not only the representation but also lies in executing the changing system structures in order to simulate the system. Therefore, a system structure needs to link the system elements to the models that represent their behavior, the relations of system elements to couplings of models.

16.4.2 Systems Entity Structures and Model Base Framework

System Entity Structure is a high-level ontology which was introduced for knowledge representation of decomposition, taxonomy, and coupling of systems (Kim, Lee, Christensen, and Zeigler 1990). It has its roots in the systems theory-based approaches to modeling and simulation (Oren and Zeigler 2012).

SES is composed of four types of elements: Entity, Aspect, Specialization, and Multi-Aspect. It can be represented as a directed and labelled tree of these elements, as in Figure 16.3. Entity is defined as an object of interest. Variables can be attached to Entities. Decomposition relationship of an entity is denoted by Aspect, while Specialization stands for its taxonomy. Specializations are designated by double lines whereas Aspects are represented by vertical single lines. The Multi-Aspect is a special kind of Aspect that is represented by three vertical lines. It stands for a multiplicity relationship that specifies that the parent entity is a composition of multiple entities of the same type.

SES is based on a clear and limited set of axioms, namely uniformity, strict hierarchy, alternating mode, valid brothers, attached variables, and inheritance (Zeigler 1984). The uniformity axiom states that any two nodes with the same labels have isomorphic subtrees.

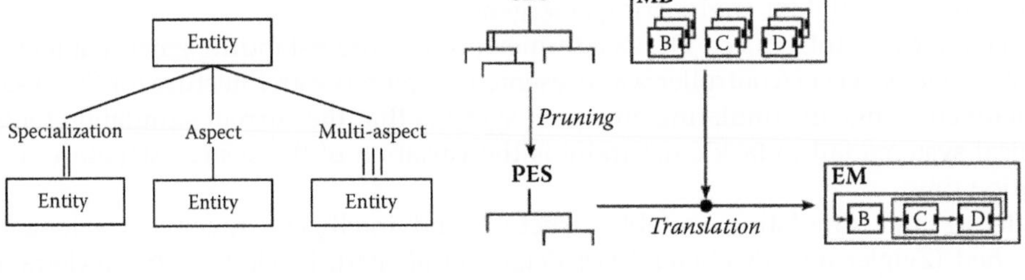

FIGURE 16.3
SES/MB Framework. (Based on Zeigler, Praehofer, and Kim 2000.)

According to the strict hierarchy axiom, a label cannot appear more than once down any path of the tree. The alternating mode axiom specifies that, if a node is Entity, then the successor is either Aspect or Specialization, and vice versa. Two brothers having the same label is prohibited by the valid brothers axiom. The attached variables axiom enforces that variable types attached to the same item shall have distinct names. Specialization inherits all variables and Aspects, according to the inheritance axiom.

Pruning is a method applied to SES for resolving the choices in Aspect, Multi-Aspect, and Specialization relations and assigning values to the variables. When multiple Aspect nodes are used to designate alternative decompositions of the system at the same hierarchical level, a particular one can be chosen in pruning based on the modeler's purpose. Specializations capture all possible variants of an entity in the same dimension, e.g., Size, Color, etc. In pruning, one entity needs to be selected from each Specialization (yielding e.g., Large_Red.). Cardinalities in Multi-Aspect relations are also specified in pruning. The outcome of pruning is a selection-free tree, which is called the Pruned Entity Structure (PES). PES specifies a particular system structure that is formed based on the parameters of pruning.

The System Entity Structure and Model Base (SES/MB) is a framework that combines the SES ontology with the classical workflow of modeling and the simulation of modular, hierarchical systems for automatic generation of an executable simulation model (Zeigler, Praehofer, and Kim 2000). Model Base is proposed as a repository to organize basic models that describe dynamic behavior. Basic models are models of atomic or coupled systems. They can be composed to an overall systems model by their input/output interfaces. Entity node attributes can be used to specify links to basic models in the MB. Pruning and translation are proposed as the steps of the pipeline. After pruning, a translation method is used to generate an executable simulation model (EM) based on the information of PES and the referenced basic models from the MB.

16.4.3 Simulating Variable System Structures with SES/MB

In the classical variable structure modeling approaches, such as (Pawletta and Pawletta 1995) or (Barros 1997), the dynamic structure configurations are modelled on the higher abstraction level but not on the system model level. Hence, they fail to provide seamless integration opportunities with the state of art simulation tools for technical systems. This chapter is based on the approach that has been presented in (Schmidt et al. 2016; Pawletta et al. 2016). It proposes extensions to the core SES/MB framework that enables a seamless integration of the approach in MATLAB/Simulink in order to support simulation of systems with variable structure and parameter configurations.

The set of all system configurations is specified in an SES. The leaf nodes define links to basic models in the MB and specify possible parameter settings in their attached attributes. Configurable dynamic models are traditionally organized in an MB.

SESVariables are proposed as an input interface for the SES/MB framework. They can specify selection of a particular system structure, including parameter settings for basic models. The selection method itself is the pruning operation which is executed based on *SESVariables*. Then, an EM can be generated based on the PES and basic models from the MB using an appropriate translation method. The output of the framework is the EM.

16.4.4 Example: Simulating a Variable SuOC Structure

The generic OC architectural concept observer/controller appoints the task to manage the negative cases and capitalize the positive cases of emergence to controller which affects the local decision rules of SuOC. The example is an extension of an application that has been introduced by The MathWorks (2015) to demonstrate features for variant modeling within MATLAB/Simulink. It is stretched to create a case where the local decision rules of SuOC are changed. The change corresponds to a varying model structure of the SuOC. The aforementioned approach is employed for simulation of variable structure SuOC in a causal block diagram based simulation tool.

Figure 16.4 presents the model structure of the SuOC which has two different controllers (ctrl), based on a linear (lc_ctrl) or a nonlinear (nc_ctrl) control structure.

The Controller of the OC System is affecting the local decision rules of SuOC elements by changing its control structure, affecting the system structure of SuOC by changing the number of available sensors, and affecting the environment that corresponds to a change in sensor measurements. Linear controller is a digital transfer function (dtfnc) and nonlinear controller is a table lookup (ltable). The available sensors are modelled as three signal generators (sg). The change in environment is then modelled as types sine, ramp, or step signals. The output is the SuOC behavior.

The control model (ctrl) can be one of two alternative submodels (lc_ctrl | nc_ctrl). Due to the varying number of possible input signals, both approaches lead to three different model structures. The minimal structure of a controller model with one input signal is illustrated with full lines. The extension for two or three input signals is pictured with dashed lines. On the other side, the model structure of the system generator (sg) depends on the number and type of included signal sources. As presented in Figure 16.5, one of the

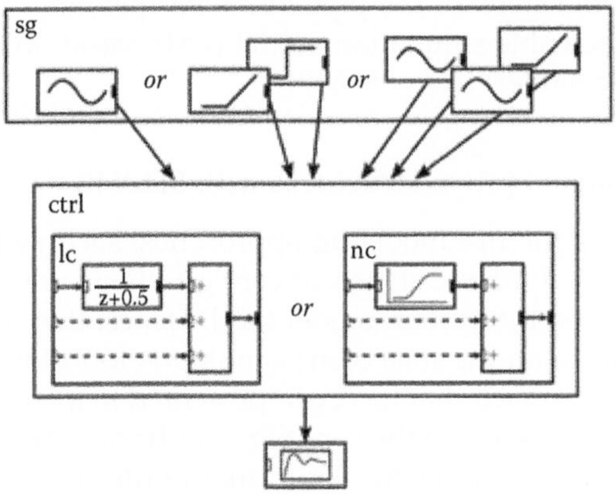

FIGURE 16.4
Example model structure.

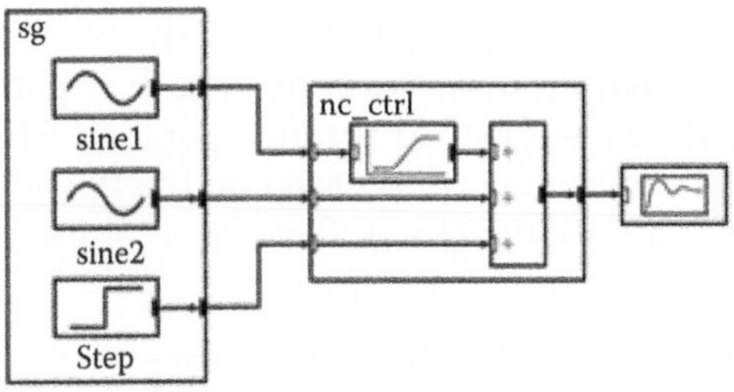

FIGURE 16.5
Structure of a system variant.

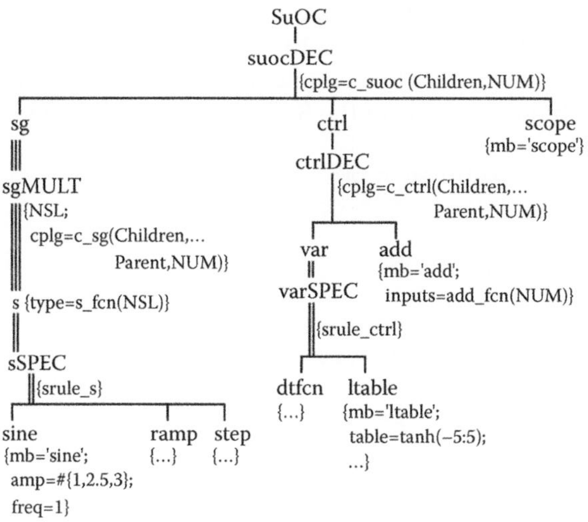

FIGURE 16.6
SES tree for the example.

possible structures for the example SoOC would be a nonlinear control structure and three signal sources, of which two are of type sine and one of type step. The number of all possible system variants even for this simple problem is more than a thousand.

Figure 16.6 presents an overall SES for the example case. *SuOC* is represented in the SES with the root node. *suocDEC* describes that the SuOC model is composed of its children, an *sg*, a *ctrl*, and a *scope*. The aspect attribute *{cplg=...}* defines the coupling relations of *SuOC*. Couplings have the following structure:

```
{'SrcEntity','FromPrt','SinkEntity','ToPrt'}
```

They can be internal couplings between children, and external input as well as external output couplings between the parent and its children.

SESVariables as the input interfaces have a global scope and are written in uppercase letters in the tree. Two example *SESVariables* are defined below:

```
SESVariables={SPEC_CTRL,NSL} with
SPEC_CTRL ∈ {'lc','nc'}
NSL ∈ {(i),(i,j),(i,j,k)|
  i ∈ {'sine[x]','ramp','step'}^
  j ∈ {'sine[x]','ramp','step'}^
  k ∈ {'sine[x]','ramp','step'}^
  x ∈ {1,2,3}}
```

SPEC_CTRL specifies the control structure to be selected and NSL gives the list with signal sources to be selected. An example value assignment to *SESVariables* is given.

```
NSL      ={'sine[1]','sine[2]','step'}
SPEC_CTRL={'nc'}
```

To express such dynamics with minimal effort and to keep a lean SES tree, the concept of *SESFunctions* has been introduced. They are executable within the simulation environment (such as MATLAB functions) and are calculated during the pruning. An example is *cplg=c_suoc(Children, NUM)* which returns the coupling relations. They can access all the SES elements and can have utilities to navigate the tree, such as *Children* and *Parent*. Below is an example in which the elements of the SuOC model are coupled.

```
function cplg=c_suoc(children,num)
%create empty data structure for couplings
cplg=cell(num+1,4);
%set variable ICs btwn sg & ctrl
for i=1:num %for 1 to num
   cplg(i,1:4)={children{1},num2str(i),...
                children{2},num2str(i)};
end
%set fixed IC btwn ctrl & scope
cplg(num+1,1:4)={children{2},'1',
                 children{3},'1' };
end
```

The children sg and ctrl are composed entities while scope is an atomic entity. Leaf node scope maps a basic system in the SES and defines with its attribute mb='scope' a corresponding link to the MB. A representative sketch of MB is given in Figure 16.7.

The node *ctrlDEC* specifies the decomposition of entity *ctrl* in the entities *var* and *add*. For both linear and nonlinear cases, the coupling relations of *ctrl* depend on the number of external inputs which again depend on the current number of signal sources. Thus, the coupling relations at *ctrlDEC* are specified by an *SESFunction* analogous to node *suocDEC*.

The node *varSPEC* is a specialization that describes an XOR selection concerning the succeeding nodes of *var*, between the entity nodes *dtfcn* and *ltable*. During pruning, the selection can be controlled by evaluating selection rules that are specified as node attribute as given below:

```
srule_ctrl={SPEC_CTRL=='lc' → dtfcn |
            SPEC_CTRL=='nc' → ltable}
```

The leaf nodes *dtfcn* and *ltable* are represented once again as a basic model. The extention also proposes parameter specifications using any built-in functions of the simulation tool, such as *tanh* in the definition *table=tanh(-5:5)* at node *ltable*.

The node *sgMULT* is a multi-aspect which specifies that it has only one succeeding entity node *(s)* and defines the number of replications of this node as an attribute. The *SESVariable NSL* is used to define a list of types for replication:

```
NSL={'sine[1]','sine[2]','step'}
```

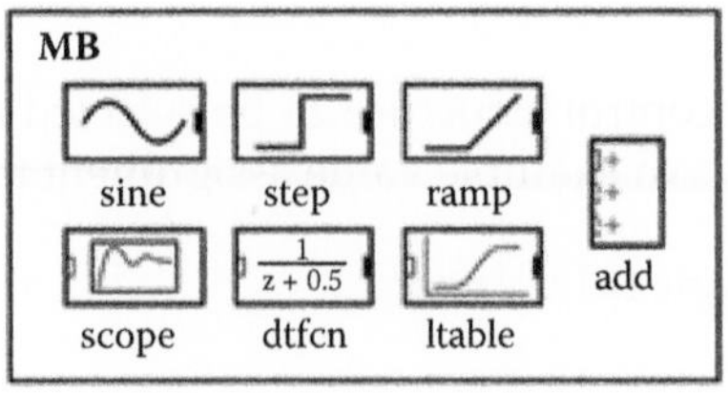

FIGURE 16.7
Model base.

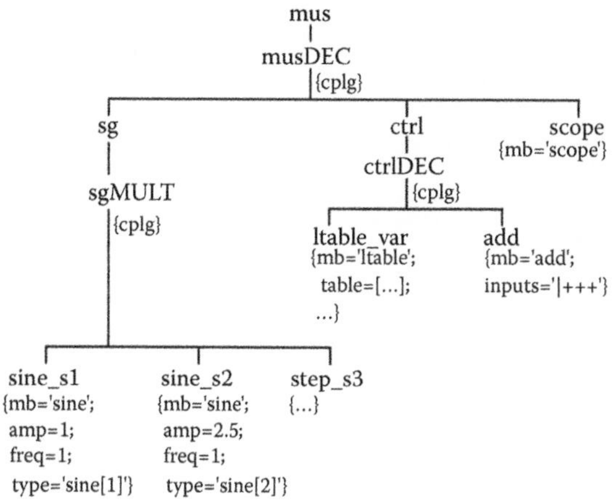

FIGURE 16.8
Pruned entity structure.

The number of replications is implicitly specified by the number of list elements. The entity *s* specifies an attribute *type* whose value o is determined by an iterator *SESFunction s_fcn.*

Based on the value of *type,* replications of entity *s* can be specialized using a succeeding specialization node *sSPEC* with the following selection rule:

```
srule_s={Parent.type=='sine[x]'  → sine[x] |
         Parent.type=='ramp'     → ramp    |
         Parent.type=='step'     → step}
```

The leaf node entities *sine, ramp,* and *step* have links to the MB and contain parameter configurations. For example, the attribute *amp=#{1,2.5,3}* of entity *sine* defines an ordered multiset for different amplitude configurations.

Pruning uses the specification of family of system structures, namely SES and value assignments to its *SESVariables,* and creates a system structure that specifies a particular system structure, namely PES. The *SESVariables*

```
NSL={'sine[1]','sine[2]','step'}
SPEC_CTRL={'nc'}
```

lead to PES that is depicted in Figure 16.8.

16.4.5 SES Toolbox for MATLAB/Simulink

The SES Toolbox (Figure 16.9) has been developed by the research group Computational Engineering and Automation (CEA) at University of Wismar. It is intended for ontology based modelling of system variants (system structures and parameter configurations) using the SES in combination with basic dynamic systems organized in an MB. With the transformation methods it provides for deriving unique system variants and for building executable simulation models for the MATLAB/Simulink environment, it is a full implementation of SES/MB framework. It can be downloaded from the following web site: *http://www.cea-wismar.de/tbx/SES_Tbx/.*

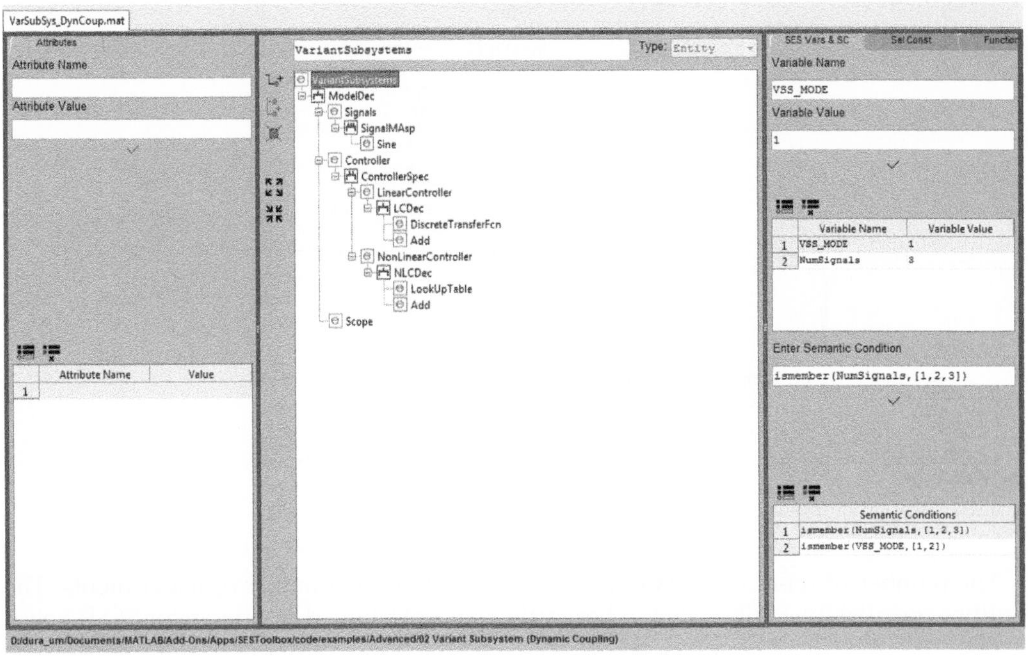

FIGURE 16.9
SES toolbox.

16.5 Outlook

This chapter presents an overview to technical systems with self-x characteristics that shows emergent behavior based on the Organic Computing approach that was developed during a German Research Foundation (DFG) priority program between 2005 and 2011. Observer/controller is being taken as reference architecture for understanding emergence in technical systems. Engineering emergence requires a control with changing the structure of the systems of interest. Simulation is now not only a tool for designing such systems but also becoming a part of them in order to supply data about dynamic system behavior. Hence, the call for approaches to simulate variable system structures is bolder than ever. Engineering emergent systems further requires such approaches to be available and seamlessly integrated with simulation tools for technical systems, such as MATLAB/Simulink.

The chapter presents a methodology that extends System Entity Structure and Model Base framework and exemplifies it with a simple case. The prototype implementation of the methodology in MATLAB/Simulink is called SES Toolbox. It is available to readers who are further interested.

References

Baker, A. 2010. Simulation-based definitions of emergence. *Journal of Artificial Societies and Social Simulation* 13(1):9.

Barros, F.J. 1997. Modeling formalisms for dynamic structure systems. *ACM Transactions on Modeling and Computer Simulation (TOMACS)* 7(4):501–515.

Branke, J., M. Mnif, C. Muller-Schloer, and H. Prothmann. 2006. Organic Computing—Addressing complexity by controlled self-organization. In *2nd International Symposium on Leveraging Applications of Formal Methods, Verification and Validation,* Paphos, Cyprus.

Brown, J.R., and Y. Fehige. 2014. Thought experiments. *Stanford Encyclopedia of Philosophy.* http://plato.stanford.edu/entries/thought-experiment (accessed on March 23, 2017).

Gianni, D., A. D'Ambrogio, and A. Tolk. 2014. *Modeling and simulation-based Systems Engineering Handbook.* CRC Taylor & Francis.

Grieves, M., and J. Vickers. 2017. Digital twin: Mitigating unpredictable, undesirable emergent behavior in complex systems. In Transdisciplinary *Perspectives on Complex Systems,* F.J. Kahlen, S. Flumerfelt, and A. Alves (Eds.), 85–113. Springer International Publishing.

Hu, X., B.P. Zeigler, and S. Mittal. 2005. Variable structure in DEVS component-based modeling and simulation. *Simulation* 81(2):91–102.

Jansen, L.J. 2015. The Benefits of Simulation-Based Education. *SIG 10 Perspectives on Issues in Higher Education* 18:32–42.

Kim, T.G., C. Lee, E.R. Christensen, and B.P. Zeigler. 1990. System Entity Structuring and Model Base Management. *IEEE Transactions on Systems, Man, and Cybernetics* 20(5):1013–1025.

Marr, B. 2017. What Is Digital Twin Technology—And Why Is It So Important?. *Forbes.* https://www.forbes.com/sites/bernardmarr/2017/03/06/what-is-digital-twin-technology-and-why-is-it-so-important (accessed on March 23, 2017).

Müller-Schloer, C.H. Schmeck, and T. Ungerer (Eds.). 2011. *Organic computing—A paradigm shift for complex systems.* Springer Science & Business Media.

NATO MSG-136. RTO Task Group on Modelling and Simulation as a Service (MSaaS) Rapid deployment of interoperable and credible simulation environments. https://www.cso.nato.int/ACTIVITY_META.asp?ACT=5642 (accessed on March 23, 2017).

Ören, T.I. 2005. Maturing Phase of the Modeling and Simulation Discipline. In *Proceedings of: ASC—Asian Simulation Conference 2005,* Beijing, P.R. China.

Ören, T.I. 2009. Modeling and Simulation: A Comprehensive and Integrative View. In *Agent-Directed Simulation and Systems Engineering,* T. Ören and L. Yilmaz (Eds.), 3–36. Wiley-Berlin.

Ören, T.I. 2010. Simulation and Reality: The Big Picture. *International Journal of Modeling, Simulation, and Scientific Computing (IJMSSC)* 1(1):1–25.

Ören, T.I. 2011. A Basis for a Modeling and Simulation Body of Knowledge Index: Professionalism, Stakeholders, Big Picture, and Other BoKs. *SCS M&S Magazine* 2(1):40–48.

Ören, T.I., and B.P. Zeigler. 2012. System Theoretic Foundations of Modeling and Simulation: A Historic Perspective and the Legacy of A. Wayne Wymore. *Simulation* 88(9):1033–1046.

Ören, T., and U. Durak. 2017. The Evolution of Simulation and Its Contributions to Many Disciplines. In *Guide to Simulation-Based Disciplines: Advancing Our Computational Future,* S. Mittal, U. Durak, T. Ören (Eds.), Springer.

Pawletta, T., and S. Pawletta. 1995. Design of an object-oriented simulator for structure variable systems. *Systems Analysis-Modelling-Simulation* 18(1):471–474.

Pawletta, T., B.P. Lampe, S. Pawletta, and W. Drewelow. 1996. An object oriented Framework for modeling and simulation of variable structure systems. In *Proc. of the SCS Summer Computer Simulation Conf.*, Portland, OR.

Pawletta, T., A. Schmidt, B.P. Zeigler, and U. Durak. 2016. Extended Variability Modeling Using System Entity Structure Ontology within MATLAB/Simulink. In *Proc. of the SCS SpringSim/ANSS 2016*, Pasadena, CA.

Richter, U., M. Mnif, J. Branke, C. Müller-Schloer, and H. Schmeck. 2006. Towards a generic observer/controller architecture for Organic Computing. *GI Jahrestagung* 93(18):112–119.

Samui, P. 2015. *Handbook of Research on Advanced Computational Techniques for Simulation-Based Engineering*. IGI Global.

Schmeck, H., C. Müller-Schloer, E. Çakar, M. Mnif, and U. Richter. 2010. Adaptivity and self-organization in organic computing systems. *ACM Transactions on Autonomous and Adaptive Systems (TAAS)* 5(3):10.

Schmeck, H. 2005. Organic computing—A new vision for distributed embedded systems. In *8th IEEE International Symposium on Object-Oriented Real-Time Distributed Computing*, Seattle, WA.

Schmidt, A., U. Durak, and T. Pawletta. 2016. Model-based testing methodology using system entity structures for MATLAB/Simulink models. *Simulation* 92(8):729–746.

Siegfried R. 2017. Call for Papers: Special Issue: Leveraging Modeling & Simulation as a Service for the Future M&S Eco-system, *Journal of Defense Modeling & Simulation: Applications, Methodology, Technology (JDMS)*. http://journals.sagepub.com/doi/pdf/10.1177/1548512914556467 (accessed on March 23, 2017).

The Mathworks. 2015. *Variant management*. http://de.mathworks.com/help/simulink/ug/variant management.html (accessed on November 12, 2015).

Thornton, M.A. 2016. *Swarm simulation*. http://markallenthornton.com/blog/swarm-simulation/ (accessed on March 23, 2017).

Tolk, A., and U. Durak. 2017. Simulation als epistemologische Grundlage für intelligente Roboter. In *Proceeding of Workshop der ASIM/GI Fachgruppen STS und GMMS*, Ulm, Germany.

Uhrmacher, A.M. 2001. Dynamic structures in modeling and simulation: A reflective approach. *ACM Transactions on Modeling and Computer Simulation (TOMACS)* 11(2):206–232.

Winsberg, E. 2015. Computer Simulations in Science. *Stanford Encyclopedia of Philosophy*. http://plato.stanford.edu/entries/simulations-science/ (accessed on March 23, 2017).

Zadeh, L.A. 1994. Fuzzy Logic, Neural Networks, and Soft Computing. *Communication of the ACM* 37(3):77–84.

Zeigler, B.P. 1984. *Multifaceted Modelling and Discrete Event Simulation*, London, UK, Academic Press Professional Inc.

Zeigler, B.P., and H. Praehofer. 1989. Systems theory challenges in the simulation of variable structure and intelligent systems. In *International Conference on Computer Aided Systems Theory*, Las Palmas, Spain.

Zeigler, B.P., T.G. Kim, and C. Lee. 1991. Variable structure modelling methodology: An adaptive computer architecture example, *Transactions of the SCS* 7(4): 291–318.

Zeigler, B.P., H. Praehoffer, and T.G. Kim. 2000. *Theory of Modelling and Simulation*. 2nd ed. Orlando, FL: Elsevier Academic Press.

17

Emergence as a Macroscopic Feature in Man-Made Systems

Moath Jarrah

CONTENTS

17.1 Introduction

Large scale systems are complex to model and simulate as systems grow in the number of their components and their interconnections. Hence a bottom-up modular approach is used to design complex systems as the computational costs have decreased. The designer of a complex system starts by modeling each component in the system according to the component's characteristics and interaction with other components. The modeler engineer puts together these components and the communication channels between them, which results in the overall system. The system and its components are created based on the knowledge of the modeling engineer. The modeler's design might need to build more systems in which a system of systems (SOS) is the final output as depicted in Figure 17.1. The entire process of modeling is based on the modeler's understanding and knowledge of the application of interest. Emergence is a new feature of the system or new knowledge about the system that arises because of dynamic interactions between the components and/or with the environment variables. In this chapter, we focus on emergence that can be detected using computer simulations.

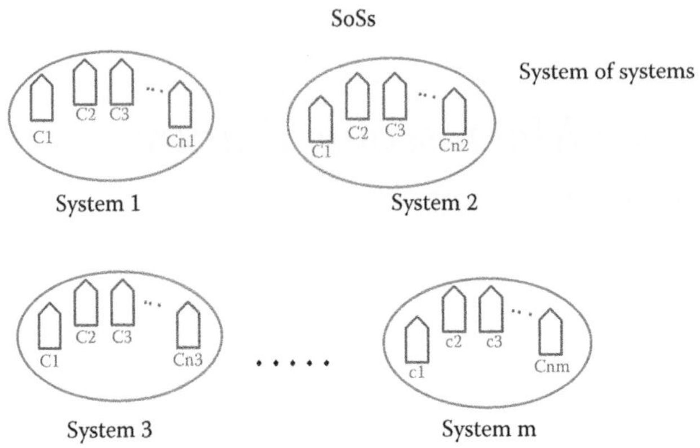

FIGURE 17.1
Different components in a system design.

Traditional equation-based approaches cannot, or are too difficult to, be formulated to solve complex systems. Hence, researchers have started to use component-based modeling or agent-based modeling because of the increasing complexity in systems. An independent component, whether it is a software or a model, can be considered as an agent where agents play the role of components [1,2]. An agent's behavior can be as simple as an if-then statement or very complex learning, adaptive, and training algorithms taken from the artificial intelligence domain. Macal and North discussed the need for agent-based modeling and simulation with four reasons [3]. First, the systems are becoming more complex in terms of their components and their interactions. Second, some systems have always been too complex to develop a model that represents them, such as economic markets. Third, data are stored at a finer level and micro-data can be modeled and simulated at a micro-level. And fourth, today's computational power, such as GPU accelerators, multi-core CPUs, cloud computing, and grid computing, is able to perform large-scale micro-level simulations. Hence, agent-based simulations (ABS) or multi-agent simulations (MAS) provide the suitable modeling and simulation environment to analyze, model, understand, and detect emergence in complex systems. Macal and North discussed that an agent is self-contained where it can function independently within its environment or with other agents. Also, it has its own characteristics, attributes, decision-making, and interaction rules at the microscopic level. Chapter 3 by Bernie Zeigler discussed the features of DEVS modeling and simulation framework where atomic and coupled models support the requirements of agent- or component-based simulation. The external, internal, and confluent functions support the behavioral foundation. In addition, local variables, states, input ports, output ports, and messages provide static and dynamic interactions capabilities. Moreover, DEVS provide variable structural capabilities for the modeler [4]. This chapter provides an example of using DEVS framework to model and simulate a man-made system.

Complex Adaptive Systems (CAS) consists of interacting agents and exhibits positive or negative features at the macroscopic level [5]. The microscopic features are the characteristics of an independent component or agent (low-level). The macroscopic features are the characteristics of the whole system. Emergence is a macroscopic feature that is observed at the whole system (high-level). For example, a car consists of many components, such as a tire, a steering wheel, a window, etc. The tire spins forward or backward which represents the tire's macroscopic features. The car's speed (e.g., 80 miles/hour) is one of the entire system's macroscopic features.

17.2 MAS Framework for Man-Made Systems

Man-made systems or system of systems are complex because of the many constituents that make up the system. As in [1] and [2], these components or constituents can be considered as agents in multi-agent simulations. Each agent has its own behavior and type at the microscopic level as shown in Figure 17.2. Two types of interactions can occur, agent to agent and agent to environment. The agent to environment interaction can be explained by the robots that were sent to clean the Fukushima Daiichi nuclear power plant in Japan after the 2011 disaster. The robots failed to perform their tasks due to the massive amount of radiation that caused failures to different components. This is a negative emergence where the man-made system (robot) failed to perform its designated task. As a result, many researchers had to study the radiation's effect on the robots in order to re-engineer them for such critical environments, such as in [6] and [7].

As shown in Figure 17.2, the microscopic level consists of many agents with different internal models. Agents can be grouped into types which are represented in the figure as different shapes. Agents of the same shape are of the same type. When the simulation starts, agents act based on their rules or as a reaction to an external event that occurs during the interaction with other agents. The higher level is the macroscopic level which describes the features of the system as a whole. Usually, the macroscopic features are the observed characteristics that measure how the system performs in reality. Weak, strong, and spooky emergences are macroscopic features of the system that are unknown beforehand. They can be positive or negative features, as discussed in Chapter 1.

17.2.1 Formalization of Emergence in the Macroscopic Features of MAS

Recent work in formalization of multi-agent simulations considers agent instances of different types. Hence, given M types of agents and N agents of each type, then an agent can be represented as A_i^j where $i \in \{1,2,3,\ldots, N\}$ and $j \in \{1,2,3,\ldots, M\}$. An agent has static and

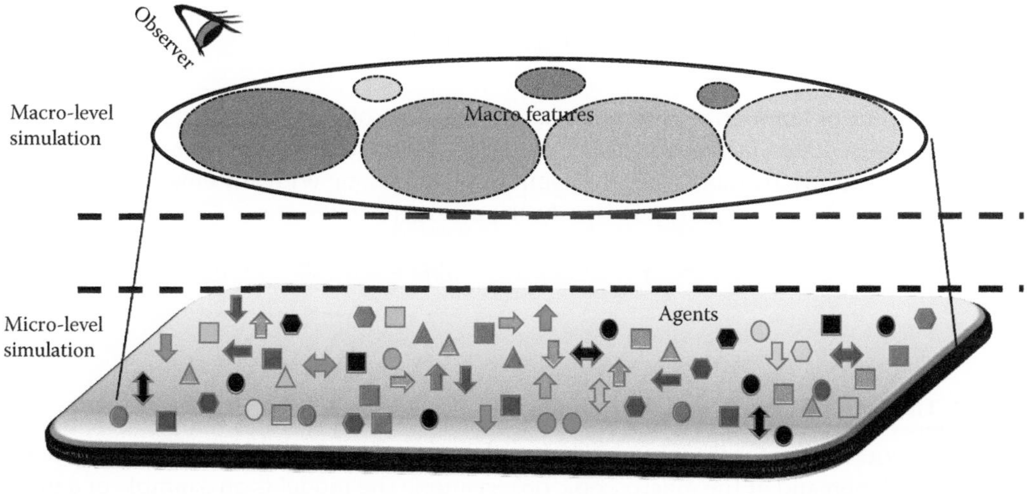

FIGURE 17.2
Micro and macro level simulation in MAS.

dynamic attributes that capture its properties and states [3,8]. An agent can be represented in terms of the attributes as shown in Equation 17.1.

$$A_i^j \rightarrow \left(statAtt_i^j, dynAtt_i^j\right) \quad (17.1)$$

The ABS or MAS system at any given simulation time t consists of a vector of all agents along with the environment state as shown in Equation 17.2 [8].

$$a(t) = \left[\left(statAtt_i^j(t), dynAtt_i^j(t)\right), s_{env}(t)\right]; \quad i \in \{1,2,3,\ldots,N\}, j \in \{1,2,3,\ldots,M\} \quad (17.2)$$

The microscopic features of the agents in the MAS simulation are denoted as L_{sum} in [9,10,11,12]; where it is defined as the sum of all individual agent behaviors without their interaction. Hence, L_{sum} and the vector $\boldsymbol{a}$ in Equation (17.2) are equivalent. At the microscopic level, after a simulation time step δt, the new state vector is a function of vector $\boldsymbol{a}$ as shown in Equation 17.3.

$$a(t+\delta t) = F(a(t), \delta t) \quad (17.3)$$

At the macro level simulation, let L_{whole} represents all possible system states due to agents' behaviors, agent-to-agent interaction, and agent-to-environment interaction. Thus,

$$L_{whole}(t+\delta t) = F(L_{whole}(t), A(t), E(t), \delta t) \quad (17.4)$$

where $A(t)$ represents the agent-to-agent interaction at the simulation time t and $E(t)$ is the agent-to-environment interaction.

After running the simulation long enough, the system reaches to equilibrium or a steady state. The macroscopic states ($L_{Equilibrium}$) at equilibrium do not change with time and can be described as in Equation 17.5.

$$L_{Equilibrium} = L_{whole}(t+\delta t) = L_{whole}(t) \quad (17.5)$$

At this point of simulation, the microscopic features of the agents change with time while the steady state macroscopic features do not. The macroscopic features consist of all states that are known by modeler of the complex system along with any emergent behavior whether it is positive or negative. Formally, the emergent states are:

$$L_{Emergent} = L_{Equilibrium} - L_{known} \quad (17.6)$$

17.2.2 The Labor Market Model Example

This subsection discusses the labor market model and shows its specifications at the microscopic simulation and at the macroscopic observation. The model is an example of a man-made system that can be used in companies. At the microscopic level, the model consists of two types of agents, a firm agent and a worker agent [8,13]. There are F number of agents of type firm and N number of agents of type worker; formally, the model's agents are:

$$\text{Agents} = A_f^1 + A_w^2 \text{ where } f \in \{1,2,3,\ldots,F\}, w \in \{1,2,3,\ldots,N\}$$

An agent of type worker has two variables that determine its microscopic behavior which are: the satisficing wage $\left(\omega_{wt}^s\right)$ at time t and the reservation wage $\left(\omega_w^r\right)$. The microscopic behavioral rules of the worker agents are:

1. A_w^2 prefers to receive ω_{wt}^s wage from the firm agent.
2. A_w^2 would accept a wage that is greater than or equal to its ω_w^r wage.

On the other hand, an agent of type firm has three microscopic properties, which are: the production amount (q_{ft}) at time t, the real offered wage (ω_{ft}) at time t, and the satisficing wage (ω_{ft}^s) at time t.

The microscopic behavioral rules of the firm agents are:

1. A_f^1 has initial values of the production amount q.
2. A_f^1 tries to force its satisficing wage $\left(\omega_{ft}^s\right)$ on any worker.

This model does not specify an agent-to-environment interaction rule. In the Fukushima Daiichi nuclear power plant case mentioned in Section 17.2, the radiation is an example of an agent-to-environment interaction. The labor market model specifies seven agent-to-agent microscopic interaction rules according to Fagiolo, Dosi, and Gabriele (2004) which are:

1. At simulation time t, firm agents determine the number of job openings.
2. Worker agents search for job openings and queue up for at least one.
3. Firm agents check their queues and negotiate with the worker agents in the queue to decide whether to hire them or not based on the wage.
4. Firm agents calculate their production using the equation, $q_{ft} = \alpha_{ft} \times n_{ft}$; where α_{ft} is the labor productivity of firm f at time t, and n_{ft} is the number of worker agents hired at time t by firm f. In addition, the price of goods and the profit for each firm are calculated (see reference [13]).
5. As a result of step 4, firm agents review their worker selection process and replace any negative profit (loss) with an average of all firm agents.
6. Based on step 5, firm and worker agents update their satisficing wages.
7. Change some of the simulation variable such as labor productivity.

The computer simulation of these two types of agents and their interactions reaches to equilibrium or steady state. This is the model's macroscopic feature that is observed by the designer. The result is that the amount of production is linearly proportional to the number of hired worker agents. In addition, the profit that is achieved by firm f at time t is calculated using Equation 17.7; where p_t is the price of the good.

$$\pi_{ft} = (p_t \times \alpha_{ft-1} - \omega_{ft}) \times n_{ft} \tag{17.7}$$

17.2.2.1 Detection

The computer simulation of the model allows the two types of the agents to interact based on the seven rules mentioned in subsection 17.2.2. At the beginning of the simulation, the observer cannot detect the amount of production and profit at the macroscopic level because many worker agents join and leave firms. After a long simulation, equilibrium is reached and the relationship becomes clear.

17.2.2.2 Classification

The equilibrium emergence that is found in the labor market can be classified as a strong emergence. The observer has not been able to observe it before simulating the model and reaching to the steady state. In addition, as shown in the description of the two types of the agents and their interactions, the model designer cannot anticipate the linearity of the relationship between the amount of production and number of workers.

17.2.2.3 Control

The designer of the market model can put minimum and maximum constraints on the number of the worker agents that are hired by a firm agent. The constraints limit or put a control on the margin of all firms' production which is directly related to the firms' profit.

17.3 Case Study of a Man-Made System (Congestion in the Internet)

The Internet is the most complex man-made system that was created. Billions of devices are connected through a communication network to allow data and information sharing and exchange. The TCP/IP network protocol is the most used protocol by the connected computing nodes. The TCP/IP protocol was initially designed in the early 1980s by the Advanced Research Projects Agency of the Department of Defense. TCP/IP is normally considered to be a 4-layer system. The 4 layers are: the application layer, the transport layer, the network layer, and the data link layer.

When the TCP/IP was first designed, the number of the connected devices was small. Now, the Internet is a large-scale system that consists of billions of devices. If many nodes are to send data to a destination through the same link, congestion occurs. Congestion is a negative emergent behavior that has emerged when many connected components with different behaviors exchange data. Manley and Cheng studied congestion for road traffic as an emergent behavior [14]. In order to solve the congestion problem in the Internet, new implementation features were added to the original TCP protocol. The TCP reference document can be found in RFC 2581 [15]. The reference specifies four congestion control protocols that are used today. The TCP/IP congestion control protocols are: slow start, congestion avoidance, fast retransmit, and fast recovery [16]. This section describes the congestion as an emergent behavior in the Internet.

When a process at a computing node (source) sends data to a destination, the process passes the data to the application layer, which in turn adds a header to the data segment. The application layer then passes the new segment (header + data) to the transport layer which in turns adds another header, and so on until it reaches to the network interface

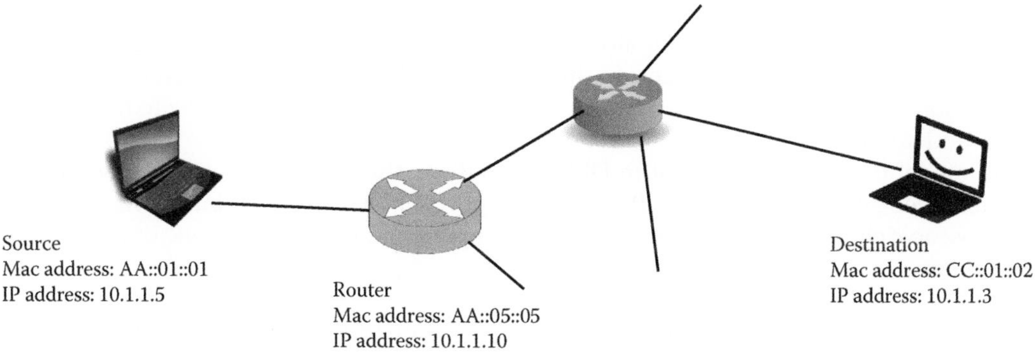

FIGURE 17.3
Routers and computing nodes in the Internet.

layer which transmits the message. This happens at the sender side which is independent of other components in the Internet system.

The Internet consists of computing nodes in addition to routers that mediate the data transfer. A router conducts its routing table in order to decide on which link or port a message should be sent in order to reach its destination. Figure 17.3 shows that routers exist as intermediate nodes between a sender and a receiver. A sender may reside in a different country or even different continent than the destination. It is the task of the routers to find out the path for a data packet to travel in order to reach to its final destination. When the router receives a data packet from a source, it strips down the Mac layer header to look for the IP header. The IP header determines the destination address. The router then checks its routing table to find the corresponding link or port that is best for the packet. There are three scenarios of the routing table which are:

1. If the destination address exists in the routing table, then the interface name will be known at which the packet is to travel.
2. If the destination address does not exist in the routing table, the IP address is converted into a network IP using a mask. Hence, the interface name will be known.
3. If the IP address is not found, then the router forwards the packet to a default interface name.

The aforementioned behavior of the router is independent of the source or the destination. If we model each of these entities as agents in MAS, we will have a sender agent, a router agent, and the destination agent. Each of these agents follows its own microscopic behavior. But the Internet is concerned with macroscopic features such as round trip time, packet loss rate, throughput, goodput, etc.

17.3.1 A Model of TCP/IP Communication in DEVS Formalism

As mentioned in the previous section, each of the sender, destination, and router components behaves independently according to their rules. Putting these components together and allowing them to interact can result in a congestion which is a negative emergence. Hence, it is reasonable to suggest that the interaction of thousands or millions of devices in the Internet can lead to congested links.

In order to model these components in DEVS formalism, we can have two types of agents or atomic models, a router atomic model and a computing atomic model. We will refer to the atomic models as agents in this subsection. This is valid as we discussed earlier in Section 17.1. The computing agent can be a sender or a receiver according to the current state and an external event.

To model the computing agent in DEVS formalism, the agent needs to have two input ports (*SendInport* and *ReceiveInport*), one output port (*Outputport*), and states. These are the microscopic characteristics of the agent. The computing agent can be in one of the following states:

- *Computing*: This state represents a normal functionality of the computing node.
- *Sending*: In a response to a process's action of sending a message to another node, the agent transits to the sending state. This is an external event that occurs on its *SendInport*. The *external* function described in Chapter 3 is responsible for this transition.
- *Applay*: This state is after the *Sending* state and adds the application header to the data segment of the message. The *internal* function described in Chapter 3 is responsible for this transition.
- *Tranlay*: This state comes after the *Applay* state and adds the transport header to the new message. The *internal* function of the computing node agent is responsible for this transition.
- *Netlay*: This state is after the *Tranlay* state and adds the network header to the new message. The *internal* function of the computing node agent is responsible for this transition.
- *Linklay*: This state is after the *Netlay* state and adds the header of the data link layer to the new message. The *internal* function of the computing node agent is responsible for this transition.
- *Transmit*: This state comes after the *Linklay* state, where the source, destination IPs, and Mac addresses are known. This state results in firing the *out* function of DEVS formalism. After this state, the agent transits into the *Computing* state to become ready for sending or receiving. The *internal* function of the computing node agent is responsible for this transition.
- *Receiving*: If the agent is in the Computing state and an external event occurs on the *ReceiveInport*, the agent enters into the *receiving* state. The *external* function of the computing node agent is responsible for this transition.
- *Unwrapping*: This state is after an internal transition from the receiving state. In this state, all the headers of the four layers are unwrapped one by one and used at each layer of the destination node. After this state elapsed time, the agent transits into the *Computing* state to become ready for sending or receiving. The *internal* function of the computing node agent is responsible for this transition.

It is worth mentioning here that the states *Sending, Applay, Tranlay, Netlay, Linklay, Receiving, Transmit,* and *Unwrapping* are non-preemptable by external events. Any external events that occur on the input ports must be queued.

At the microscopic level, the router has a receiving buffer of size x, a sending buffer of size y, IP address, and a Mac address. It has multiple input ports for receiving messages

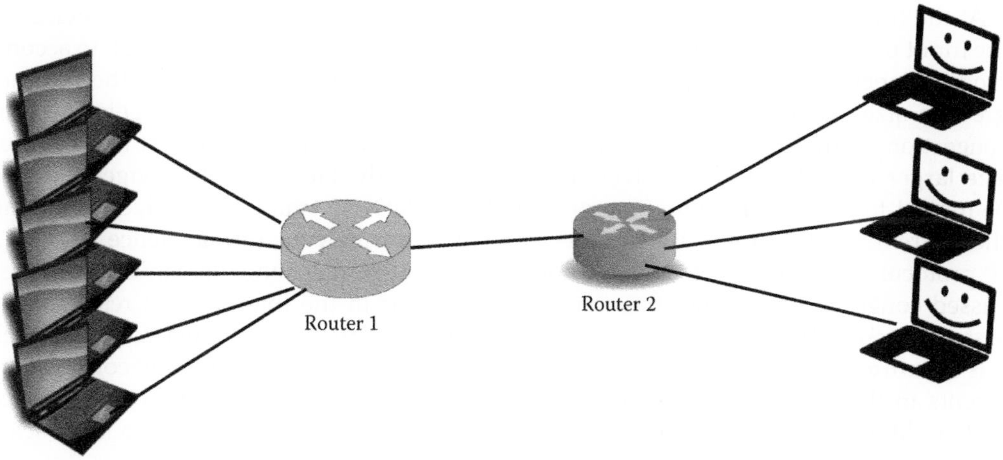

FIGURE 17.4
Input and output ports of the agents at the microscopic level.

and multiple output ports for sending out messages. The number of distinct ports depends on the network topology. For example, Router 1 in Figure 17.4 has five input ports and one output port. Router 2 has one input port and three output ports. Nevertheless, the physical devices of today's technology allow bidirectional connections (duplex) where a port can be used for both sending and receiving. In DEVS formalism, a port must be determined as to whether it is an input or output port (unidirectional or simplex).

A modeler can model the router agent to have three states, which are:

- *Ready*: This state represents a normal functionality of the router node as being waiting to receive messages on its input ports to perform rerouting.
- *Reroute*: In a response to receiving a message on its inputs, the router gets the message and finds the destination address. Then it checks its internal routing table to find which output port is the appropriate one to reach the destination. The *external* function of the router agent is responsible for this transition.
- *Sending*: After determining the appropriate port to send a message, the agent transits to this state causing the out function to execute and the message to be sent on the output port. The *internal* function of the router agent is responsible for this transition.
- After the *Sending* state, the agent goes back to the *Ready* state waiting for more messages to be rerouted. The *internal* function is responsible for this transition.

The aforementioned characteristics of the computing agent and router agent in DEVS formalism are known as L_{sum} or the vector $\boldsymbol{a}$ in Equation (17.2). L_{sum} was defined as the sum of all individual agent behaviors without their interaction in subsection 17.2.1.

Coupling is used in DEVS to connect different agents. We can have N agents of type computing and M of type router. Then we add coupling between the input ports and the output ports to establish a specific network topology. The modeling engineer can add more and more agents resulting in a large scale multi-agent simulation.

At the time when the TCP/IP was first designed, the number of connected devices to the input ports of a router was small. The receiving buffer can be large enough to accommodate concurrent messages that arrive on the input ports. However, as the Internet grew in size and millions of devices were plugged in, the input buffer gets full and congestion occurs on many links. Replacing a router with a new one that has a buffer size larger than the old one solves the problem locally. However, the congestion will be moved to another neighboring router with a relatively small buffer size. Changing the entire infrastructure of the Internet is costly and time-consuming. Hence, the TCP/IP protocol was modified to include congestion control mechanisms. Smith (2011) discussed congestion and other phenomena that appear in the Internet as a result of the system complexity. Congestion is a phenomenon that appears at the macroscopic level. It is a feature of the entire Internet system as a result of interaction of many components (agents in the simulation domain). It is clear that congestion is a negative emergence that could have been detected if MAS was used to model and simulate the large-scale Internet system.

17.3.1.1 Detection

A multi-agent simulation using DEVS formalism of the TCP/IP protocol requires having many agents of the two types: computing nodes and routers. When building a large scale simulation to depict what we find in the Internet, many computing agents can be sending messages to one destination. The DEVS simulation can find out that the number of messages might overwhelm the input buffer of a router on the path. This results in dropping of packets, raising the congestion problem. The observer can detect the congestion phenomenon at the macroscopic level by increasing the rate at which the atomic computing models send messages.

17.3.1.2 Classification

The congestion is classified as a strong emergence. When first designing and developing the TCP/IP protocol, designers were not aware of the massive scale of today's Internet or the congestion problem. By using MAS in DEVS formalism, designers can add more and more atomic computing and routers agents in order to detect that congestion can occur. As a result, DEVS formalism could have helped designers to avoid the man-made system failures or shutdowns. Congestion is a negative strong emergence.

17.3.1.3 Control

In order to control the congestion between two atomic models, DEVS formalism provides a rich environment of coupling, decoupling, and state changing to alter the behavior of a router atomic model. A DEVS modeler can use coupling between the receiver and sender to allow the receiver to inform the sender to increase or decrease the sending rate. This is similar to the sliding window that was added to the TCP/IP. Also, an overwhelmed intermediate atomic model can inform others using its output ports to route messages to other neighbors. This prevents specific links (coupling) from becoming congested. By using DEVS formalism, a designer can experiment with many different options and observe the network performance.

17.4 Conclusion

The microscopic level of a man-made system specifies each agent's or component's behavior and action rules independently. However, the agents exist together with other agents in a complex man-made system. Agents interact and affect each other's behavior, resulting in features for the entire system. The system's features are described at a macroscopic level. Modeling and simulation using DEVS formalism can help engineers in discovering the macroscopic features of an entire complex system. The macroscopic features can be known ahead of time (at the design time), or can emerge as a new knowledge about the system, such as link congestion in the case of the Internet. This chapter reviewed different efforts in formalization of emergence at the simulation macroscopic level. It also discussed how DEVS formalism provides features to describe agents at the microscopic level, and the simulation can result in showing macroscopic features. The labor market model and TCP/IP protocol in the Internet were discussed to explain emergence by examples.

References

1. Eric Bonabeau. (2002). Agent-based modeling: Methods and techniques for simulating human systems. In Proceedings of the National Academy of Sciences. 99(3): 7280–7287.
2. Brian Heath, Raymond Hill, and Frank Ciarallo. (2009). A Survey of Agent-Based Modeling Practices (January 1998 to July 2008). Journal of Artificial Societies and Social Simulation 12(4): 9.
3. Charles M. Macal and Michael J. North. (2009). Agent-based modeling and simulation. In Winter Simulation Conference (WSC '09). 86–98.
4. Xiaolin Hu, Xiaolin Hu, Bernard P. Zeigler, and Saurabh Mittal. (2005). Dynamic Reconfiguration in DEVS Component-based Modeling and Simulation. Journal Simulation. 81(2): 91–102.
5. Maarika Teose, Kiyan Ahmadizadeh, Eoin O'Mahony et al. (2011). Embedding system dynamics in agent based models for complex adaptive systems. In Proceedings of the Twenty-Second International Joint Conference on Artificial Intelligence. 3: 2531–2538.
6. Shimeng Li, Y. F. Zheng, J. Qiu, and Lei (Raymond) Cao. (2015). Testing of harmonic drive degrading under radiation environment. International Journal of Mechatronics and Automation, Vol. 5, No. 2/3, pp. 69–79, 2015.
7. Shimeng Li, Adib Samin, J. Qiu, Y. F. Zheng, and Lei (Raymond) Cao. (2015). Study on radiation induced performance degradation of BLDC motor in robot servo systems. International Journal of Mechatronics and Automation, Vol. 5, No. 2/3.
8. Zehong Hu, Meng Sha, Moath Jarrah, Jie Zhang, and Hui Xi. (2016). Efficient computation of emergent equilibrium in agent-based simulation. In Proceedings of the Thirtieth AAAI Conference on Artificial Intelligence (AAAI '16). AAAI Press 2501–2508.
9. A. Kubí. (2003). Toward a Formalization of Emergence. In *Artificial Life*, vol. 9, no. 1, pp. 41–65.
10. Yong Meng Teo, Ba Linh Luong, and Claudia Szabo. (2013). Formalization of emergence in multi-agent systems. In Proceedings of the 1st ACM SIGSIM Conference on Principles of Advanced Discrete Simulation (SIGSIM PADS '13). ACM, New York, NY, USA, 231–240.
11. Claudia Szabo and Yong Meng Teo. (2015). Formalization of Weak Emergence in Multiagent Systems. ACM Transactions on Modeling and Computer Simulation, vol. 26, no. 1, 6.
12. C. Szabo, Y. M. Teo, and G. K. Chengleput. (2014). Understanding complex systems: Using interaction as a measure of emergence. Proceedings of the Winter Simulation Conference 2014, Savannah, GA, pp. 207–218.

13. G. Fagiolo, G. Dosi, and R. Gabriele. (2004). Matching, bargaining, and wage setting in an evolutionary model of labor market and output dynamics. Advances in Complex Systems. 7(02): 157–186.
14. E. J. Manley and T. Cheng. 2010. Understanding road congestion as an emergent property of traffic networks. In Proceedings of 14th World Multi-Conference on Systemics, Cybernetics and Informatics. 25–34.
15. M. Allman, V. Paxson, and W. Stevens. (1999). TCP Congestion Control, RFC 2581.
16. Reginald D. Smith. (2011). The Dynamics of Internet Traffic: Self-Similarity, Self-Organization, and Complex Phenomena. Journal of Advances in Complex Systems, 14, 6 pp. 905–949.

18

Monterey Phoenix—Behavior Modeling Approach for the Early Verification and Validation of System of Systems Emergent Behaviors

Kristin Giammarco and Mikhail Auguston

CONTENTS

18.1 Introduction

This chapter teaches how to detect, classify, predict, and control emergent behaviors in System of Systems (SoS) aided by modeling and simulation (M&S) with Monterey Phoenix (MP), a behavior modeling approach and tool for system, software, and process architecture. In each of three examples from different domains, these activities are illustrated as a well-partitioned but collaborative effort between humans and automated tools that produces useful insight into the emergent behaviors of a system of interest. The sections that follow make the case for a key realization from this research, which is that in order to improve the state of the art in detection, classification, prediction, and control of emergent behaviors during the early stages of design, we must adhere consistently to certain old and new heuristics. In brief, these heuristics are: separate behaviors and interactions, model system behaviors and environment behaviors, formalize models for automatic execution, properly allocate each task to a human or to a machine, and use abstraction and refinement to manage large models. Consistent implementation of these modeling heuristics results in substantially improved SoS models that contain far more emergent behaviors to be detected, classified, predicted, and controlled—manually at first, then with increasing automation as we equip our modeling tools with behavior pattern templates to scan across all system behavior models. The automated behavior pattern templates, a main discussion of Chapter 19, formalize value judgments assigned and controlled by humans (e.g., whether an emergent behavior is favorable or unfavorable).

After the motivation is established in Section 18.2, basic behavior modeling concepts for engineering emergence are presented in Section 18.3, followed by an overview of the Monterey Phoenix (MP) approach in Section 18.4. The approach is illustrated with findings of emergent behaviors in student-developed MP models in Section 18.5. As shown in Section 18.6, a major finding from this research is that certain emergent behaviors of a SoS can be detected, classified, predicted, and controlled through early M&S. The chapter concludes in Section 18.7 with a summary of conclusions and next steps.

The concepts of detection, classification, prediction, and control of emergent behaviors are used throughout this chapter, defined in Chapter 1, and reprinted here for reference:

Detection: The initial discovery of the presence/existence of emergence.

Classification: Four types of emergence have been identified to date:

1. Simple: According to Page (2009), "*simple emergence* is generated by the combination of element properties and relationships and occurs in non-complex or 'ordered' systems."
2. Weak: Page (2009) describes "*weak emergence* as expected emergence which is desired (or at least allowed for) in the system structure. However, since weak emergence is a product of a complex system, the actual level of emergence cannot be predicted just from knowledge of the characteristics of the individual system components."
3. Strong: "The term '*strong emergence*' is used to describe unexpected emergence; that is, emergence not observed until the system is simulated or tested or, more alarmingly, until the system encounters in operation a situation that was not anticipated during design and development. Strong emergence may be evident in failures or shutdowns." Guide to the Systems Engineering Body of Knowledge.
4. Spooky: "An emergent property that is inconsistent with the known properties of the system's components. The property is not reproduced in any model of the system, even one with complexity equal to that of the system itself, even one that appears to be precisely simulating the system itself in all details." Rainey and Tolk (2015).

Prediction: Based upon initial detection of the existence of emergence, then being able to postulate the potential future states of emergence.

Control: Specification of a modeling and simulation technique, possibly in concert with another analytical technique, used either to manage negative cases of emergence or to capitalize on positive cases of emergence.

18.2 Motivation

Consider some of the following scenarios, and their potential consequences:

1. An order processing system enters a waiting state after a transaction is cancelled (Pilcher 2015).
2. A first responder administers rescue medication to an unconscious patient, unaware that the medication was already administered (Bryant 2016).
3. The International Space Station is unaware of a hazardous condition within a supply spacecraft as that spacecraft approaches to dock (Nelson 2015).
4. An Unmanned Aerial Vehicle (UAV) on a search and track mission reaches a return-to-base condition, then finds and begins to track a new target (Revill 2016).
5. A UAV on a humanitarian assistance and disaster relief mission reports acceptable system status, then the operator suddenly commands the UAV to abort the mission without provocation (Reese 2017 on Beaufait, Constable, and Jent 2017).

The finders of these scenarios were students ranging from high school to graduate school, with little to no behavior modeling experience. They each used the lightweight formal methods framework for behavior modeling described herein to model their respective problem and generate a set of scenarios, some of which were valid, and some of which were invalid, such as the examples given above. Despite the wide range of application domains, each of the scenarios exhibits some SoS-level emergent behavior that was not anticipated by the modeler or any of the model's stakeholders.

Detecting, classifying, predicting, and controlling emergent behavior in SoS is a very new endeavor in the field of engineering, and demands differences in thinking about how we model system behavior. Many definitions for the concept of a system exist, but none so far have the precision needed to explain the mechanics of how interacting systems exhibit emergent behavior. Consider, for example, the following definitions for the term "system":

- "A system is a construct or collection of different elements that together produce results not obtainable by the elements alone. The elements, or parts, can include people, hardware, software, facilities, policies, and documents; that is, all things required to produce systems-level results. The results include system level qualities, properties, characteristics, functions, behavior and performance. The value added by the system as a whole, beyond that contributed independently by the parts, is primarily created by the relationship among the parts; that is, how they are interconnected." (INCOSE 2017)
- "A system is a set of entities and their relationships, whose functionality is greater than the sum of the individual entities." (Crawley, Cameron & Selva 2016)

The following definition of SoS is also relevant for consideration:

- "A system of systems (SoS) brings together a set of systems for a task that none of the systems can accomplish on its own. Each constituent system keeps its own management, goals, and resources while coordinating within the SoS and adapting to meet SoS goals." (SEBoK 2017)

First, if a model of a system or SoS is to exhibit emergent behaviors, the model itself must have the properties described in these definitions:

- "functionality is greater than the sum of the individual entities" (Crawley, Cameron & Selva 2016).
- "elements that together produce results not obtainable by the elements alone" (INCOSE 2017).
- "a task that none of the systems can accomplish on its own" (SEBoK 2017).

Next, we must define more precisely what is meant by "greater than the sum" or that which is "not obtainable by the elements alone." The first definition provides a key clue for where to focus our engineering efforts for emergence: "the relationship among the parts; that is, how they are interconnected." In between the parts or systems are the numerous possible interactions among the parts or systems, which are either permitted or prohibited by the design to arise. The multitude of different possible interactions is what "results" in "greater than the sum" expression of functionality.

We define *emergent behavior* as all behaviors permitted to arise from individual behaviors and interactions among systems or their components. Interactions may be specified as a set of constraints imposed on behaviors of parts, which are assembled together using available stakeholder requirements into a component, a system, or a SoS. Emergent SoS behavior, in principle, does not contain the individual system behaviors. When we assemble systems together into a SoS, we can only prevent some of the individual system behaviors from occurring. The task is now to recognize and specify the composite behaviors remaining in the assembled SoS, as a subset of the plain Cartesian product of system behaviors. Section 18.4 goes into the mechanics of this process; for now, suffice it to say that a key activity of engineering emergence is that of engineering interactions.

If the possible behaviors of a system are not controlled with any natural or engineered constraints, it is likely that given enough repetition and time, the system will eventually perform every one of its possible behaviors. This idea is implicit in Murphy's famous law: "If there are two or more ways to do something, and one of those ways can result in a catastrophe, then someone will do it" (Götz 2017). Experience has shown that interactions among and within people and technology give rise to many possible combinations of behaviors, which can be characterized as "positive" or "negative" emergence (Mittal & Rainey 2015) (Zeigler 2016). Engineers and system stakeholders want their designed systems to exhibit "positive" emergent behaviors, and to suppress or exclude "negative" emergent behaviors. These positive and negative behaviors become more difficult to predict as the numbers of systems and possible system behaviors increase.

In summary, to engineer emergence, we must think in terms of relaxing or restricting control over component or system interactions. We must view each system as having its own behaviors that will express unless we deliberately constrain those behaviors out of the design. To tease out ahead of time what those unwanted behaviors are, however, requires that some old and new modeling concepts be employed to expose them during modeling and simulation.

18.3 Behavior Modeling Concepts for Engineering Emergence

This section summarizes some fundamental concepts for engineering emergence observed by the authors over a decade of behavior modeling research. Employment of these concepts allows more emergent behaviors to appear in system models, where their discovery is less costly as compared with making the same discovery in the operational system. The example models shown later in this chapter adhere to the following principles.

18.3.1 Separate Behavior and Interactions

The foremost concept of engineering emergence is that of *behavior.* Ultimately, stakeholder satisfaction is dependent on the system's or SoS's behavior. Behavior is a term that is commonly associated with the social sciences, but in engineered systems, behavior has a broader definition. It is the set of sequential, concurrent, alternate, optional, and iterating events that occur in or are shared by engineered technology, physical phenomena in the natural environment, people, organizations, other biological life, and anything else that can be described in terms of a process or as a timeline of events. System behavior is the overall action of a system subject to the dependencies among its components and between those components and the system's environment. Likewise, SoS behavior is the overall action of a SoS subject to the dependencies among its constituent systems and the SoS environment. Interactions are used to capture dependencies between components or systems, as in the example of Figure 18.1, which illustrates a User and a System interacting in a basic authentication sequence.

A main challenge in engineering is to expose undesired behaviors and interactions early during design, to control their presentation in the actual system. Engineering diagrams such as Figure 18.1 typically account for desired system behaviors and possibly for some off-nominal alternatives. The more difficult task is to account for *all* possible alternative behaviors permitted by a design, including all disruptions to desired sequences. Though nominal and off-nominal use case scenarios may be abundant on well-funded programs, these still represent only a subset of possible system behaviors. To steer emergent system behaviors, we need more than a collection of use cases that describe how we want a system to behave under *known* conditions. We need a comprehensive set of use cases that also include combinations of events that *no one conceived in advance.* Such a feat requires a fundamental change in the structure of system behavior models.

In order to expose both desired *and* undesired emergent behaviors, we must model system behaviors separately from system interactions. Applying this separation enables room for expansion of each behavior model into a fuller description of each system's behavior (Giammarco 2017). All examples provided in this chapter employ this concept. Figure 18.2 recasts the systems from Figure 18.1 to show each system's ordered activities in separate boxes. Interactions between the systems are shown as horizontal arrows that cross the swim lanes.

The system interactions (horizontal arrows) may be considered as constraints on the two independent models for User and System, as shown at the top of Figure 18.3. The interactions in the center of this figure operate as constraints that may be included or excluded

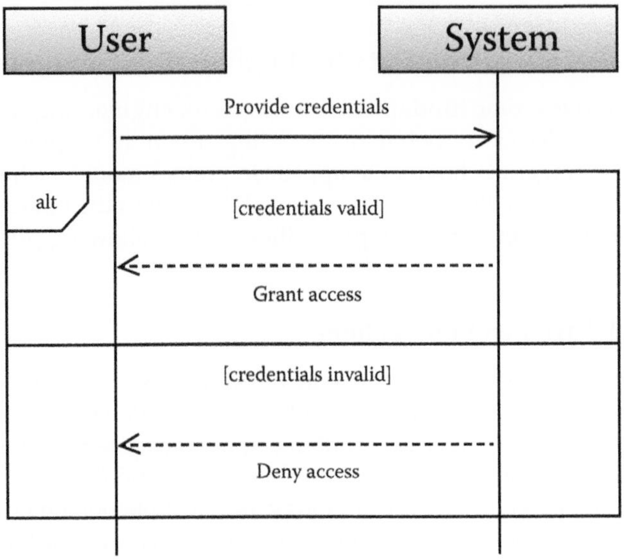

FIGURE 18.1
A User requests access to a System. The System grants the User access if the supplied credentials are valid; otherwise the System denies the User access (Source: Giammarco 2017).

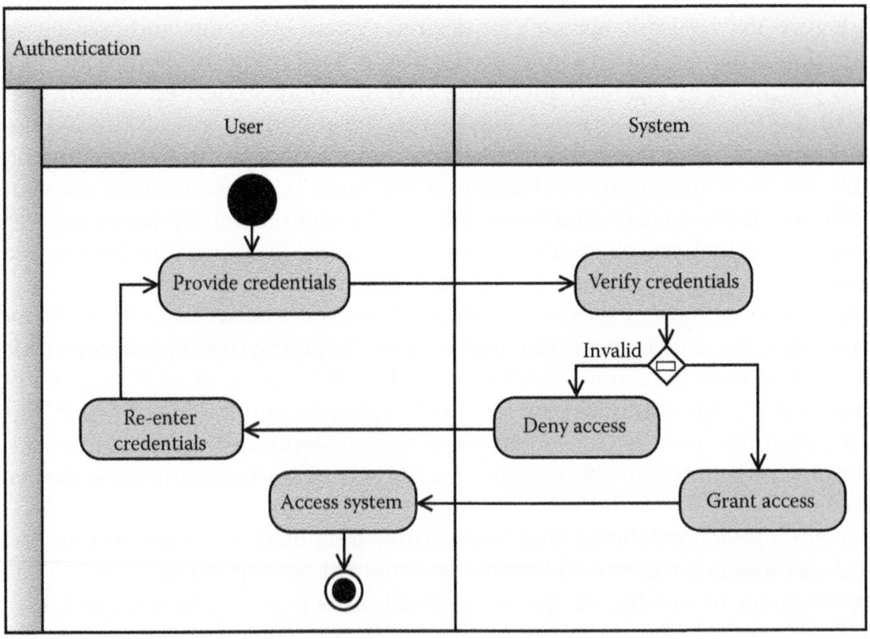

FIGURE 18.2
User behaviors and system behaviors are in separate swim lanes, left and right, respectively. Precedence dependencies (interactions) among activities inside the User and the System are represented by horizontal arrows that cross swim lanes (Source: Giammarco 2017).

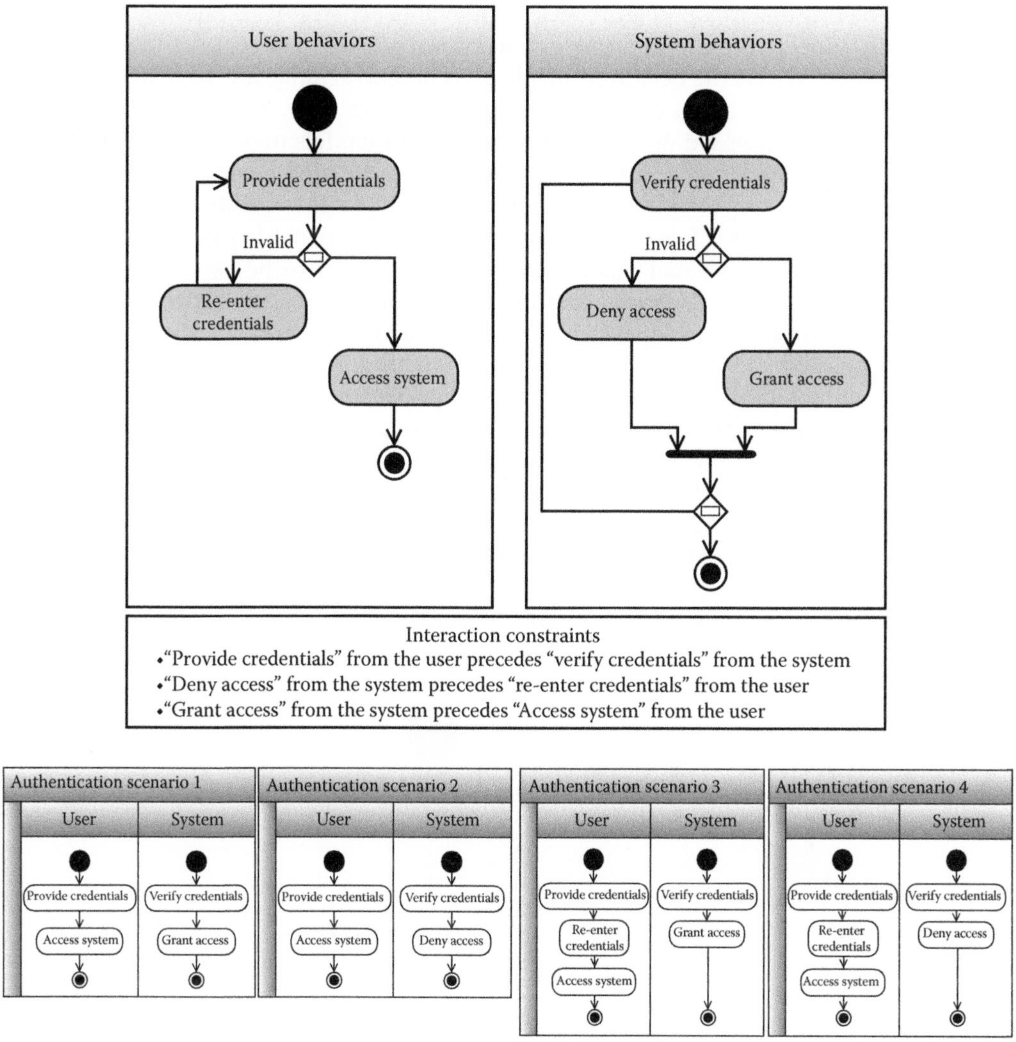

FIGURE 18.3
Top and center: Behaviors of the User and System, and interactions between them are logically separated (After: Giammarco 2017). Bottom: Four possible instances of emergent SoS behavior when the constraints in the center are not imposed.

at the will of the modeler, to allow or to remove unwanted emergent behaviors from the User and System models. When these interaction constraints are lifted, more behavior combinations for User and System *operating together as a SoS* are observed. Figure 18.3 (bottom) shows the various possible outcomes of the SoS model, absent of all constraints, from left to right: the User accessing the System after being granted access, the User accessing the System after being denied access, the User accessing the System after re-entering valid credentials, and the User accessing the System after re-entering invalid credentials. Each of these activity combinations provokes consideration about cases in which each scenario would be favorable or unfavorable. The emergent behaviors exposed by lifting the interaction constraints cause the stakeholders to think through potential scenarios that may not otherwise have been considered, because of the suppressive effect that the constraints had on them in the original model (Figure 18.2). Removal of constraints is one way to coax the emergent behaviors to present themselves (Giammarco 2017).

Separation of concerns is the design principle used here to arrange a system model so that each system in the SoS has a behavior specification of its own. System interactions are separate constraints imposed upon the system behavior models in order to drive the path selected in each separate behavior model. The result is the SoS emergent behavior; that is, all the behaviors permitted in the systems subject to the constraints. From Vaneman's and Jaskot's definition of a SoS, "… independent and task-oriented systems … integrated into a larger systems construct" (2013), system tasks ought to be separately modeled. Furthermore, to support *independent* and *integrated* systems per this definition, behavior and interactions are modeled separately, then integrated to generate the emergent behaviors that are possible to expose through logical modeling. Many emergent behaviors are masked by current modeling approaches, which over-constrain the models. This symptom is alleviated by (1) separating system behaviors, grouping them by system performing them, and (2) separating interactions among those systems, and using those interactions as constraints on the separate system models.

To find emergent behaviors, it is critical to practice clear allocation of behaviors to systems or physical components, rather than develop a blended process of behaviors that does not specify which systems do which activities, as exemplified in Figure 18.4. The blended approach does not employ separation of system behaviors and system interactions,

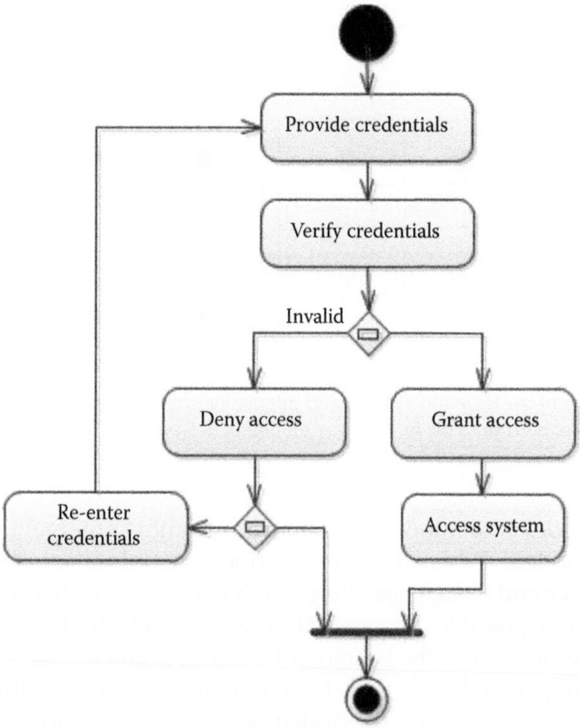

FIGURE 18.4
A blended process model shows activities in different systems, introducing ambiguity about which systems are performing which activities (Source: Giammarco 2017). It is better to have a separate model for each system, and use the interaction constraints to merge separate behavior models into integrated SoS behavior models.

and leaves the model vulnerable to misinterpretation concerning who is doing what. Furthermore, developing a set of separate system models and interactions among those models informs the development of SoS requirements.

18.3.2 Model System Behaviors and Environment Behaviors

A system should be considered in the context of other systems in the environment in which it operates, as suggested in (ISO 2011), as illustrated in Figures 18.1–18.3, and as commonly practiced in disciplined modeling efforts. What should be more commonly practiced is modeling of environment-system behaviors as thoroughly as the system-under-design's behaviors. Systems and SoS operate within an environment that has its own behaviors. All these behaviors together produce emergent behaviors. To increase the number of emergent behaviors exposed, the environment behaviors must also be modeled in as much detail as the system-under-design. We must conceive alternate events that could occur in the system's environment, concurrent with system-under-design behaviors. Figure 18.5 depicts an extension of the example in Figure 18.3, showing additional possible behaviors (User walks away and System terminates the session). Applying this concept does not

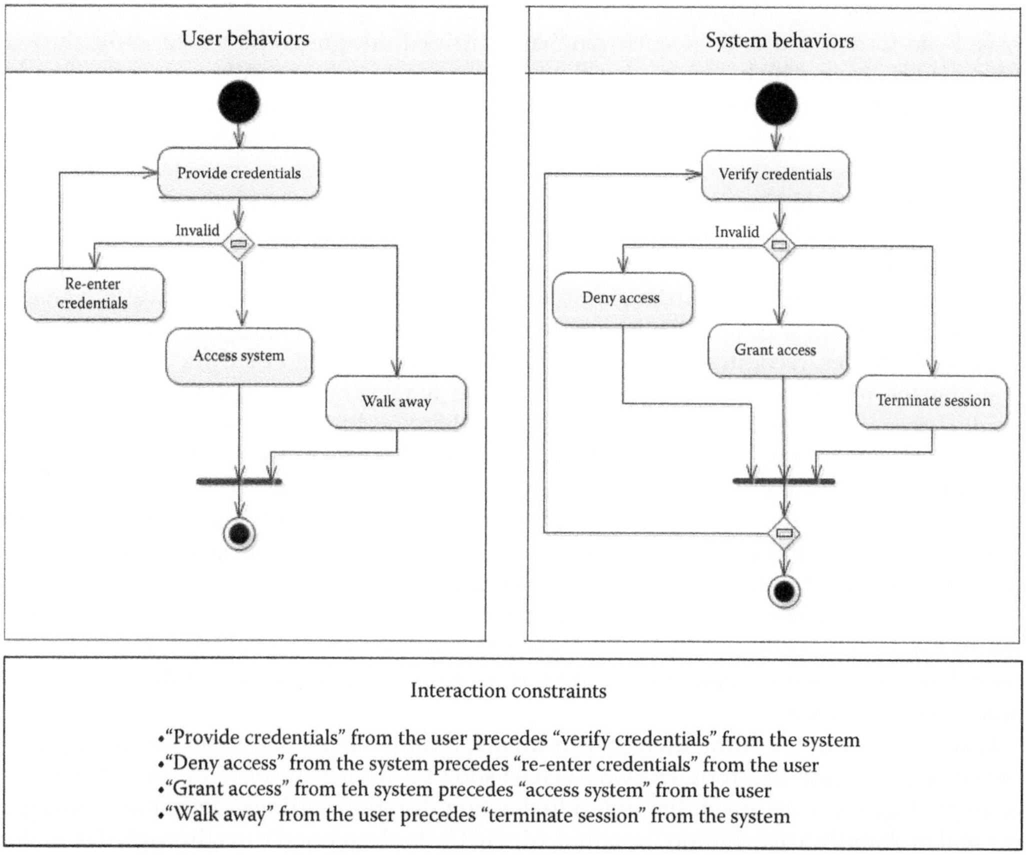
User behaviors
Provide credentials
Invalid
Re-enter credentials
Access system
Walk away
System behaviors
Verify credentials
Invalid
Deny access
Grant access
Terminate session
Interaction constraints
•"Provide credentials" from the user precedes "verify credentials" from the system
•"Deny access" from the system precedes "re-enter credentials" from the user
•"Grant access" from teh system precedes "access system" from the user
•"Walk away" from the user precedes "terminate session" from the system

FIGURE 18.5
The User model (the System's environment) is extended with an additional possible User behavior "Walk away." This addition prompts an idea for a new System activity, "Terminate Session." The interaction between these new activities would require a timing attribute (After: Giammarco 2017).

eliminate the possibility of errors of omission, which is a problem common to all models that are, by definition, an imprecise representation of reality. The objective of this concept is to conceive as many potential behaviors in the system *and in the system's environment* as is possible.

For human-designed systems, modeling anticipated system and environment behaviors and interactions reduces the risk of unwanted emergent behaviors occurring during system operation. The possible behaviors and interactions of Figures 18.1–18.3, and 18.5 were simple enough to determine manually, but larger models quickly exceed human cognitive capacities for accurate and complete determination of emergent behaviors. The next concept therefore focuses on the use of automated tools for modeling system and environment behaviors and interactions.

18.3.3 Formalize Models for Automatic Execution

Modeling component activities of systems and SoS increases our understanding of emergent system behavior. Once the number of possible behaviors and interactions grows beyond the human capacity for short term memory (Miller 1956), our biological processors need to be supplemented with synthetic processors (automated tools) for help with defining, inspecting, analyzing, acting on, and preserving design decisions. Models that have been formalized and executed enable nontrivial designs to be tested early, during lifecycle phases in which changes are the least expensive to make. A *formal* model should have one and only one possible interpretation. Formalization of system behavior descriptions helps expose errors, remove ambiguity from less formal descriptions, and makes models machine-readable so they can be executed in simulation. Simulating a formal model helps with testing and debugging of the logic and design, using tools to unravel the emergent behaviors out across a timeline (e.g., discrete event simulation). Such simulations may be focused on the logic alone, or be accompanied by attributes such as durations, probabilities, costs, resource utilizations, and other parameters of interest to design analysis. Informal models rely on reviewers to expose tacit assumptions without automated support; however, recognizing the presence of assumptions can be difficult to do when viewing the model using the same lens through which the assumptions were initially made—an example of Albert Einstein's saying that "problems cannot be solved by the same level of thinking that created them." "Having the capability to automatically turn a model over in different ways to view behaviors and interactions, from different angles and over time, has proven to be useful in the exposure of not only modeling errors, but genuinely unexpected emergent behaviors that otherwise may not have been identified until they occurred in the actual system" (Giammarco 2017).

The behaviors shown in Figure 18.3 are formalized for simulation in Figure 18.6. Note that the informal feedback loops from Figure 18.3 are described in a way that enables the model to be executed in simulation using scripts to specify dependency of the exit logic on access being granted.

Using discrete event simulation on this model, a timeline may be generated for a possible outcome (Figure 18.7, top). The Gantt chart shows a Total Time of 12.21 seconds for one example login activity, given the individual action duration settings (determined using a normal probability distribution function). Monte Carlo simulation runs through the model a specified number of times (e.g., 100), using the probabilities inside the OR blocks to make branch selections (Figure 18.7, bottom). The Time Tree Map visualizes the relative activity durations in the model over many runs, and shows that longest activities are typically

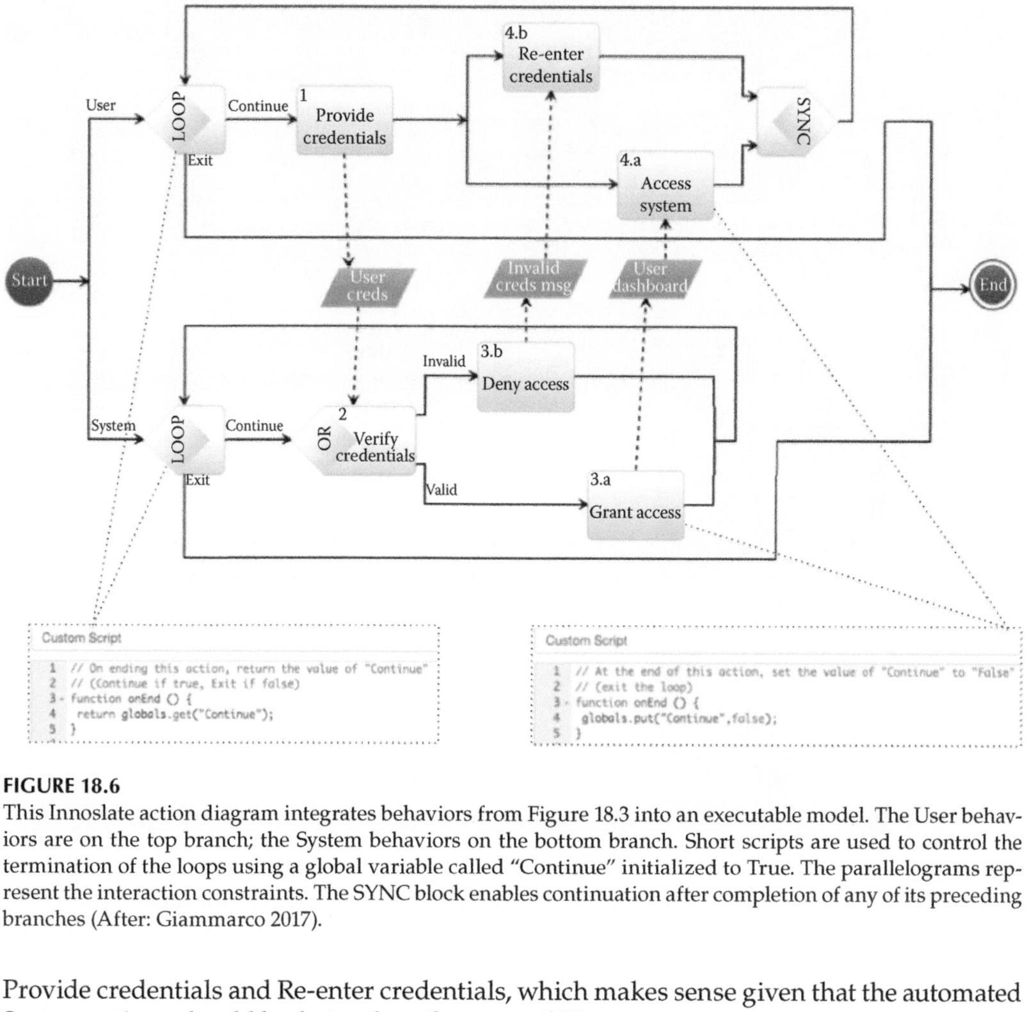

FIGURE 18.6
This Innoslate action diagram integrates behaviors from Figure 18.3 into an executable model. The User behaviors are on the top branch; the System behaviors on the bottom branch. Short scripts are used to control the termination of the loops using a global variable called "Continue" initialized to True. The parallelograms represent the interaction constraints. The SYNC block enables continuation after completion of any of its preceding branches (After: Giammarco 2017).

Provide credentials and Re-enter credentials, which makes sense given that the automated System actions should be faster than the manual User actions.

The larger SoS models grow, the more essential automated tools with simulation become for conducting verification and validation of the behavior logic. Since humans are error-prone when required to process large amounts of information, it is important to decide which tasks ought to be assigned to a human and which tasks ought to be assigned to an automated tool.

18.3.4 Properly Allocate Each Task to a Human or to a Machine

The concepts earlier presented demonstrate expression and suppression of emergent behaviors using constraints. Many of today's modeling methods over-constrain models so that they contain a small subset of possible behaviors (Giammarco & Auguston 2013) (Auguston et al. 2015). This over-constraining of models is a byproduct of inefficient human/machine task allocation. Behavior models were first developed with drawing tools, from which standard notations evolved. Now, more and more automated tools for behavior modeling contain a simulation capability, providing the means to shift investment of human capital from

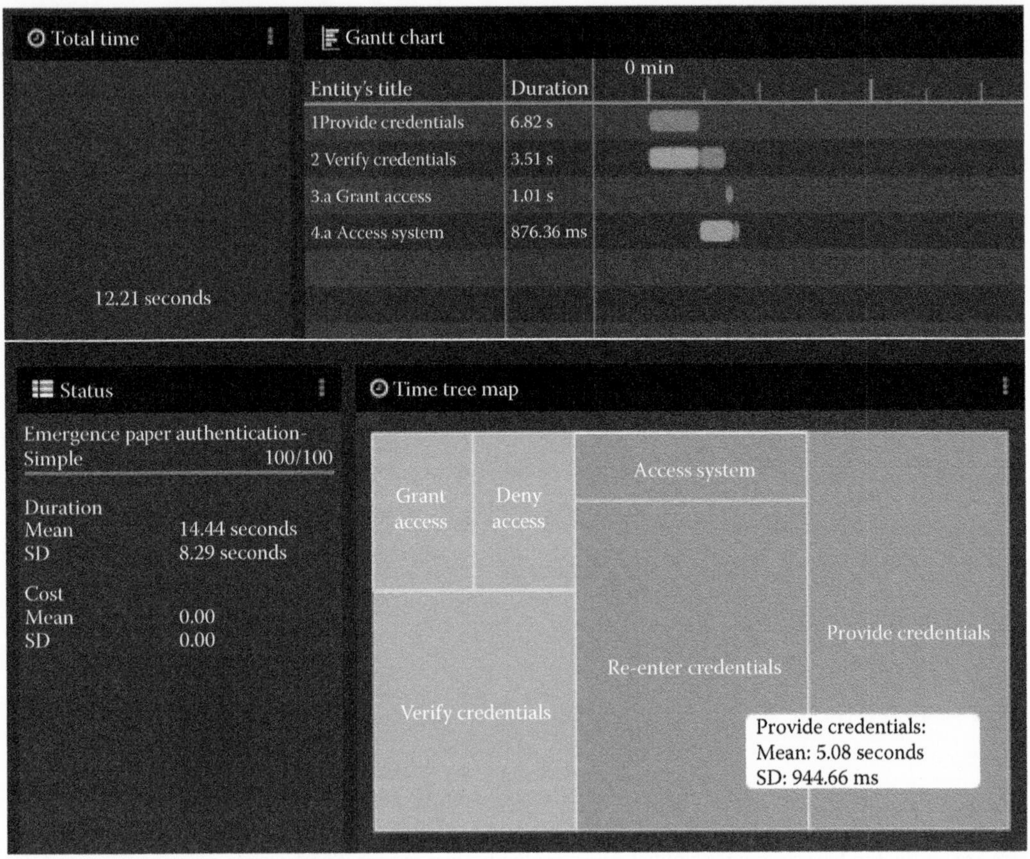

FIGURE 18.7
User and System behaviors executed in discrete event simulation (top), and in Monte Carlo simulation (bottom). For these runs, the OR block probability settings were 0.80 for selection of the valid branch and 0.20 for selection of the invalid branch.

drawing tasks to behavior rule specification, followed by review and inspection of automatically generated diagrams. The human modeler should focus on behavior specification, verification, and validation activities, determining which emergent behaviors to keep and which to reject with constraints. Using this approach, the human has more time to assess the value of each emergent behavior (favorable or unfavorable), when the heavy lifting of scenario generation is delegated to an automated tool (Auguston 2009a, 2009b). Section 18.4 presents Monterey Phoenix (MP) (Giammarco & Whitcomb 2017) (Giammarco & Giles 2017) (Auguston et al. 2015) (Giammarco & Auguston 2013), which has a fast, accurate, and cost-effective event trace generator that computes all possible use case scenario variants up to a specified scope. Using this task allocation, human intellect is used for *description* and *validation* of emergent behaviors, and automated tools are used (in between description and validation) for the *generation* of emergent behaviors (Giammarco & Whitcomb 2017). Relieved of the manual labor of detecting emergent behaviors, humans will have more time to determine the best ways to classify, predict, and control emergent behavior, using their innate talents for reasoning about alternative behaviors for interacting systems. With the advent of new tools, modeling tasks being conducted within organizations must be re-evaluated for allocation to a human or to a machine, and assigned appropriately.

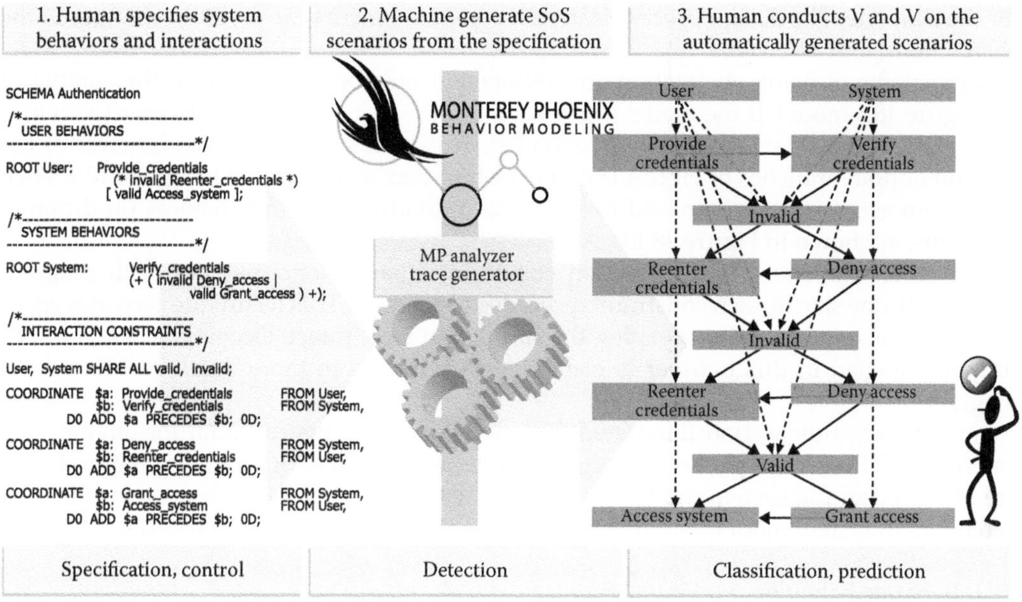

FIGURE 18.8
A modeler specifies behavior and interaction separately (left), uses MP to automatically generate SoS scenarios (center), then inspects the resulting scenarios (right). The example on the right is one out of six possible SoS scenarios at scope 3 (After: Giammarco 2017).

Figure 18.8 illustrates the human-authored MP model capturing the logic in Figure 18.3 (left), followed by automatic scenario generation (detection) using the MP Analyzer (center), then followed by inspection of detected emergent behaviors in the generated scenarios (right), where emergent behaviors may be classified, predicted, and controlled through modifications to the specification. Here, the feedback loops in Figure 18.3 are formalized using the optional loop (`* invalid Reenter_credentials *`), and the loop only occurs if access is denied. This implementation logic differs from the Innoslate logic by using Jackson's (2006) Small Scope Hypothesis to control the number of loop iterations.

18.3.5 Use Abstraction and Refinement to Manage Larger Models

Applying the concepts of abstraction and refinement enhances the readability of models, especially as the number of systems and activities grows. Component hierarchies can be used to develop corresponding behavior hierarchies to partition an otherwise large, flat model into a collection of smaller, more legible diagrams or models. Hiding the detailed inner-workings of components is a strategy long used to keep the contents of a model well organized and manageable. High-level models ignore implementation details, including specific hardware components and software algorithms, to focus attention on the architecture prior to solution details. A process of stepwise refinement gradually elaborates on a high-level model, and helps to maintain clear delineations between design levels. Components or activities that appear higher up in the hierarchy are more abstract, whereas components and activities appearing at lower levels are more refined. Whether one is working from the bottom-up or from the top-down, it is advisable to verify and validate the design at the working level, preferably with automated tools, before leaving one level for the next one. The design process is iterative, so refinements can result in changes to the abstractions, and vice versa.

The benefits of using abstraction and refinement may be illustrated on the content of the Figure 18.6 model. If the model in Figure 18.6 were to grow larger with more activities and actors, it may start to become overloaded. To address this issue, the contents of the top and bottom branches of Figure 18.6 may be bundled, as illustrated in Figure 18.9. Then each main activity is decomposed to show the individual system behaviors on different diagrams, as shown in Figure 18.10.

Both diagrams show the interaction constraints (parallelograms), but each diagram shows only one side of each constraint (precedes or follows). This technique provides space to expand each model, while keeping the simulation logic intact. Decomposition can continue to be used in this manner to elaborate on activities in more detail on lower level diagrams.

In summary, this section illustrated five modeling concepts essential for engineering emergence. These concepts are a product of a decade of behavior modeling research and are observed in the system models exemplified later in the chapter. To expose emergent behaviors in system models, employ the following practices (Giammarco 2017):

1. Separate behaviors and interactions,
2. model system behaviors and environment behaviors,
3. formalize models for automatic execution,
4. properly allocate each task to a human or to a machine, and
5. use abstraction and refinement to manage large models.

The next section describes an approach and language that uses these concepts to help stakeholders detect, classify, predict, and control emergent behaviors via system modeling and simulation.

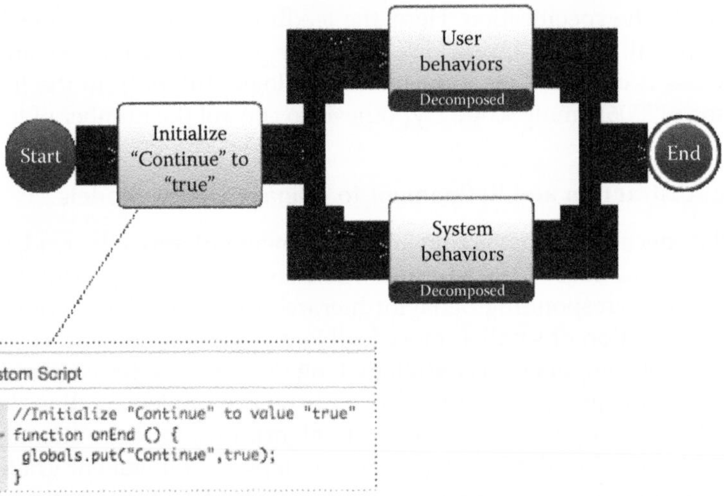

FIGURE 18.9
The concurrent User behaviors and System behaviors are bundled using abstraction. The activity that initializes the "Continue" variable to "true" is also shown here. At the end of the "Initialize 'Continue' to 'true'" action, the global variable "Continue" is set to "true."

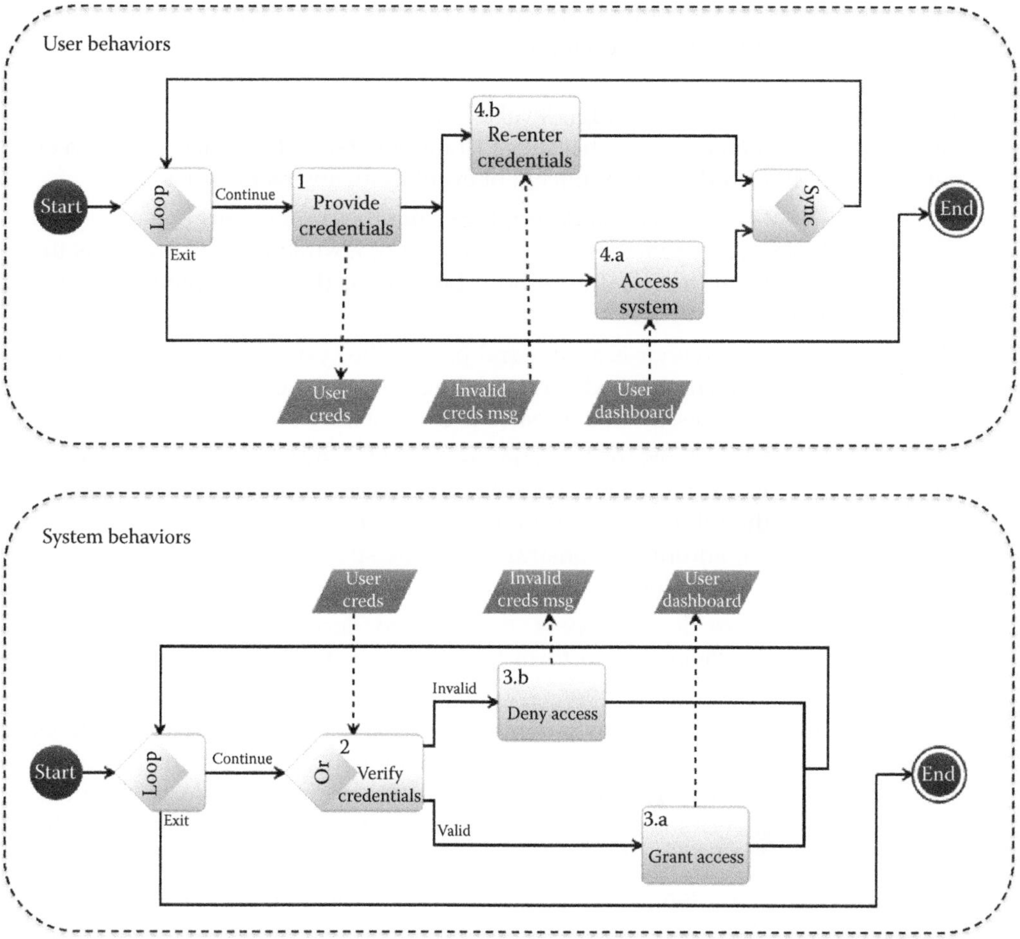

FIGURE 18.10
The concurrent User behaviors and System behaviors are refined on separate diagrams. Each diagram shows only one end of each interaction constraint.

18.4 An Introduction to Monterey Phoenix

Monterey Phoenix (MP) is a framework for software system architecture and business process (workflow) specification based on behavior models (Auguston, 2009a, 2009b, 2014, 2018). Architecture plays a role as a bridge between requirements and implementation of a system, and represents a stepwise refinement in the design process with specific objectives and stakeholders. Business process often represents a part of the system's requirements, and interactions between the system and its environment are an integral part of architecture specification.

MP provides an executable architecture model, and behavior modeling is at the core of the MP approach. In common architecture models, the main elements are components (representing the functionality) and connectors (representing the information flow between components). In MP manual, the main concepts are activities and coordination between activities, based on the following principles:

- A view of the architecture as a high-level description of possible system behaviors, emphasizing the behavior of subsystems (components) and interactions between subsystems. MP introduces the concept of event as an abstraction of activity.
- The separation of the interaction description from the components behavior is an essential MP feature. It provides for a high level of abstraction and supports the reuse of architectural models. Interactions between activities are modeled using event coordination constructs.
- The environment's behavior is an integral part of the system architecture model. MP provides a uniform method for modeling behaviors of the software, hardware, business processes, and other parts of the system.
- The event grammar models the behavior as a set of events (event trace) with two basic relations, where the PRECEDES relation captures the direct dependency abstraction, and the IN relation represents the hierarchical relationship. Since the event trace is a set, additional constraints can be specified using set-theoretical operations and predicate logic.
- The MP architecture description is amenable to deriving multiple views (or abstractions), and provides a uniform basis for specifying both structural and behavioral aspects of a system.
- MP models are executable. MP tools support automated and exhaustive (for a given scope) scenario generation for early system architecture verification. The Small Scope Hypothesis (Jackson 2006) states that most flaws in models could be demonstrated on relatively small counterexamples. Testing and debugging of architecture is supported by assertion checking for verifying the behavior properties.
- Humans understand examples better than formal models. Scenarios (or use cases) generated by MP can facilitate communication with system's stakeholders.
- MP framework can assist in unifying UML activity and sequence diagrams, and statechart notations. MP may be used in addition to the existing architecture definition tools and methodologies.

MP is intended for the use of lightweight Formal Methods in software and system architecture design and maintenance. It provides an ecosystem for sanity checking tools, reusable architecture patterns, reusable assertions, queries, and tools for extracting architecture views. A public online demo with MP editor, trace generator, and trace visualization is available at http://firebird.nps.edu/.

The purpose of any model is to help answer questions. MP model is designed to answer questions about a system's behaviors, including such aspects as structure of behavior, dependencies between actions involved in the behavior, constraints on behaviors, simple queries about behavior, as well as to provide a source for different visualizations or views of behaviors.

The MP behavior model is based on the concept of *event* as an abstraction of activity. The event has a beginning and an end, and may have duration (a time interval during which the action is accomplished).

The behavior of a system is modeled as a set of events with two binary relations defined for them: precedence (PRECEDES) and inclusion (IN)—the *event trace*. One action is required to precede another if there is a dependency between them, e.g., the Send event should precede the Receive event. Events may be nested, when a complex activity contains a set of other activities. Defining one of these basic relations for a pair of events represents an important design decision. Usually system behavior does not require a total ordering of events. Both PRECEDES and IN are partial ordering relations. If two events are not ordered, they may overlap in time; in particular, they may occur concurrently.

18.4.1 Event Grammar

The structure of possible event traces is described by an event grammar. A grammar rule specifies structure for a particular event type (in terms of IN and PRECEDES relations) and has a form

```
A:      pattern_list;
```

where A is an *event type* name and pattern_list is composed from event patterns. Event types that do not appear in the left-hand part of rules are considered atomic and may be refined later by adding corresponding rules.

An instance of an event trace satisfying the grammar rule can be visualized as a directed graph with two types of edges (one for each of the basic relations). Events are visualized as boxes, and basic relations as arrows. Figure 18.11 outlines the event patterns for use in the grammar rule's right-hand part. Here B, C, D stand for event type names or event patterns.

Sequence denotes ordering of events under the PRECEDES relation. The grammar rule A: B C means that an event a of the type A contains ordered events b and c matching B and C, correspondingly, and relations b IN a, c IN a, and b PRECEDES c hold. A grammar rule may contain a sequence of several events, like A: B C D.

Pattern (+ B +) may be used to denote a sequence of one or more events B, and {+ B +} denotes a set of one or more events B. In all cases, it is assumed that iterated event instances are unique. Event patterns may use iteration to describe repeated behavior patterns. An

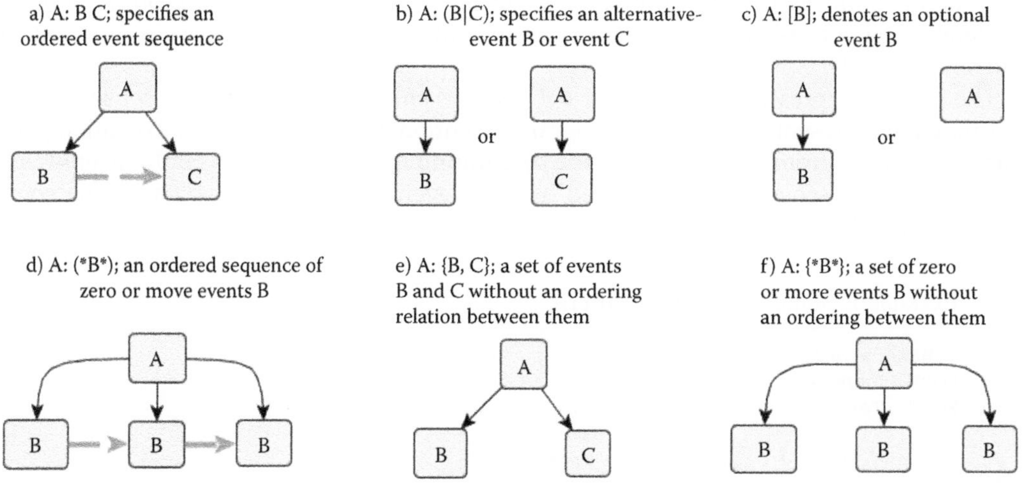
a) A: B C; specifies an ordered event sequence
A
B
C
b) A: (B|C); specifies an alternative-event B or event C
A
B
or
A
C
c) A: [B]; denotes an optional event B
A
B
or
A
d) A: (*B*); an ordered sequence of zero or move events B
A
B
B
B
e) A: {B, C}; a set of events B and C without an ordering relation between them
A
B
C
f) A: {*B*}; a set of zero or more events B without an ordering between them
A
B
B
B

FIGURE 18.11
Event patterns and examples of event traces.

event grammar is a graph grammar for directed acyclic graphs of vertices (events) with edges representing relations IN and PRECEDES.

An event trace represents an example of a particular execution of the system or a use case, especially if the behavior of the environment is included. Event traces can be effectively derived from the event grammar rules and then adjusted and filtered according to the composition operations and constraints in the schema. This justifies the term *executable architecture model*. For a given MP schema it is possible to obtain all valid event traces up to a certain limit. Usually such a limit (*scope*) is set as the upper limit on the number of iterations in grammar rules (recursion can be limited in a similar way, but current Firebird implementation does not allow explicit or implicit recursion in event grammar rules). For many purposes, a modest limit of 3 iterations will be sufficient. This process of generating and inspecting event traces for the schema is similar to the traditional software testing process.

Since it is possible to automatically generate all event traces within the given scope for MP model, it provides for exhaustive testing—a feature usually not available for traditional software testing. Careful inspection of generated traces (scenarios/use cases) may help developers to identify undesired behaviors. Usually it is easier to evaluate an example of behavior (particular event trace) than the generic description of all behaviors (the schema). The Small Scope Hypothesis (Jackson 2006) states that most errors can be demonstrated on relatively small counterexamples. Assertion checking and SAY clauses in MP make the event trace inspection easier.

18.4.2 Behavior Composition

The behavior of a particular system is specified as a set of possible event traces using a *schema*. The purpose is to define the structure of event traces in terms of IN and PRECEDES relations using event grammar rules and other constraints. A schema contains a collection of events called *roots*, representing the behaviors of parts of the system (components and connectors in common architecture descriptions), *composition operations* specifying interactions between these behaviors, and additional constraints on behaviors.

There is precisely one instance of each root event in a trace. A schema can contain auxiliary grammar rules for composite event types used in other rules. In principle, schema may define both finite and infinite traces, but most analysis tools for reasoning about a system's behavior assume that a trace is finite. Only finite event traces are considered in current MP tools.

The schema represents instances of behavior (event traces) in the same sense as Java source code represents instances of program execution. Just as a particular execution path can be extracted from a Java source code by running it on a JVM, a particular event trace specified by MP schema can be derived from the event grammar rules by applying composition operations and constraints.

Example 1. Simple interaction pattern.

```
SCHEMA simple_message_flow
ROOT  Task_A:     (* send *);
ROOT  Task_B:     (* receive *);

COORDINATE   $x: send         FROM Task_A,
             $y: receive FROM Task_B
      DO     ADD $x PRECEDES $y; OD;
```

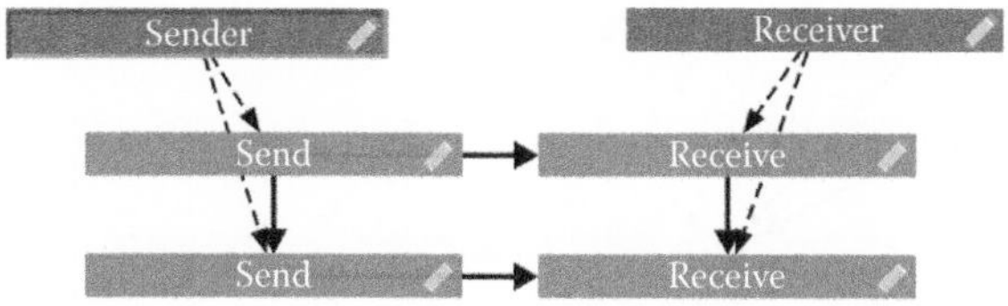

FIGURE 18.12
Example of a composed event trace for the `simple _ message _ flow` schema (scope 2).

The composition operation **COORDINATE** coordinates behaviors of two root events sending and receiving messages. This trace operation takes two root event traces and produces a modified event trace (merging behaviors of **Task_A** and **Task_B**) by adding the **PRECEDES** relation for the selected **send** and **receive** pairs. Essentially it is a loop performed over the structure of coordinated events, hence the **DO - OD** notation for the loop body.

This **COORDINATE** composition uses event selection patterns to specify subsets of root traces that should be coordinated. The **send** pattern identifies the set of events selected from **Task_A**. The default for event selection in the coordination source within COORDINATE preserves the order of events generated during the derivation process. The default means that the ordering of events in selected sets will remain "as is."

Both selected event sets should have the same number of elements (**send** events from the first trace and **receive** events from the second), and the pair coordination follows their ordering (synchronous coordination), i.e., first **send** is paired with first **receive**, second with the second, and so on. MP variables $x and $y provide access to the pair of events matching the selection pattern within each iteration. The **ADD** operation completes the behavior adjustment, specifying additional **PRECEDES** relation for each pair of selected events. Behavior specified by this schema is a set of matching event traces for **Task_A** and **Task_B** with the modifications imposed by the composition. If the numbers of events in the selected event sets (coordination sources) are not equal, the coordination operation fails to produce a resulting trace.

Figure 18.12 gives a sample of event trace satisfying the schema **simple_message_flow**. It resembles a UML sequence diagram's "swim lanes."

18.4.3 Data Items as Behaviors

Data items in MP are represented by actions (events) that may be performed on that data. This principle follows the Abstract Data Type (ADT) concept introduced in (Liskov & Zilles 1974).

Example 2. Data flow.

```
SCHEMA Data_flow
  ROOT Writer: (* ( working | writing ) *);
  /* writing events should precede reading */
  ROOT File:   (+ writing +) (* reading *);
   Writer, File SHARE ALL writing;
  ROOT Reader: (* ( reading | working ) *);
   Reader, File SHARE ALL reading;
```

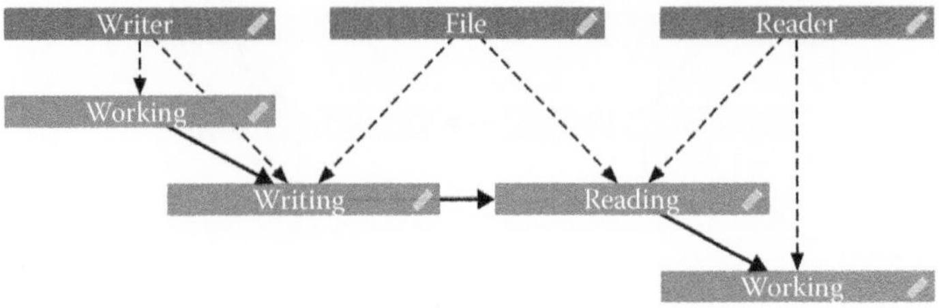

FIGURE 18.13
An example of composed event trace for the `Data_flow` schema.

The behavior of **`File`** requires **`writing`** events to be completed before **`reading`** events, and there should be at least one **`writing`** event. The **`SHARE ALL`** composition operations ensure that the schema admits only event traces where corresponding event sharing is implemented.

Event sharing is defined as following (here X, Y are events, Z is an event type, and FROM is a transitive closure of IN).

`X, Y SHARE ALL  Z  ≡ { v: Z | v FROM X } = { w: Z | w FROM Y }`

Event sharing is yet another way of behavior coordination. It is assumed that shared events may appear within an event at any level of nesting. In Figure 18.13, the shared events **`writing`** and **`reading`** are shown to have IN relations with multiple roots.

At the architectural level, data items are inputs or outputs of activities and are modeled as operations that may be performed on them. This is a simple and uniform concept.

18.4.4 Behavior of the Environment

The following example demonstrates how to integrate the behavior models of an environment and a system that operates in this environment. The **`ATM_Withdrawal`** schema specifies a set of possible interactions between the **`Customer`**, **`ATM_System`**, and **`Date_Base`**.

Example 3. Withdraw money from ATM.

```
SCHEMA ATM_withdrawal
ROOT Customer:     (*  insert_card
                         ( identification_succeeds
                           request_withdrawal
                          ( get_money | not_sufficient_funds )          |

                          identification_fails                          )
                   *);

ROOT ATM_system:   (*     read_card
                           validate_id
                          ( id_successful
                                check_balance
                                ( sufficient_balance
                                      dispense_money                    |
                                         unsufficient_balance     )     |
                            id_failed                                   )
                   *);
```

```
COORDINATE        $x: insert_card              FROM Customer,
                      $y: read_card                  FROM ATM_system
                      DO     ADD $x PRECEDES $y;            OD;

COORDINATE        $x: request_withdrawal       FROM Customer,
                      $y: check_balance              FROM ATM_system
                      DO     ADD $x PRECEDES $y;            OD;

COORDINATE        $x: identification_succeeds  FROM Customer,
                      $y: id_successful              FROM ATM_system
                      DO     ADD $y PRECEDES $x;            OD;

COORDINATE        $x: get_money                FROM Customer,
                      $y: dispense_money             FROM ATM_system
                      DO     ADD $y PRECEDES $x;            OD;

COORDINATE        $x: not_sufficient_funds     FROM Customer,
                      $y: unsufficient_balance       FROM ATM_system
                      DO     ADD $y PRECEDES $x;            OD;

COORDINATE        $x: identification_fails     FROM Customer,
                      $y: id_failed                  FROM ATM_system
                      DO     ADD $y PRECEDES $x;            OD;

ROOT Data_Base:   (*  validate_id [ check_balance ] *);
Data_Base, ATM_system SHARE ALL validate_id, check_balance;
```

An event trace generated from the schema can be considered as a use case example. The event trace on Figure 18.14 can be viewed also as an analog of UML sequence diagram's "swim lanes" for Customer and ATM_system interactions.

The concept of environment in an architecture model includes behavior of other systems, hardware, business processes, and any other behaviors, which are not part of the

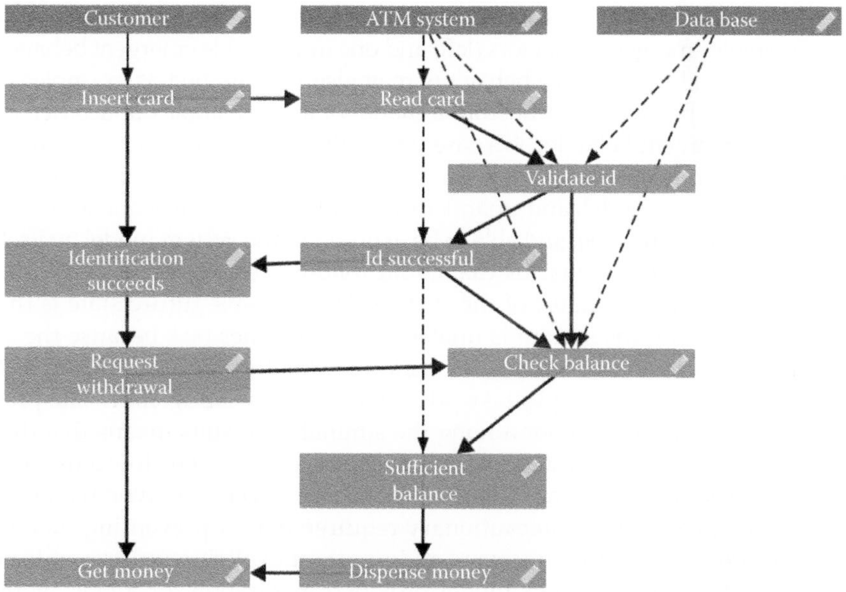

FIGURE 18.14
An example of event trace for `ATM _ withdrawal` schema.

system under consideration, but may interact with it. In particular, this approach may be of use for analyzing emergent behaviors of System of Systems, when the model of SoS is composed from the models of its components.

18.5 Examples of Emergent Behaviors in MP Models

This section provides examples of emergent behaviors discovered according to the definition of emergent behavior provided in Section 18.2: "all behaviors permitted to arise from individual behaviors and interactions among systems or their components." The Monterey Phoenix models in this section were developed by high school to graduate school students. In no case did any of the students anticipate the exemplified emergent behaviors found upon inspection of the scenarios generated from their own models. The description of each example is presented in terms of the concepts of detection, classification, prediction, and control, which are defined in the chapter introduction.

18.5.1 Order Processing System Model

An order processing system model was developed to demonstrate conversion of an OV-6b State Transition Description into an MP behavior model by modeling states and transitions as events and interactions (Pilcher 2015).

Pilcher first *detected* all possible emergent behaviors by generating a set of MP event traces from her system behavior model, using the MP Analyzer on firebird.nps.edu, up to a scope of 2. The detection process, once a very manually intensive job to do thoroughly (Giammarco, Giles & Whitcomb 2017), is now automatic and scope-complete with the MP Analyzer, using finite models of infinite behaviors authored by the human.

Pilcher then *classified* each behavior in the set as favorable or unfavorable. Figure 18.15 shows two favorable emergent behaviors (left) and one unfavorable emergent behavior (right). Both the favorable and unfavorable behaviors may also be classified as examples of simple emergence, since the behaviors result from a model of a non-complex and ordered system. The unfavorable emergent behavior is a case where the order processing system returns to a waiting state after cancellation of an order, which was not explicit in the source data that Pilcher used to build the model, and unanticipated, but indeed a consequence of the formal specification of the system alone, and able to be exposed in the system model by itself.

Pilcher then *predicted* the scenarios that could follow such emergent behavior by postulating the potential future states of the system. One possible future state is that a customer or employee using the system is unable to complete her task because the system is hanging indefinitely, unable to reset from the error. Another possible future state could involve a nefarious actor seeking to take advantage of the waiting state. The presence of this unfavorable emergent behavior among the simulation results means that the design, as modeled, permits this unwanted state. If such a scenario were to occur in a "safety critical" or "cyber secure" system, this may indeed be a very unfavorable behavior for the system owner, prompting a precautionary requirement for preventing such behavior. The prediction process therefore uncovers an important explicit requirement for how the system should *not* behave. Pilcher's model provided inspiration for a new requirement: "The Order Processing System shall end all started transactions in either the Cancelled or Delivered state" (Giammarco & Giles 2017).

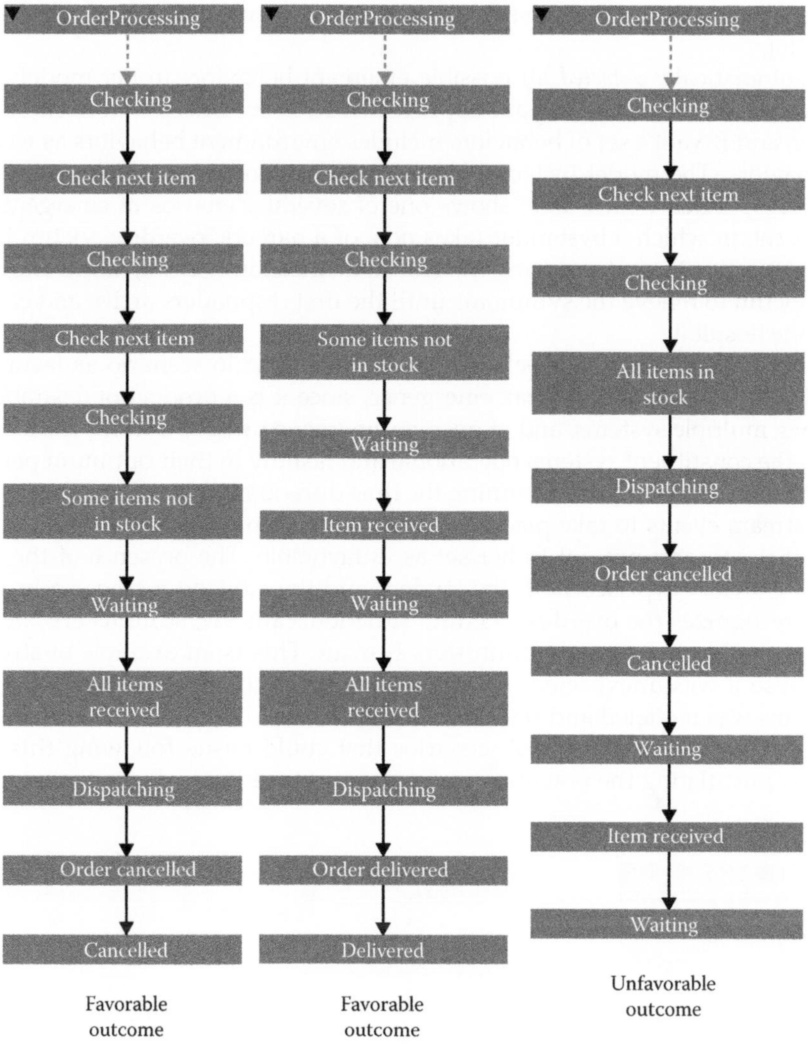

FIGURE 18.15
Two favorable event traces, left (ending in cancelled or delivered) and one unfavorable trace, right (ending in waiting) detected using the MP Analyzer (Source: Pilcher 2015). These are examples of simple emergence since it was possible to expose them in the system model without modeling interactions in the system's environment.

The presence of such an unfavorable scenario during M&S using MP offers a specific example of undesired system behavior having cost and vulnerability implications should it occur during system operations. Noting "that questionable result," Pilcher "identified design errors resulting in a revised design" (2015). Pilcher thus *controlled* the unfavorable emergent behavior by rearranging the logic in the MP model of the system under design to preclude its occurrence.

18.5.2 First Responder Process Model

A first responder process model was developed to assess safety issues pertaining to a proposed process for layperson administration of a rescue medication called Narcan (naloxone) as constrained by existing regulations governing first responder behaviors (Bryant 2016).

Bryant automatically *detected* all possible emergent behaviors in her model at a scope of 1 using the MP Analyzer. Bryant applied more of the concepts in Section 18.2 than did Pilcher, and Bryant's set of behaviors includes environment behaviors as well as composite behaviors. The model included behaviors for the overdose victim, the bystander, and a first responder. Figure 18.16 shows one of several scenarios of emergent behavior in Bryant's set, in which a bystander takes note of a narcotic overdose victim in respiratory distress, calls 911, and, carrying the rescue medication Narcan, administers it to the overdose victim to relieve the symptoms until the first responders arrive and can take the victim to the hospital.

Bryant *classified* the emergent behavior in the Figure 18.16 scenario as favorable. This scenario is also an example of weak emergence, since it is a product of desirable interactions among multiple systems, and of positive emergence, since "it fulfils the SoS purpose and keeps the constituent systems operational and healthy in their optimum performance ranges" (Mittal & Rainey 2015), assuming the time durations of the upstream events enable the downstream events to take place as shown. With help from her advisor, Bryant classified another scenario present in her set as unfavorable. The presence of the following emergent behavior surprised both the student and the advisor: a scenario in which the bystander recognizes the overdose victim's situation, calls 911, administers Narcan, then the first responder arrives and administers Narcan. This is an example of strong emergence because it was unexpected and not evident until the possible interactions among these systems was modeled and simulated using MP.

Bryant and her advisor *predicted* scenarios that could ensue following this emergent behavior by postulating the potential future states of the system. In particular, the victim

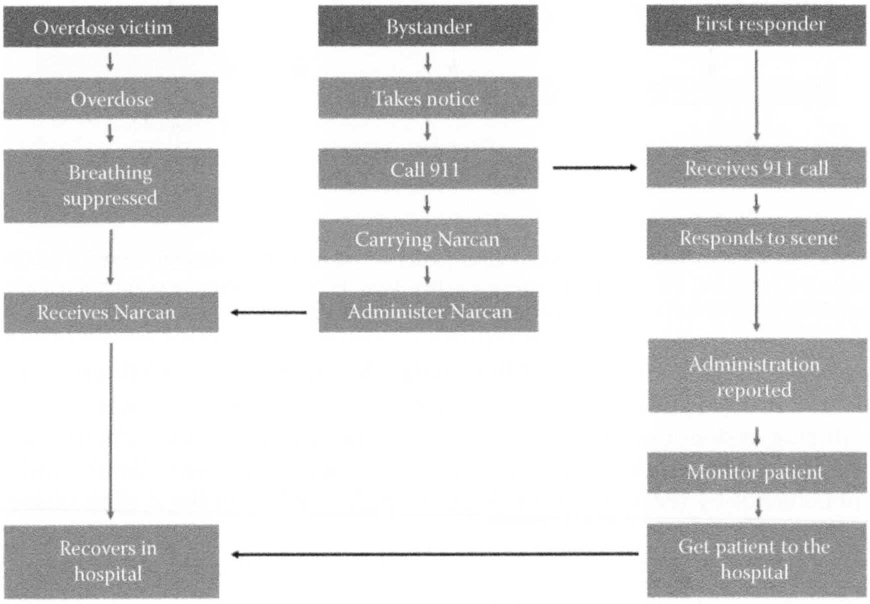

FIGURE 18.16
This favorable scenario, in which all actors are behaving according to the expected process (Bryant 2016), is an example of weak positive emergence.

is already at risk for common and severe side effects from the first Narcan administration. It follows that the victim may be placed at increased risk by a second administration of the drug. Reflecting on the predicted alternative scenarios that could follow the emergent behavior shown in Figure 18.17, this scenario could be classified as an example of negative emergence when it does not satisfy the purpose of the SoS and produces "undesired behaviors such as load hot-swapping and cascaded failures" (Zeigler 2016, citing Mittal & Rainey 2015). Possible cascaded failures triggered by delivering Narcan include severe system-wide failures ("side effects") such as seizures, coma, and death (Cunha, 2016). As such, it is a reasonable precaution to try to prevent such a scenario from occurring by accident in reality. There are, however, cases in which such a scenario may be deemed necessary by the first responders, in the event that the victim is still not breathing upon their arrival and the risk of not administering the drug again outweighs the risks of re-administration. In this case, the scenario would be classified as strong positive emergence. Therefore, whether a scenario is classified as positive or negative highly depends on the details surrounding the scenario, and having such scenarios exposed well in advance of their real manifestation affords mission planners time to consider and make decisions concerning the optimal policies and procedures in advance instead of in real time.

For the purposes of her senior capstone project, Bryant facilitated *control* of the emergent behavior by modifying her model with additional events "administration reported" and "administration NOT reported," to help subject matter experts envision what additional controls should be applied to prevent negative emergence they may not have otherwise considered, such as accidental overdosing of Narcan resulting in excessive side effects, costs in time and dollars, legal implications, and protocols for interacting with bystanders

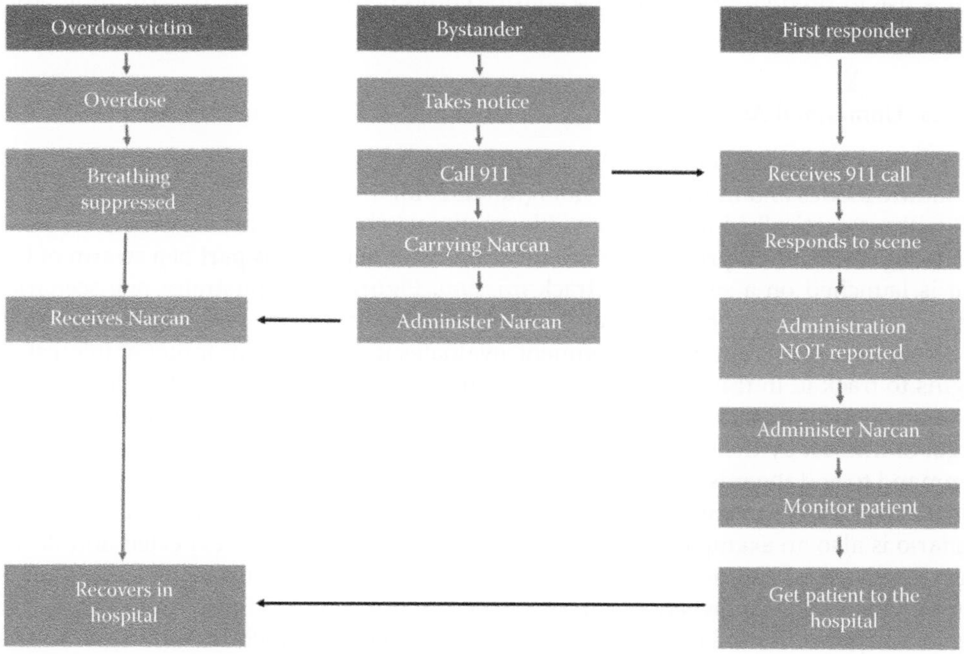

FIGURE 18.17
This unfavorable scenario, in which two doses of Narcan are administered—once by the bystander and then by the first responder (Bryant 2016)—provides a canvas for predicting a range of strong positive or negative emergence, depending on the details surrounding these events.

carrying Narcan. As a result of the prediction activity, a modification to the process was proposed, for including a way for the bystander to mark the victim by means of a medical bracelet that would be included with the Narcan kit, which can be used to indicate dose amount and time administered so first responders have that information, regardless of the presence of the bystander or others to communicate verbally with the first responders when they arrive. Bryant's MP model therefore provided inspiration for the following requirement: "Any Bystander who administers Narcan to an Overdose Victim shall place a band around the Overdose Victim's wrist that indicates the amount and time of the Narcan dose administered" (Giammarco & Giles 2017). Such a requirement, if adhered to, would provide first responders with information to which they otherwise may not have had access.

The model can be improved by adding some interactions that are not shown, such as event "overdose" in the Overdose Victim preceding event "takes notice" in the Bystander. It is interesting to note that detection, classification, prediction, and control of emergent behaviors in models was conducted in spite of the presence of such modeling errors. It seems to be advantageous to look for emergent behaviors in models with specification gaps, since these models have not yet been "locked down" with design constraints that would suppress the very behaviors we would want to consider. Therefore, MP behavior models need not be, and perhaps should not be, thoroughly verified before beginning the validation process, to make the best use of MP's ability to permute through all combinations of system behaviors far more comprehensively than a human would do, and with far better accuracy. Bryant concludes, "By inspecting the MP event traces, unexpected events, including miscommunication and patient response, can be identified and procedures refined before errors occur in real life" and "MP's use in predicting overdose scenarios could be beneficial to determining advantages to the medical field" (2016).

18.5.3 Unmanned Aerial Vehicle on a Search and Track Mission Model

An Unmanned Aerial Vehicle (UAV) on a search and track mission was modeled to look for failure modes and failsafe behaviors (Revill 2016).

Revill automatically *detected* all possible emergent behaviors in his UAV model at a scope of 1 using the MP Analyzer. In this model, the UAV is depicted as part of a swarm of UAVs that is launched on a search and track mission. Figure 18.18 illustrates one scenario in which the swarm operator commands mission commencement, and a UAV then detects an object signature from the environment, evaluates it, resolves it as target of interest, and begins to track it. In this scenario, the UAV communicates the information to the swarm operator, who then assesses the validity of the target. The target is determined to be a valid target of interest by the operator. After some time elapses, the operator issues a recovery command to end the mission.

Revill classified the emergent behavior in the Figures 18.18 scenario as favorable. This scenario is also an example of weak positive emergence, since it is expected and desired. Then Revill found another scenario in the set that clearly stood out to him as unfavorable, one in which a UAV reached a bingo fuel condition (just enough fuel to return to base), began an egress to the landing site, then spotted a target, which it then began to track. The object is evaluated by the swarm operator and determined to not be a target of interest, and issues a recovery command to end the mission (Figure 18.19). This is an example of strong emergence, because it was unexpected and discovered only when the possible interactions of the systems in the scenario were simulated in MP.

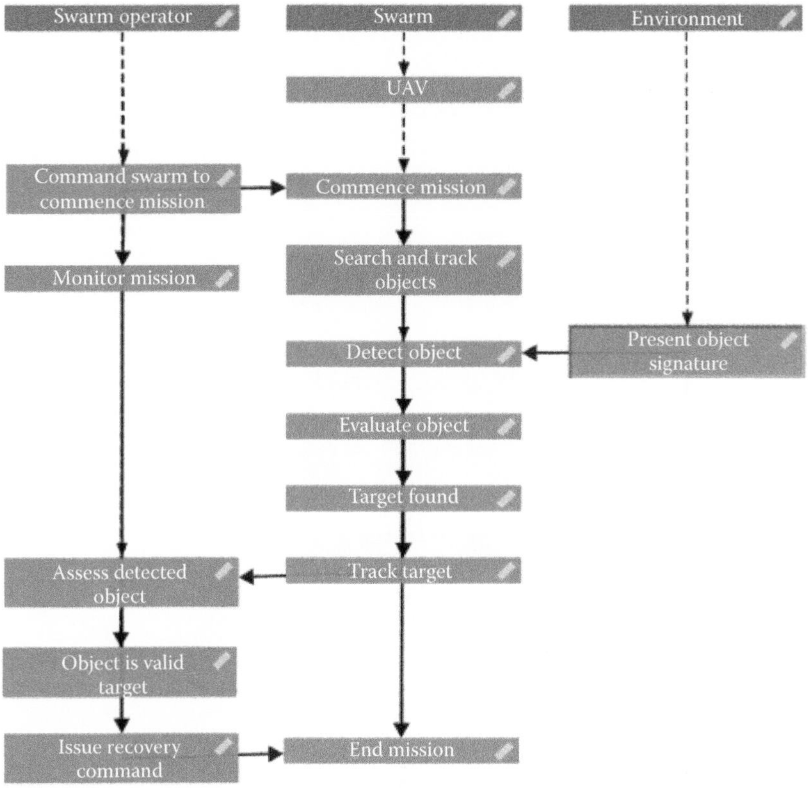

FIGURE 18.18
A favorable scenario where all systems are exhibiting expected behavior according to the operational process (Revill 2016), and an example of weak positive emergence.

Revill *predicted* that if this scenario were to play out in reality, the UAV may be forced to land at an alternative landing site, or potentially be lost altogether. Considering the different outcomes, he envisioned scenarios in which the risk of alternative landing/losing the UAV was worth the additional time on station, and in which the additional time on station was not worth the risk of alternative landing/losing the UAV. Therefore, this is an example of strong emergence that can be positive or negative, depending on details such as timing considerations and UAV disposability.

The strong emergent behavior in Revill's model can be *controlled* by modifying his model with constraints that disallow scenarios in which the event "Track Target" follows event "Bingo Fuel." This is the strictest course of action and corresponds to the requirement "A UAV shall only track targets found before reaching bingo fuel conditions" (Giammarco & Giles 2017). Several alternative requirements that are not as restrictive may also be considered as a result of this finding, such as: "A UAV that has reached a bingo fuel condition shall request permission from the Swarm Operator to track any new targets found," and "A UAV that has found a possible target after reaching bingo fuel shall relay the last known location of the target to the Swarm Operator, then continue to return to base" (Giammarco & Giles 2017). Running further simulations of this scenario with time durations, probabilities of occurrence, and other quantitative attributes would further help to determine which requirement is most appropriate for the system under design.

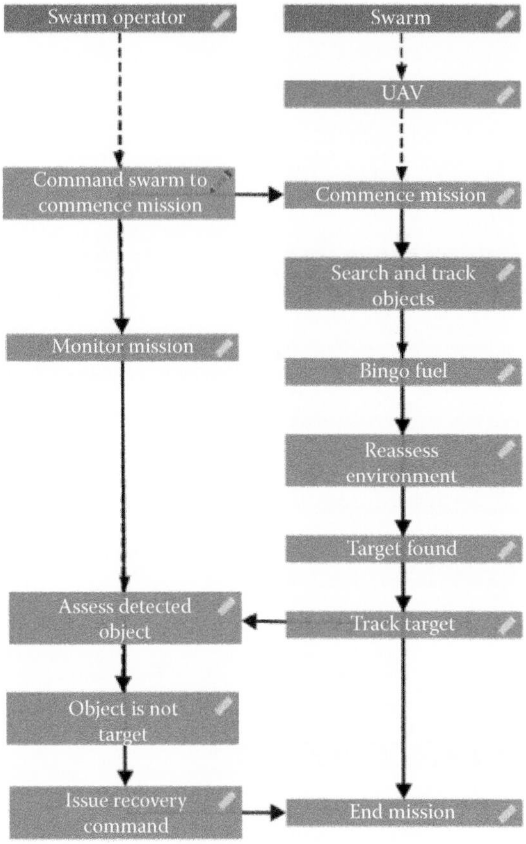

FIGURE 18.19
This unfavorable scenario, in which a target is found and tracked after bingo fuel conditions are reached (Revill 2016), is an example of strong emergence that may be positive or negative, depending on the exact quantities of resources such as time and fuel. Note that in this scenario, the object detected was determined not to be a target of interest after all.

In all of these scenarios, the discovery of the possible emergent behaviors in advance of experiencing them in reality affords designers and other stakeholders some time to consider which behaviors should be admitted, and which ones should be prohibited through the introduction of design constraints.

18.6 Detection, Classification, Prediction, and Control of Emergent Behavior

The three examples across different domain applications show how each modeler conducted emergent behavior detection, classification, prediction, and control through early modeling and simulation with MP. Table 18.1 summarizes the analysis of these examples.

The MP tool automatically *detected* unanticipated behaviors through modeling and simulation of all possible combinations of system behaviors and interactions permitted by the

TABLE 18.1

Summary of Emergent Behavior Analysis

Example	Figure	Detection	Classification	Prediction	Control
Pilcher's Order Processing System	18.15 left	Automatic and scope-complete with MP	Simple positive emergence	Order Cancelled	–
	18.15 middle		Simple positive emergence	Order Delivered	–
	18.15 right		Simple negative emergence	Order hangs in a Waiting state: Customer inconvenience; employee inconvenience; Cyber security vulnerability	Behavior logic modification in system model to prevent sequences that end in Waiting state
Bryant's First Responder Process	18.16	Automatic and scope-complete with MP	Weak positive emergence	Overdose Victim recovers after Bystander administers Narcan	–
	18.17		Strong positive emergence	Overdose Victim recovers after Bystander and First Responder administer Narcan	Add details to the model to be explicit about requirements to ensure this outcome
			Strong negative emergence	Overdose Victim at risk from lack of communication between Bystander and First Responder	Add details to the model to be explicit about requirements to mitigate this risk
Revill's UAV Mission	18.18	Automatic and scope-complete with MP	Weak positive emergence	Valid target detected and tracked	–
	18.19		Strong positive emergence	UAV is successfully recovered after tracking an object of interest after bingo fuel	Add details to the model to be explicit about requirements to ensure this outcome
			Strong negative emergence	UAV forced to emergency land / crash after tracking an object of interest after bingo fuel	Add details to the model to be explicit about requirements to mitigate this risk

model. The modeling concepts discussed in Section 18.3 are what enabled these emergent behaviors to present in simulation so that a human analyst can inspect them.

As seen in the discussion of the examples, detection of emergent behaviors is a task done automatically by MP, followed by the application of human intellect during classification to interpret the automatically generated scenarios and prediction of the broader implications and consequences therein. The human then codifies control using event ordering and constraints in the MP Analyzer to automatically reject unwanted emergent behaviors from the allowable behavior instances of the design. The event ordering and constraints, once tested for their effectiveness, become formal requirements to prevent such behaviors in the real system.

During inspection, the emergent behaviors may be *classified* as favorable or unfavorable, simple, weak, strong, or spooky, and positive or negative. Favorable and unfavorable are terms that convey a general value judgment by a human. Favorable generally corresponds to positive emergence, and unfavorable generally corresponds to negative emergence; however, the favorable/unfavorable terms refer to initial value judgments that take place before the scenarios are thoroughly analyzed and more formally classified as exhibiting positive or negative emergence. There may also be cases where positive emergence exists and stakeholders disagree whether all elements of the scenario are favorable or unfavorable (from their viewpoint). Further work is indicated to resolve this taxonomy.

Negative emergent behaviors attract great attention and desire for more understanding for when and how they can occur. Using the MP approach, positive emergent behaviors are the ones that remain once the design is purged of the negative emergent behaviors by rearranging the order of events, imposing deliberate design constraints, or otherwise altering the behaviors or interactions.

Once emergent behaviors have been detected and classified, MP is used as a canvas upon which humans may paint predictions about when and how these behaviors may occur, and whether they would result in positive or negative emergence. Using tools capable of assigning event attributes such as durations and probabilities, as shown in Figure 18.7, would provide quantitative data to inform decisions about whether and how to *control* the emergent behaviors.

Emergent behaviors are controlled through modification of the individual behavior models themselves, or when this is not possible (as is often the case with independently owned and operated systems), through relaxing or restricting interaction constraints that create dependencies between the individual behavior models. Engineering emergence in SoS is thus largely a practice of engineering interactions. In addition to controlling the order of events inside of systems, human architects can control emergent behaviors among different systems by introducing interaction constraints that prune them out of the set of allowable behaviors, or when that is not possible, by planning for their occurrence through early detection, classification, and prediction and designing some failsafe behaviors to control unwanted emergent behaviors.

18.7 Conclusions and Way-Ahead

A major finding from this research is that certain emergent behaviors of a SoS can be detected, classified, predicted, and controlled through early modeling and simulation with MP. To engineer emergence, we must think in terms of relaxing or restricting control over component or system interactions. We must view each system as having its own behaviors that will express unless we deliberately constrain those behaviors out of the design. To enable expression of emergent behaviors in simulation, the following modeling concepts should be employed: separate behaviors and interactions, model system behaviors and environment behaviors, formalize models for automatic execution, properly allocate each task to a human or to a machine, and use abstraction and refinement to manage large models.

The key to exposing emergent behaviors using MP lies in its ability to automatically generate a scope-complete set of event traces from separate system specifications. The Small Scope Hypothesis assures that most unwanted behaviors will be possible to detect that way. This capability automates the detection process, leaving humans with more time for the classification, prediction, and control activities requiring human intellect and judgment. Emergent behaviors are controlled through modifications to the behavior and interaction specifications. The essential features of MP enabling control of emergent behaviors are: ability to formalize interactions within the system and between the system and its environment at an appropriate level of abstraction; automated generation of exhaustive sets of scenarios within a given scope; and tools for the analysis of generated scenarios supporting assertion checking, event trace visualization, and annotation.

The ease with which the emergent behaviors were detected, classified, predicted, and controlled by students at various levels of education suggests that MP is user friendly for practitioners who have basic skills in logic and logical thinking, and is suitable for application in the engineering emergence problem space. Next steps in this research are to develop a formal classification taxonomy characterizing different types of emergent behavior and use cases for such classification, to specify MP-compatible profiles for graphical languages such as the System Modeling Language (SysML) and the Lifecycle Modeling Language (LML) that will demonstrate more fully how to implement the modeling concepts in section 18.3, and to enable set based design tools with MP as a comprehensive use case scenario set generator that accounts for event properties such as durations and probabilities.

References

Auguston, M. 2009a. Monterey Phoenix, or How to Make Software Architecture Executable, OOPSLA'09/Onward conference, Orlando, Florida, OOPSLA Companion, October 2009, pp. 1031–1038.

Auguston, M. 2009b. Software Architecture Built from Behavior Models, ACM SIGSOFT Software Engineering Notes, 34:5.

Auguston, M. 2014. Behavior models for software architecture, NPS Technical Report NPS-CS-14-003, November 2014, http://calhoun.nps.edu/handle/10945/43851.

Auguston, Mikhail. 2018. System and Software Architecture and Workflow Modeling Language Manual. Naval Protgraduate School, Monterey, CA. https://wiki.nps.edu/display/MP/.

Auguston, Mikhail, Kristin Giammarco, W. Clifton Baldwin, Ji'On Crump, and Monica Farah-Stapleton. 2015. "Modeling and Verifying Business Processes with Monterey Phoenix." In *2015 Conference on Systems Engineering Research*, 44:345–53.

Beaufait, James, Anthony Constable, and Daniel Jent. "Modeling UAS HADR in Monterey Phoenix." SI4022 Term Paper. Monterey, CA. June 2017.

Bryant, Jordan. "Using Monterey Phoenix to analyze an alternative process for administering Naloxone." Capstone Research Project, Science and Math Academy, Aberdeen, MD. June 2016. Retrieved September 17, 2016 from http://www.scienceandmathacademy.com/academics/srt4/student_work/2016/bryant_jordan.pdf.

Crawley, Edward, Bruce Cameron, and Daniel Selva. System Architecture: Strategy and Product Development. Hoboken, NJ: Pearson Higher Education, 2016, p. 9.

Cunha, John P. "Narcan Side Effects Center." [Online]. http://www.rxlist.com/narcan-side-effects-drug-center.htm. Accessed July 24, 2017.

Giammarco, Kristin. "Practical Modeling Concepts for Engineering Emergence in Systems of Systems." Proceeding of the 12th Annual System of Systems Engineering Conference, Waikoloa, HI, June 18–21, 2017.

Giammarco, Kristin and Clifford A. Whitcomb. "Comprehensive use case scenario generation: An approach and template for modeling system of systems behaviors." Proceedings of the 12th Annual System of Systems Engineering Conference. Waikoloa, HI: June 18–21, 2017.

Giammarco, Kristin and Katy Giles. "Verification and validation of behavior models using lightweight formal methods." Proceedings of the 15th Annual Conference on Systems Engineering Research. Redondo Beach, CA: March 23–25, 2017.

Giammarco, K., and M. Auguston. "Well, you didn't say not to! A formal systems engineering approach to teaching an unruly architecture good behavior." Procedia Computer Science, 20, (2013): 277–282.

Götz, Andreas. "The Ultimate Collection of Murphy's Laws." [Online]. Available: http://murphyslaws.net/. Accessed February 27, 2017.

International Council on Systems Engineering (INCOSE). "What Is Systems Engineering?: Definition of a System." [Online]. Available: http://www.incose.org/AboutSE/WhatIsSE. Accessed February 27. 2017.

International Organization for Standardization. ISO/IEC/IEEE 42010:2011, "Systems and Software Engineering—Recommended Practice for Architectural Description of Software-Intensive Systems."

Jackson, D. 2006. Software Abstractions: Logic, Language, and Analysis, Cambridge, Massachusetts: The MIT Press.

Liskov, B., and S. Zilles. 1974. Programming with abstract data types, ACM SIGPLAN Notices, Vol. 9 Issue 4, pp. 50–59.

Miller, George A. "The Magical Number Seven, Plus or Minus Two: Some Limits on Our Capacity for Processing Information." Psychological Review, Vol. 63 (1956): 81–97.

Mittal, S. and L. Rainey. Harnessing emergence: The control and design and emergent behavior in system of systems engineering. In: *SummerSim: summer simulation multi-conference 2015*, Chicago, USA, 26–29 July, 2015.

Nelson, Cassie. "Modeling a spacecraft communication system using Monterey Phoenix: A systems engineering case study." Master's Project, Stevens Institute of Technology, Hoboken, NJ. November 2015.

Page, S.E. 2009. *Understanding Complexity*. The Great Courses. Chantilly, VA, USA: The Teaching Company.

Pilcher, Joanne D. "Generation of department of defense architecture framework (DODAF) models using the Monterey Phoenix behavior modeling approach." Master's Thesis, Naval Postgraduate School, September 2015.

Reese, Cody. "A Model-Based Evaluation of Unmanned Aerial Systems for Humanitarian Assistance and Disaster Relief Operations." SI4022 Term Paper. Monterey, CA. June 2017.

Revill, Michael B. "UAV swarm behavior modeling for early exposure of failure modes." Master's Thesis, NPS, September 2016.

SEBoK authors. 2017. "System of Systems (SoS)," in BKCASE Editorial Board. 2016. *The Guide to the Systems Engineering Body of Knowledge (SEBoK)*, v. 1.8. R.D. Adcock (EIC). Hoboken, NJ: The Trustees of the Stevens Institute of Technology. *Released 27 March 2017, http://sebokwiki.org/wiki/Systems_of_Systems_(SoS)#Definition_and_Characteristics_of_Systems_of_Systems* (accessed 12 July 2017).

Vaneman, W., and R. Jaskot. 2013. "A Criteria-Based Framework for Establishing System of Systems Governance." Proceedings of the 7th Annual International IEEE Systems Conference, pp. 491–496. Orlando, FL: April 15–18, 2013.

Zeigler, Bernard P. "A note on promoting positive emergence and managing negative emergence in systems of systems." The Journal of Defense Modeling and Simulation 13, no. 1 (2016): 133–136.

19

A Model-Based Approach to Investigate Emergent Behaviors in Systems of Systems

John J. Quartuccio and Kristin M. Giammarco

CONTENTS

19.1 Introduction

The International Council on Systems Engineering (INCOSE) Vision 2025 suggests that the system engineer will conduct "...sensitivity and uncertainty analysis to analyze a system design from all relevant perspectives across the entire life cycle. While adding fidelity to models, adapting modeling formalisms, and combining multiple concurrent modeling efforts, systems engineers will be able to perform increasingly detailed trade studies and analyses" [39].

This chapter seeks to implement a part of this vision by proposing a methodology that develops a behavior model architecture suitable for evaluating potential emergent properties of a System of Systems (SoS) using lightweight formal methods. MP was inspired by model checking of software structures, and specifically by the Small Scope Hypothesis [43] and the associated Alloy Analyzer environment [42]. Consistent with these concepts, MP finds all possible outcomes of the model, within the scope of execution, while the Small Scope Hypothesis suggests that most problems in the architecture can be found with only a few iterations of the model. The execution environment supports assertion checking, trace visualization, annotation of each instance of behavior (or trace), and post-execution analysis. The authors propose a methodology to build a model, illustrated by an example problem. Analysis of detection, classification, prediction, and control of emergent behavior is conducted by the model developer, post execution of the model as outlined in the Discussion section.

19.1.1 The Language of Behavior Models

In order to present a cohesive representation of a system architecture in this chapter, a significant discussion on the concepts of abstraction, behavior, activity, events, modeling, and simulation become necessary.* These definitions may be somewhat unique to the proposed methodology, become relevant within the context of a systems demonstrating positive emergence.

19.1.1.1 Behaviors, Activities, Actions, and Events

Monterey Phoenix (MP), developed by Auguston [14,15], models behavior as a set of events and is proposed as a means to develop a system of systems architecture. A behavior relies upon an operator performing a particular action on some object or operand. This action is considered an event. The following paragraphs list definitions for behavior, activity, action, and event from the Merriam-Webster dictionary, with alternate uses removed for clarity, as follows:

> **behavior** [8]
> 1b: anything that an organism does involving action and response to stimulation
> c: the response of an individual, group, or species to its environment ...
> 3: the way in which something functions or operates ...

From this definition, a behavior involves an action in response to some stimulation.

> **activity** [7]
> 1: the quality or state of being active : behavior or actions of a particular kind ...
> 2: vigorous or energetic action ...
> 3: natural or normal function ...

An activity relates to some set of actions or functions.

> **action** [6]
> 1b: the accomplishment of a thing usually over a period of time, in stages, or with the possibility of repetition
> c: actions plural : behavior, conduct ...
> 2: an act of will ...
> 3: the bringing about of an alteration by force or through a natural agency - the action of water on rocks...
> 7a: an operating mechanism
> b: the manner in which a mechanism or instrument operates – a drill's twisting action...

Here, we see that an action brings about temporal relationships, including some period of time, implying a beginning and an end.

> **event** [9]
> 1c: a postulated outcome, condition, or eventuality ...
> 2a: something that happens : occurrence ...
> 5: a subset of the possible outcomes of an experiment

* This section derived from a conversation and email with Dr. Mikhail Auguston [13], Naval Postgraduate School.

Starting with these general definitions, concise uses of these terms need to be employed in order to develop the concepts of behaviors within some technical or engineering construct. To that end, let us first define a **behavior** as a set of events or actions. Behaviors are typically represented by multiple systems interacting with the environment and among the systems. There is also typically some sort of observable outcome (e.g., formation flight).

The *temporal* relationships among events is of critical importance to a behavior. A particular event or any be detectable or observable entity has a beginning and an end. This allows us to then discuss the concept of *precedence*. Precedence enables ordering of multiple events, since a particular events may occur before another, establishing a precedence relationship. Alternatively, multiple events may occur concurrently or simultaneously, without a precedence relationship. This lack of precedence is also a temporal property, but in this case, the events are independent of each other. Both the existence of a precedence relationship or the non-existence of a precedence relationship are temporal attributes of an event.

An event also has an additional relationship, *inclusion*. This relates the hierarchy of events, such that all event must have a root or source. One may think of a radio that would either send or receive a message. The event under consideration is the send or receive action, the operator is the radio, and the operand is the message. And so one can say that the send and receive events stem from, or are included in, the radio. Complex events may be described within multiple subordinate events that stem from a composite event. For example, there may be many functions necessary for the radio to send or receive, and so the operator and operand remain the same, but the send and receive events may be made up of many sub-events. Within this chapter, actions and events are treated synonymously, though when used in reference to the MP code, the term "event" will be used exclusively. More simply, an activity is a higher level abstraction of an event. Following a defined process is an example of an activity.

Key points:

1. A **behavior** is a set of events or actions, typically leading to some observable end point.
2. An **event** has a beginning and an end, with temporal relationship(s) among multiple events. A temporal relationship may include the existence of precedence between events, or the lack of precedence among events. Concurrent events exhibit a lack of precedence.
3. An **event** also has an additional attribute, inclusion, relating the events within a hierarchy.
4. An **action** is considered as an individual event.
5. An **activity** is a higher level abstraction of an event or multiple sequence of events. For example, a process may be considered as an activity.

19.1.1.2 Abstraction, Models, and Simulation

Once again, let us start with the Merriam-Webster dictionary results for definitions of abstract, models, and simulation; again only the relevant definitions are listed as follows:

abstract [5]
1a: disassociated from any specific instance–an abstract entity

model [10]
3: structural design–a home on the model of an old farmhouse
4: a usually miniature representation of something–a plastic model of the human heart; also: a pattern of something to be made
5: an example for imitation or emulation...
11: a description or analogy used to help visualize something (such as an atom) that cannot be directly observed
12: a system of postulates, data, and inferences presented as a mathematical description of an entity or state of affairs; also: a computer simulation (see simulation 3a) based on such a system climate models...

simulate [11]
2: to make a simulation of (something, such as a physical system)

simulation [12]
1: the act or process of simulating... 3a: the imitative representation of the functioning of one system or process by means of the functioning of another–a computer simulation of an industrial process b: examination of a problem often not subject to direct experimentation by means of a simulating device

The concept of an abstraction is critical to any discussion of a model. Simply put, a model represents only those functions necessary for a particular purpose; the model is a representation, not the actual entity. As an example, a training system may need to instantiate a radio in order to provide instruction to aviators. However, the representation of the radio does not need to perform the thousands of functions of an operational radio; it only simulates a finite number of instances of the operational system to the user.

Many are familiar with the popular quote from Box that "all models are wrong, but some are useful" [21]. Buede developed a consistent thought as he defines a **model** as "any incomplete representation of reality, an abstraction" [23, p. 75]. He further elaborates that a model may take on the following forms:

- *a physical representation* such as an aircraft wind tunnel model, used to identify aerodynamic properties to be projected to the final configuration,
- *a mathematical representation* such as a random number generator or a physics-based simulation of aerodynamic loads, and
- *a mental representation* that may or may not agree with a physical representation, a mathematical representation, or estimation of reality.

Buede's key point is that the *"essence of a model is the question or set of questions that the model can reliably answer"* [23]. Though a model is an incomplete abstraction of reality, yet the model must enable a *complete set of answers to the questions of interest.* Thus in order to represent behaviors, a model needs to identify all aspects of these behaviors including both positive and negative interactions.

Consistent with those definitions, consider the following: "A model is a representation of something. It captures not all attributes of the represented thing, but rather only those seeming relevant. The model is created for a certain purpose and stakeholders" [1].

A *simulation* is any particular instance of a behavior model. Using the previous examples, a simulation of the wind tunnel aerodynamic model is one run of the wind tunnel, consisting of the initial set-up description and final results. The simulation of a physics-based computer model is one execution, based on a set of inputs, and the computed results.

As we shall investigate behavior models in the next section, these too have particular results based on execution of the model.

> ***Key points:***
>
> 1. A model is an abstraction of the physical or instantiated system, developed in order to represent essential elements of the design.
> 2. The model must represent a complete set of behaviors, and analytically derive a means to enable intended behaviors, while restricting unintended behaviors.
> 3. A simulation is any instance of behavior extracted from the model.

19.1.1.3 A Behavior Model

We have established that a behavior is a set of events or actions leading to some observable end-point, and a model is an abstraction of the physical or instantiated system, developed in order to represent essential elements of the design.

A behavior model in the analytical or computational realm consists of an algorithm that employs a formal language to derive a set of events, representing essential interactions and relationships held within the system. This approach avoids the specification of parameters of the design for as long as possible, enabling concentrated effort to evaluate the underlying and intrinsic interactions within the system and among interactions with other systems, the user, or the environment. The temporal (precedence) and hierarchical (inclusion) properties of behavior formulate the essence of the model.

A formal language, based in set theory, is fundamental to the description of a behavior model. As opposed to a natural language, with possibilities for misinterpretation, a formal language is logically complete, consistent, and verifiable. Monterey Phoenix (MP) employs such a language, explicitly for the purpose of defining and modeling behaviors. As a light weight formal method, models developed within MP rely upon intrinsic investigation by the developer for verification in lieu of formal proofs. This approach encourages an interactive methodology for development, outlined later in this chapter.

> ***Key points:***
>
> 1. A behavior model employs a formal language to derive a set of events representing essential interactions and relationships within the design.
> 2. This analysis is performed at a level of abstraction that isolates the logical behavior of the design.

In execution, MP generates all possible outcomes of the model. Within MP, these instances are called traces (or use cases) and are based on the scope of execution of the model. Simple models contain relatively few interactions and few alternative possibilities; and so a measure of complexity may be associated with the number of interactions and number of traces of the model.

MP employs the Small Scope Hypothesis developed by Jackson [43], positing that most software failures can be found with relatively few iterations of the code. Modeling objects, as opposed to behaviors, Jackson also encoded this concept through the Alloy Analyzer [42].

MP models behaviors of a design, as the developer systematically defines the events, adds temporal and hierarchical interactions, applies constraints, evaluates the results, and adjusts the definition as needed.

Incorporating a concise language, MP employs event grammar rules to define the model. Within these grammar rules, the schema, or code, separates the definition of the behavior sequence of a system from the definition of the interactions among systems. In this way, MP uniquely separates the definition of the behaviors within a system from the definition of the interactions of the events among multiple systems, the user, and an environment [17,33]. This construct allows the model developer flexibility and control in describing the intended behavior.

19.1.1.4 A Taxonomy of Computer-Based Models

Within the construct of computer modeling and simulation of dynamics behaviors and interactions within a system, a proposed taxonomy of models and simulation including behavior models, system dynamics modeling, discrete-event modeling, and agent-based modeling, is summarized in Table 19.1 and discussed in the following outline [20,23]. The latter three types require some level of parameterization, such as speed, range, timing, mass properties, and band-width in order to project the system dynamics. Behavior modeling produces a logically-sound architecture and therefore it should be conducted prior to more detailed analysis.

1. *Behavior models* are developed at a high level of abstraction, prior to populating parameters to the model. Monterey Phoenix (MP) is a behavior model, employing lightweight formal methods, that develops all possible traces, or use cases, within

TABLE 19.1

Types of Modeling and Simulation

Type	Applications
Behavior models	Early system architectures
Logic-based	Cross-domain analysis of all behaviors
Event-based models	Production and manufacturing
Sequenced by events	Transportation systems
	Logistics
Agent-based models	Organizational behavior
Sequenced by an agent	Animal behavior
	Biological systems
	Crowd dynamics
	Search patterns
System dynamics models	Flight dynamics
Physics, time-based	Weapon separation dynamics & trajectory
	Computational fluid dynamics
	Dynamic structural loading
Hybrid models	Autonomous platforms
	Combined effects

a given scope of execution [17]. Behavior models focus on the essence of the interactions within a system and with external interfaces, producing a logically sound architecture. MP is available for anyone to use at http:\\firebird.nps.edu.

2. *Discrete-event models* are used for analysis of transactions such as those encountered in financial institutions, logistics and shipping companies, and other behavioral representations. While temporal properties remain integral to the model, the progression of the simulation relies upon completion of a sequence of events. Credit card transactions, commercial airlines transportation, and commercial cargo and shipping systems are examples of processes that are well suited for discrete event modeling. Many aspects of systems of systems are tied to such discrete events. The Discrete Event System Specification (DEVS) formulation was introduced by Zeigler [62], enabling a mathematical representation of system dynamics based on event sequences.
3. *System dynamic models* are used to simulate classic time-based problems in engineering. Many of these models simplify the system dynamics to a set of first or second order differential equations, code these as difference equations with a variable or fixed time-step, and solve these equations using a numerical method such as Runge-Kutta or other applicable algorithm [61]. Most engineering disciplines use system dynamic models to study phenomena and their applications to include areas of study within dynamics of structures, particle dynamics, flight dynamics, and computational fluid dynamics, with application to many scientific fields. Interestingly, systems theory as applied to other fields also gain benefit from this method, including economics, sociology, and psychology. MATLAB® and Simulink® [61] and AnyLogic® [20] are commercially available packages that model system dynamics. Higher order languages such as any instantiation of C or Java™ can also be used to develop these models.
4. *Agent-based models* are more recently developed than system dynamic models and discrete event models. The agent-based formulation describes the individual parameters or variables within the agent, the expected interactions between agents, and the overall context of the environment including number of dimensions, number of agents, and constraints of the problem. AnyLogic® is a commercially available platform for agent-based modeling [20].
5. *Hybrid models* combine aspects of two or more types of models.

Key point:
Behavior modeling forms the essence of system interactions and produces a logically sound architecture. From this basis, further parameterized description and follow-on modeling of event-based, agent-based, system dynamics, or hybrid types may be conducted.

19.1.2 Further Definitions Applicable Systems Engineering

19.1.2.1 System

A general definition of a *system*, supported by International Council on Systems Engineering (INCOSE), International Organization for Standardization (ISO), International Electrotechnical Commission (IEC), and Institute of Electrical and Electronics Engineers (IEEE).

These definitions apply to "engineered" systems that have specifically and intentionally been developed by mankind, as opposed to natural systems [46]. From this perspective a system is defined as "a combination of interacting elements organized to achieve one or more stated purposes" [41], where the elements may include "hardware, software, firmware, people, information, techniques, facilities, services, related natural artifacts and other support elements" [18].

Crawley et al. define a system as "a set of entities and their relationships, whose functionality is greater than the sum of the individual entities" [25]. The authors further indicate that in this definition the system includes both "entities and their inter-relationships." This construct of a system has a natural extension to concept of Systems of Systems (SoS); in this case each entity is a system.

19.1.2.2 Model-Based Systems Engineering

The INCOSE defines *Model-Based Systems Engineering (MBSE)* as

> ... the formalized application of modeling to support system requirements, design, analysis, verification, and validation activities beginning in the conceptual design phase and continuing throughout development and later life cycle phases [40].

19.1.2.3 System of Systems

INCOSE further defines a *System of Systems (SoS)* as a system of interest "whose elements are managerially and/or operationally independent systems. These inter-operating and/or integrated collections of constituent systems usually produce results unachievable by the individual systems alone. Because an SoS is itself a system, the systems engineer may choose whether to address it as either a system or as an SoS, depending on which perspective is better suited to a particular problem" [40].

The Defense Acquisition Guide (DAG) adds context to this definition that is directly consistent with the ODUSD Systems Engineering Guide for Systems of Systems [53]. The DAG reads as follows:

> Most DoD capabilities today are provided by an aggregation of systems often referred to as System of Systems (SoS). A SoS is described as a set or arrangement of systems that results when independent and useful systems are integrated into a larger system that delivers unique capabilities. For complex SoS, the inter-dependencies that exist or are developed between and/or among the individual systems being integrated are significantly important and need to be tracked. Each SoS may consist of varying technologies that matured decades apart, designed for different purposes but now used to meet new objectives that may not have been defined at the time the systems were fielded. [59, p. 167]

Dahmann et al. indicated that the system of systems framework and processes "push systems thinking beyond the traditional arena of new system development and acquisition to address the reality of today's system challenges of integrating and evolving existing systems to meet changing needs" [28]. Within the continuum of types of systems of systems, consideration has been given to basic categories of managerial control to include "directed, collaborative, virtual, and acknowledged" [47,48,27], an emphasis on the importance of acknowledged systems have "recognized capability needs, management, and SE at the SoS level as well as autonomous objectives, management, and technical development approaches of the systems which contribute to the SoS capability objectives" [28].

19.1.2.4 Complex Systems and Emergence

Mitchell defined a *Complex System* as "a system in which large networks of components with no central control and simple rules of operation give rise to complex collective behavior, sophisticated information processing, and adaptation via learning or evolution" [51, p. 13].

Taking into account the "self-organizing" aspects of unique "emergent" or collective behavior that can occur with no internal or external command and control of the individual elements or agents, Mitchell also proposed a second definition of a *Complex System* as "a system that exhibits nontrivial emergent and self-organizing behaviors" [51, p. 13].

Complexity is closely aligned with Chaos Theory [35,60,58] and its implications apply to many disciplines, including economics, psychology, sociology, medicine, and organization dynamics; within the traditional sciences of physics, chemistry, biology, and mathematics; and within engineered systems including network theory, computer science, integrated systems, and in general, systems of systems. Characteristics associated with complexity include emergence and self-organization. These aspects cannot be deduced by decomposing the system to its smallest components, but they are tied to the individual system behaviors, and are associated with interactions among the systems and their environment. Aspects that also contribute to those collective behaviors include, dependence on initial conditions, diversity within the population of agents, and the environment itself [60,58,24,50,51]. Considering these definitions, a complex system is special sort of system of systems that demonstrates a property known as emergence.

Zeigler identified the importance of both positive and negative emergence, and is a primary motivation for this text [63]. Crawley et al. relate complex systems and emergence, including the concept of positive and negative as follows:

> Complex systems have behaviors and properties that no subset of their elements have. Some of these are deliberately sought as the product of methodical design activity. While achieving these behaviors, the designers often accept certain undesirable behaviors or side effects. In addition, systems have unanticipated behaviors commonly called emergent. Emergent behaviors may turn out to be desirable in retrospect, or they may be undesirable. [26]

Emergence is exhibited through collective behavior and self-organization of a system of systems that cannot be deduced by the investigation of an individual constituent system, element, or agent. This behavior depends not only upon the properties of an individual constituent system, but the collective behavior emanates from tightly coupled interactions among the individuals within the group [60,58,24,50,51].

Maier [49] defined an emergent property as "a property possessed by an assemblage of things that is not possessed by any members of the assemblage individually" [49, p. 21]. Maier [49] further outlined types of emergence, summarized in Table 19.2. Fromm [31] had also developed a classification of emergence, summarized in Table 19.3, based on the level of interaction of constituent systems within the whole system. There exists some level of inconsistency between these descriptions; the former description may be better suited for an early assessment of predictability of the system response, while the latter description lends itself to a computational method that can enable automated classification based on the system response.

TABLE 19.2

Types of Emergence, Derived from Maier [49, p. 21–22]

Type	Emergent Property Mapping to Modeling and Simulation
Simple	Behavior readily predicted, model abstracted with lower complexity than the actual system
Weak	Behavior consistent with known properties of the system and readily and reproduced by simulation; however, system interactions must be included in the model
Strong	Behavior consistent with known properties, but unable to reliably predict where [or when] emergent properties occur
Spooky	Behavior inconsistent with known properties of the system components, even with model equivalent to complexity of the system

TABLE 19.3

Classification of Emergence, Indicating a Particular Designation Based on the Level of Interaction and Feedback, Developed by Fromm [31]

Class	Title	Description
Type I	*Simple*	Interaction of systems producing some global response, but without feedback to the systems.
Type II	*Weak*	Inclusive of the interactions of Type I, but with feedback from the global response to the systems.
Type III	*Multiple*	Inclusive of Type II, but with multiple systems.
Type IV	*Strong*	Inclusive of Type III, with the interaction of these multiple systems with each other creating additional global responses.

19.1.2.5 Architecture

As defined by Crawley et al., an architecture is "an abstract description of the entities of a system and the relationship between those entities" [26]. Perhaps more succinctly, we may think of an architecture as a "function enabled by form" [25]. For this chapter, an architecture is generally used as a means to describe the topology and semantics of the model description.

19.1.3 Systems Engineering through MBSE

The objective of a system of systems developer extends the traditional systems engineering approaches of requirements definition, functional analysis, and design synthesis in order to attain objectives not possible by a single system [19,46,23]. The developer of a System of Systems (SoS) needs to ensure that all possible behaviors of the design have been thoroughly investigated and verified, prior to product release to the end-user. Therefore, the developer faces a risk that inherent or possibly latent characteristics may cause unexpected or unwanted outcomes in the system of systems. The design needs to exhibit desired behaviors to perform a particular function and also constrain undesired behaviors, avoiding hazardous or damaging outcomes.

Model-Based Systems Engineering (MBSE) enables a means to capture the interactions and connectivity among components within a system of systems. As outlined by Delligatti, in order to implement an MBSE approach, the user needs to employ a language, method, and tool [29]. The Object Modeling Group (OMG) definition of Unified Modeling Language (UML) [2] and the associated dialect of Systems Modeling Language (SysML) [3] provide a means for standard definition of syntax, semantics, and exchange of behavior models and

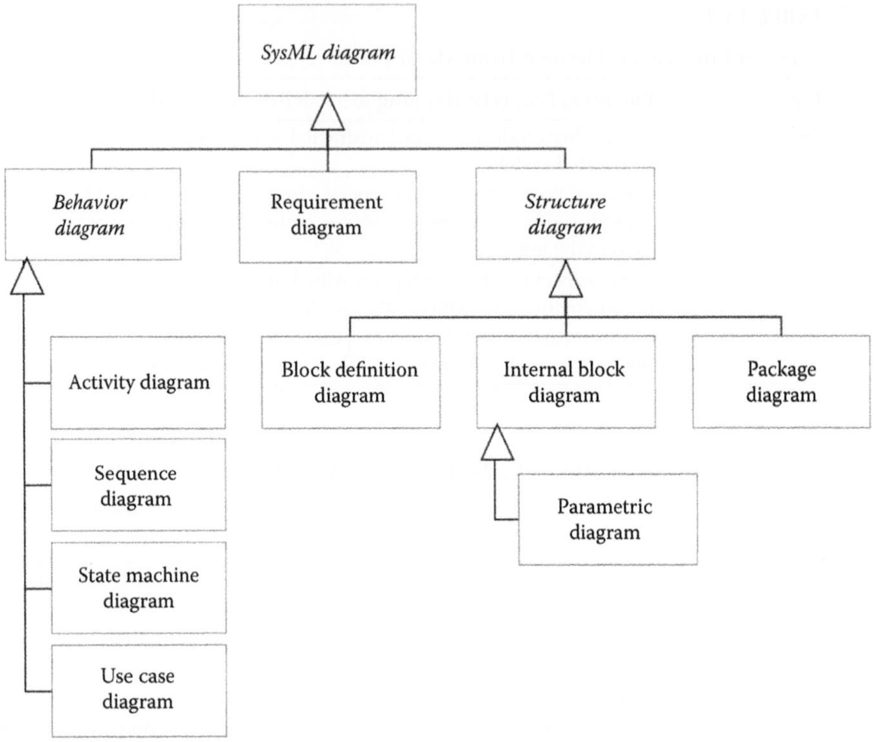

FIGURE 19.1
SysML taxonomy, where the open arrow indicates a "type of" the indicated block. This figure was derived from the SysML reference documentation [3] p. 194.

associated diagrams, including activity diagrams, sequence diagrams, state charts, and use cases. The elements SysML describe the model structure, behavior, requirements, and constraints using a semi-formal modeling language that aligns the grammar, or a set of rules used to depict the system. As illustrated in Figure 19.1, SysML depicts the following aspects of the system, [3]:

- Structure or hierarchy
- System behaviors
- System requirements
- Parametric values

The objective of the SoS developer is to drive the all behaviors to be with consistent and predictable results, thereby gaining the benefits of positive emergence, while limiting or constraining the design to avoid unwanted emergence [63].

Key point:
UML and SysML provide a coherent language for system diagrams; however, these definitions are not intended to evaluate the content of the architectures that they represent.

The model developer needs to consider the following questions in coordination with developing a SysML model:

- What are all of the potential outcomes of the system under development?
- What factors affect the probability of gaining a desired system outcome?
- What negative properties are inherent in the architecture design?
- What is the potential for re-use of successful architectures?
- What constraints are necessary for the system?
- What requirements are derived from the system model?

WHAT IS THE PURPOSE OF THIS CHAPTER?

The methodology and examples derived in this chapter seek to answer critical questions posed by the System of Systems (SoS) developer by implementing lightweight formal methods as a means to fundamentally define system behaviors, to methodically simulate the system outcomes, and then to subsequently evaluate the results.

19.2 Background

Model-Based Systems Engineering (MBSE) structures form a foundation of engineering principles that developers have begun to apply to System of Systems (SoS) by applying Systems Modeling Language (SysML) structures [38,37,54]. Behavior models derived from Monterey Phoenix (MP) provide a logical foundation for the system behavior architectures [32] as a means to capture intrinsic nature of the system.

19.2.1 Enabling Technologies That Implement Emergent Behavior within a System of Systems

Advances in digital computing through exponential decrease in the size of integrated circuits [52] has enabled an unprecedented level of interaction among systems. For over fifty years, Moore's Law, indicating the number of transistors per square inch doubles every year, has been expanded as a general rule for integrated circuits, even as computing architecture reach the atomic level. Advances have enabled responsive sensors and interfaces among systems possible, thereby enabling behaviors that use advances in sensor systems, communication, and computing power.

- *Sensor systems*: Electro-Optical/Infra-Red (EO/IR), Light Detection and Ranging (LiDAR), Radio Frequency (RF), and Global Position System (GPS) have been implemented in a wide range of sensors across many platforms including autonomous vehicles.
- *Data links and communication*: Networked sensor systems via RF data links and other communication forms enable advances in localization and simultaneous fusion of multiple physical locations.
- *Computing power*: Enabling behaviors with real-time algorithms running on distributed multiple systems.

19.2.2 Potential Sources of Unfavorable Emergent Behavior

- *Information latency*: Operating on latent data may be induced by filtering, physical responses, communication protocol, external sources, or environmental conditions.
- *Hardware response*: Physical limitations or responses of the physical system unanticipated by the developer, nonlinear failure, progressive structural changes due to cyclical loading to the structure, or state changes of the any component or transport utility (e.g., liquid to gas of a cooling function) may cause unanticipated response of the system.
- *Environmental conditions* may include off-nominal conditions of heat, cold, power availability, atmospheric changes that impact critical phenomena of the system.
- *Communication delay or failure* may include the network construct, interference in radio frequencies, saturation, and network management, network contention, or other imposed protocol.
- *Human user* induced failures due to poor design may impact timing, response, and throughput; thereby introducing errors.
- *Design constraints* as limitations to the system behavior may inhibit both favorable and unfavorable aspects of the system and become critical factors within a SoS design.

19.2.3 Analysis of Emergent Behaviors within Systems of Systems

Developers of System of Systems (SoS) have employed a range of simulation methods as a means to identify and characterize emergent properties. These means have often employed agent-based models or hybrid models that incorporate both agent-based parameters along with a time-space simulation of a dynamic model. Mobile robotics have this type of hybrid response, such that the interaction of systems is dependent upon the spatial relationship of multiple agents. The decision logic employed within these analyses often includes a Markov chain type of logic representation. A simple example of a Markov chain may be described as a system that may be in either state *A* or state *B*, and the alternatives are whether to remain in the existing state or to move to the alternate state. As illustrated in Figure 19.2, the researcher assigns probabilities to these alternatives.

Examples of these analysis include the following:

Pentland [55] represented human decision behaviors employing a series of Markov chain models, such that the overall behavior represented as multiple model dynamics, with these decision points broken down as a series of prototypical steps.

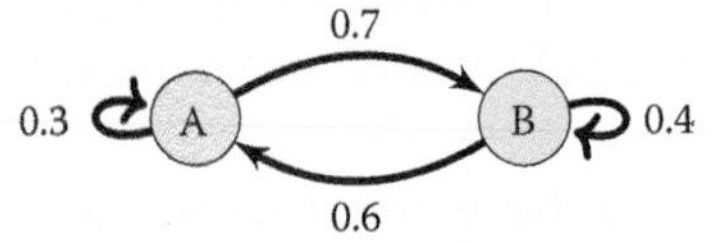

FIGURE 19.2
Example of a simple Markov chain, consisting of states *A* and *B*, where the probability of moving from state *A* to state *B* is 0.7, while the probability of remaining in state *A* is 0.3. Conversely, the probability of moving from state *B* to state A is 0.6, while the probability of remaining in state *B* is 0.4.

Singh [57] developed a model, representing complex adaptive behaviors of an Unmanned Air Vehicle (UAV) swarm using the NetLogo™ simulation environment, employing an agent-based simulation model. The model represented continuous-time response of each system acting as an agent in the swarm, and then described the interaction of the systems within the swarm. These interactions were dependent upon the spatial separation of each UAV. Singh identified emergent behaviors based on the a particular system response and classified the emergence using the taxonomy proposed by Holland [36] and classification proposed by Fromm [31], summarized in Table 19.3.

Kantert [44] developed an agent-based simulation, but employed graphical means of analysis as a means to identify norms of behaviors. By establishing norms, an instance of non-compliance also relates to non-normative behavior, thus, posing a means for identifying negative emergence. The application is intended for real-time evaluation of cyber-security.

These examples illustrate methods employed for analysis of emergent behaviors within systems of systems. Each of these examples relied upon some architecture describing the interactions and behaviors of the systems. The remainder of this chapter outlines a methodology of how behavior analysis using Monterey Phoenix (MP) can be used to derive the underlying architectures of theses types of systems. This methodology emphasizes the behaviors at a high level of abstraction such that subsequent modeling using agent-based, discrete-time based, system dynamics, or hybrid approaches will proceed employing a sound underlying architecture. Specifically, behavior modeling identifies constraints that are necessary to implement the designed system. Analysis of the model also provides insight to the probability of obtaining desired results of the system.

19.3 Methodology

Monterey Phoenix (MP) [16,33,34] provides explicit definition of the system's behavior architecture and may thereby expose potential flaws in logic that can show up by building and analyzing the architecture. The MP construct separates the descriptions of behavior within the system and the interaction among the systems, environment, and users, giving the developer flexibility in the design of the architecture.

The Small Scope Hypothesis, derived by Jackson [43], postulates that most errors in computer code can be found within a relatively small number of iterations or execution cycles. The small scope hypothesis and the associated Alloy Analyzer [42] provided inspiration to apply this practice from the software domain to the systems engineering domain with Monterey Phoenix and the MP-Firebird Analyzer [33].

Key points:

1. Monterey Phoenix (MP) exhaustively identifies all possible outcomes of the behavior architecture, within the scope of execution. These results expose patterns of behavior leading to the need for analytic tools to interactively identify these patterns, both positive and negative.

2. MP applies the Small Scope Hypothesis [43] such that most inconsistencies in the model are found within relatively few iterations.
3. MP builds a model at a high level of abstraction, prior to adding parameters to the system description.
4. Evaluating the model with MP early in the development enables the use of a mature structure for follow-on simulations employing event-based models, agent-based models, time-based physical models including system dynamics, or hybrids of these methods.

19.3.1 Using the MP-Firebird Analyzer

MP uses pseudo-code, or schema, in order to instantiate particular behaviors within the model. Each section of code is written and edited in the designated text-box. The scope of execution controls the number of iterations computed in accordance with the schema. Upon execution, diagrams of an exhaustive list of traces are produced.

The MP execution environment is available through MP-Firebird analyzer, illustrated in Figure 19.3. MP-Firebird supports the import of user-defined code as well as readily accessible examples. The environment also supports export of code and diagrams.

The following are fundamental concepts of Monterey Phoenix (MP):

- MP employs a powerful language, or pseudo-code, that formally describes the behaviors within the system. The complete model is referred to as a schema.
- All blocks in the system are considered as events.
- Root events initiate the hierarchy of subordinate or child events. The subordinate events are established by an *inclusion* relationship stemming from the root event.
- Events also exhibit *precedence* relationships both within a root hierarchy and coordinated among multiple root hierarchies.

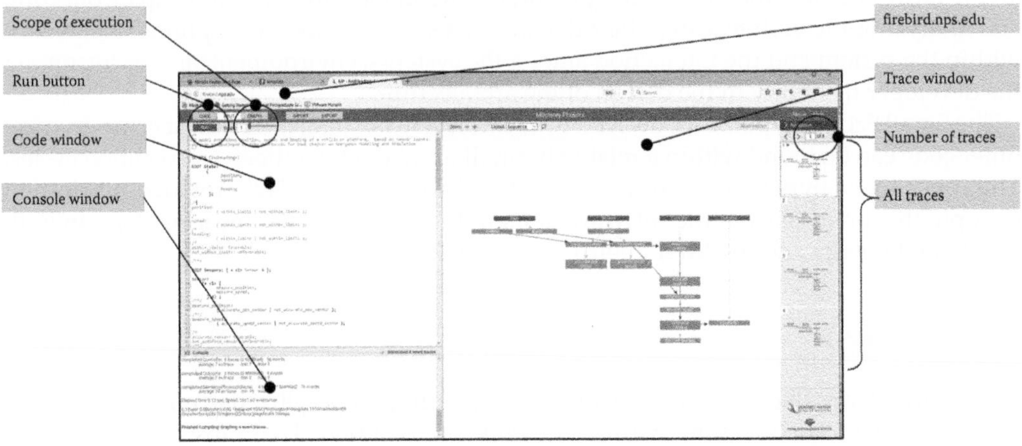

FIGURE 19.3
The Monterey Phoenix (MP) model environment, illustrating the code window, scope of execution, the current trace, and all traces produced by executing the schema. MP is available for anyone to use at firebird.nps.edu.

- MP separates the definition of behaviors within the system from interaction with other systems, users, or the environment through coordination of inter-related events.
- Atomic events are comprised with no constituent event, such that these events have a parent event in the hierarchy, but no children.
- Composite events are comprised with one or more constituent events, such that thee events have both a parent event in the hierarchy and at least one child event.
- Execution of MP exhaustively produces all possible outcomes in accordance with the schema and within the scope of execution.

Key point:
Execution of the MP schema produces all possible outcomes within the scope of execution, thereby enabling a means to identify emergent behaviors.

NOTE: Throughout this chapter, all figures that were derived from MP employ the key shown in Figure 19.4, unless otherwise noted on the figure.

19.3.2 Methodology Description

Six steps of the proposed methodology to build behavior models that support System of Systems (SoS) development is shown in Figure 19.5. Subject Matter Expert (SME) input is fundamental to successful implementation of the model, ensuring that critical aspects of the design are addressed in the model. At nearly all of the steps, the model is executed by the developer such that only small changes are implemented at a time. Then the effects are immediately analyzed within the MP-Firebird environment. This immediate feedback, produced in the execution environment, encourages interaction and engagement by the developer with the central engagement of a SME. Also at each step, the developer considers whether criteria is met to complete the step and then decides whether to proceed to the next step, repeat work in the current step, or go back to any previous step. The following sections describe the methodology as a step-by-step possess to develop an MP model, one set of commands at a time.

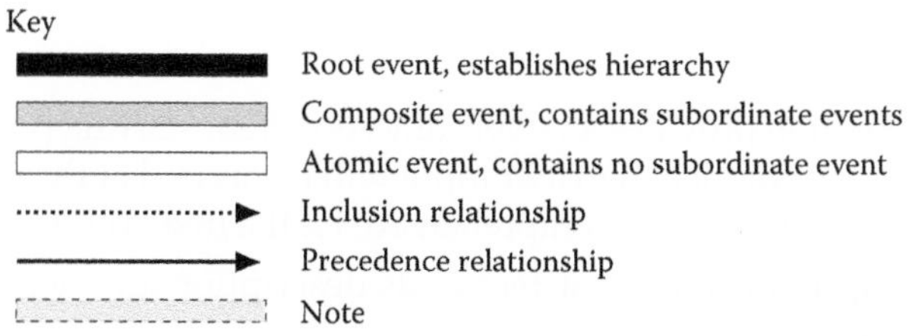

FIGURE 19.4
MP diagram key.

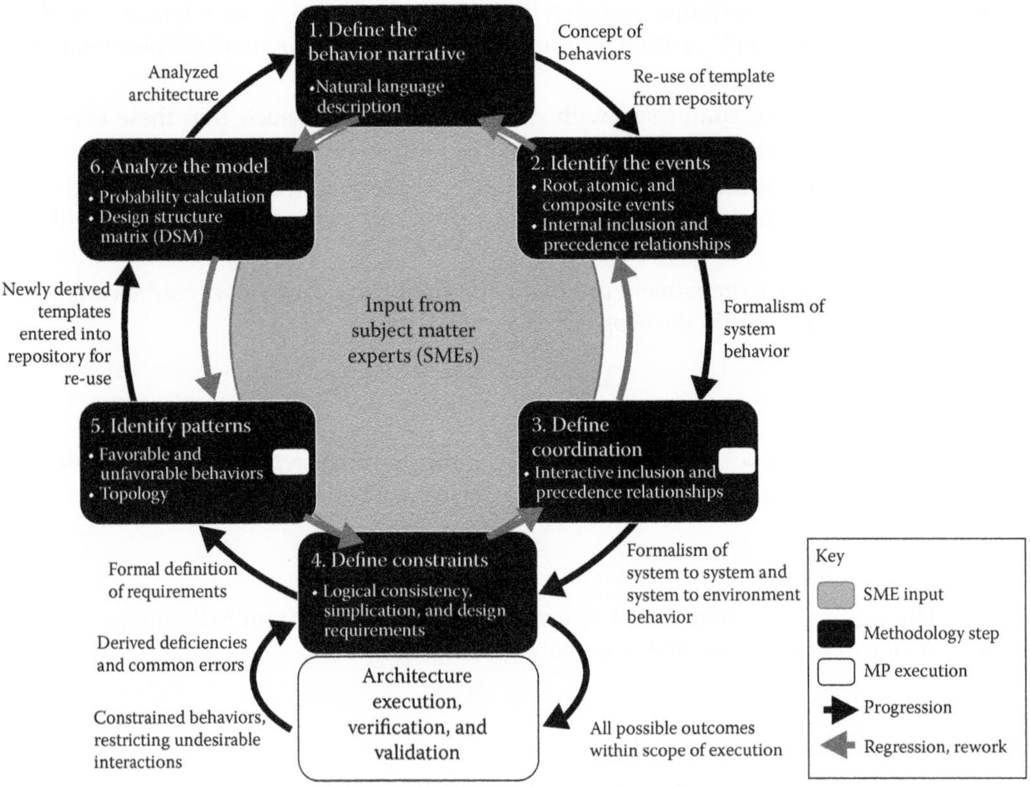

FIGURE 19.5
A methodology to mature System of Systems (SoS) behavior modeling using Monterey Phoenix (MP).

19.3.2.1 Step 1: Develop a Narrative of the Behavior

It is typical that the model developer is not the end-consumer of the model and associated analysis, and so SME input is critical to the process. The SME may be an end-user of the product, the funding sponsor of the project, or perhaps the developer of a legacy or predecessor system. The model developer begins the methodology by interviewing the relevant SMEs necessary to formulate a complete, analyzable, internally consistent, and elegant description, in accordance with qualities of great models [4].

Criteria: This step is complete when the SMEs have described the relevant behaviors, constraints, assumptions, and limitations of the SoS. A written and approved document is the recommended practice, especially if significant resources are spent in order to develop the model.

Decision: The developer needs to decide to proceed to next step, repeat this step, or go back to a previous step.

Proceed if documentation provides an internally consistent starting point, then proceed with the model development. Further input will be needed at the next steps, and so the description will may not be comprehensive for the first cycle of the methodology.

Repeat if the description is not consistent in its description and definitions; clarification will be necessary.

Go Back if the SME input cannot be reconciled; consideration needs to be given to address conflicts with sponsoring authority, as necessary.

Key point:
Subject Matter Expert (SME) input is critical to modeling a system architecture.

19.3.2.2 Step 2: Identify MP Events

A simple example was used throughout each of the next sections in order to illustrate how to describe events in MP, as shown in Figure 19.6. This model includes two root events A and B, with corresponding atomic events "task_a" and "task_b." Root events are typically the environment, a user, a system, or a component of the system. Subsequent events are typically actions taken by the root.

The inclusion relationship is represented with the dotted line. The model was executed with scope of two, producing two traces. Trace 1 has a single atomic event for both roots A and B, while trace 2 has two atomic events for root A and a single atomic event for root B.

The code for this model is listed below, where the curly brackets "{...}" indicate concurrent events, and the plus sign, "+," surrounding task_a indicates that the number of events is at least one and up to the scope of execution. Since this schema was executed at a scope of two, the output includes a trace with one "task_a" event and a second trace with two "task_a" events. Both traces have a single "task_b" event. Each root description ends with a semicolon, ";".

```
SCHEMA simple
ROOT A:
{ + task_a + };
ROOT B:
    task_b   ;
```
MP code

Criteria: This step is complete when the relevant environment, users, systems, and components are defined as root events, and all functions are captured as atomic or composite events.
Decision: Proceed to next step, repeat this step, or go back to previous step, as follows:

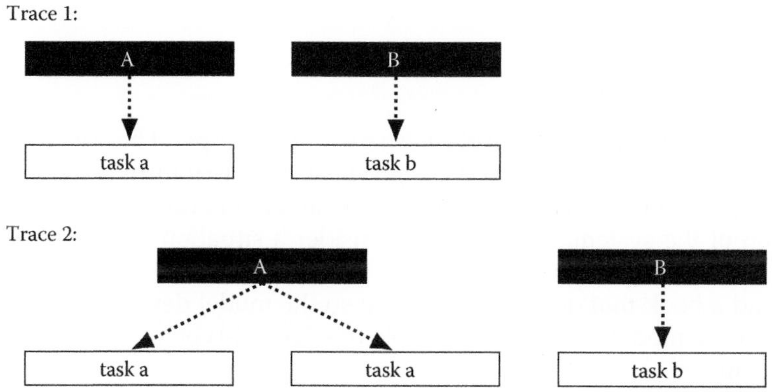

FIGURE 19.6
A simple Monterey Phoenix (MP) model illustrating root events A and B, with corresponding atomic events "task_a" and "task_b". This inclusion relationship is represented with the dotted line. The model was executed with scope of two, producing two traces. Trace 1 has a single atomic event for both roots A and B, while trace 2 has two atomic events for root A and a single atomic event for root B.

Proceed when coordination with the SMEs affirm that all events are captured.

Repeat if the SMEs identify significant errors.

Go Back if fundamental problems exist in the narrative.

19.3.2.3 Step 3: Identify Coordination

Coordination establishes precedence relationships between events within roots A and B. In the simple illustration we can establish that in all cases, "task_a" precedes "task_b", as illustrated in Figure 19.7, with the code listed below the figure. The "DO…OD" command establishes a loop for asynchronous coordination so that all occurrences of "task_a" precede "task_b." Events may also be shared by multiple roots using a "SHARE ALL" command.

```
COORDINATE $b: task_b
DO COORDINATE
        $a: task_a
    DO ADD $a PRECEDES $b; OD; OD;
                                    MP code
```

Criteria: This step is complete when coordination across all root events are complete.
Decision: Proceed to next step, repeat this step, or go back to previous step.

Proceed if the developer and SMEs agree with the coordination structure.

Repeat if errors are found.

Go Back if the root events need to be restructured.

> ***Key point:***
> Monterey Phoenix (MP) separates the behavior definition within a system, modeled as a root event, from the interaction among the System of Systems (SoS), modeled as multiple root events.

19.3.2.4 Step 4: Define Constraints

At least three types of constraints are of concern to the developer. These include logical constraints, simplification constraints, and design constraints that form system requirements.

A ***logical-type*** of constraint inhibits certain behaviors in order to maintain a realizable representation of the system. For example, consider a situation where a book may exist or not exist, and a student shall either read or not read the book. It is impossible for the student to read a book that does not exist, and so the model developer must restrict this sequence from the model. This is an example of a logical-type of constraint to be implemented in the model.

The developer also needs to consider the impact of irregular activities within the model, especially in situations where emergent behaviors may exist, but are not obvious. Consider a situation where a passenger train can either stop at a station or not stop at a station; and a passenger at the station can either board the train or not board the train. The developer may decide to restrict the condition where the passenger boards the train if it has not stopped.

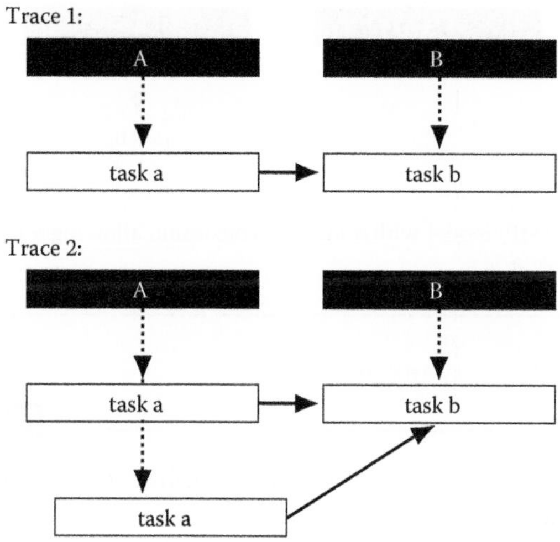

FIGURE 19.7
A simple Monterey Phoenix (MP) model with coordination between atomic events from root A and B. Coordination from task a to task b for both traces are illustrated as solid lines, representing precedence relationships. As such, task a must be accomplished before task b.

Under most situations, this restriction is reasonable; however, if searching for irregular interactions, the constraint will inhibit the behavior that is of interest to the developer. Irregular behavior may include the activity of a person illegally boarding a moving train. If restricted, the developer would not identify this potentially emergent event. Therefore, the developer needs to ensure that the model represents all areas of concern, by not overly-constraining the model.

A ***simplification-type*** of constraint is conducted at the discretion of the developer in order to concentrate the effort of the model where significant impact is to be expected. For example, if a student completes an assignment and turns it in to the professor, and the professor is responsible to grade all assignments that are turned in, a reasonable assumption is that the student's assignment is graded by the professor. This is a simplification-type of constraint that enables a comprehensive structure of the model, but it will limit the results to areas of greater interest to the developer.

A ***design-type*** of constraint establishes requirements of the system in order to ensure that sufficient boundaries are placed on the interaction of relevant events. For example, if a developer was interested in finding a means to ensure that vehicle operators remain alert while driving their car, it would not benefit the developer to target children under twelve years old for the methods or intervention. Considering a physical system, if an aircraft is executing a landing mode, it would not make sense for the vehicle to simultaneously attempt to re-fuel from another aerial platform. This would be a typical design-constraint such that certain behaviors are limited to particular modes or states of the platform.

Implementing constraints for our simple example: if the schema were executed without the coordination of precedence relationships, the developer would find the occurrence of a trace shown in Figure 19.8, where "task_b" occurs without a logical predecessor of "task_a". A read function cannot occur before a write function, nor a receive function before a send function. Clearly in these cases, a logical-type of constraint is needed for the model. The following code ensures that the number of "trace_a" events is greater than the number of "trace_b" events.

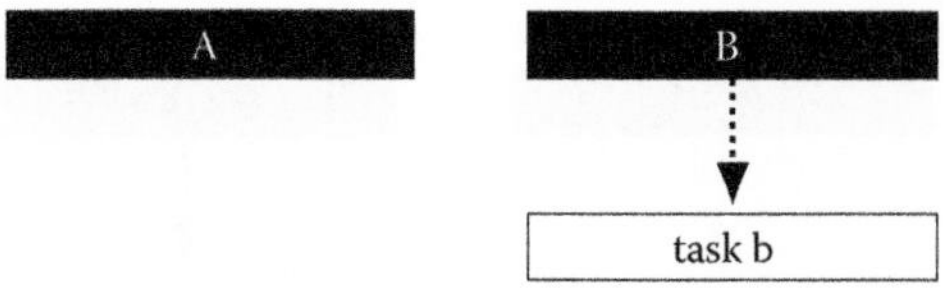

FIGURE 19.8
A simple Monterey Phoenix (MP) model with a missing constraint, allowing a trace in which "task_b" exists without a corresponding "task_a."

```
ENSURE ( #task_a >= #task_b);
```
MP code

Criteria: This step is complete when all logical, simplification, and design constraints are implemented in the schema.
Decision: Proceed to next step, repeat this step, or go back to previous step.

Proceed when the developer and SMEs agree that the constraints are an appropriate representation.

Repeat if constraints are missing or overly used, restricting desirable or actual behavior.

Go Back if any part of the event structure needs to be reworked in order to satisfy the needed behaviors.

Key points:

1. Constraints may be used to identify requirements that are necessary for enabling desirable behavior and restraining undesirable behavior.
2. Over-use of constraints may unintentionally hide intrinsic behaviors of the architecture necessary to identify emergence.

19.3.2.5 Step 5: Identify Patterns

Identifying patterns in the MP results, using the assertion checking functions along with the "MARK" and "SAY" commands, has several purposes that are critical to the system developer, including the following:

- Quickly investigate large sets of output that can easily reach in the tens of thousands of traces.
- Ensure that all traces contain only expected output, avoiding strong and spooky forms of emergence [49].
- Develop a repository of effective architectures that can be employed in future designs, including instances of emergence, thereby capturing positive lessons learned.
- Develop a repository of ineffective or problematic architectures, including unwanted instances of emergence, thereby capturing negative lessons learned.

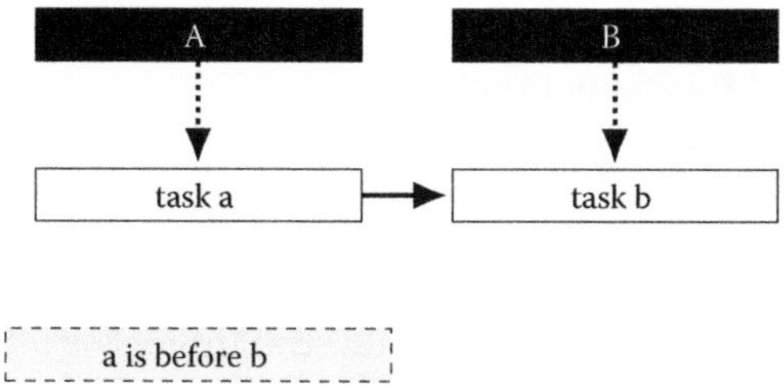

FIGURE 19.9
A simple Monterey Phoenix (MP) model with assertion checking such that each trace that satisfies the condition that "a is before b" is marked and labeled. This capability establishes the means to identify templates, or patterns, in the results.

Figure 19.9 illustrates how the assertion checking function is used in MP. The code listed below the figure checks for conditions where "task_a" occurs before "task_b." If the condition is met, the trace is marked and annotated with the "SAY" command. Each of these checks can be formulated as templates of unique behavior in the model. Long term objectives include aligning a repository of these templates to enable re-use of working and validated architectures, or as importantly as abandoned ineffective or problematic architectures.

```
IF EXISTS
  $alpha : task_a,
  $beta:  task_b
    (
        $alpha BEFORE $beta
    )
THEN MARK; SAY("a is before b");
FI;
                                              MP code
```

Criteria: This step is complete when templates are associated desirable and undesirable behaviors of the model.
Decision: Proceed to next step, repeat this step, or go back to previous step.

Proceed when the developer and SMEs agree that the templates have been effectively implemented.

Repeat if significant portions of the output cannot be resolved as expected behavior.

Go Back if additional constraints are needed in the model to avoid unwanted patterns.

19.3.2.6 Step 6: Evaluate the Results

Several types of analyses may be of interest to the developer, using the results from executing the MP model, including the following:

- Probability of any trace or outcome
- A summary of all interactions in the form of a Design Structure Matrix (DSM) or N-squared Diagram

- The throughput of data flow
- The sequencing and timing of events
- Requirements on computing power

For the purpose of this chapter, just the first two types of analyses will be explored as follows:

Analysis of probability of each trace may be straight-forward for a very simple model, but the developer needs to ensure that the fundamental principles of statistics are followed closely. Not all architectures lend themselves to a viable solution. The authors worked out certain conditions where an MP model can be used to define a Bayesian belief-network, and then probabilities of each trace could derived from a given set of assumptions [56]. The conditions include the following:

1. Each trace is an instance of the architecture and is represented as a directed graph without loops. This is a given for MP.
2. Each trace has the same topology, such that a consistent Bayesian belief network may be applied to all traces. This is not typical for many models, since the scope tends to change the topology, representing additional iterations. And so the developer needs to maintain a single scope for all traces in the output.
3. Constraints need to be written as conditional probabilities in order to be integrated with the belief network.
4. An approach is defined to prorate probability after multiple constraints are applied to the model. Constraints tend to eliminate large segments of the belief network.
5. Of note, the impact of an overly-constrained model will become apparent when attempting to implement a prorating scheme.

Key point:
Probability calculations may uncover overly-constrained models that inhibit realistic modes of behavior.

A Design Structure Matrix (DSM) can be derived directly from the MP model. The DSM, also called an N-squared diagram, forms a matrix by taking every event along the row and column, and then identifying the interaction of events within the matrix. The simple example DSM is shown in Table 19.4. As outlined by Browning [22] and Eppinger and Browning [30], DSMs have been used to identify spatial requirements, energy flow, information flow, and material flow across systems. These are helpful to the developer

TABLE 19.4

A simple Design Structure Matrix (DSM), derived from the Monterey Phoenix (MP) model as illustrated in Figure 19.7

FROM	TO	A	task_a	B	task_b
		1	2	3	4
A	1	0	3	0	0
task_a	2	0	0	0	3
B	3	0	0	0	2
task_b	4	0	0	0	0

because they support an additional form of review, particularly when the developer may not be aware of a particular behavior in the system. Consider the following activities:

1. Inspection of rare occurrences of interactions between events. These may have slipped the notice of the developer, especially with large data sets.
2. Investigation of common occurrences of desirable behaviors in the model,
3. Comparison of results with different scopes of execution, ensuring that no unexpected events occur.
4. Re-ordering of events so that clusters of common exchanges can be centralized and support system level instantiation during development [22,30].

Key point:
A Design Structure Matrix (DSM) may help the developer to identify negative emergence by investigating rarely exercised interactions in the model.

Criteria: This step is complete when evaluation criteria is calculated or analyzed (e.g., probability of each trace, probability of success, effect of scope on execution, or other parameter).
Decision: Proceed to next step, repeat this step, or go back to previous step.

Proceed in order add detail or complexity to the behavior narrative for another full cycle of the methodology.

Repeat if added detail is needed in the analysis, such as executing the existing model at higher scope in order to identify any unique occurrences in the data.

Go Back if the events, coordination, or constraints need to be revised.

WHEN IS THE MODEL COMPLETE?

The model is complete when requirements are identified in terms of constraints and the model has provided sufficient detail to formulate a simulation of system behavior. Types of models that form the basis of simulation include behavior models, physics-based models, agent-based models, discrete event models, or a combination of these methods.

19.4 Example Problem

The objectives of this example problem are to illustrate that a behavior model, defined in a high level of abstraction, can effectively define essential system behaviors in order to support the investigation of emergence within systems of systems. These behaviors are characterized by event interactions within the system functions, and from system to system. Emergent behaviors are derived from these sources of interaction, and so the developer needs to ensure that only expected behaviors are employed in the system by implementing constraints. Analysis of the results includes deriving every possible outcome and using computational tools to investigate all instances of the output.

19.4.1 Example Description

A generic air vehicle or platform with a behavior of formation flight serves as a means to apply the proposed methodology, stepping through building the MBSE model, MP formulation, evaluation of the results, and evolution of the design architecture. As illustrated in Figure 19.10, a platform or vehicle develops formation flight by deriving its current state (position, speed, and heading), communicates that information with neighboring platforms, derives a navigation command, and then attains a flight position. Formation flight is the desired emergent behavior, but if a failure occurs, the vehicle behavior reduces to solo-flight.

19.4.2 Methodology Applied to the Example Problem

The First Cycle of the Methodology develops the behaviors needed to determine the platform position, speed, and heading, thereby enabling navigation of a single platform.
The Second Cycle of the Methodology develops the behaviors associated with communication between two platforms.
The Third Cycle of the Methodology combines the previous two segments to define the behaviors needed for formation flight.

19.4.3 Segment 1: Single Platform Determination of Current Heading

19.4.3.1 Segment 1 Narrative

This segment develops a model of the platform determination of position, speed, and heading based on single or multiple position and speed sensors. The diagram and coordinate system are illustrated in Figure 19.11. Also as illustrated in Figure 19.12, sensors measure the platform position and velocity, then implement an algorithm to determine position, speed, and heading. Separation of the platform state, measurement sensors, and controller is fundamental to the model.

19.4.3.2 Segment 1 Events

Events included in the State are the actual position and speed of the platform within time and space. The "State" is a root event since it has no predecessor, and the "position" and "speed" are atomic events since they have no events at a lower level of indenture. There is no precedence relationship between the position and speed and so this concurrence nature is captured by using the "{" and "}" characters with the events separated by a comma. Figure 19.13 illustrates these relationships, with the code listed following the figure.

```
ROOT State:
    {
            position,
            speed
    };
```
MP code

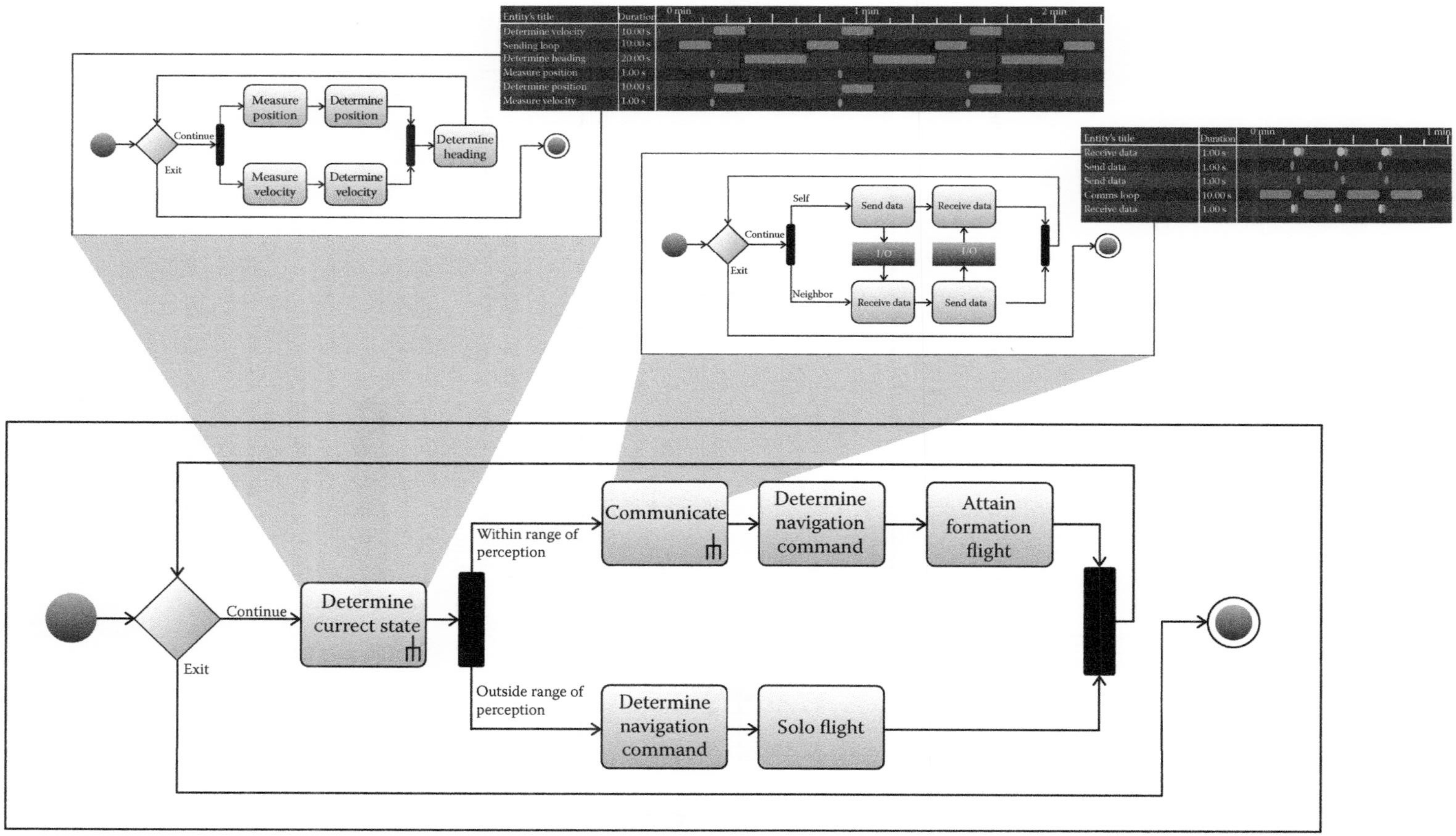

FIGURE 19.10
Formation Flight Activity Diagram showing discrete event simulation for heading determination and communication sub-tasks. Illustration of these activity diagrams and discrete-event simulations were developed in Innoslate®, www.innoslate.com.

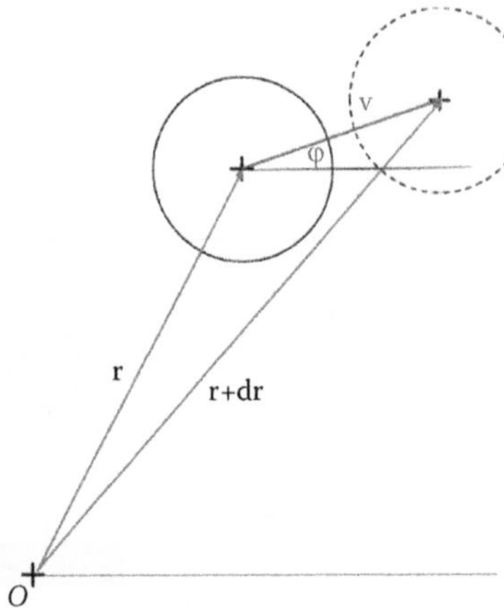

FIGURE 19.11
Coordinate system for a single platform with basic navigation parameters, consistent with that developed by Konle [45].

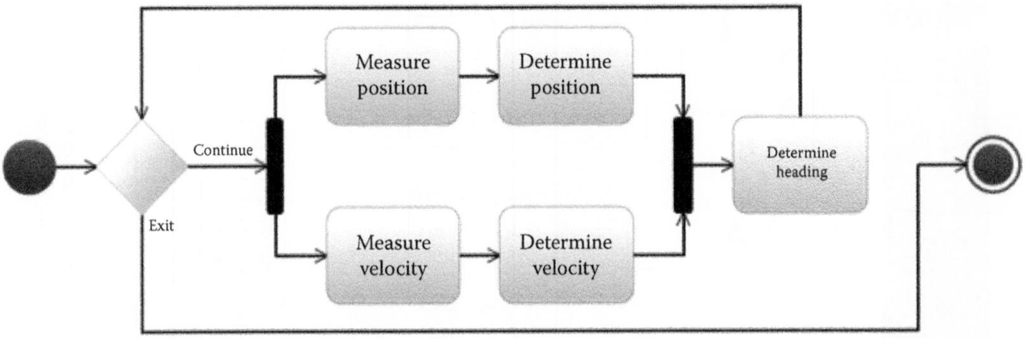

FIGURE 19.12
Determination of the current state of a single platform (e.g., air vehicle), indicating the use of sensors to measure position and velocity, then an algorithm to determine position, speed, and heading. Illustration of this activity diagram was developed in Innoslate®, www.innoslate.com.

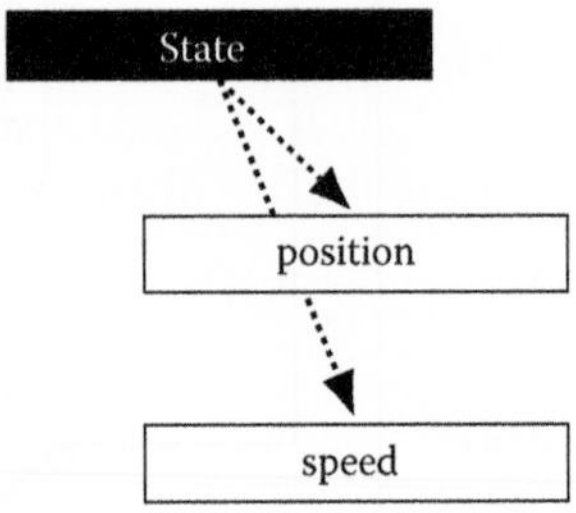

FIGURE 19.13
Events associated with the vehicle state, indicating position and speed.

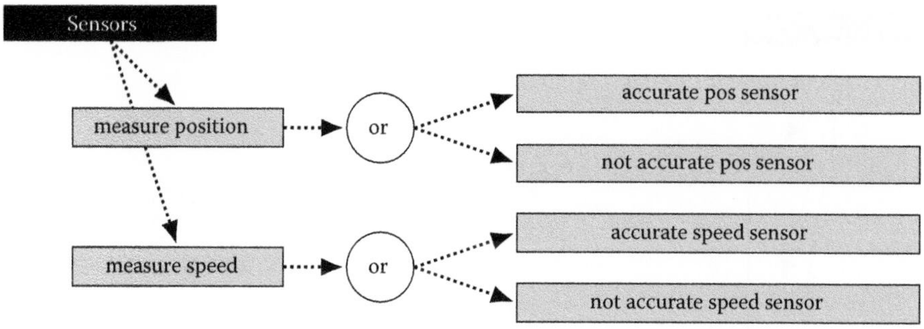

FIGURE 19.14
Sensors of a single platform, including position and speed as direct measurements. Each sensor may be accurate or inaccurate in its measurements.

Events included in the Sensors are the measurement of position and speed. Here, the "Sensors" event is a root event, while the "measure position" and "measure speed" events are composite events since each of these are divided further to account for the attribution of either an "accurate sensor" or a "not accurate" sensor. These accuracy attributes are also composite events composed of either favorable or unfavorable semantics used later in the model derivation. Figure 19.14 illustrates these relationships, following with the code listing. Notice that the alternatives in the code are composed by each alternative in the parenthesis, separated by the pipe character, "|".

```
ROOT Sensors:
       {
           measure_position,
           measure_speed,
       };

measure_position:
           ( accurate_pos_sensor |
             not_accurate_pos_sensor);

measure_speed:
           ( accurate_speed_sensor |
             not_accurate_speed_sensor);
                                        MP code
```

Events included in the Controller are the determination of position, speed, and heading. The "Controller" is the root event, while "determine position," "determine speed," and "determine heading" are composite events. Once again each of these events is followed by an accuracy attribute. Also, each of these attributes is a composite event since it is designated as favorable or unfavorable (see Section 19.4.3). Figure 19.15 illustrates these relationships, following with the code listing. Notice that a precedence relationship exists for both the position and speed determination to occur prior to the heading determination.

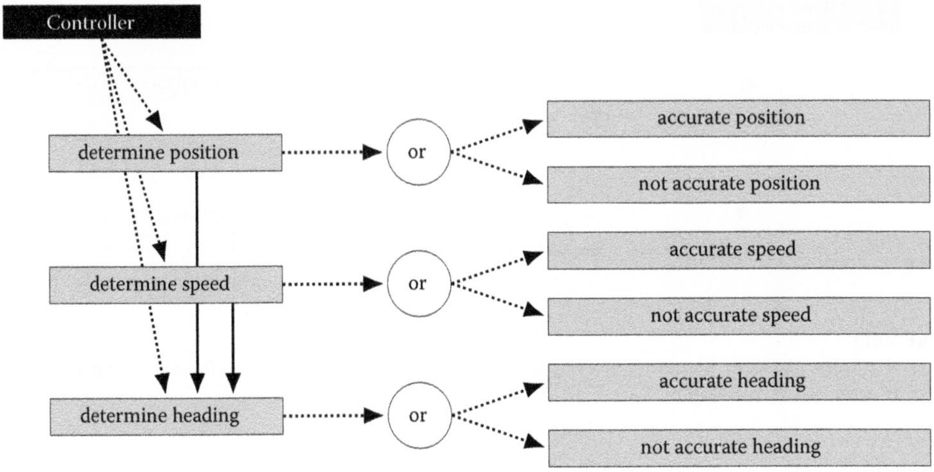

FIGURE 19.15
Events associated with the controller. The controller determines the platform position, speed, and heading. Each calculation by the controller may be accurate or inaccurate. The position and speed are not dependent upon each-other. The heading depends upon the position and speed as indicated by the precedence relationships (solid lines).

```
ROOT Controller:
       {
           determine_position,
           determine_speed
        }
           determine_heading

determine_position:
          ( accurate_position |
               not_accurate_position);

determine_speed:
          ( accurate_speed |
               not_accurate_speed);

determine_heading:
          ( accurate_heading |
               not_accurate_heading);
                                    MP code
```

Events included in the Outcome are the results of either a failed or successful execution of the controller heading command. The result may be considered as the physical enactment of the command, such that the control surfaces and propulsion of the UAV physically respond to the command from the controller. The structure consists of a root event of "Outcome," a composite event "current state," and atomic events of "successful" or "failed," as shown in Figure 19.16.

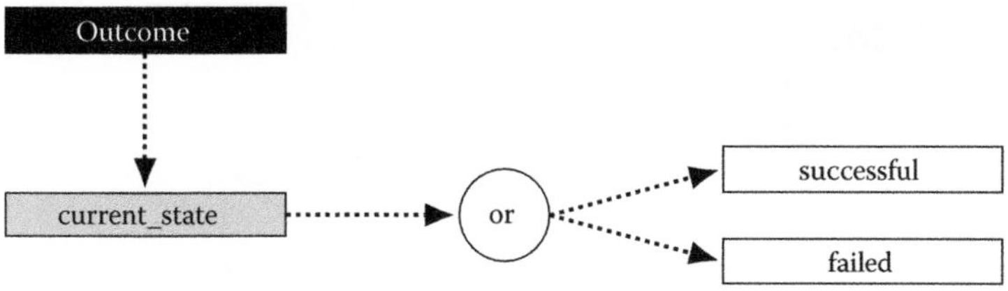

FIGURE 19.16
Events associated with the outcome. The UAV either successfully implements the heading command or fails to implement the command. The Outcome is the root event, and the navigation command is a composite event comprised of a successful or failed atomic event.

```
ROOT Outcome:
      current_state;

current_state:
    ( successful| failed);
```
MP code

19.4.3.3 Segment 1 Coordination

MP defines the interaction of events across the swim-lanes of each root separately from the definition of interactions within the root hierarchy, thereby enabling flexibility for the developer in developing these interactions. These interactions across root events are considered coordination. For our example problem in segment 1, coordination includes interactions from the state to the sensors, from the sensors to the controller, and from the controller to the outcome.

Coordination of the State to the Sensors is established by precedence relationships between the actual position and speed to the measured position and speed, and indicated by solid lines with arrows. Figure 19.17 illustrates these relationships, following with the code listing.

```
COORDINATE $a: position
     DO COORDINATE<!>
            $m: measure_position
     DO ADD $a PRECEDES $m; OD; OD;

COORDINATE $a: speed
     DO COORDINATE<!>
            $m: measure_speed
     DO ADD $a PRECEDES $m; OD; OD;
```
MP code

Coordination of the Sensors to the Controller is established by precedence relationships between the measurement of the position and speed to the determination of position and speed. With only one set of position and speed sensors, this may appear redundant or unnecessary to introduce into the model; however, this structure supports multiple position and

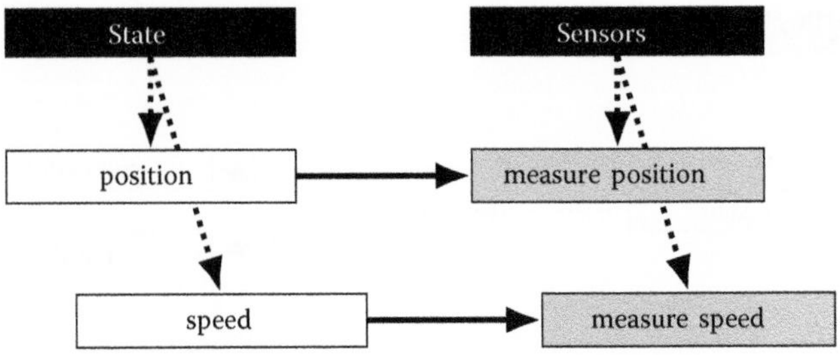

FIGURE 19.17
Interaction 1: Coordination of the state to the sensor measurements of speed and position. The vehicle state (position and speed) precedes the measurement of the state.

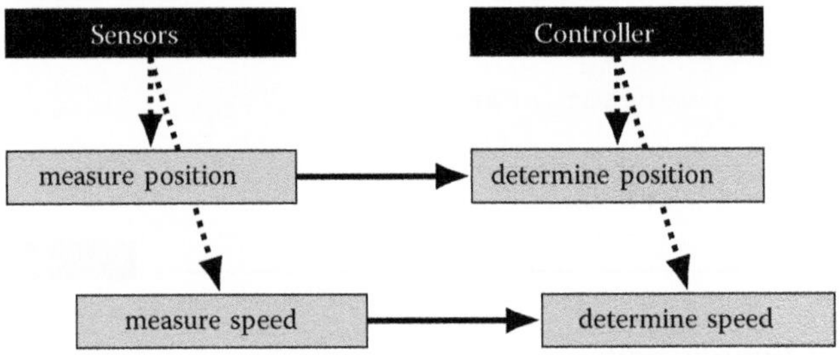

FIGURE 19.18
Interaction 2: Coordination of the sensor measurements of speed and position to the determination of speed and position. The measurement precedes the determination of each property.

speed sensors adapted to the model. Notice that the position and speed processing remains concurrent. Figure 19.18 illustrates these relationships, following with the code listing.

```
COORDINATE $d: determine_position
     DO COORDINATE
            $m: measure_position
         DO ADD $m PRECEDES $d; OD; OD;

COORDINATE $d: determine_speed
     DO COORDINATE
            $m: measure_speed
         DO ADD $m PRECEDES $d; OD; OD;
                                        MP code
```

Coordination of the Controller to the Outcome is established by the determination of heading leading to an outcome identified as a composite event, the current state, with sub-events of successful or a failure as previously described. Figure 19.19 illustrates these relationships, following with the code listing.

```
COORDINATE
    $t: determine_heading FROMC ontroller
  DO COORDINATE
    $o: current_stateFROM Outcome
  DO ADD $t PRECEDES $o  ; OD; OD;
                                        MP code
```

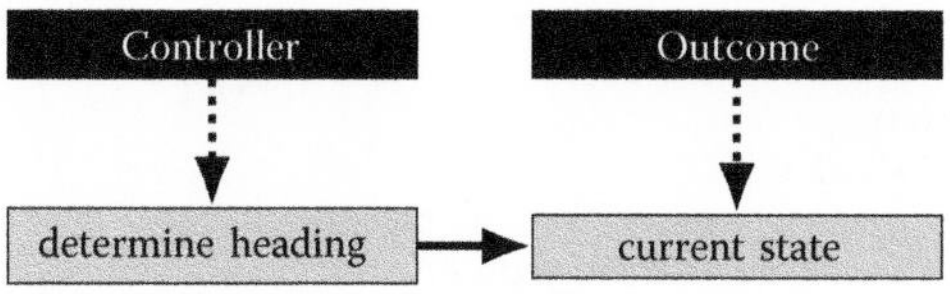

FIGURE 19.19
Interaction 3: Coordination of the controller to the outcome, whether success or failure.

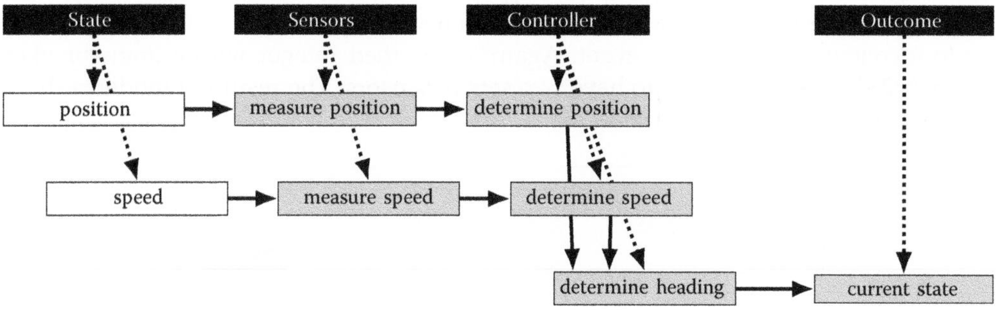

FIGURE 19.20
The basic MP model to determine the position, speed, and heading of the vehicle.

The basic MP model including the behaviors within each root event and the coordination of behaviors between the root events is shown in Figure 19.20.

Adding sample iterations to the basic MP model is easy to achieve within MP. By simply adding the + character before and after the sensor events within parentheses, the maximum number of samples of the position and speed sensors will match the scope of execution. The code listed below also includes a specified integer within angle-brackets, such as "<2>", indicating two samples of each sensor, as illustrated in Figure 19.21.

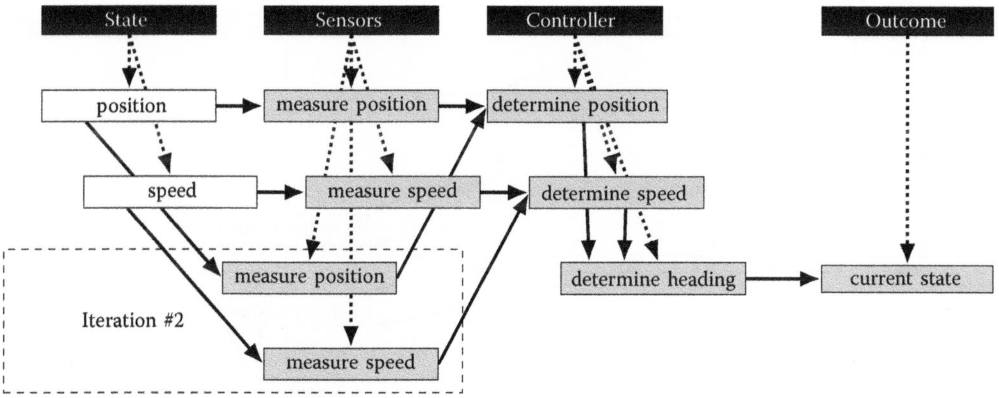

FIGURE 19.21
The basic MP model with two iterations of the sensors.

```
Sensor:
    (+ <2> {
          measure_position,
          measure_speed,
      } +) ;
```
MP code

Adding multiple sensors to the basic MP model is even easier than adding sample iterations. As shown in the following code, changing the root event from "Sensor" to "Sensors" and using the curly brackets "{…}" to indicate concurrency changes each Sensor description to a concurrent composite event. Again, a specified integer within angle-brackets, such as "<2>", causes all traces to have two sets of sensors. The resulting model is shown in Figure 19.22, followed by the code.

```
ROOT Sensors: { + <2>Sensor+ };
```
MP code

19.4.3.4 Segment 1 Constraints

Constraint 1: Determining the heading leads to a successful outcome and the corollary. This is a *simplification-type* of constraint that enables separate event for the outcome of the model.

```
ENSURE ( #accurate_heading     == 1 ->
             #successful == 1);
ENSURE ( #not_accurate_heading== 1 ->
             #failed        == 1);
```
MP code

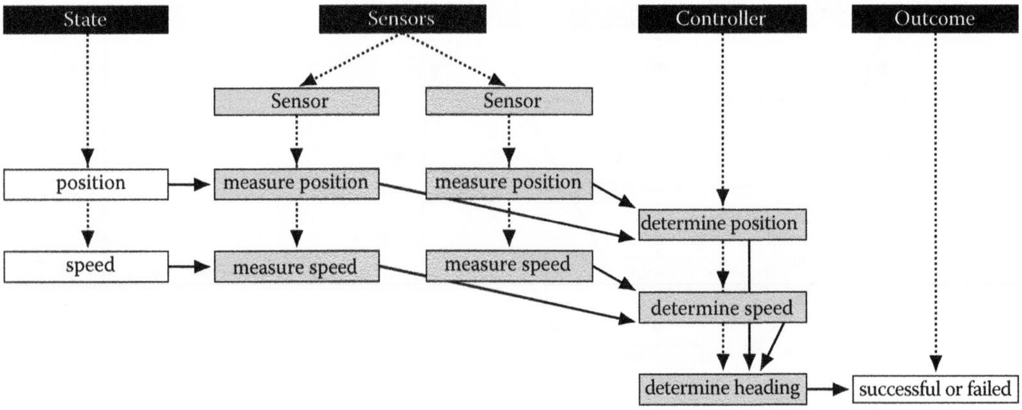

FIGURE 19.22
The basic MP model with two sets of sensors.

Constraint 2: If all position sensors are inaccurate, then the determination of position by the controller is inaccurate, and if all speed sensors are inaccurate, then the determination of speed by the controller is inaccurate. This is a ***logical-type*** of constraint since inaccurate sensors cannot produce an accurate measurement. The following code is listed for the case where two position sensors and two speed sensors are used.

```
ENSURE ( # Sensor *2 ==
            #not_accurate_pos_sensor              ->
                # not_accurate_position == 1 );
ENSURE ( # Sensor *2 -
            #not_accurate_speed_sensor == 0->
                # not_accurate_speed      == 1 );
                                                MP code
```

Constraint 3: At least one accurate measurement of position leads to an accurate determination of the position, and at least one accurate measurement of speed leads to an accurate determination of speed. This is a ***design-type*** of constraint and forms a requirement for the algorithm within the Controller that determines the position and speed of the platform.

```
ENSURE ( #accurate_pos_sensor >= 1 ->
            #accurate_position == 1 );
ENSURE ( #accurate_speed_sensor >=  1 ->
            #accurate_speed    == 1 );
                                          MP code
```

Constraint 4: If both position and speed are determined accurately by the controller, then the algorithm within the controller accurately determines the heading, and the corollary that if either the position or the speed are inaccurate, then the heading is inaccurate. This is a ***design-type*** of constraint leading to a requirement for the controller to consistently produce a correct heading, given accurate speed and position inputs.

```
ENSURE ( #accurate_position +
             #accurate_speed == 2 ->
                   #accurate_heading== 1 );
ENSURE ( #accurate_position +
            #accurate_speed < 2   ->
                   #accurate_heading== 0 );
                                          MP code
```

For the basic model with one set of sensors and one iteration, running the model with no constraints resulted in sixty-four traces, or possible outcomes; conversely, adding four constraints resulted in only four traces after running the model. Figure 19.23 illustrates the effects of constraints for the basic model, and Table 19.5 lists additional results, increasing the number of sensors and number of iterations.

19.4.3.5 Segment 1 Patterns

As demonstrated in the last section, the number of outcomes possible with even simple models becomes exceptionally large. Identification of patterns and assertion checking help the developer to ensure that all aspects of the model have been investigated sufficiently.

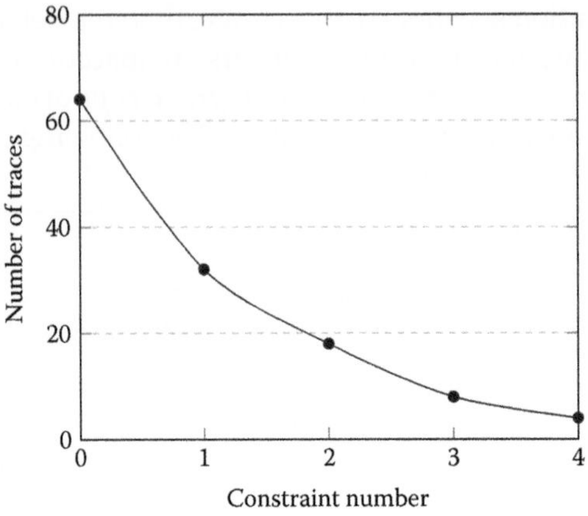

FIGURE 19.23
The effect of constraints on the number of traces derived from the MP model for segment 1, determination of heading.

Four templates were developed to ensure the expected outcomes. These templates, based on the topology of Figure 19.24 and illustrated in Figures 19.25 through 19.28, include all sensors functioning accurately leading to a successful outcome, speed sensor failing, position sensor failing, and both sensors failing. The code for template #1 is listed below. Notice that the "MARK" and "SAY" commands provide immediate feedback to the developer so that all traces can quickly be evaluated. Implementing these templates showed that all traces complied with one and only one template, including those with increased numbers of sensors or numbers of samples.

```
IF EXISTS
/*     from Sensors     */
    $S1: accurate_pos_sensor,
    $S2: accurate_speed_sensor
/*     from Controller     */
    $C1: accurate_position
    $C2: accurate_speed
    $C3: accurate_heading,
/*     from Outcome     */
    $O1: successful
    (
    $S1 BEFORE $C1
    AND
    $S2 BEFORE $C2
    AND
    $C1 BEFORE $C3
    AND
    $C2 BEFORE $C3
    AND
    $C3 BEFORE $O1
    )
THEN
    MARK;
    SAY("T1: Favorable measurements");
FI;
                                        MP code
```

TABLE 19.5

Number of Traces without and with Constraints for Multiple Sensors and Multiple Samples of Each Sensor

Number of Sensors (position and speed)	Zero Constraints		Active Constraints	
	One Data Sample	Two Data Sample	One Data Sample	Two Data Samples
1	64	256	14	16
2	256	4,069	16	256
3	1024	65,536	64	4,096

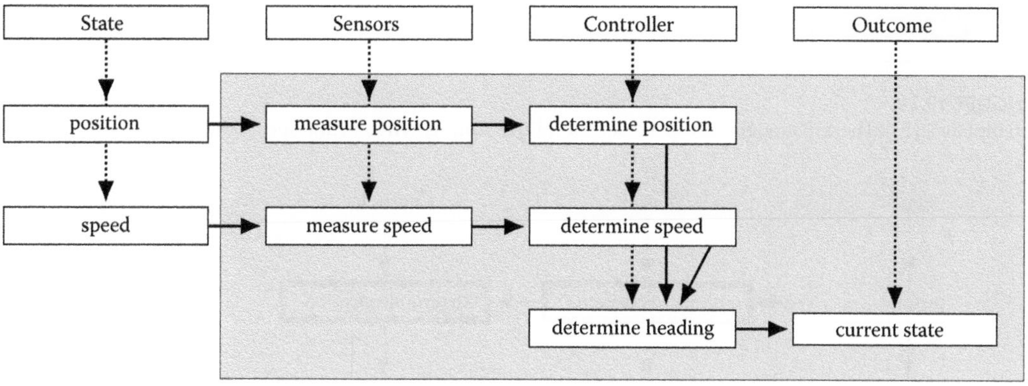

FIGURE 19.24
The basic MP model topology. Notice that the topology of the model remains consistent for all traces.

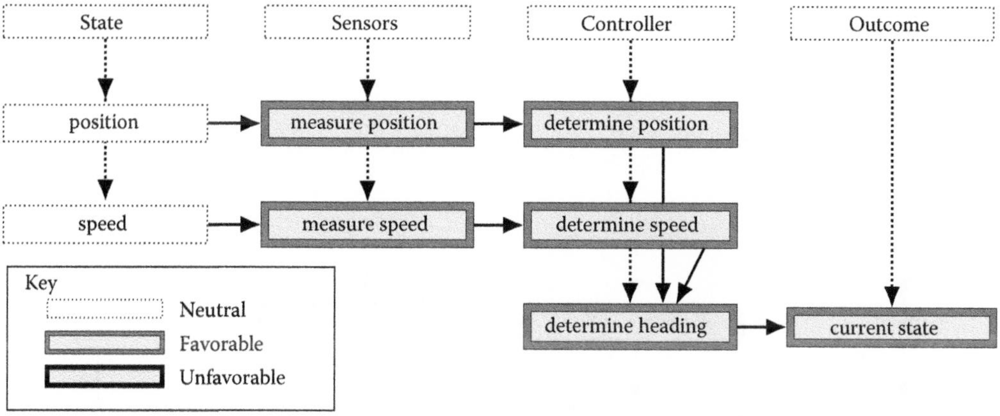

FIGURE 19.25
Template 1 (T1): The MP model with a successful outcome, determining position, speed, and the current state.

Key point:
Semantics added to the topology enable an algorithm within MP to search for and identify behavior patterns.

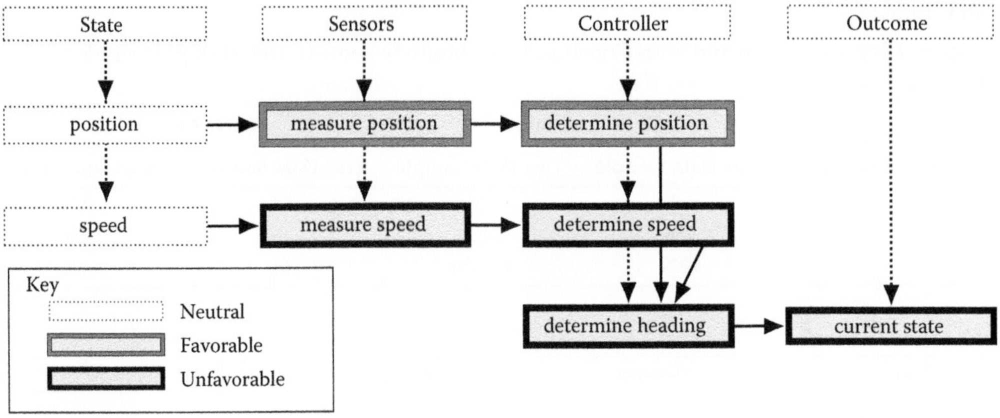

FIGURE 19.26
Template 2 (T2): The MP model with speed sensor failure, failing to determine speed and heading.

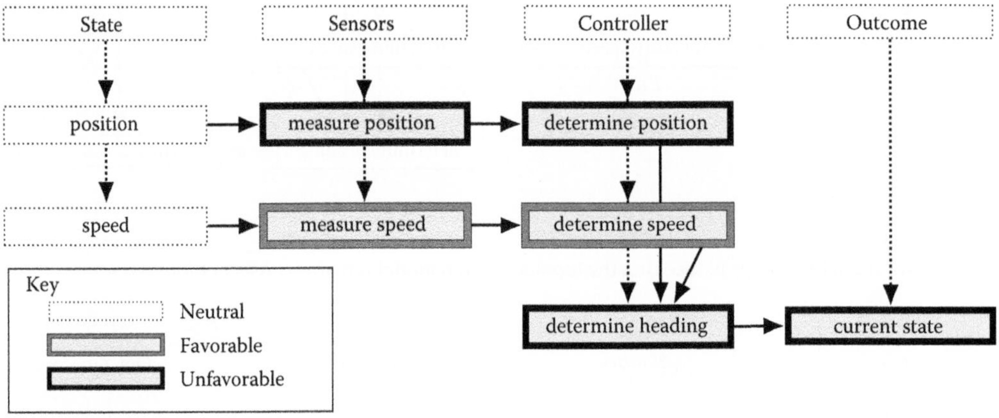

FIGURE 19.27
Template 3 (T3): The MP model with position sensor failure, failing to determine position and heading.

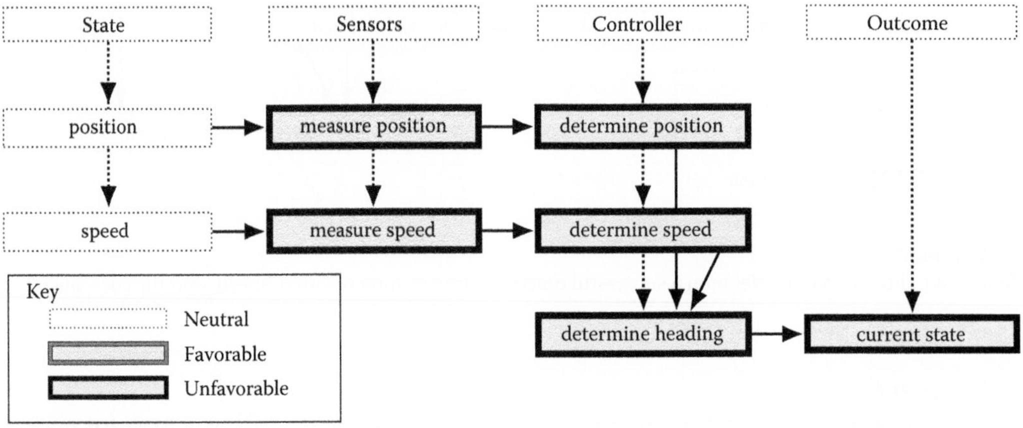

FIGURE 19.28
Template 4 (T4): The MP model with all sensor failure.

The following code adds attributes of "favorable" and "unfavorable" to the sensors and controller events. These semantics added to the model topology assist the developer to immediately recognize visual patterns that are also realized in MP code.

```
accurate_pos_sensor : favorable ;
not_accurate_pos_sensor : unfavorable ;

accurate_speed_sensor : favorable ;
not_accurate_speed_sensor : unfavorable ;

accurate_position : favorable ;
not_accurate_position : unfavorable ;

accurate_speed : favorable ;
not_accurate_speed : unfavorable ;

accurate_heading : favorable ;
not_accurate_heading : unfavorable ;
                                        MP code
```

19.4.3.6 Segment 1 Evaluation

Calculation of probability of successfully determining the platform state, including the position, speed, and heading is derived in this section. The MP model produces a directed graph without loops, and so this topology directly supports a belief network, as illustrated in Figures 19.29 and 19.30.

Table 19.6 outlines the definitions of the indices, consistent with Figure 19.30. The constraints limit the number of valid traces from sixteen to just four, as listed in Table 19.7. Furthermore, because of the constraints, the valid traces, those with probability greater than zero, have a probability of one for events C, D, E, and F, also listed in Table 19.8.

$$P\left(c_k \middle| a_i^{(1)}, a_q^{(2)}\right) = P\left(d_m \middle| b_j^1, b_r^2\right) = P\left(e_n \middle| c_k, d_m\right) = P\left(f_p \middle| e_n\right) = 1 \tag{19.1}$$

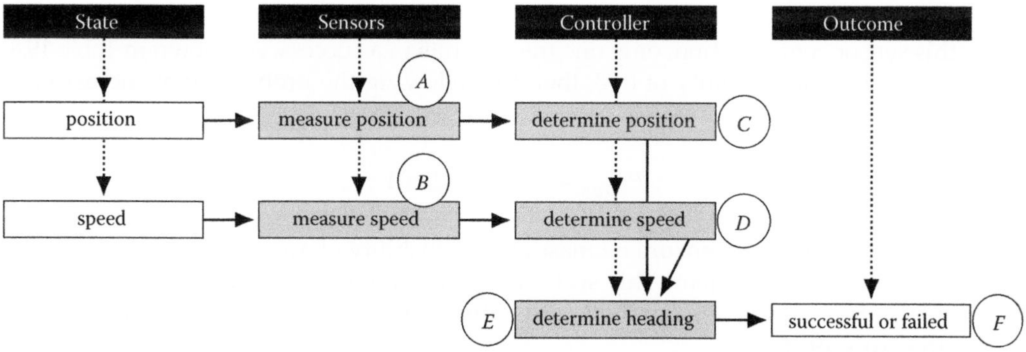

FIGURE 19.29
The relationship to a probability network is straightforward since the MP model produces a directed graph without loops. The nodes A through F, representing MP events, form a Bayesian-belief network. The constraints add conditional probabilities to the belief network.

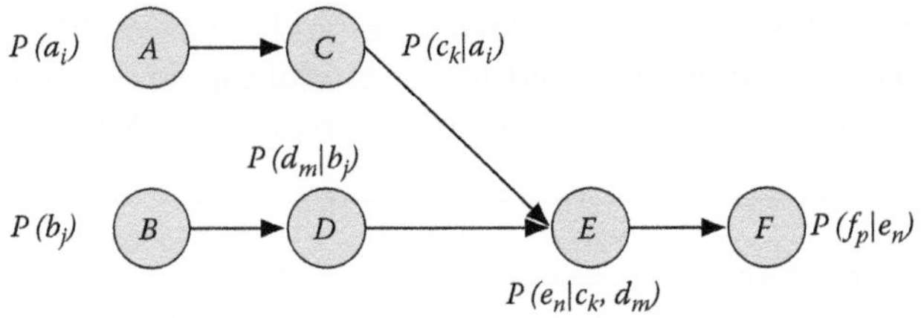

FIGURE 19.30
Bayesian belief network derivation from the MP model, showing independent probabilities for the sensors in nodes A and B and conditional probabilities for nodes C through F.

TABLE 19.6
Belief Network Indices for the Heading Determination Model, with One Set of Sensors

Node	P	Index
A. measure_position:	$P(a_i)$	$i \in \{1, 2\}$ $i = 1$: accurate $i = 2$: not accurate
B. measure_speed:	$P(b_j)$	$j \in \{1,2\}$ $j = 1$: accurate $j = 2$: not accurate
C. determine_position:	$P(c_k)$	$k \in \{1, 2\}$ $k = 1$: accurate $k = 2$: not accurate
D. determine_speed:	$P(d_m)$	$m \in \{1, 2\}$ $m = 1$: accurate $m = 2$: not accurate
E. determine_heading:	$P(e_n)$	$\text{n} \in \{1, 2\}$ $n = 1$: accurate $n = 2$: not accurate
F. navigation_command:	$P(f_p)$	$p \in \{1,2\}$ $p = 1$: successful $p = 2$: failed

For this sensor configuration, only one trace results in a success. As shown in Table 19.8, if all sensors have a reliability of 0.98, then the following the probability of success is as follows:

$$P_{\text{success}} = P_{\text{trace\_1}} = 0.9604 \tag{19.2}$$

Adding a second set of sensors increases the probability of accurately determining the desired heading of the system. This section outlines the derivation of the probability of success with these added sensors. Table 19.9 outlines the definitions of the indices, consistent with Figure 19.31.

As with the single set of sensors, the constraints limit the number of valid traces from 256 to 16, as listed in Table 19.10. Furthermore, because of the constraints, the valid traces,

TABLE 19.7

Constraints Written in Terms of Conditional Probabilities

Constraint 1:	***If*** the heading is accurate, $n = 1$, ***then*** have a successful outcome, $p = 1$, and the corollary ***if*** the heading is not accurate $n = 2$, ***then*** have a failed outcome, $p = 2$. $P(f_{p=1} \mid e_{n=1}) = 1, P(f_{p=2} \mid e_{n=1}) = 0$ $P(f_{p=1} \mid e_{n=2}) = 0, P(f_{p=2} \mid e_{n=2}) = 1$
Constraint 2:	***If all*** position sensor(s) are not accurate, $i = 2$, ***then*** the position determination is not accurate, $k = 2$, and ***if all*** speed sensor(s) are not accurate $j = 2$, ***then*** the speed determination is not accurate, $m = 2$. $P(c_{k=1} \mid a_{i=2}) = 0, P(c_{k=2} \mid a_{i=2}) = 1$ $P(d_{m=1} \mid b_{j=2}) = 0, P(d_{m=2} \mid b_{j=2}) = 1$
Constraint 3:	***If any*** of the position sensor(s) are accurate, $i = 1$, ***then*** the position determination is accurate, $k = 1$, and ***if any*** of the speed sensor(s) are accurate $j = 1$, ***then*** the speed determination is accurate, $m = 1$. $P(c_{k=1} \mid a_{i=1}) = 1, P(c_{k=2} \mid a_{i=1}) = 0$ $P(d_{m=1} \mid b_{j=1}) = 1, P(d_{m=2} \mid b_{j=1}) = 0$
Constraint 4:	***If both*** the position and speed determinations are accurate,$k = 1$ and $m = 1$, ***then*** the heading determination is accurate, $n = 1$, and the corollary ***if either*** the position or speed determination is not accurate $k = 2$ or $m = 2$, ***then*** the heading determination is not accurate, $n = 2$. $P(e_{n=1} \mid c_{k=1}, d_{m=1}) = 1, P(e_{n=2} \mid c_{k=1}, d_{m=1}) = 0$ $P(e_{n=1} \mid c_{k=2}, d_{m=2}) = 2, P(e_{n=2} \mid c_{k=2}, d_{m=2}) = 1$ $P(e_{n=1} \mid c_{k=1}, d_{m=2}) = 2, P(e_{n=2} \mid c_{k=1}, d_{m=2}) = 1$ $P(e_{n=1} \mid c_{k=2}, d_{m=1}) = 2, P(e_{n=2} \mid c_{k=2}, d_{m=1}) = 1$

TABLE 19.8

Belief Network Results For the Heading Determination Model, with One Set of Sensors

Trace	Template	P_{trace}	$P(a_i)$	$P(b_j)$	$P(c_k \mid a_i)$	$P(d_m \mid b_j)$	$P(e_n \mid c_k, d_m)$	$P(f_p \mid e_n)$
1	T1	0.9604	$P(a_{i=1}) = 0.98$	$P(b_{j=1}) = 0.98$	$P(c_{k=1}) = 1$	$P(d_{m=1}) = 1$	$P(e_{n=1}) = 1$	$P(f_{p=1}) = 1$
2	T2	0.0196	$P(a_{i=1}) = 0.98$	$P(b_{j=2}) = 0.02$	$P(c_{k=1}) = 1$	$P(d_{m=2}) = 1$	$P(e_{n=2}) = 1$	$P(f_{p=2}) = 1$
3	T3	0.0196	$P(a_{i=2}) = 0.02$	$P(b_{j=1}) = 0.98$	$P(c_{k=2}) = 1$	$P(d_{m=1}) = 1$	$P(e_{n=2}) = 1$	$P(f_{p=2}) = 1$
4	T4	0.0004	$P(a_{i=2}) = 0.02$	$P(b_{j=2}) = 0.02$	$P(c_{k=2}) = 1$	$P(d_{m=2}) = 1$	$P(e_{n=2}) = 1$	$P(f_{p=2}) = 1$

those with probability greater than zero, have a probability of one for events C, D, E, and F, also listed in Table 19.10.

$$P\left(c_k \middle| a_i^{(1)}, a_q^{(2)}\right) = P\left(d_m \middle| b_j^1, b_r^2\right) = P\left(e_n \middle| c_k, d_m\right) = P\left(f_p \middle| e_n\right) = 1 \tag{19.3}$$

Successful traces are those that have at least one accurate position sensor and one accurate speed sensor so that the position and speed are determined to be accurate, thereby satisfying Template #1. This condition is enforced by Constraint #3, defined in Section 19.4.3.

Let N be the number of traces that satisfy Template 1, then from Table 19.10, we find that nine traces satisfy this restriction, namely traces 1, 2, 3, 5, 6, 7, 9, 10, and 11. Thus the

TABLE 19.9

Belief Network Indices for the Heading Determination Model, with Two Sets of Sensors

Node	P	Index
$A^{(1)}$: measure_position, sensor 1:	$P\left(a_i^{(1)}\right)$	$i \in \{1, 2\}$ $i = 1$: accurate $i = 2$: not accurate
$A^{(2)}$: measure_position, sensor 2:	$P\left(a_q^{(2)}\right)$	$q \in \{1, 2\}$ $q = 1$: accurate $q = 2$: not accurate
$B^{(1)}$: measure_speed, sensor 1:	$P\left(b_j^{(1)}\right)$	$j \in \{1,2\}$ $j = 1$: accurate $j = 2$: not accurate
$B^{(2)}$: measure_speed, sensor 2:	$P\left(b_r^{(2)}\right)$	$r \in \{1, 2\}$ $r = 1$: accurate $r = 2$: not accurate
C: determine_position:	$P(c_k)$	$k \in \{1, 2\}$ $k = 1$: accurate $k = 2$: not accurate
D: determine_speed:	$P(d_m)$	$m \in \{1,2\}$ $m = 1$: accurate $m = 2$: not accurate
E: determine_heading:	$P(e_n)$	$n \in \{1, 2\}$ $n = 1$: accurate $n = 2$: not accurate
F: navigation_command:	$P(f_p)$	$p \in \{1, 2\}$ $p = 1$: successful $p = 2$: failed

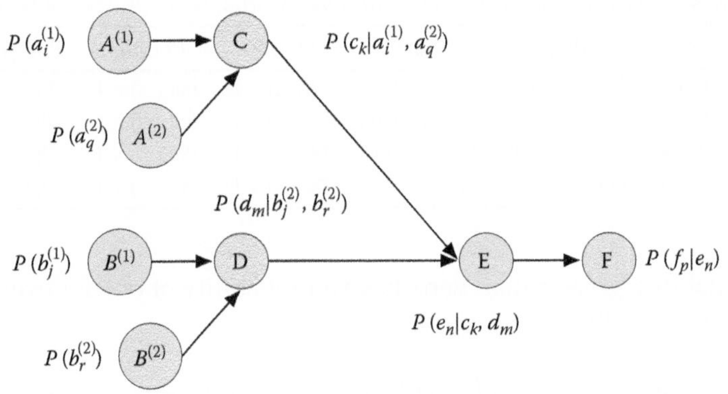

FIGURE 19.31
Bayesian belief network derivation from the decision model, indicating conditional probabilities for each node.

probability of successfully determining the heading of the platform is found by summing these individual trace probabilities, as follows:

$$P_{success} = \sum_{n=1}^{N} P_{trace\_n} = 0.9992 \tag{19.4}$$

TABLE 19.10

Belief Network Results for the Heading Determination Model, with two Sets of Sensors

Trace	Template	P_{trace}	$P\left(a_i^{(1)}\right)$	$P\left(a_q^{(2)}\right)$	$P\left(b_j^{(1)}\right)$	$P\left(b_r^{(2)}\right)$
1	T1	0.9223682	$P\left(a_{i=1}^{(1)}\right)=0.98$	$P\left(a_{q=1}^{(2)}\right)=0.98$	$P\left(b_{j=1}^{(1)}\right)=0.98$	$P\left(b_{r=1}^{(2)}\right)=0.98$
2	T1	0.0188238	$P\left(a_{i=1}^{(1)}\right)=0.98$	$P\left(a_{q=1}^{(2)}\right)=0.98$	$P\left(b_{j=1}^{(1)}\right)=0.98$	$P\left(b_{r=2}^{(2)}\right)=0.02$
3	T1	0.0188238	$P\left(a_{i=1}^{(1)}\right)=0.98$	$P\left(a_{q=1}^{(2)}\right)=0.98$	$P\left(b_{j=2}^{(1)}\right)=0.02$	$P\left(b_{r=1}^{(2)}\right)=0.98$
4	T2	0.0003842	$P\left(a_{i=1}^{(1)}\right)=0.98$	$P\left(a_{q=1}^{(2)}\right)=0.98$	$P\left(b_{j=2}^{(1)}\right)=0.02$	$P\left(b_{r=2}^{(2)}\right)=0.02$
5	T1	0.0188238	$P\left(a_{i=1}^{(1)}\right)=0.98$	$P\left(a_{q=2}^{(2)}\right)=0.02$	$P\left(b_{j=1}^{(1)}\right)=0.98$	$P\left(b_{r=1}^{(2)}\right)=0.98$
6	T1	0.0003842	$P\left(a_{i=1}^{(1)}\right)=0.98$	$P\left(a_{q=2}^{(2)}\right)=0.02$	$P\left(b_{j=1}^{(1)}\right)=0.98$	$P\left(b_{r=2}^{(2)}\right)=0.02$
7	T1	0.0003842	$P\left(a_{i=1}^{(1)}\right)=0.98$	$P\left(a_{q=2}^{(2)}\right)=0.02$	$P\left(b_{j=2}^{(1)}\right)=0.02$	$P\left(b_{r=1}^{(2)}\right)=0.98$
8	T2	0.0000078	$P\left(a_{i=1}^{(1)}\right)=0.98$	$P\left(a_{q=2}^{(2)}\right)=0.02$	$P\left(b_{j=2}^{(1)}\right)=0.02$	$P\left(b_{r=2}^{(2)}\right)=0.02$
9	T1	0.0188238	$P\left(a_{i=2}^{(1)}\right)=0.02$	$P\left(a_{q=1}^{(2)}\right)=0.98$	$P\left(b_{j=1}^{(1)}\right)=0.98$	$P\left(b_{r=1}^{(2)}\right)=0.98$
10	T1	0.0003842	$P\left(a_{i=2}^{(1)}\right)=0.02$	$P\left(a_{q=1}^{(2)}\right)=0.98$	$P\left(b_{j=1}^{(1)}\right)=0.98$	$P\left(b_{r=2}^{(2)}\right)=0.02$
11	T1	0.0003842	$P\left(a_{i=2}^{(1)}\right)=0.02$	$P\left(a_{q=1}^{(2)}\right)=0.98$	$P\left(b_{j=2}^{(1)}\right)=0.02$	$P\left(b_{r=1}^{(2)}\right)=0.98$
12	T2	0.0000078	$P\left(a_{i=2}^{(1)}\right)=0.02$	$P\left(a_{q=1}^{(2)}\right)=0.98$	$P\left(b_{j=2}^{(1)}\right)=0.02$	$P\left(b_{r=2}^{(2)}\right)=0.02$
13	T3	0.0003842	$P\left(a_{i=2}^{(1)}\right)=0.02$	$P\left(a_{q=2}^{(2)}\right)=0.02$	$P\left(b_{j=1}^{(1)}\right)=0.98$	$P\left(b_{r=1}^{(2)}\right)=0.98$
14	T3	0.0000078	$P\left(a_{i=2}^{(1)}\right)=0.02$	$P\left(a_{q=2}^{(2)}\right)=0.02$	$P\left(b_{j=1}^{(1)}\right)=0.98$	$P\left(b_{r=2}^{(2)}\right)=0.02$
15	T3	0.0000078	$P\left(a_{i=2}^{(1)}\right)=0.02$	$P\left(a_{q=2}^{(2)}\right)=0.02$	$P\left(b_{j=2}^{(1)}\right)=0.02$	$P\left(b_{r=1}^{(2)}\right)=0.98$
16	T4	0.0000002	$P\left(a_{i=2}^{(1)}\right)=0.02$	$P\left(a_{q=2}^{(2)}\right)=0.02$	$P\left(b_{j=2}^{(1)}\right)=0.02$	$P\left(b_{r=2}^{(2)}\right)=0.02$

The model readily supports analysis through investigation of a Design Structure Matrix (DSM), as shown in Tables 19.11 and 19.12. Each event associated with the model is listed on both the rows and columns of the matrices. Each instance of an interaction that is tabulated in the corresponding block is updated. Direction of the interaction that is captured from the set of rows to the columns. Simple inspection of these tables indicates that every relationship found with two sets of sensors also exists with a single set of sensors. As an example, look at the row that reads "determine position" to the column "determine heading" four times for the single set of sensors, and sixteen times for the dual set of sensors. Since no new information occurs with increasing the scope, or number of sensors, the developer can have confidence that the model has a stable response.

19.4.4 Segment 2: Communication among Multiple Platforms

19.4.4.1 Segment 2 Narrative

As illustrated in Figures 19.32 and 19.33, the communication function depends significantly upon the relative position of both platforms. The model takes this range into account by enabling successful communication only when this condition is satisfied. The concepts of "self" and "neighbor" are introduced in this segment, enabling the model of multiple platforms.

TABLE 19.11

Design Structure Matrix (DSM) Summarizing all Four Traces for the Single Set of Position and Speed Sensors

From	To	State	Position	Speed	Sensors	Sensor	measure_position	accurate_pos_sensor	measure_speed	accurate_speed_sensor	Controller	determine_position	accurate_position	determine_speed	accurate_speed	determine_heading	accurate_heading	Outcome	Successful	not_accurate_speed_sensor	not_accurate_speed	not_accurate_heading	Failed	not_accurate_pos_sensor	not_accurate_position
		1	2	3	4	5	6	7	8	9	10	11	12	13	14	15	16	17	18	19	20	21	22	23	24
State	1	0	4	4	0	0	0	0	0	0	0	0	0	0	0	0	0	0	0	0	0	0	0	0	0
Position	2	0	0	0	0	0	4	0	0	0	0	0	0	0	0	0	0	0	0	0	0	0	0	0	0
Speed	3	0	0	0	0	0	0	0	4	0	0	0	0	0	0	0	0	0	0	0	0	0	0	0	0
Sensors	4	0	0	0	0	4	0	0	0	0	0	0	0	0	0	0	0	0	0	0	0	0	0	0	0
Sensor	5	0	0	0	0	0	4	0	4	0	0	0	0	0	0	0	0	0	0	0	0	0	0	0	0
measure_position	6	0	0	0	0	0	0	2	0	0	0	4	0	0	0	0	0	0	0	0	0	0	0	2	0
accurate_pos_sensor	7	0	0	0	0	0	0	0	0	0	0	0	0	0	0	0	0	0	0	0	0	0	0	0	0
measure_speed	8	0	0	0	0	0	0	0	0	2	0	0	0	4	0	0	0	0	0	2	0	0	0	0	0
accurate_speed_sensor	9	0	0	0	0	0	0	0	0	0	0	0	0	0	0	0	0	0	0	0	0	0	0	0	0
Controller	10	0	0	0	0	0	0	0	0	0	0	4	0	4	0	4	0	0	0	0	0	0	0	0	0
determine_position	11	0	0	0	0	0	0	0	0	0	0	0	2	0	0	4	0	0	0	0	0	0	0	0	2
accurate_position	12	0	0	0	0	0	0	0	0	0	0	0	0	0	0	0	0	0	0	0	0	0	0	0	0
determine_speed	13	0	0	0	0	0	0	0	0	0	0	0	0	0	2	4	0	0	0	0	2	0	0	0	0
accurate_speed	14	0	0	0	0	0	0	0	0	0	0	0	0	0	0	0	0	0	0	0	0	0	0	0	0
determine_heading	15	0	0	0	0	0	0	0	0	0	0	0	0	0	0	0	1	0	1	0	0	3	3	0	0
accurate_heading	16	0	0	0	0	0	0	0	0	0	0	0	0	0	0	0	0	0	0	0	0	0	0	0	0
Outcome	17	0	0	0	0	0	0	0	0	0	0	0	0	0	0	0	0	0	1	0	0	0	3	0	0
Successful	18	0	0	0	0	0	0	0	0	0	0	0	0	0	0	0	0	0	0	0	0	0	0	0	0
not_accurate_speed_sensor	19	0	0	0	0	0	0	0	0	0	0	0	0	0	0	0	0	0	0	0	0	0	0	0	0
not_accurate_speed	20	0	0	0	0	0	0	0	0	0	0	0	0	0	0	0	0	0	0	0	0	0	0	0	0
not_accurate_heading	21	0	0	0	0	0	0	0	0	0	0	0	0	0	0	0	0	0	0	0	0	0	0	0	0
Failed	22	0	0	0	0	0	0	0	0	0	0	0	0	0	0	0	0	0	0	0	0	0	0	0	0
not_accurate_pos_sensor	23	0	0	0	0	0	0	0	0	0	0	0	0	0	0	0	0	0	0	0	0	0	0	0	0
not_accurate_position	24	0	0	0	0	0	0	0	0	0	0	0	0	0	0	0	0	0	0	0	0	0	0	0	0

TABLE 19.12

Design Structure Matrix (DSM) Summarizing All Sixteen Traces for Two Sets of Position and Speed Sensors

	To	State	Position	Speed	Sensors	Sensor	measure_position	accurate_pos_sensor	measure_speed	accurate_speed_sensor	Controller	determine_position	accurate_position	determine_speed	accurate_speed	determine_heading	accurate_heading	Outcome	Successful	not_accurate_speed_sensor	not_accurate_speed	not_accurate_heading	Failed	not_accurate_pos_sensor	not_accurate_position
From		1	2	3	4	5	6	7	8	9	10	11	12	13	14	15	16	17	18	19	20	21	22	23	24
State	1	0	16	16	0	0	0	0	0	0	0	0	0	0	0	0	0	0	0	0	0	0	0	0	0
Position	2	0	0	0	0	0	32	0	0	0	0	0	0	0	0	0	0	0	0	0	0	0	0	0	0
Speed	3	0	0	0	0	0	0	0	32	0	0	0	0	0	0	0	0	0	0	0	0	0	0	0	0
Sensors	4	0	0	0	0	32	0	0	0	0	0	0	0	0	0	0	0	0	0	0	0	0	0	0	0
Sensor	5	0	0	0	0	0	32	0	32	0	0	0	0	0	0	0	0	0	0	0	0	0	0	0	0
measure_position	6	0	0	0	0	0	0	16	0	0	0	32	0	0	0	0	0	0	0	0	0	0	0	16	0
accurate_pos_sensor	7	0	0	0	0	0	0	0	0	0	0	0	0	0	0	0	0	0	0	0	0	0	0	0	0
measure_speed	8	0	0	0	0	0	0	0	0	16	0	0	0	32	0	0	0	0	0	16	0	0	0	0	0
accurate_speed_sensor	9	0	0	0	0	0	0	0	0	0	0	0	0	0	0	0	0	0	0	0	0	0	0	0	0
Controller	10	0	0	0	0	0	0	0	0	0	0	16	0	16	0	16	0	0	0	0	0	0	0	0	0
determine_position	11	0	0	0	0	0	0	0	0	0	0	0	12	0	0	16	0	0	0	0	0	0	0	0	4
accurate_position	12	0	0	0	0	0	0	0	0	0	0	0	0	0	0	0	0	0	0	0	0	0	0	0	0
determine_speed	13	0	0	0	0	0	0	0	0	0	0	0	0	0	12	16	0	0	0	0	4	0	0	0	0
accurate_speed	14	0	0	0	0	0	0	0	0	0	0	0	0	0	0	0	0	0	0	0	0	0	0	0	0
determine_heading	15	0	0	0	0	0	0	0	0	0	0	0	0	0	0	0	9	0	9	0	0	7	7	0	0
accurate_heading	16	0	0	0	0	0	0	0	0	0	0	0	0	0	0	0	0	0	0	0	0	0	0	0	0
Outcome	17	0	0	0	0	0	0	0	0	0	0	0	0	0	0	0	0	0	9	0	0	0	7	0	0
Successful	18	0	0	0	0	0	0	0	0	0	0	0	0	0	0	0	0	0	0	0	0	0	0	0	0
not_accurate_speed_sensor	19	0	0	0	0	0	0	0	0	0	0	0	0	0	0	0	0	0	0	0	0	0	0	0	0
not_accurate_speed	20	0	0	0	0	0	0	0	0	0	0	0	0	0	0	0	0	0	0	0	0	0	0	0	0
not_accurate_heading	21	0	0	0	0	0	0	0	0	0	0	0	0	0	0	0	0	0	0	0	0	0	0	0	0
Failed	22	0	0	0	0	0	0	0	0	0	0	0	0	0	0	0	0	0	0	0	0	0	0	0	0
not_accurate_pos_sensor	23	0	0	0	0	0	0	0	0	0	0	0	0	0	0	0	0	0	0	0	0	0	0	0	0
not_accurate_position	24	0	0	0	0	0	0	0	0	0	0	0	0	0	0	0	0	0	0	0	0	0	0	0	0

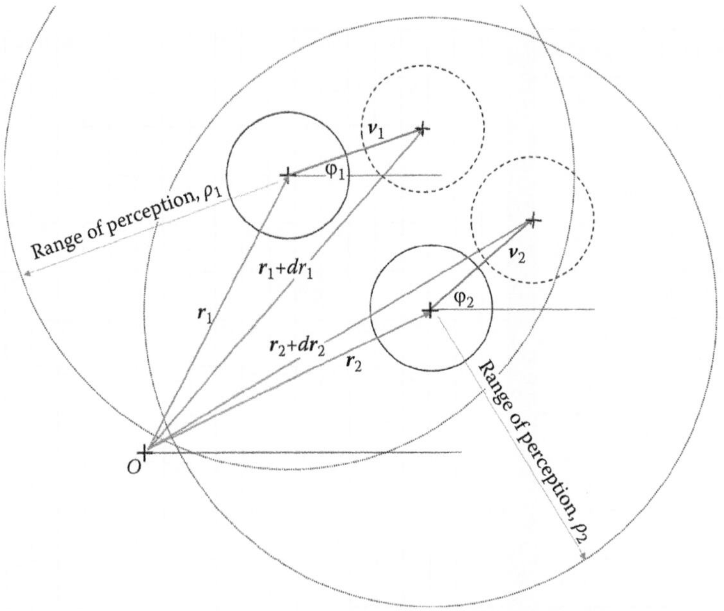

FIGURE 19.32
Two platforms illustrated within the range of perception of each platform. These platforms are now able to employ a formation relative to one another when simultaneously within this range.

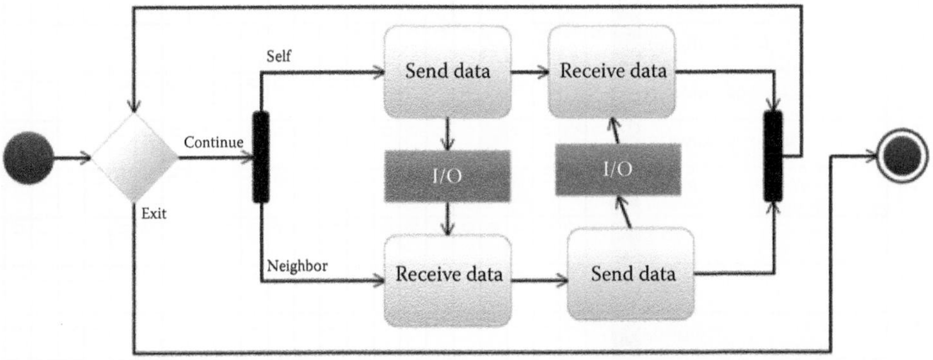

FIGURE 19.33
Communication activity diagram. Illustration of this activity diagram was developed in Innoslate®, www.inno-slate.com.

19.4.4.2 Segment 2 Events

The root events for this segment include the Environment, Self, and Neighbor. The Environment includes the range, or distance between the platforms, which may be either within range of the communication device (e.g., radio), or not within the range of the communication device. Each platform (Self and Neighbor) has zero or more occurrences, as indicated by the asterisk character.

```
SCHEMA platformCommunication

ROOT Environment:
        range;
range:
    (   within_range| not_within_range );

ROOT Self:
{ ( * send *) ,
  ( * receive * ) };

ROOT Neighbor:
 { ( * send *) ,
   ( * receive * ) };
```
MP code

19.4.4.3 Segment 2 Coordination

Two types of interactions occur between the environment and the platforms (self and neighbor) and also between the platforms. Both establish precedence relationships as follows:

Interaction 1: The range of perception falling within a limit is a pre-cursor to the ability of the platform to receive a communication. This establishes a precedence relationship as follows:

```
COORDINATE $e: within_range
    DO COORDINATE
            $r: receive
    DO ADD $e PRECEDES $r; OD; OD;
```
MP code

Interaction 2: A precedence relationship also exists between send and receive function for both self and neighbor, which can be written as follows:

```
COORDINATE $r: receive FROM Neighbor,
            $s: send     FROM Self
        DO ADD $s PRECEDES $r; OD;

COORDINATE $r: receive FROM Self,
            $s: send     FROM Neighbor
        DO ADD $s PRECEDES $r; OD;
```
MP code

19.4.4.4 Segment 2 Constraints

Two constraints are associated with the mode, both of which are of the ***logical-type***.

Constraint 1: *If* the platforms are out of range, then there can be no receipt of communication. This is a ***logic-type*** of constraint that that recognizes the environmental boundary of the system.

```
ENSURE ( #not_within_range == 1
                 -> #receive == 0);
```
MP code

Constraint 2: *If* there is no send, *then* there is no receive. This is a ***logical-type*** of constraint that restrains an impossible outcome in the model.

```
ENSURE ( #send FROM Self
                  >= #receive FROM Neighbor);
ENSURE ( #send FROM Neighbor
                  >= #receive FROM Self);
```
MP code

19.4.4.5 Segment 2 Patterns

Two basic patterns are associated with the communication model: (1) send from self and received by neighbor, and (2) send from neighbor and received by self, as illustrated in the output of the model execution in Figures 19.34 through 19.38. These patterns may occur singly, simultaneously, or not at all. It is significant to note that Figure 19.38 is the only trace that is out of range, and communication is inhibited; thus constraint #1 was effective in limiting any occurrences of communication while out of range.

The MP code is listed below for the first pattern, such that the communication is from self to the neighbor.

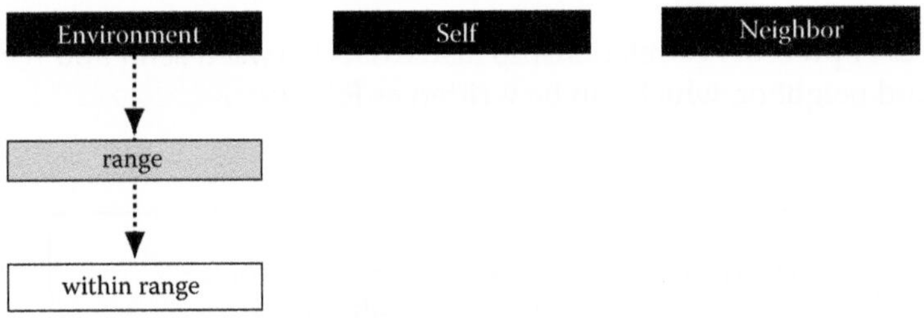

FIGURE 19.34
MP results for platform communication, with scope 1, trace 1. The platforms are within range, but no send-receive interaction occurs between the platforms, and so no template is satisfied in this trace.

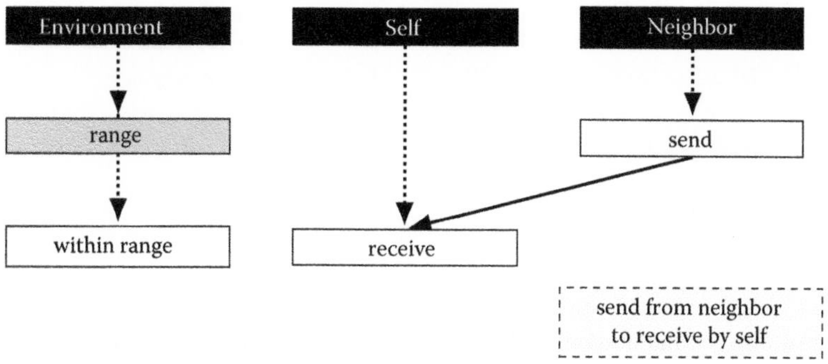

FIGURE 19.35
MP results for platform communication, with scope 1, trace 2. The platforms are within range, and the neighbor sends content to self, satisfying Template #2.

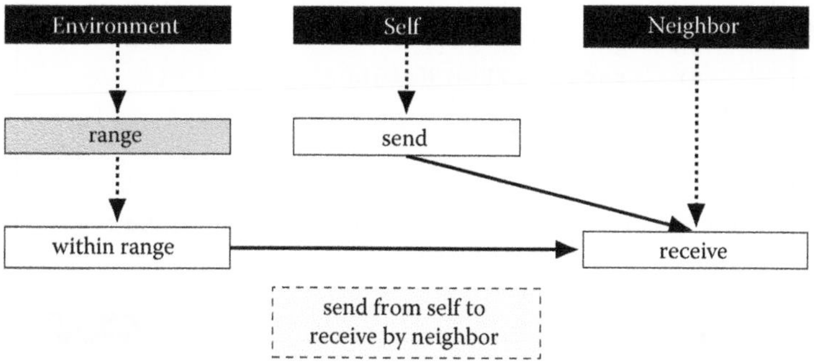

FIGURE 19.36
MP results for platform communication, with scope 1, trace 3. The platforms are within range, and self sends content to the neighbor, satisfying Template #1.

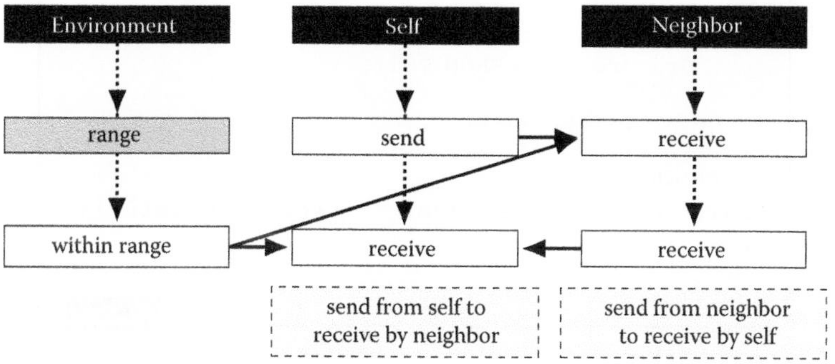

FIGURE 19.37
MP results for platform communication, with scope 1, trace 4. The platforms are within range, and both self and neighbor sends content to each other, satisfying Template #1 and #2.

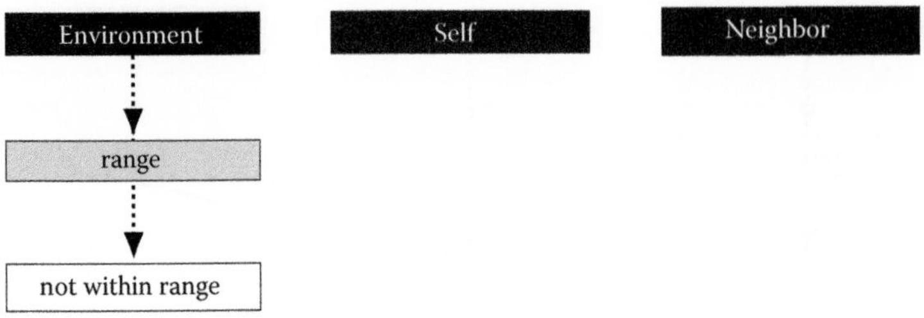

FIGURE 19.38
MP results for platform communication, with scope 1, trace 5. The platforms are out of range, with no send-receive interaction between the platforms. This is the only trace in which the platforms are out of range. No template is associated with this trace.

```
/*Template 1: Send from self is before receive
                    from neighbor*/
IF EXISTS
    $s:  send FROM Self,
    $r:  receive FROM Neighbor
    (
        $s BEFORE $r
    )
THEN MARK;
SAY("send from self to receive by neighbor");
FI;
                                          MP code
```

The MP code for Template 2 is listed below, such that the communication is from the neighbor to self.

```
/*Template 2: Send from neighbor is before
                    receive from self*/
IF EXISTS
    $s:  send FROM Neighbor
    $r:  receive FROM Self
    (
        $s BEFORE $r
    )
THEN MARK;
SAY("send from neighbor to receiveby self");
FI;
                                          MP code
```

19.4.4.6 Segment 2 Evaluation

Evaluation of the probability of successful communication is dependent on many factors, and this analysis considers only the range between the platforms. Other considerations include environmental weather conditions, power availability, and antenna performance. Nonetheless, the structure of the model is diagrammatic of these additional factors.

Certain assumptions were made in order to consider the probability of communication of the completed model in Section 19.4.5. The DSM for scope of execution equal to one and two are shown in Tables 19.13 and 19.14, respectively. Figure 19.39 illustrates the number of traces accomplished for a scope up to 4. Notice that the DSM structures remain consistent, with no new relationships evident with the increase the scope of execution, other than the subsequent "send" and "receive" events.

TABLE 19.13

Design Structure Matrix (DSM) for Platform Communication, with Scope 1

From	To	Environment	Self	Neighbor	Range	within_range	not_within_range	Receive	Send
		3	1	2	4	5	6	7	8
Environment	3	0	0	0	5	0	0	0	0
Self	1	0	0	0	0	0	0	2	2
Neighbor	2	0	0	0	0	0	0	2	2
Range	4	0	0	0	0	4	1	0	0
within_range	5	0	0	0	0	0	0	4	0
not_within_range	6	0	0	0	0	0	0	0	0
Receive	7	0	0	0	0	0	0	0	0
Send	8	0	0	0	0	0	0	4	0

TABLE 19.14

Design Structure Matrix (DSM) for Platform Communication, with Scope 2

From	To	Environment	Self	Neighbor	Range	within_range	not_within_range	Receive	Send
		1	4	5	2	3	8	6	7
Environment	1	0	0	0	10	0	0	0	0
Self	4	0	0	0	0	0	0	9	9
Neighbor	5	0	0	0	0	0	0	9	9
Range	2	0	0	0	0	9	1	0	0
within_range	3	0	0	0	0	0	0	18	0
not_within_range	8	0	0	0	0	0	0	0	0
Receive	6	0	0	0	0	0	0	6	0
Send	7	0	0	0	0	0	0	18	6

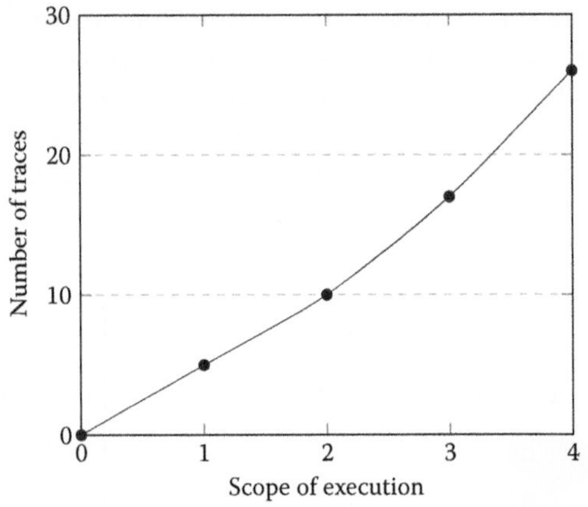

FIGURE 19.39
The effect of scope on the number of traces derived from the MP model, shown scope up to four. Each change in scope increases the number of send-receive cycles possible within the model.

19.4.5 The Combined Formation Flight Mode

The combined formation flight model incorporates the previously described segments to determine the platform heading and communication between platforms in order to determine a navigation command. Recall that determining the current state is based on measuring the speed and position in order to derive a heading as developed in Segment 1. Similarly, the communication model relies upon the platforms being within range to support communication. With those segments defined, the developer is able to derive a combined model that supports formation flight.

19.4.5.1 Formation Flight Narrative

Formation flight relies upon the function of a particular platform to sense its current state (including position, speed, and heading), to attempt to communicate (send and receive) that information to and from neighboring platforms, determine a navigation command, and then to reach an outcome that supports formation flight or resorts to solo flight. Figure 19.40 outlines a basic structure, as an SysML diagram.

19.4.5.2 Formation Flight Events

From the narrative, the events of the platform are to "sense" the state (position, speed, and heading) of its own platform, to "communicate" that information with neighboring platforms, and to "navigate" based on the relationship of the self and the neighboring platforms. Figures 19.41 and 19.42 illustrate these relationships, followed respectively by the associated MP code.

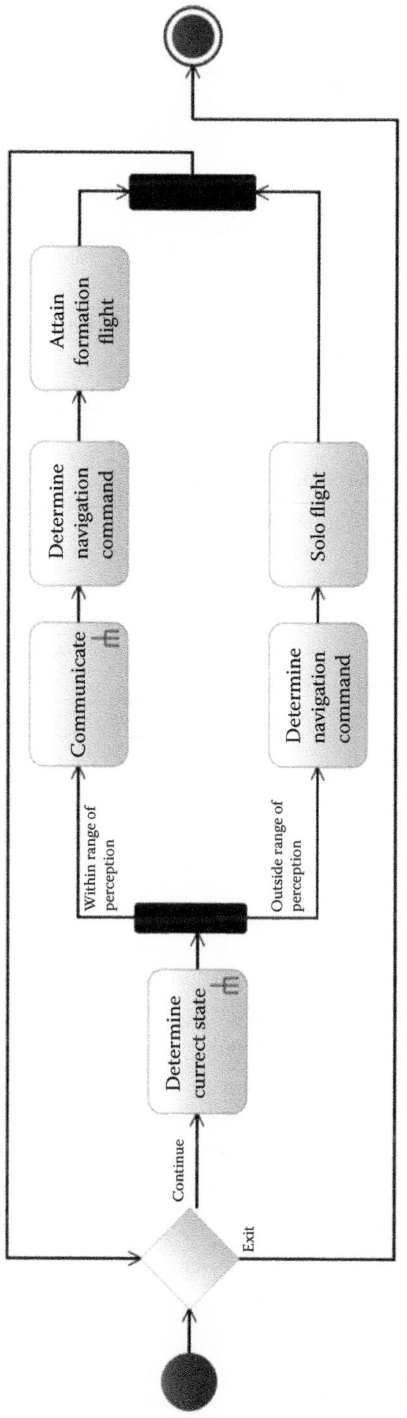

FIGURE 19.40
Formation flight activity diagram. Illustration developed in Innoslate®, www.innoslate.com.

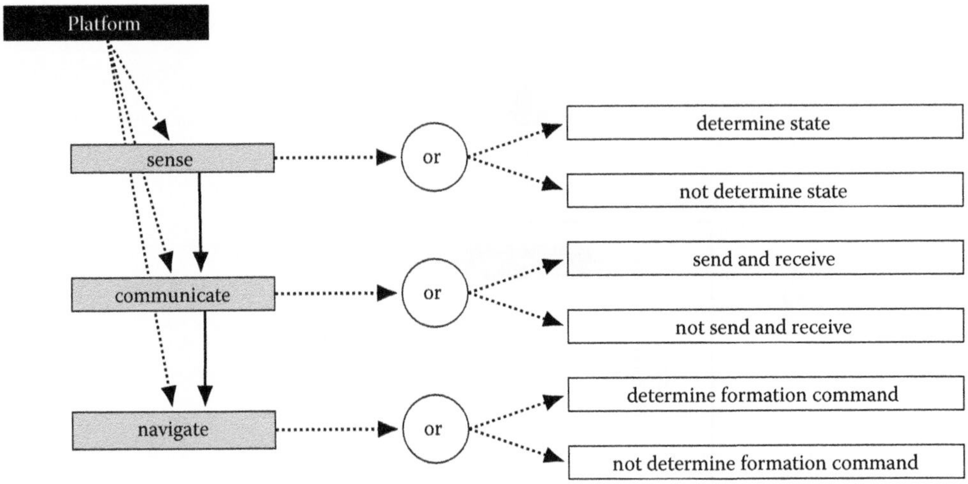

FIGURE 19.41
Formation flight events associated with the platform or vehicle.

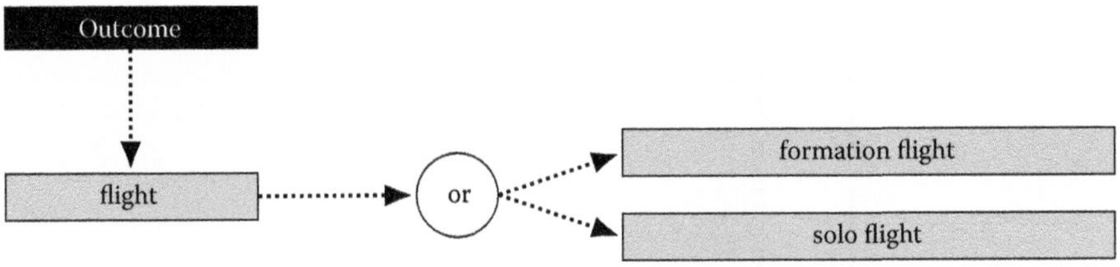

FIGURE 19.42
Formation flight events associated with the outcome.

```
SCHEMA formationFlight

ROOT Platform:
                sense
                communicate
                navigate;

sense:
   (  determine_state|
          not_determine_state)
;

communicate:
   (  send_and_receive|
          not_send_and_receive)
;

navigate:
   (  determine_formation_command|
          not_determine_formation_command)
;
                                   MP code
```

```
ROOT Outcome:
        flight;

flight:
      ( formation_flight|solo_flight );
```
MP code

19.4.5.3 Formation Flight Coordination

Coordination of this model is very simple: the output of the command by the platform relates directly to the outcome of the formation flight. This code is listed below:

```
COORDINATE$n: (determine_formation_command
             | not_determine_formation_command),

             $f: (formation_flight
                 | solo_flight  )
    DO ADD $n PRECEDES$f; OD;
```
MP code

19.4.5.4 Formation Flight Constraints

The constraints reduce the number of viable traces from sixteen to four, as illustrated in Figure 19.43, and listed below.

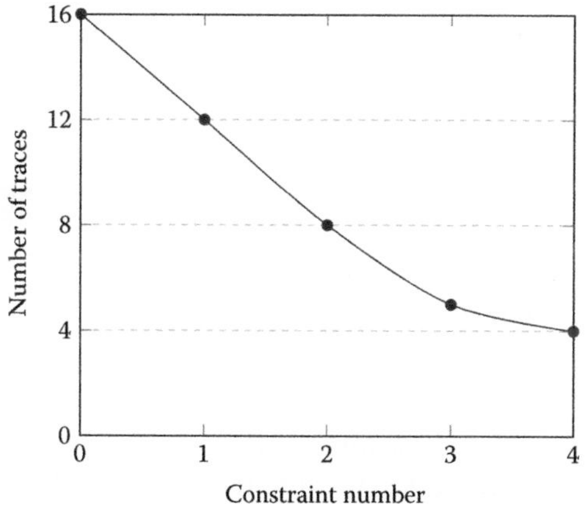

FIGURE 19.43
The effect of constraints on the number of traces derived from the MP model for formation flight model. Running the model with no constraints resulted in sixteen traces, or possible outcomes; conversely, adding four constraints resulted in only four traces after running the model.

Constraint 1: *If* the state is not known, *then* there is no data to send. This is a ***logical-type*** of constraint that prevents incorrect alternatives from forming in the model.

```
ENSURE ( #not_determine_state == 1 ->
              #not_send_and_receive == 1);
```
MP code

Constraint 2: *If* there is no communication, *then* the navigation algorithm cannot determine a formation flight command. This is also a ***logical-type*** of constraint since the platform needs to account for neighboring platform positions, speed, and heading in order to adjust its own formation.

```
ENSURE    not_send_and_receive == 1 ->
           #not_determine_formation_command
             == 1);
```
MP code

Constraint 3: *If* no formation command is issued, *then* solo-flight commands are executed. This is a ***design-type*** of constraint for the navigation algorithm to implement, establishing a requirement for the system.

```
ENSURE ( #not_determine_formation_command == 1
          -> #solo_flight== 1);
```
MP code

Constraint 4: *If* the state is determined and communication is established, *then* the navigation algorithm determines a formation command. This also is a ***design-type*** of constraint that leads to a requirement for the navigation algorithm.

```
ENSURE (#determine_state+ #send_and_receive
          == 2 ->
           #determine_formation_command == 1);
```
MP code

19.4.5.5 Formation Flight Patterns

Patterns exposed executing the formation flight model are shown in Figures 19.44 through 19.47. The patterns include successful formation, failure in flight though a correct command was given, failure in communication, and failure in deriving the current state leading to all subsequent events failing. Identifying the pattern in MP is listed in the following code for the first template.

```
/*Template 1: Successful formation flight*/
/**/
IF EXISTS DISJ

$s:  determine_state
$c:  send_and_receive
$f:  determine_formation_command
$o:  formation_flight

    (
        $s BEFORE $c
        AND
        $c BEFORE $f
        AND
        $f BEFORE $o
    )

THEN MARK;
/*
SAY("T1: formation flight achieved");
/**/
FI;
```

MP code

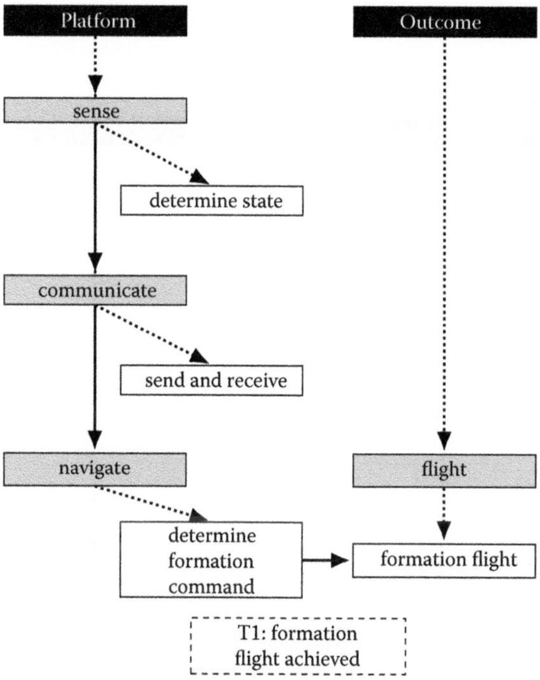

FIGURE 19.44
Formation flight trace 1, successful formation achieved.

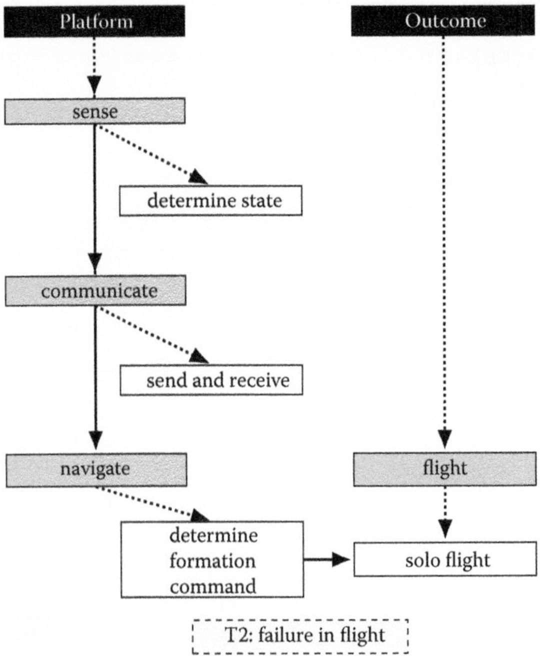

FIGURE 19.45
Formation flight trace 2, correct command given but a failure occurred in flight, resulting in an inability to attain formation.

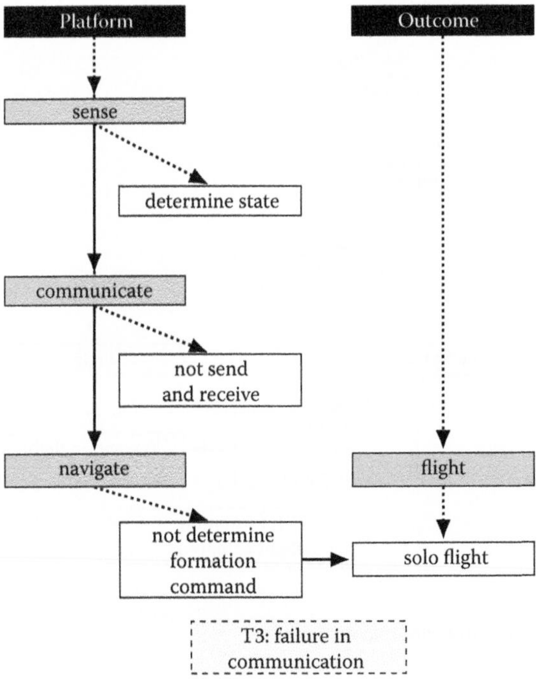

FIGURE 19.46
Formation flight trace 3, failure in communication.

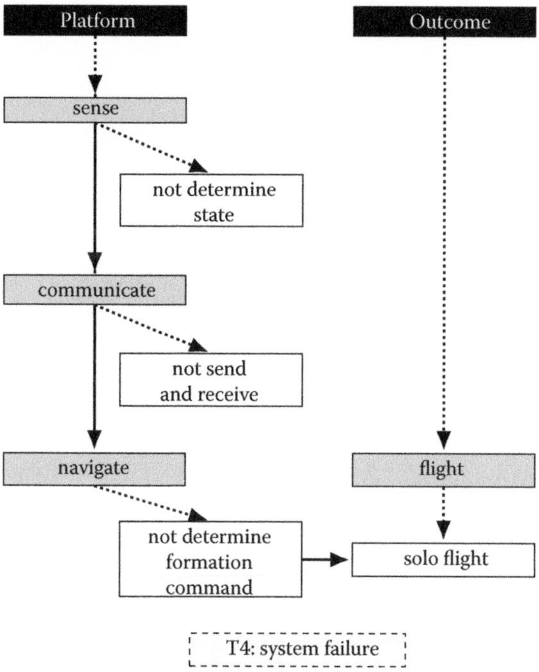

FIGURE 19.47
Formation flight trace 4, failure in determining the state and a ripple-down effect to the following events.

19.4.5.6 Formation Flight Evaluation

Calculation of probability of success is also readily derived from the model. The topology for a belief-network flows directly from the MP model, as illustrated in Figures 19.48 and 19.49. Node A represents the platform sensing function developed in Segment 1, node B represents the communication function developed in Segment 2, node C represents the navigation command function, and node D represents the outcome flight state. The probability for node A to be successful was derived extensively in the paragraphs above, and so for the purpose of this example, we shall use $P(a) = .996$. The probability of communication depends upon many factors, including the relative position of the platforms, environmental factors, and interference with the signal. Therefore, for the purpose of this

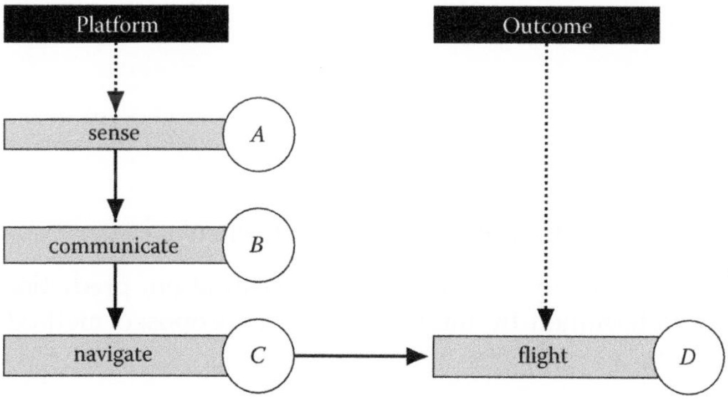

FIGURE 19.48
Topology of the formation flight model, showing the relationship to a Bayesian belief network.

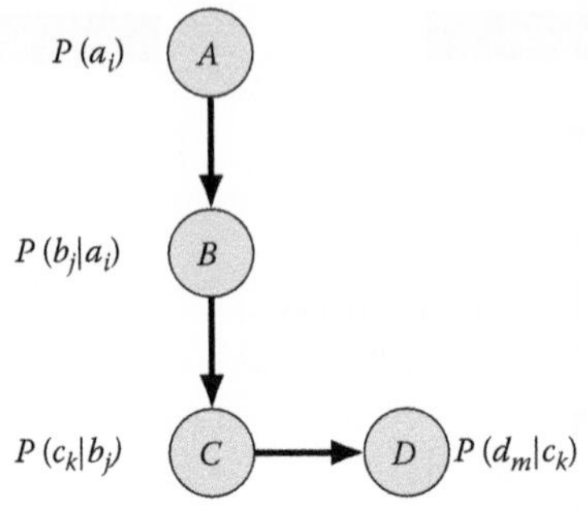

FIGURE 19.49
Bayesian belief network derivation from the formation-flight model.

TABLE 19.15

Formation-Flight Probability of Success. All Calculations Are for a Successful Sequence, but Adding Communication Iterations to Improve Performance

$P_{success}$	$P(a)$	$P(b\|a)$	$P(c\|b)$	$P(d\|c)$
0.7194	0.9992	$(1 - 0.2^1) = 0.8$	1	0.9
0.8633	0.9992	$(1 - 0.2^2) = 0.96$	1	0.9
0.8921	0.9992	$(1 - 0.2^3) = 0.992$	1	0.9
0.8978	0.9992	$(1 - 0.2^4) = 0.9984$	1	0.9

exercise, let us consider that $P(b) = .8$ is a reasonable boundary for the problem. And then, we can derive the possible effect of using multiple measurements, such that $P(b) = (1–0.8^n)$, where n is the number of consecutive measurements. Given correct inputs, the navigation algorithm will derive the correct navigation command, and so we can use $P(c|b) = 1$. And finally, though a correct command may be given, the outcome is not always successful due to material failures, environmental conditions, or other interference, and so let us use $P(d|c) = 0.9$. Given these assumptions, Table 19.15 lists the results indicating that the probability of successfully reaching formation flight is between 0.72 and 0.9.

Design Structure Matrix (DSM) for the formation flight model is shown in Table 19.16. The model incorporates no effect for changes in scope since the number of parameters remains constant. Nonetheless, the DSM holds no surprises for this relatively linear model.

19.5 Discussion

19.5.1 Detection, Classification, Prediction, and Control

This section outlines the analysis of detection, classification, prediction, and control of emergent behavior determined by implementing the proposed methodology, and exercised by the examples.

TABLE 19.16

Design Structure Matrix (DSM) Derived from the Formation Flight Model

From \ To		1 Platform	2 Sense	3 determine_state	4 Communicate	5 send_and_receive	6 Navigate	7 determine_formation_command	8 Outcome	9 Flight	10 formation_flight	11 solo_flight	12 not_send_and_receive	13 not_determine_formation_command	14 not_determine_state
Platform	1	0	4	0	4	0	4	0	0	0	0	0	0	0	0
Sense	2	0	0	3	4	0	0	0	0	0	0	0	0	0	1
determine_state	3	0	0	0	0	0	0	0	0	0	0	0	0	0	0
Communicate	4	0	0	0	0	2	4	0	0	0	0	0	2	0	0
send_and_receive	5	0	0	0	0	0	0	0	0	0	0	0	0	0	0
Navigate	6	0	0	0	0	0	0	2	0	0	0	0	0	2	0
determine_formation_command	7	0	0	0	0	0	0	0	0	0	1	1	0	0	0
Outcome	8	0	0	0	0	0	0	0	0	4	0	0	0	0	0
Flight	9	0	0	0	0	0	0	0	0	0	1	3	0	0	0
formation_flight	10	0	0	0	0	0	0	0	0	0	0	0	0	0	0
solo_flight	11	0	0	0	0	0	0	0	0	0	0	0	0	0	0
not_send_and_receive	12	0	0	0	0	0	0	0	0	0	0	0	0	0	0
not_determine_formation_command	13	0	0	0	0	0	0	0	0	0	0	2	0	0	0
not_determine_state	14	0	0	0	0	0	0	0	0	0	0	0	0	0	0

19.5.1.1 Detection

By its implementation, MP automatically provides detection of emergent behaviors by investigating the output traces for particular interactions of interest. And so, for all cases, the traces provide the evidence of emergence. For the examples described in the previous section, the UAVs display emergent behavior of formation flight by sensing the environment for position and heading, communicating position with neighboring platforms, deriving a navigation and formation flight command relative to neighboring platforms, and executing a prescribed formation. Figure 19.50 illustrates the positive emergence of each segment of the model, including the combined model representing formation flight.

19.5.1.2 Classification

Using the taxonomy outlined by Maier [49], simple, weak, and strong classifications are possible outcomes for a system analyzed within the methodology framework, as outlined in Table 19.2. Recall from this table that simple emergence is characterized by readily predictable outcomes, weak emergence is characterized by dependency upon interactions within the model, and strong emergence includes dependencies that are difficult to predict as to where or when they occur.

The goal of the developer is to enable positive emergence within the simple and weak classifications such that the behavior is readily predictable, consistent with known properties, and enabled through interactions among the systems and the environment. The developer needs to manage systems that encounter the strong classifications that occur without reliable predictability. Unexpected environmental conditions may cause a system to fall into this category.

Table 19.17 reflects the classification of each of the example problems. For Segment 1: Determine current state, the emergent behavior depends upon the sensor function and the interaction of the sensors with the environmental state and the UAV controller. Therefore, this segment demonstrates weak emergence because of the dependency upon the interaction of the sensors. For Segment 2: Communication, the behavior depends upon the interaction among the UAVs, designated as self and neighbor. However, in this case, there is also dependency upon the range of the sensors and unknown or undetermined environmental effects that could inhibit the receive function. Therefore, these interactions could lead to strong negative emergence since these dependencies would make it difficult to reliably predict whether communication can occur. Under favorable conditions, positive emergence would be dominant, but under unfavorable conditions, the emergence would have a strong negative emergence level, impacting the ability to communicate. Finally, for the Combined Model: Formation Flight, the classification of strong is appropriate because of the dependency of the communication among the individual UAVs which has already been identified as strong.

Maier also proposed the additional field of *spooky* that would function in ways that are inconsistent with known properties. Since it is not possible to model an unknown behavior, it is thus not possible to have some prescribed outcome within this category. On the contrary, this category serves the developer as motivation to derive some reasonable representation such that the behavior would be consistent with known properties and thereby fall into one of the simple, weak, or strong classifications.

19.5.1.3 Prediction

Two methods for prediction of emergence were used: (1) calculation of the probability of behaviors, and (2) investigation of the Design Structure Matrix (DSM). By employing the first

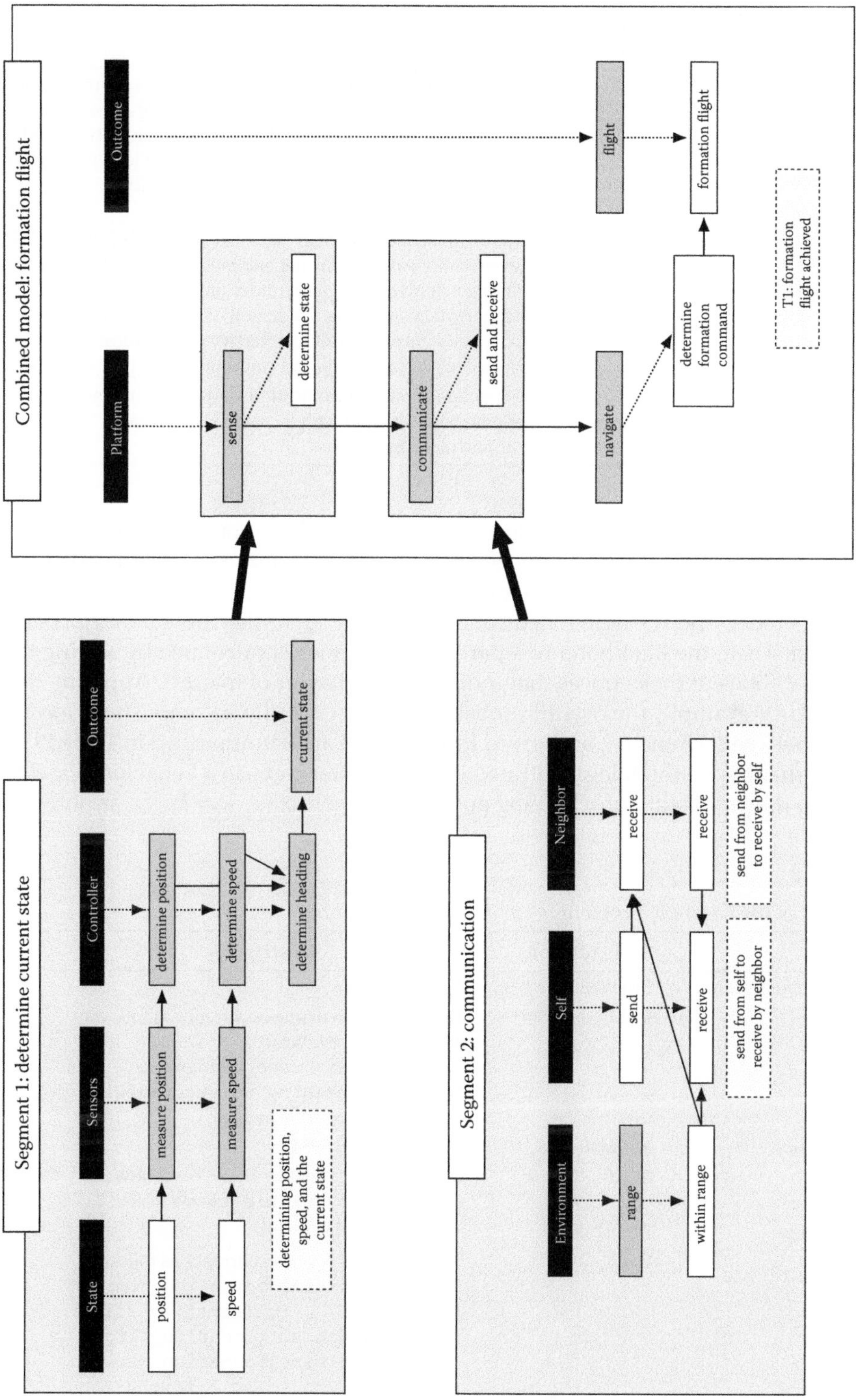

FIGURE 19.50
Combined Model Emergence indicating the dependence of determining the current state and communication in order to successfully establish formation flight as a form of positive emergence. Automatic detection of these conditions enabled by assertion checking of the model, and labeling the results according to the search template.

TABLE 19.17

Classification of the Example Problem

Segment	Classification	Description
Determine current state	Weak	Interaction of sensors with environment provides inputs to the controller and so this interaction supports the level of weak emergence. Incorrect inputs lead to incorrect calculations and negative emergence. Conversely, correct inputs lead to effective determination of the UAV state (position, velocity, and heading), thereby enabling positive emergence.
Communication	Strong	Environmental conditions may limit likelihood of receiving communication, and so this dependency supports the potential for strong negative emergence under unfavorable environmental conditions. Prediction of negative emergence also depends upon controlling effective communication through design constraints and communication performance.
Formation flight	Strong	The dependency on communication supports strong negative emergence since it is difficult to predict unfavorable environmental conditions.

method, the probability calculations enable a trace-by-trace calculation of the likelihood of occurrence of each trace. Behaviors of interest were identified through an automated query of the model by assertion checking. This query identifies templates reflecting patterns of behavior in the model, and the automated function readily identifies these behaviors and labels the output. Then, the likelihood of a particular behavior is calculated by adding the associated probabilities of those traces that indicate the behavior of interest. Applying this analysis to the UAV example, the overall probability of successfully accomplishing formation flight falls between 0.7 and 0.9, as derived in Table 19.15 and summarized in Table 19.18. This illustrates that the methodology outlined in this chapter can use a behavior model to derive insight to the probability of achieving positive emergence, $P_{positive} = P_{success}$, as outlined

TABLE 19.18

Predictions of the Example Problem

Segment	Prediction	Description
Determine current state	Probability	$P_{success} = 0.9992$

	DSM	Three instances of one occurrence as follows:
		`determine_heading to accurate_heading`
		`determine_heading to successful`
		`Outcome to successful`

Communication	Probability	$P_{success} = 0.8$ to 0.9984
	DSM	One instance of one occurrence as follows: `range to not_within_range`
Formation flight	Probability	$P_{success} = 0.7$ to 0.9

	DSM	Four instances of one occurrence as follows:
		`determine_formation_command to formation_flight`
		`Flight to formation_flight`
		`determine_formation_command to solo_flight`
		`Sense to not_determine_state`

in Table 19.15. Conversely, the probability of negative emergence is the result of the traces that do not satisfy a successful outcome, represented as $P_{negative} = 1 - P_{success}$.

The second method used to predict emergence was accomplished by employing the DSM or otherwise known as the N-squared diagram. Since MP calculates all possible traces and all possible interactions of the system and associated events, the DSM captures all interactions within a single matrix. A simple post-processing routine identifies the number of times each possible interaction occurs, thereby populating the DSM. Infrequent occurrences may be a source to investigate for unintended or unfavorable emergence since the developer may unintentionally overlook these interactions.

As an illustration, Table 19.18 reflects both prediction methods, including the probabilities for a successful outcome and instances of a single occurrence of a relationship for each of the segments of the example problem. These methods of prediction apply to the simple, weak, or strong emergence since a representative model and resulting simulation is necessary to apply the prediction methods.

19.5.1.4 Control

In all cases outlined by the examples, the control of emergent properties is implemented by the constraints applied to the system as outlined in Table 19.19. Types of constraints include *logical, simplification,* and *design*. Logical constraints enable the model to reflect achievable interactions; as an example, a system cannot receive communication if it were not previously sent. Simplification constraints allow a robust model structure or topology, while simplifying the output for the purpose of analysis. An example of simplification includes the idea that every instance of a correct calculation will lead to a successful outcome.

TABLE 19.19

Controls of the Example Problem, Defined as Logical, Simplification, or Design Constraints

Example	#	Type	Description
Heading	1	Simplification	An accurate heading leads to a success and its corollary that an inaccurate heading leads to a failure.
	2	Logical	Inaccurate sensors lead to inaccurate determination of position and speed.
	3	Design	One accurate measurement leads to an accurate determination for both position and speed sensors. This indicates that the design of the controller needs to have a determination of accurate and inaccurate sensors, forming a design constraint.
	4	Design	When both position and speed are accurate, the heading is accurate. This forms a design constraint for the controller.
Communication	1	Logical	If the platforms are out of range, no platforms can receive communication.
	2	Logical	If there is no send, then there is no receive.
Formation flight	1	Logical	If the position, speed, and heading is not known, then there is no information to send and receive.
	2	Logical	If there is no communication, then formation flight cannot be commanded.
	3	Design	If there is no formation command, execute solo-flight. This is a design constraint for the navigation controller.
	4	Design	If the state is determined and communication established, then a formation command is determined. This is a design constraint for the controller.

Controls can be applied only to those segments of the model that the developer can affect, and so simple and weak emergence is readily effected by controls. Strong emergence may be impacted by limiting the effects of negative emergence through constraints or by improving design reliability and performance. For instance, in our example, we identified that the communication can be affected by environmental conditions outside the control of the developer. However, the developer may be able to influence the design performance and reliability of the radio or receiver. In this way, strong negative emergence is still possible, but it may be less likely to occur.

19.6 Conclusion

This chapter illustrated that a model described in a high level of abstraction can effectively define system behaviors at a fundamental level. These behaviors are characterized by internal event interactions comprising the system-level functions, and by external interactions from system to system, system to environment, or system to user. Emergent behaviors are derived from these sources of interaction, and so the developer needs to ensure that only expected behaviors are employed in the system. The user shapes the behaviors by applying implementing constraints and executing the architecture to evaluate the results of all potential interactions. Analysis includes deriving every possible outcome and using computational tools to investigate all instances of the output. The example problem outlined in this chapter demonstrated the following assertions:

- A behavior model description using the formal methods from Monterey Phoenix (MP) exposes the fundamental essence of behavior and can be used to derive efficient architectures that produce well-understood behaviors of a system of systems.
- Analysis of behaviors at a higher level of abstraction captures the inherent behaviors of the system and the interactions among the systems, emphasizing the fundamental source of emergent behaviors. This abstraction enables the developer to focus on the core area of interest and the essence of behaviors. This analysis is to be conducted prior to adding detailed parameters to the model. These parameters may be necessary for discrete-event simulation, agent-based simulation, dynamic simulation, or a hybrid simulation; therefore, behavior analysis should precede these types of simulation.
- Within the scope of execution, all possible permutations of model output are derived and analyzed within the execution environment. This provides a level of assurance that the system architecture that complies with the model definition and constraints, and that the system under development exhibits only known behaviors.
- MP-Firebird provides the developer with an interactive simulation environment to test whether constraints and coordination of events have the desired impact on the model. Using assertions to propose a query and then marking the associated traces enables the user to quickly identify the model response.
- Templates of prototypical behaviors are supported by the query and mark commands, enabling an automated means to parse the resulting traces and characterize the positive and negative behaviors of the model.

- Constraint types include logical limitations, simplification decisions, and design requirements. Overly-constraining the model could eliminate essential outcomes and failure to identify emergent properties of the system.
- The execution environment supports analysis of the results including probability calculation and automated population of a Design Structure Matrix (DSM).
- The developer can ensure the existence of desirable behaviors and limit undesirable behaviors through constraints. MP supports positive emergence verification by the powerful means to define model behaviors and interactions through lightweight formal methods, developing appropriate constraints, and deriving all possible outcomes of the architecture within the scope of execution. Negative emergence can be detected by assertion checking and by investigating whether mistakes and omissions reside in model.

References

1. What is a model?, March 2013. http://model-based-systems-engineering.com/2013/03/15/what-is-a-model/(accessed July 5, 2017).
2. OMG Unified Modeling Language (UML), March 2015. http://www.omg.org/spec/UML/2.5/PDF/.
3. OMG Systems Modeling Language (SysML), May 2017. http://www.omg.org/spec/SysML/1.5/.
4. Qualities of great models. Massachusetts Institute of Technology, Short Course titled Model-Based Systems Engineering: Documentation and Analysis, 2017. Accessed June 11, 2017 from https://mitprofessionalx.mit.edu/courses/course-v1:MITProfessionalXSysEngx31T2017/.
5. Abstract, (n.d.). Retrieved June 10, 2017, from https://www.merriam-webster.com/dictionary/abstract.
6. Action, (n.d.). Retrieved June 10, 2017, from https://www.merriam-webster.com/dictionary/action.
7. Activity, (n.d.). Retrieved June 10, 2017, from https://www.merriam-webster.com/dictionary/activity.
8. Behavior, (n.d.). Retrieved June 9, 2017, from https://www. merriam-webster.com/dictionary/behavior.
9. Event, (n.d.). Retrieved June 10, 2017, from https://www.merriam-webster.com/dictionary/event.
10. Model, (n.d.). Retrieved June 9, 2017, from https://www.merriam-webster.com/dictionary/model.
11. Simulate, (n.d.). Retrieved June 9, 2017, from https://www. merriam-webster.com/dictionary/simulate.
12. Simulation, (n.d.). Retrieved June 9, 2017, from https://www. merriam-webster.com/dictionary/simulation.
13. M. Auguston, June 2017. phone conversation and email of June 8, 2017.
14. Mikhail Auguston. Monterey Phoenix, or how to make software architecture executable. In *Proceedings of the 24th ACM SIGPLAN Conference Companion on Object Oriented Programming systems languages and applications*, pages 1031–1040. ACM, 2009.
15. Mikhail Auguston. Software architecture built from behavior models. *ACM SIGSOFT Software Engineering Notes*, 34(5):1–15, 2009.
16. Mikhail Auguston. Behavior models for software architecture. Technical Report NPS-CS-14-003, Monterey, California. Naval Postgraduate School, 2014.

17. Mikhail Auguston. *Monterey Phoenix, System and Software Architecture and Workflow Modeling Language Manual (version 2)*. Naval Postgraduate School, Monterey California, 2 edition, April 2016. https://wiki.nps.edu/display/MP/MP+Crash+Course.
18. BKCASE. The guide to the systems engineering body of knowledge (sebok), v 1.4. BKCASE is managed and maintained by the Stevens Institute of Technology Systems Engineering Research Center, the International Council on Systems Engineering, and the Institute of Electrical and Electronics Engineers Computer Society, 2015 (accessed July 11, 2015).
19. Benjamin S. Blanchard, Wolter J. Fabrycky, and Walter J. Fabrycky. Systems engineering and analysis, volume 4. Prentice Hall Englewood Cliffs, NJ, 1990.
20. Andrei Borshchev. The Big Book of Simulation Modeling: Multimethod Modeling with AnyLogic 6. AnyLogic North America, 2013.
21. George E.P. Box. Science and statistics. Journal of the American Statistical Association, 71(356):791–799, 1976.
22. Tyson R. Browning. Applying the design structure matrix to system decomposition and integration problems: A review and new directions. *IEEE Transactions on Engineering Management*, 48(3):292–306, 2001.
23. Dennis M. Buede. *The engineering design of systems: Models and methods*. John Wiley & Sons, 2nd edition, 2009.
24. C. Calvano and P. John. Systems engineering in an age of complexity. *IEEE Engineering Management Review*, 32(4):29–37, 2004.
25. Edward Crawley, Bruce Cameron, and Daniel Selva. *System architecture: Strategy and product development for complex systems*. Prentice Hall Press, 2015.
26. Edward Crawley, Olivier De Weck, Christopher Magee, Joel Moses, Warren Seering, Joel Schindall, David Wallace, Daniel Whitney et al. The influence of architecture in engineering systems. 2004.
27. J.S. Dahmann and K.J. Baldwin. Understanding the current state of U.S. defense systems of systems and the implications for *systems engineering. In Systems Conference, 2008 2nd Annual IEEE*, pages 1–7, April 2008.
28. J.S. Dahmann, A.S. Bhatti, and M. Kelley. Importance of systems engineering in early acquisition. In *Systems Conference, 2009 3rd Annual IEEE*, pages 110–115, March 2009.
29. Lenny Delligatti. SysML distilled: *A brief guide to the systems modeling language*. Addison-Wesley, 2013.
30. Steven D. Eppinger and Tyson R Browning. *Design structure matrix methods and applications*. MIT Press, 2012.
31. Jochen Fromm. Types and forms of emergence. *arXiv preprint* nlin/0506028, 2005.
32. Kristin Giammarco. Architecture modeling software analytics—Model quality and maturity using automated tools. In *System of Systems Engineering (SoSE), 2017 12th International Conference on*, 2017.
33. Kristin Giammarco and Mikhail Auguston. Well, you didn't say not to! a formal systems engineering approach to teaching an unruly architecture good behavior. *Procedia Computer Science*, 20(0):277–282, 2013. Complex Adaptive Systems.
34. Kristin Giammarco and Mikhail Auguston. Monterey Phoenix main principles and advantages, viewed July 16, 2015. https://wiki.nps.edu/display/MP/Main+Principles+and+Advantages.
35. James Gleick. Chaos: *Making a new science*. Viking Penguin, Inc., 1987.
36. O. Thomas Holland. Taxonomy for the modeling and simulation of emergent behavior systems. In *Proceedings of the 2007 Spring Simulation Multiconference—Volume 2*, SpringSim '07, pages 28–35, San Diego, CA, USA, 2007. Society for Computer Simulation International.
37. Jianpeng Hu, Linpeng Huang, Xuling Chang, and Bei Cao. A model driven service engineering approach to system of systems. In *Systems Conference (SysCon), 2014 8th Annual IEEE*, pages 136–145, March 2014.
38. T.V. Huynh and J.S. Osmundson. A systems engineering methodology for analyzing systems of systems using the systems modeling language (sysml), 2006.

39. INCOSE. World in motion: Systems engineering vision 2025. *International Council on Systems Engineering*, San Diego, CA, USA, 2014.
40. INCOSE. *Systems Engineering Handbook: A Guide for System Life Cycle Processes and Activities, Version 4*. International Commitee on Systems Engineering, Wiley, 2015.
41. ISO/IEC/IEEE. *Systems and Software Engineering—System Life Cycle Processes*. International Organization for Standardization (ISO), International Elecrotechnical Commission (IEC), Institute of Electrical and Electronics Engineers (IEEE), Geneva, Switzerland, 2015. 15288:2015.
42. Daniel Jackson. Alloy: A lightweight object modelling notation. *ACM Trans. Softw. Eng. Methodol.*, 11(2):256–290, April 2002.
43. Daniel Jackson. *Software Abstractions: Logic, language, and analysis*. MIT Press, 2012.
44. Jan Kantert, Sven Tomforde, Melanie Kauder, Richard Scharrer, Sarah Edenhofer, Jörg Hähner, and Christian Müller-Schloer. Controlling negative emergent behavior by graph analysis at runtime. *ACM Transactions on Autonomous and Adaptive Systems* (TAAS), 11(2):7, 2016.
45. Wolfgang Konle. Track heading and speed estimation. In *GI Jahrestagung (2)*, pages 785–791. Citeseer, 2010.
46. A. Kossiakoff and W. Sweet. *Systems Engineering Principles and Practice*. Wiley, 2003.
47. Mark W. Maier. Architecting principles for systems-of-systems. *INCOSE International Symposium*, 6(1):565–573, 1996.
48. Mark W. Maier. Architecting principles for systems-of-systems. *Systems Engineering*, 1(4):267–284, 1998.
49. Mark W. Maier. The role of modeling and simulation in system of systems development. *Modeling and simulation support for system of systems engineering applications*, pages 11–41, 2015.
50. John H. Miller and Scott E. Page. *Complex adaptive systems: An introduction to computational models of social life: An introduction to computational models of social life*. Princeton University Press, 2009.
51. Melanie Mitchell. *Complexity: A guided tour*. Oxford University Press, 2009.
52. Gordon E. Moore. The future of integrated electronics. *Electronics*, 1965.
53. ODUSD(A&T)SSE. Systems engineering guide for systems of systems. Office of the Deputy Under Secretary of Defense for Acquisition and Technology, Systems Software Engineering, August 2008. Version 1.0.
54. Xing Pan, Baoshi Yin, and Jianmi Hu. Modeling and simulation for SoS based on the DoDAF framework. In Reliability, Maintainability and Safety (ICRMS), *2011 9th International Conference on*, pages 1283 –1287, June 2011.
55. Alex Pentland and Andrew Liu. Modeling and prediction of human behavior. *Neural computation*, 11(1):229–242, 1999.
56. John Quartuccio, Kristin Giammarco, and Mikhail Auguston. Deriving probabilities from behavior models defined in Monterey Phoenix. *In System of Systems Engineering (SoSE), 12th International Conference on*, 2017.
57. Shweta Singh, Shan Lu, Mieczyslaw M. Kokar, and Paul A. Kogut. Detection and classification of emergent behaviors using multi-agent simulation framework (wip). In *Proceedings of the Symposium on Modeling and Simulation of Complexity in Intelligent, Adaptive and Autonomous Systems*, page 3. Society for Computer Simulation International, 2017.
58. Steven H. Strogatz. *Sync: How order emerges from chaos in the universe, nature, and daily life*. Hyperion, 2003.
59. Defense Acquisition University. Defense acquisition guide (DAG). 19:1248, June 2013.
60. Mitchell M. Waldrop. *Complexity: The emerging science at the edge of order and chaos*. Simon and Schuster, 1993.
61. Oleg A. Yakimenko. *Engineering computations and modeling in MATLAB/Simulink*. American Institute of Aeronautics and Astronautics, 2011.
62. Bernard P. Zeigler. *Theory of Modeling and Simulation*. John Wiley, 1976.
63. Bernard P. Zeigler. A note on promoting positive emergence and managing negative emergence in systems of systems. *The Journal of Defense Modeling and Simulation: Applications, Methodology, Technology*, 13(1):133–136, 2016.

20

InterDyne: A Simulation Method for Exploring Emergent Behavior Deriving from Interaction Dynamics

Christopher D. Clack and Leo Carlos-Sandberg

CONTENTS

20.1 Introduction

Emergent Behavior in Systems of Systems may derive from a variety of different low-level behaviors, including (i) behavior *within* subsystems; and (ii) behavior *between* subsystems. We are interested in the latter—in the way that emergent behavior can arise in a System of Systems as a result of the dynamics of interaction between subsystems.

Our leading case study for interaction dynamics in a System of Systems is the financial markets, which exhibit a range of complex and undesirable emergent behavior. For example, in the U.S. markets "Flash Crash" of May 6, 2010 (CFTC and SEC 2010), market prices became disconnected from rational valuations, and irrational prices and frenzied trading in one market rapidly spread to many other markets, causing extreme price crashes and spikes. For the U.S. financial markets this was "one of the most turbulent periods in their history" (Kirilenko et al. 2014).

The financial markets are an interesting subject for study from a System of Systems perspective since they have a highly complex and interactive structure. Each individual market (such as the equities market, the commodities market, the derivatives market, and the foreign exchange market) is itself a complex System of Systems comprising a network of entities such as traders, brokers, dealers, inter-broker dealers, exchanges, clearing houses, investment funds, retail investors, and so on. Each of these markets works in a slightly different way (for example, equities are traded on one or more public exchanges, whereas there is no central exchange for foreign exchange and instead there is a network of foreign exchange dealers quoting slightly different prices).

These markets are further interlinked by traders who operate in multiple markets simultaneously. For example, a trader might manage risky trades in one market by entering into risk-reducing trades in other markets. Further linkage is provided by traders who exploit arbitrage opportunities (such as transient inconsistencies in price) between markets.

20.1.1 Emergent Behavior Deriving from Interaction Dynamics

It is well known in the field of physics that very complex behavior can emerge from the *interactions* between objects, even between a small number of simple components. Consider, for example, the three-body problem (Sundman 1912), the behavior of coupled oscillators (Wilberforce 1896), and feedback oscillation in electronics (Barkhausen 1935). An everyday example of undesirable behavior arising from interaction in a System of Systems occurs within an audio amplifier when the loudspeaker output system interacts with the microphone input system; the interaction causes a feedback loop, resulting in a high-pitched oscillation.

With all of the above examples, the high-level emergent behavior in the System of Systems arises from the interaction between the Systems, not from the Systems themselves (other than the Systems possessing properties that permit them to become engaged in interaction—such as mass, momentum, inertia, energy, and the ability to translate sound into electricity and vice versa). In each example, if the interaction between Systems were removed, the high-level behavior of the System of Systems would disappear: for example, a single body if not acted upon by a force is stationary or moves with constant velocity (Newton 1687), a single oscillator alone exhibits simple movement, and an audio amplifier does not normally emit a high-pitched oscillation if the microphone System is separated from the loudspeaker System.

Within physics, interactions might be physical collisions or the effect experienced by a component when it is acted upon by a field, such as an electric, magnetic, or gravitational field. Within the financial markets, trading by humans has almost completely been replaced by computer-directed trading, with computer algorithms making decisions on what, and when, to trade; thus, in the financial markets interactions primarily occur between computer programs, and the medium for such interactions is the passing of messages between computers (such as a computer in an investment bank sending an order to a computer at a stock exchange).

Day and Huang (Day and Huang 1990) have demonstrated how interactions between two simple but different trading strategies and a market maker can cause complex emergent features of stock market prices such as alternating periods of rising ("bull market") and falling ("bear market") with sudden switching between the two at irregular intervals. Further, Lyons (Lyons 1997) has shown how a feedback loop can emerge between foreign exchange dealers, causing them to repeatedly transfer inventory between themselves. Our aim is to develop a framework for modelling and analyzing emergent behavior that arises from the dynamics of interaction, and in the context of our case study to analyze behavior that may increase risk to the stability of the financial markets.

20.1.2 Feedback Loops

A good example of complex interaction dynamics in the financial markets System of Systems is the creation of feedback loops, where a System observes in its input some value that derives from its own output. For example, falling market prices can cause traders to sell (to prevent further loss) and this selling causes prices to fall further (which provokes further selling).

Some feedback loops may have a benign effect whereas others may be malignant; we call these, respectively, *stabilizing* and *destabilizing feedback loops* (Clack et al. 2014). Control systems routinely use stabilizing feedback loops to ensure an output signal stays close to a desired reference signal—see for example (Skogestad and Postlethwaite 2007, Zames 1966, Horowitz 1959).

The process by which an output value is transformed into an observed input value may be complex and transitive (e.g., it may involve intermediate processing by one or more other subsystems); this is illustrated in Figure 20.1 below, where interactions between four subsystems are depicted by directional arrows—one direct feedback loop is depicted by

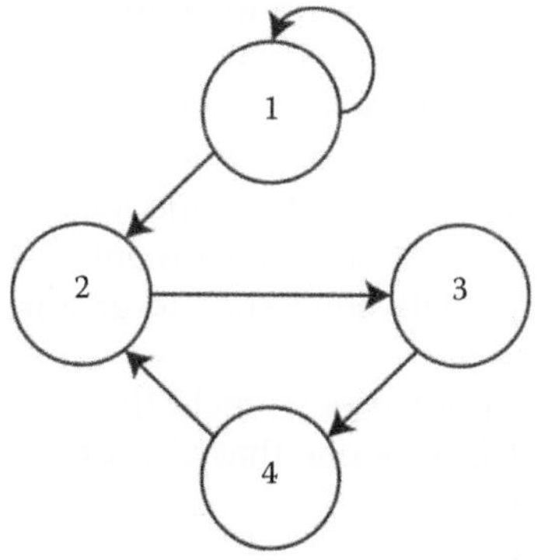

FIGURE 20.1
Four subsystems, with one direct feedback loop and one indirect feedback loop.

an arrow from subsystem 1 to itself, and one indirect feedback loop can be discovered by following the arrows from subsystem 2 to subsystem 3, to subsystem 4, and back to subsystem 2. A practical example of a direct feedback loop would be a trading algorithm that issues orders to a stock exchange where the size and price of those orders is based on the previously issued order (or perhaps on a history of previously issued orders). Alternatively, the size and price of an issued order may depend on the current size of the trader's inventory—when each order is executed at the exchange this will cause the trader's inventory to be changed, and this will affect the next order issued. Notice how in these two examples the feedback loop does not exist at a single point in time but rather *across* time—the trader's inventory is affected by orders filled at a previous time, which are due to orders issued at a time before that, and whose sizes were affected by the size of the trader's inventory at a further previous time. The fact that a feedback loop between Systems extends across time does not make it less real (nor, potentially, less powerful) but it can make it more difficult to analyze.

When constructing a System of Systems model of a financial market, the behavior of the subsystems will be specified and initial conditions will be set. In such a model, feedback loops may exist *ab initio*, and may potentially be discoverable by analysis of the specification and initial conditions. We call this a *static analysis* of the System of Systems specification, because it does not require the behavior of the Systems to be traced (or simulated) in time. Different kinds of static analysis may be required to identify different kinds of feedback loop between Systems—for example, identification of a circular loan relationship between banks, where each is modelled as a separate System, might require a different analysis to identification of a feedback effect on the calculation of order sizes by a trader System.

Alternatively, feedback loops might be created dynamically as a result of the changing behavior of the Systems specified in the System of Systems model (for example, a new loan relationship may create a feedback loop where there was none previously). The role of the content of messages should not be underestimated, since feedback between Systems may for example depend upon the sizes of transactions and whether one System is buying or selling. Similarly, dynamic effects may involve several Systems, and feedback may depend upon repeated patterns of messages occurring between several Systems.

To identify and understand such dynamically-created feedback loops requires analysis of the dynamic behavior of the different Systems in the System of Systems model—for example, it requires the values of different functions and state variables to be traced from one time step to the next, and it may require inspection of the history of messages (including their data content) sent from one System to another. We call this a *dynamic analysis* of the System of Systems model.*

As part of dynamic analysis, we may also wish to investigate whether feedback loops between Systems might over time increase or decrease their effect on various—typically high-level—properties of the System of Systems model, and whether such feedback loops might increase or decrease their number of constituent Systems, break apart to create two or more smaller feedback loops between separate groups of Systems, join together, or disappear.

For a given model of a System of Systems, including its specification and initial conditions, we define a *static feedback loop* as one that always exists, with unchanging size and

* It can be difficult to derive this information from a static inspection of the specification and initial conditions. Indeed, if our models are viewed as executable specifications some properties are undecideable (Turing 1937), yet observations of the dynamics of behavior during simulation can provide valuable insight.

effect. We expect that a static feedback loop is potentially discoverable by a static analysis of the model.* By contrast we define a *dynamic feedback loop* to be any feedback loop that is not static: it may be transient, may have changing size or effect, and might not be detectable via static analysis. We expect that the identification of a dynamic feedback loop would require a dynamic analysis of the model, and we anticipate that dynamic feedback loops might be highly unpredictable and difficult to identify, analyze, and understand. These definitions and behaviors are introduced and discussed further in (Clack et al. 2014).

20.2 Research Focus

Our initial interest in the financial markets derived from reading the reports of the U.S. Commodity Futures Trading Commission (CFTC) and the U.S. Securities and Exchange Commission (SEC) on the "Flash Crash" of May 6, 2010 (CFTC and SEC 2010), coupled with early observations from academics (Easley et al. 2011) and from industry practitioners (Nanex 2010a). We noticed that much of the approach and reasoning that was employed in the analyses immediately following the Flash Crash did not take into account the System of Systems nature of the financial markets; the issue of emergent behavior was not discussed, and although the evidence presented in the CFTC/SEC report described several feedback loops we felt that there was insufficient attention paid to the market impact of low-level dynamic interaction, including these feedback loops. Our research therefore seeks to improve the conceptual understanding of financial markets as a System of Systems with emergent behavior arising from interaction dynamics.

20.2.1 Fine-Grained Microstructure Approach

To investigate the interaction dynamics involved in the Flash Crash, it is necessary to model message-passing at a very fine-grained level. For example, it is necessary to model the timing of the arrival of trader's orders at an exchange, because the sequence in which orders are processed by the exchange might support or defeat a feedback loop. It is necessary not only to model message-passing but also to run experiments to collect and observe the *precise orderings* of messages and their content. Furthermore, several reports (CFTC and SEC 2010, Nanex 2010b) have mentioned the existence of communication delays and it is therefore necessary to model the effects of varying delays. Thus, our task has been to model the detailed microstructure of the financial markets at the level of the passing between entities of individual messages such as orders, confirmations, and market data.

20.2.2 Discrete Time

One challenge in modelling the interaction behavior of financial markets is the linking of the extremely fast behavior of discrete-time computers (and the automated trading and matching algorithms that they run) with the comparatively slow human observation of market behavior.

* Such loops may be intended or unintended.

When viewed at human timescales, the financial markets may appear to operate in continuous time. However, at a fine-grained level of detail all computer operations are effected in discrete time dictated by the change in voltage of a system clock (a chip that emits an extremely precise square-wave oscillating voltage—for example, a 2.7GHz chip oscillates 2,700,000,000 times per second). The passing of messages between two computers is a communication between two discrete-time systems linked by a transmission system; the transmission system itself typically comprises a sequence of cables and intermediary devices, where the intermediary devices may operate in discrete time yet the cables do not operate in discrete time. Despite the continuous-time nature of some parts of the transmission system, a message will only be received by a computer at a discrete time determined by the receiving computer's system clock (a message arriving earlier will not be processed until the next triggering edge of the clock voltage).

Discrete time is also used in coarse-grained models, where in different models an individual time step may represent an intraday period, a day, or a month. An important characteristic of discrete-time models is that each time step represents an equal amount of time; the extent of each time step may be defined (e.g., as a number of seconds, minutes, hours or days), though in some models (e.g., (Huang et al. 2012)) the extent is left unstated in the model and may then be instantiated in a numerical simulation.

20.3 Method

Our approach is to model in discrete time at the finest level of detail appropriate for an experiment, and then to simulate that model for sufficiently many time steps to explore the high-level behavior of interest (this naturally requires computational efficiency). For our purposes, the time gaps between events are just as important as the events themselves, since for example excessive time gaps may represent information delays that have been suggested by some observers to be contributory to emergent behavior such as the Flash Crash (CFTC and SEC 2010, Nanex 2010b). Thus, we require a discrete-time model rather than a discrete-event model; furthermore, our interest is focused on the *interaction* between entities rather than on the entities themselves, and we require the ability to observe these interactions in detail.

We have considered several different approaches to modelling interaction. At first we explored using a process calculus (Milner 1999) to model explicit communication between subsystems (Rötzer 2012), but we found this frustrating since the very low level of specifying communication steps hindered expression of the higher-level algorithms that initiate or respond to that communication. What we sought for exploring interaction dynamics was a technique that occupies a "Goldilocks position" that neither models at a level of abstraction that is too high (e.g., probabilistic modelling) nor at a level that is too low (e.g., process calculus).

An agent-based model (Chan et al. 2010) provides an attractive fit to modelling interaction dynamics in a System of Systems, since each subsystem can be modelled by a separate agent or by a group of agents with varying levels of detail, with private communication within a group; furthermore, messaging and discrete time are handled naturally, and numerical solutions can be obtained for large problems that are analytically intractable. However, we were initially wary of this approach because of the limitations that cause agent-based models to be not widely accepted by economists, as described for example by

Gould et al. (Gould et al. 2013) and Leombruni and Richiardi (Leombruni and Richiardi 2005). For our purposes, the most important limitations appear to be the following:

- Although each agent is fully specified, the model as a whole may lack a formal definition.
- It can be difficult to track how a specified input parameter affects the output, and parameter-estimation may be achieved in a way that is not representative of all possible outcomes the model can produce.
- Finding a set of agent rules that produces a specific high-level behavior provides no guarantee that it is the only set of rules to do so.

Leombruni and Richiardi (Leombruni and Richiardi 2005) resolved the first of these problems by providing a formal specification of an agent-based model in terms of a set of recurrence relations. This has the added advantage that the recurrence relations are easily understandable by non-programmers. They also showed how to ameliorate the second problem by analysis of the sensitivity of outcomes to parameter selections. The third problem remains, and is shared by many models,* yet such models are of great value in hypothesis formulation where they can be used to disprove a statement of the form "This high-level behaviour cannot occur."

20.3.1 Recurrence Relations

Leombruni and Richiardi (Leombruni and Richiardi 2005) have shown how a discrete-time agent-based model of the dynamic microstructure of financial markets can be expressed as a set of recurrence relations.† In their formulation, each of n agents is well described by a state variable given by $x_{i,t}$ where i is the agent identity ($i \in 1, \dots, n$) and t is time, and where each $x_{i,t+1}$ is describable by a recurrence relation as follows:

$$x_{i,t+1} = f_i(x_{i,t}, x_{-i,t}; \alpha_i)$$

In the above equation, the value $x_{i,t+1}$ at time $t+1$ may depend on its previous value $x_{i,t}$ at time t. The state evolution function $f_i()$ may be different for each agent, and each agent may have a bespoke parameter α_i. The parameter $x_{-i,t}$ refers to the states at time t of all agents other than i. Note that the recurrence relation describes a state at time $t+1$ only in terms of states at a *previous* time t (since it is not possible to know future events) and that initial values $x_{i,0}$ are assumed to exist.

Macro-level properties of the model can always be solved by iteratively solving each term $x_{i,t}$. This provides a formal definition of the model that is accessible by domain experts, that expressly indicates the dependency of one subsystem on another (i.e., if $x_{3,t}$ occurs in the definition of $x_{1,t+1}$ then we say that $x_{1,t+1}$ depends on $x_{3,t}$) and that is amenable to static analysis (e.g., it would be possible to perform an automatic dependency analysis). It is also possible to reason about information delays if we generalize the model to $x_{i,t+1} = f_i(x_{i,t}, x_{i,-t}, x_{-i,t}, x_{-i,-t}; \alpha_i)$ where the subscript $-t$ refers to all times previous to t. Thus, it is for example possible to express a dependency of $x_{1,t+1}$ on $x_{3,t-4}$.

* We recall Box's dictum that "All models are wrong; some models are useful." (Box 1976).

† Recurrence relations (and difference equations) are widely used in economics and finance—for example, see (Day and Huang 1990, Huang et al. 2012, Rosu 2009).

Although recurrence relations are naturally recursive, this does not mean that destabilizing feedback loops necessarily exist in a model that uses recurrence relations. For example, the relation $x_{t+1} = x_t$ represents a harmless static value. By contrast, the relation $x_{t+1} = x_{t+1}$ is badly formed (upon evaluation it loops forever). Nor is it true that the only destabilizing feedback loops are those that are specificly coded ("hard-wired") into a recurrence relation, since an unintended and unidentified feedback loop may exist between a group of several recurrence relations.

There are many benefits that accrue from the use of a set of recurrence relations to model market behavior, and such a model is very amenable to static analysis. However, a set of recurrence relations is not as well suited to dynamic analysis:

- although dependencies are explicit, viewing interaction as the sending and receiving of messages is less explicit; and
- although an individual recurrence relation provides a time history of values for a given subsystem, we do not have a time history of messages between two selected subsystems.

For the analysis of interaction dynamics, we wish to investigate not only the history of the states of an agent (i.e., the sequence $x_{i,0}, x_{i,1}, \ldots$ for every agent i) but also the history of messages between every pair of agents. For example, if $x_{1,t+1}$ depends on its previous state at time t and on the state of agent x_2 at time $t-3$ (i.e., the dependency set is $\{x_{1,t}, x_{2,t-3}\}$) then the history of dependencies at each time step ($t = 1, t = 2, t = 3, t = 4, t = 5, \ldots$) for x_1 is given by the sequence $\{x_{1,0}, x_{2,-2}\}, \{x_{1,1}, x_{2,-1}\}, \{x_{1,2}, x_{2,0}\}, \{x_{1,3}, x_{2,1}\}, \{x_{1,4}, x_{2,2}\}, \ldots$ and if we say $x_{i,t}$ is undefined for $t < 1$ (i.e., no dependency exists) we have: $\{\}, \{x_{1,1}\}, \{x_{1,2}\}, \{x_{1,3}, x_{2,1}\}, \{x_{1,4}, x_{2,2}\}, \ldots$

Finally we note that where the function $f_i()$ contains conditional statements, e.g., where the state $x_{i,t+1}$ is conditioned on some value $x_{j,t}$, then the set of dependencies may be conditional and in the general case the dependency set becomes more difficult to analyze.*

By contrast, with an agent-based simulator it is typically a simple matter to investigate the interaction dynamics by observing actual messages transmitted from one agent to another.

20.3.2 Two Views

Ideally we would like to perform both static analysis of a System of Systems model expressed as a set of recurrence relations and dynamic analysis of an agent-based model of the same System of Systems. We consider these to be two complementary views of the same System of Systems. Eventually, we would like to be able to express the System of Systems as a set of recurrence relations and then generate the agent-based model automatically; if this could be done in a way that preserves the semantics of the specification, then we would be confident that the static and dynamic analyses are being performed on models of the same System of Systems.

A numerical simulator for an agent-based model of interaction dynamics for System of Systems has been developed and will be presented in the following section. This simulator

* It is similar to the problem of determining for a function $g(x,y)$ whether evaluation of that function will necessarily require evaluation of its arguments x and y (Clack and Peyton Jones 1985a, Clack and Peyton Jones 1985b).

(called "InterDyne" (Clack 2016)) has been operational for some time and has been used to model a variety of financial Systems of Systems and to explore emergent behavior in those systems.

We have also performed static analyses on specifications of financial Systems of Systems expressed as a set of recurrence relations. Creating the link between the two views is the focus of current research, and we are exploring the approach of using a sequence of correctness-preserving transformations that would make incremental modifications starting with the set of recurrence relations and ending with agent code that can be used in the InterDyne simulator.

20.4 The InterDyne Simulator

InterDyne is an agent-based simulator for the exploration of interaction dynamics in complex Systems of Systems. Although InterDyne is a general-purpose simulator that can be used in a variety of fields it has primarily been used to investigate models of financial markets. In this section we summarize and discuss the detailed information provided in the InterDyne User Manual (Clack 2016).

The structure of an InterDyne simulation is straightforward, with one or more agents passing messages via a "Simulator Harness" as illustrated in Figure 20.2, and with each subsystem in the System of Systems model being represented by a single agent or by a collection of agents. The primary output from InterDyne is a trace file, suitable for post-hoc analysis and interactive visualization to explore the antecedents of emergent behavior.

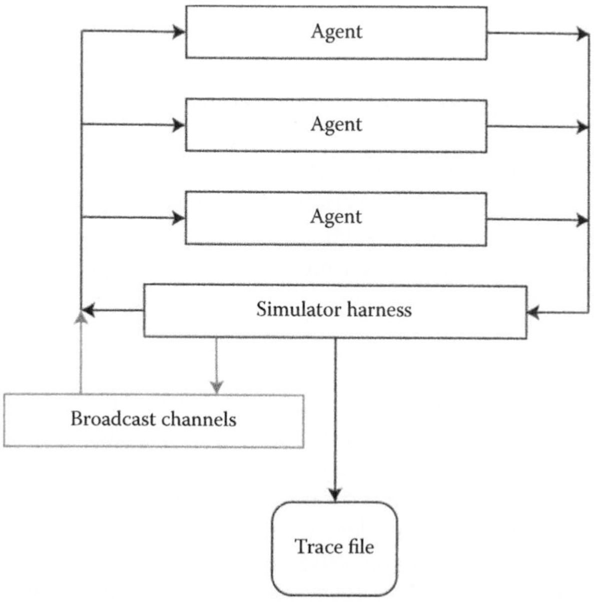

FIGURE 20.2
InterDyne simulation with three agents.

InterDyne is run by executing the function "`sim`" applied to appropriate arguments (see Figure 20.4). The arguments include:

- the number of time steps for the simulation,
- a list of runtime arguments: these are (key, value) pairs that are made available to every agent in the system, and
- a list of information about each agent: namely, a two-tuple containing the name of the function that implements the agent, and a list of broadcast channel identifiers (see Section 20.4.3) to which the agent will listen.

Each agent is uniquely identified by its position in that list of agent information—the first agent has identifier 1, the second has identifier 2, and so on. Identifier 0 is reserved for the simulator "harness" function that mediates messaging, controls the passage of time during InterDyne simulation, and sends output to a trace file. The agent identifiers are used to specify the source and destination of all one-to-one messages.

Each agent receives a sequence of packets containing messages and outputs a sequence of packets containing messages; at each time step an agent both consumes one input packet and generates one output packet. Each output packet contains zero or more messages (of varying type—see Section 20.4.3), and each input packet contains three items: (i) the current time, (ii) a collection of zero or more direct messages, and (iii) a collection of zero or more broadcast messages. Each agent only receives a direct message if its agent identifier is in the recipient field of the message (i.e., agents do not see messages sent to other agents), and only receives broadcast messages sent to broadcast channels to which this agent has subscribed. Broadcast messages support sending the same message to a large number of other agents, such that the recipients can be defined and changed in the set-up of the experiment rather than in the agent code.

20.4.1 InterDyne Time

InterDyne operates in discrete time, and simulation experiments are executed for a defined number of time steps. InterDyne does not specify what period of real time corresponds to each time step; this is a matter for the experimenter to define, according to the requirements of the model and its simulation, and a time step is typically set to be the smallest required resolution with all other times being integer multiples of that time. The experimenter may also choose not to give such a definition; for example, where the behavior being investigated is known to be or assumed to be rate-independent.

20.4.2 InterDyne Agents

InterDyne supports a heterogeneous set of communicating agents, and each agent can be modelled with a different level of detail. For example, a System of Systems model might comprise two agents: one that generates outputs dependent on a statistical distribution together with a second agent that is a thorough implementation of a complex trading algorithm with considerable internal complexity.

At each time step an agent must consume one inbound message packet and must generate one outbound message packet. Each packet may contain multiple messages and inbound packets are different to outbound packets (see later example). If an agent does not have any messages to receive or send at a given time step, then it will either receive

```
f st args ((t, msgs, bcasts):rest) myid = []:(f st args rest myid)
```

FIGURE 20.3
A simple agent function.

or generate an empty packet. Optionally an agent may distinguish between an output packet that is empty by mistake and an output packet that is empty by design—it does this by generating an output packet containing the distinguished empty message called a "Hiaton."*

InterDyne agents are typically (but not always) written in two parts: (i) a "wrapper" function that manages the consumption of inbound messages, the generation of outbound messages, and the update of local state, and (ii) a "logic" function that is called by the wrapper function and which calculates the messages to be sent. The "wrapper" function is the agent function, and the "logic" function is a subsidiary function with restricted scope so that it can only be invoked by the wrapper function.

Figure 20.3 illustrates a simple agent wrapper that does nothing and does not call a logic function, written in the programming language Haskell:† at each time step it reads an inbound item, and creates an empty outbound item (the inbound item is not used). An inbound packet is represented as a Haskell three-tuple (discussed below), and a sequence of packets is represented as a list of packets.

In Figure 20.3 the agent function "`f`" is recursively defined and loops once per time step. A typical Haskell syntax is to write the name of the function, followed by names for its arguments, to the left of the equals sign: to the right of the equals sign is the expression that calculates the value returned by the function. In this case, the function "`f`" takes four arguments: `st` is a local state variable (in this example it is never inspected and never changed), `args` is a copy of the runtime arguments (every agent is passed a copy of all the runtime arguments), and the penultimate argument is the list of inbound packets written as a structural pattern so that individual parts can be named: the first available inbound packet is a three-tuple and the remainder of the list (the packets that will be available in the future) is given the name "`rest`"; the components of the first inbound packet are (i) the current time "`t`"; (ii) a list "`msgs`" of all one-to-one messages sent to this agent to be received in this time step; and (iii) a list "`bcasts`" of all broadcast messages available at this time step on all the broadcast channels (see Section 20.4.3) to which this agent is subscribed. The last argument `myid` is the identifier of this agent (which should never be changed).

The output of the function is a Haskell list of outbound packets, one per time step, and each outbound packet is a list of messages (including both one-to-one messages and broadcast messages). Thus, the output comprises a first list item (an outbound packet—in this case the empty list "`[]`") connected via the Haskell operator ":" to the remainder of the list items (given by the result of the recursive call). In the recursive call, the three-tuple "`(t, msgs, bcasts)`" is missing from the third argument (only the tail of the list, called "`rest`" in this example, is used), and so in the next recursive invocation the function `f` will observe at the front of the list the inbound packet corresponding to the following time step.

Figure 20.4 illustrates two versions of the function `sim` running an InterDyne simulation with three agents. In both examples the simulator runs for 60 time steps. The second

* Faustini (Faustini 1982) asserts that this term is due to W. Wadge and E. Ashcroft.

† The current version of InterDyne (and therefore the agent code) is written in the programming language Haskell (Hudak et al. 1992). This aligns with our view of modelling a System of Systems as an executable specification (Turner 1985).

```
exampleExperiment1
 = do
   sim 60 [] agents
   where
   agents  = [ (traderWrapper,    []),
               (brokerWrapper,    []),
               (exchangeWrapper, [])  ]

exampleExperiment2
 = do
   sim 60 [convert] (map snd agents)
   where
   convert = generateAgentBimapArg agents
   agents  = [ ("Trader",   (traderWrapper,    [1] )),
               ("Broker",   (brokerWrapper,    [3] )),
               ("Exchange",(exchangeWrapper, [2,3]))  ]
```

FIGURE 20.4
Two simple code examples for running an InterDyne simulation.

argument is a list of runtime arguments—empty for the first example, and the single argument `convert` in the second example. The argument `convert` is an application of the library function `generateAgentBimapArg` to `myagents` which results in a runtime argument containing a function that converts a name to an agent identifier and vice versa.* The third argument is a list of agents and broadcast channels on which they will listen—in the first example there are no subscriptions. In the second example the first agent subscribes to broadcast channel 1, the second subscribes to channel 3, and the third subscribes to channels 2 and 3.†

These examples omit many details—for example they do not give the definitions for the agent functions `traderWrapper`, `brokerWrapper`, and `exchangeWrapper`; they do not illustrate how to define an output file for the results; nor how to use names instead of integers for broadcast channels; nor how to specify the communications topology and the delays that should be applied to each communication link (see Section 20.4.4). They do however indicate the parsimonious style that can be achieved when using InterDyne.

20.4.3 InterDyne Interaction and Messages

InterDyne supports message-based communication between agents as the only form of interaction. Messages are directed (a message is sent from one agent to one or more others), and where two agents need to send messages to each other simultaneously this is modelled as two separate directed messages. InterDyne supports a wide range of messages from the very simple (for which examples are given below) to the very complex (for example, detailed messages between subsystems of a financial system using the

* Messages (see Section 20.4.3) can be sent using agent identifiers or agent names. To achieve the latter, the user includes in the list of runtime arguments a function that when executed will convert an agent identifier into a name or vice versa. Runtime arguments are available to all agents, and the user can change the mapping each time `sim` is called.

† In the second example, the Haskell function "map" takes two arguments, a function and a list, and applies the function to every item in the list. The function "snd" when applied to a structure such as "("Trader", (traderWrapper,[1]))" returns the second component, which is "(traderWrapper, [1])".

industry-standard Financial Information eXchange protocol). The messaging system is of central importance to InterDyne since messages are the medium for interaction between subsystems.

This communication can be one-to-one (a private message from one agent to another) or one-to-many (a broadcast message which can be read by all other agents). Both of these messaging types are needed in modelling financial markets; for example, an exchange would send a broadcast message to all members updating them on the latest executed trades, whereas a trader would send a private message to an exchange to place an order. Broadcast messages are directed to a broadcast channel, and agents subscribe to zero or more broadcast channels at the start of a simulation.

All messages comprise (i) a tag to indicate the type of message being sent; (ii) a pair of integers that indicate either the identifiers of the sending and receiving agents (for one-to-one messages) or the sending agent identifier and the receiving broadcast channel identifier (for broadcast messages); and (iii) the message data. The message types (and associated tags) include both generic types and a few domain-specific types. Examples include:

- `Message (1, 2) data`—a one-to-one message, from agent 1 to agent 2, typically sending a list of (key, value) pairs
- `Ordermessage (3, 4) data`—a domain-specific one-to-one message, from agent 3 to agent 4, sending data that represents an order (typically sent to a subsystem that models an exchange)
- `Datamessage (3, 1) data`—a one-to-one message, from agent 3 to agent 1, where the data component is a string
- `Debugmessage (1, 4) data`—a one-to-one message, from agent 1 to agent 4, where the data component is a string (for debugging purposes)
- `Broadcastmessage (3, 1) broadcastdata`—a broadcast message, sent from agent 3 to broadcast channel 1, containing data that will be received by all agents that have previously subscribed to broadcast channel 1

If a message is sent to agent identifier 0, it receives special treatment. Agent identifier 0 is reserved for the simulator harness, and a message sent to identifier 0 is printed to an output file. This is the primary mechanism for defining the output of the simulator. Currently the main output file is the "trace file" that records all messages sent to identifier 0 except for those with type `Datamessage`—these messages are instead output to a comma-separated file, and this provides a mechanism for structured output that can be viewed as a spreadsheet.

Figure 20.5 illustrates a simple agent wrapper that sends a debug message to the output trace file (via the simulator harness) at every time step. It does not call a logic function:

```
f st args ((t, msgs, bcasts) : rest) myid
     = [m]  : (f st args rest myid)
       where
       m = Debugmessage (myid,0) ("Debug "++(show t))
```

FIGURE 20.5
A simple agent wrapper.

20.4.4 InterDyne Delays and Topology

An important aspect of interaction dynamics is the role played by communication delays. In our discrete-time models there is always a minimum communication delay in that we assume a message sent at time step *t* will not be received until time step *t+1*. Where greater time delays are required, these could be implemented in a variety of ways; for example, an agent could implement code that explicitly puts some or all messages into a delay queue prior to processing, or explicit delay agents could be used. Having explored a number of different mechanisms, we have chosen to extend InterDyne to support (but not mandate) the use of delay information that is passed to the simulator as an optional runtime argument.

The delay information provides a unique delay value (as an integer multiple of time steps) for each directed communication path between two agents. A specified delay applies to any messages using that interaction path, whether the message is defined as being one-to-one or one-to-many. Since the communication paths are directed, asymmetric delays can be specified between two agents: for example, messages sent from agent A to agent B may have a greater delay than messages sent from agent B to agent A. Passing the delay information as a runtime argument provides a uniform mechanism where the amounts of delay between different agents can be varied systematically without requiring agent code to be changed.

This approach also provides an opportunity to define the topology of connections between InterDyne agents: if messages are not permitted from Agent A to Agent B. then the communication delay for that path could be set either to an abort error message or to a delay that is longer than the expected length (in time steps) of the experiment.* This defines the interaction topology of the simulation as a directed graph, with the agents being the nodes of that graph and the communication paths being the edges.

The interaction topology can be used as a validation tool (to test a complex experiment to determine whether a subsystem is sending messages as expected), but normally is used as part of the mapping from model to simulation, to reflect a semantic interpretation of the subsystems of the modelled System. For example, as illustrated in Figure 20.6, an investment bank subsystem might be modelled as two agents—a sales agent and a trader agent; the sales agent may communicate with several client subsystems and pass client orders to the trader agent for execution, and the trader agent would send orders to a stock-exchange subsystem, yet the sales agent may be prohibited from sending/receiving messages directly to/from the exchange, and the trader may be prohibited from sending/receiving messages directly to/from the clients. This provides a way to express a degree of hierarchy of systems and subsystems in an overall System of Systems.†

To define delays and interaction topology, two runtime arguments must be passed to the `sim` function (both arguments must be present, or none):

1. the name of a function that takes two agent identifiers (integers that uniquely specify the start point and end point of an interaction) and returns an integer delay which is taken to be a number of time steps; and
2. the maximum delay in the system.

* Optionally an error message can also be sent to the trace file without aborting.

† InterDyne is not fully hierarchical—an agent cannot itself be another InterDyne simulation.

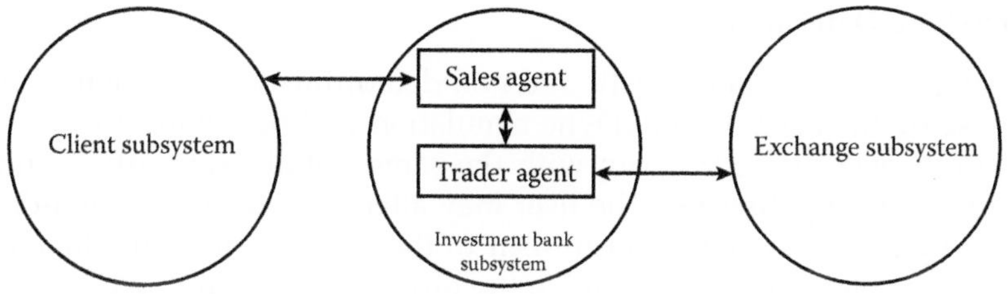

FIGURE 20.6
Subsystems, agents, and permitted communication paths.

The user has complete freedom to define the delay function, including sending trace messages and raising errors if an attempt is made to send a message on a path that is not permitted. The simulator harness (agent identifier 0) detects the presence of these two runtime arguments and, if they are present, uses the function to calculate the required delay for every sent message (agents normally ignore these two runtime arguments). In the current implementation, the delay function cannot be conditioned on other factors such as time or local state, and can therefore only implement static delays (which, of course, could be changed for different runs of an experiment). To implement dynamically-varying delays the user would create a delay agent to act as an intermediary on a given communication path.

Figure 20.7 gives a simplified example for three agents: runtime arguments are (key, value) pairs: the maximum delay argument has a string key `"maxDelay"` and value `maxDelay`, and is passed as a runtime argument using the tag `Arg` (a Haskell constructor), whereas the delay function has the key string `"DelayArg"` and uses the Haskell constructor `DelayArg`. The maximum delay is a whole number of time steps—in this case three.

After setting the delays as illustrated in Figure 20.7, one-to-one messages are automatically delayed by the stated number of time steps (additional to the minimum communication delay of one time step), and broadcast messages are split into separate messages for each recipient and delayed by the amount stated for each communication path (these "routed broadcasts" are received in the list of normal messages in an agent's input).

```
exampleExperiment
= do
  sim 60 args agents
  where
  args      = [ (Arg   (Str "maxDelay", maxDelay)),
                (DelayArg (Str "DelayArg", delay)) ]
  maxDelay  = 3
  delay x y = if ((x = 1) && (y = 2)) then 1 else
              if (x = 2) then 0 else
              if ((x = 3) && (y = 2)) then 3 else
              error ("bad message: from "++show x++" to "++show y)
  agents    = [ (traderWrapper, [1] ),
                (brokerWrapper, [3] ),
                (exchangeWrapper, [2,3]) ]
```

FIGURE 20.7
Simplified code to set communication delays.

20.4.5 InterDyne Determinism

InterDyne supports both deterministic and non-deterministic simulations. Unless specified otherwise by the user, an InterDyne simulation will be deterministic and provide the same results every time it is run with the same initial values (this can be helpful in determining causal pathways). The user may alternatively express non-determinism in two ways: by including non-determinism in the code for an individual agent, and/or by instructing InterDyne to randomize the order in which multiple inter-agent messages are received at each time step (the alternative is that where multiple messages are received in one time step they are processed in order according to their agent identifier). Randomizing message arrival is effective where an agent receives more than one message from another agent in one time step, and can be useful to remove suspected systematic bias when exploring dynamic behavior (for example, a feedback loop might be manifest if a subsystem representing a stock exchange always processes orders from one trader subsystem first, yet may disappear when this systematic bias is removed). The re-ordering of messages is managed in the same way every time the simulator is run and is therefore repeatable (if different behavior is required, different randomization seeds can be used on each run).

20.5 Case Study

To illustrate the use of InterDyne to explore interaction dynamics, we investigate the emergent phenomenon of "Hot Potato" trading, as observed for example in Foreign Exchange (Lyons 1997) and Futures (CFTC and SEC 2010, Kirilenko et al. 2014) markets.

We define Hot Potato trading as the *repeated* passing of inventory imbalances (i.e., trading) between dealers (market makers) (Lyons 1997), and we define Hot Potato Instability as the repeated passing of inventory imbalances between dealers when market supply and demand are zero. Hot Potato trading starts when a dealer with excess inventory sells to another dealer: if that causes the second dealer to have excess inventory, the second dealer sells to another dealer; if the inventory returns to the first dealer, a cycle is created and repeats until non-dealer trades remedy the imbalances. Hot Potato trading may be prompted by repeated market supply or demand: by contrast Hot Potato Instability may have been initially triggered by market supply or demand but thereafter repeats without further input from the market (it is thereafter a self-exciting oscillation).

We start with a simple problem (we call it "the *n-dealer* problem" for Hot Potato trading on a limit order book (Clack 2013)): given n dealers (market makers) who, being averse to inventory risk, employ a threshold strategy to avoid excessively high or low inventory, and each of which submits orders to an exchange running a limit order book, in the context of a market with supply s and demand d, under what circumstances can those dealers be compelled to enter into Hot Potato Instability?*

* The *n-dealer* problem for Hot Potato trading in an order-book market is important since this phenomenon was observed at the heart of the Flash Crash (CFTC and SEC 2010).

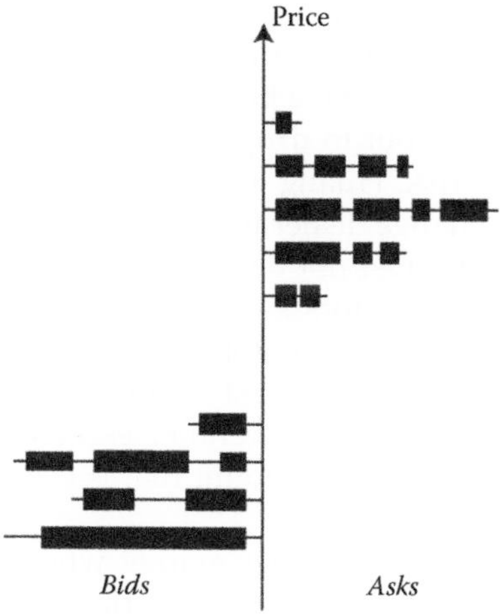

FIGURE 20.8
A limit order book stores limit orders, ordered by price: "bids" to buy, and "asks" to sell, a quantity at a price. At each price there is a queue of orders sorted by arrival time. An arriving "buy" market order is matched against the lowest-priced "ask" with earliest arrival time; a trade executes at that price for the smaller of the "buy" and "ask" quantities—any remaining "ask" quantity stays in position in the queue, and any remaining "buy" quantity is matched against the next lowest-priced "ask." This is mirrored when a "sell" market order is matched against "bid" limit orders.

To explore this problem, we consider a System of Systems comprising the following subsystems, each implemented as a single agent:*

- A subsystem that implements the behavior of a financial exchange operating a central limit order book (see Figure 20.8), e.g., for the trading of equities (shares) in a single company.
- *n* subsystems, each of which implements the behavior of a dealer that both buys and sells equities and makes a profit from the difference between the buy price and the sell price. Each dealer will implement an inventory threshold policy that controls the types, sizes, and prices of orders sent to the exchange.

Market supply and demand prior to the Hot Potato behavior are implemented by proxy—i.e., by the setting up of initial inventories for the dealers. So the research question is whether it is possible to trigger Hot Potato Instability (which requires no further market supply or demand) by some prior pattern of market behavior that sets the dealers' inventories in such a way that the dealers enter into unstable trading between themselves.

The interaction topology of this System of Systems is defined such that the exchange can send and receive messages to and from all dealers (and vice versa) but none of the dealers can interact with each other directly.

Each dealer must implement a threshold policy for inventory control. Dealers are typically subject to multiple risks, with the two predominant risks being inventory risk (for

* This System of Systems is highly simplified. In reality there would for example be a very large number of different kinds of traders and market makers, and there would be multiple exchanges, trading equities for multiple companies. Furthermore, we assume that the market makers implement identical policies, though in reality they would differ.

example, if too high an inventory were kept, then its value might suddenly diminish if the market price were suddenly to drop) and adverse selection risk (for example, that a trader who sells to the dealer might have better information than the dealer and might know that the market price is about to drop). In the *n-dealer* problem only one type of risk is considered—inventory risk. Huang et al. (Huang et al. 2012) have shown that an optimal way for a Foreign Exchange dealer to manage inventory risk is to use a dynamic threshold, where the inventory must not be permitted to exceed a threshold value that may vary according to dynamic factors such as the size of the spread (the difference between buy and sell prices). When inventory does not exceed a given threshold, the dealer submits both a "bid" quote and an "ask" quote and profits from the spread when both orders are executed. By contrast, if inventory exceeds a given threshold, a market order is issued to another dealer for an amount that is exactly equal to the amount that exceeds the threshold.

The Huang et al. model assumes a dealer market where prices are constrained by a dominant dealer, and so there is no control of the prices or sizes of the quotes issued by the dealer whose inventory is optimized. By contrast, in our model we have an order-book market and we will provide more subtle inventory control by allowing the market maker to vary both the prices and sizes of its "bid" and "ask" limit orders; for example, as inventory approaches (but is not in excess of) the upper threshold, the market maker may make the size and price of its "bid" orders unattractive to sellers while making the size and price of its "ask" orders more attractive to buyers (the combined effect of which should be to reduce the dealer's inventory).

Given the greater control afforded to market makers in an order-book market, it is not obvious how Hot Potato behavior could be triggered. Given that a market maker whose inventory is just below the upper threshold could opt to issue no bid orders at all* or could issue a bid for a price that is below other bids and therefore is unlikely to be executed, how could such a market maker be induced to buy inventory that it does not want to buy?

20.5.1 Time

Our model assumes that all subsystems have synchronized clocks with no time drift between subsystems. All communications delays are assumed to be equal.

We let the minimum quantum of time be the total amount of time taken for data to be communicated, for calculations to be made, and for a new datum to be generated: in this simple model we assume that this is equal for all subsystems interacting via any communication path.

At the start of time t, a subsystem can access the values of other subsystems at time $t-x$, where $x>=1$, and generates its own datum for time t. This datum will then be available to other subsystems at time $t+y$, where $y>=1$.

20.5.2 Static Analysis

In this section we show how static analysis can be performed on a recurrence-relation model of the defined System of Systems. Further detail, and a slightly different presentation, can be found in (Clack et al. 2014).

* Although this may come at a cost: several exchanges offer a rebate to dealers who agree always to have a bid and an ask submitted to the order book; if no bid or ask were received, an exchange might remove the rebate.

20.5.2.1 The Exchange Subsystem

We present recurrence relations to define the dynamic behavior of the two order books of bids and asks respectively (*ob_bids* and *ob_asks*), of the market orders that have been executed (*xsells* and *xbuys*), and of the limit orders that have been executed (*xbids* and *xasks*). These recurrence relations assume appropriate definitions of operators "⊕", "∅", "⊗", and "§" as explained below:

- $(x \oplus y)$ incorporates limit orders contained in "y" into the order book "x" using the correct price-time sequence (Gould et al. 2013). The result is an order book.
- $(x \varnothing y)$ takes the market orders contained in "y" and calculates which limit orders in the order book "x" should be executed against those market orders and should therefore be removed from "x". The result is an order book.
- $(x \otimes y)$ calculates which of the market orders contained in "y" will be executed against the limit orders held in order book "x". The result is a set of market orders, and if either x or y are empty or undefined, the result is the empty set *{}*.
- $(x \oplus y)$ calculates which of the limit orders in the order book "x" will be executed against the market orders held in "y". The result is a set of limit orders, and if either x or y are empty or undefined, the result is the empty set *{}*.

The empty order book is given by *ob_empty*. To simplify the model, we assume that all limit orders are good for only one time step—if they are not executed within one time step of being added to the order book, they are discarded (this requires the dealers to issue new limit orders at each time step).

Given the above operators, Figure 20.9 illustrates the required recurrence relations. Notice that $xsells_t$, for example, depends on ob_bids_t which in turn depends on lo_bids_{t-1}: in calculating the set of executed sells at the start of time t, the exchange uses the bids that were issued at the start of the previous timestep t–1. The market orders *mo_sells* and *mo_buys* and limit orders *lo_bids* and *lo_asks* are the combined orders that are submitted to the exchange by the dealers.

$$
\begin{aligned}
ob\_bids_t &= (ob\_empty \oplus lo\_bids_{t-1}) \varnothing mo\_sells_{t-1} \\
ob\_asks_t &= (ob\_empty \oplus lo\_asks_{t-1}) \varnothing mo\_buys_{t-1} \\
xsells_t &= ob\_bids_t \otimes mo\_sells_{t-1} \\
xbuys_t &= ob\_asks_t \otimes mo\_buys_{t-1} \\
xbids_t &= ob\_bids_t \oplus mo\_sells_{t-1} \\
xasks_t &= ob\_asks_t \oplus mo\_buys_{t-1} \\
mo\_sells_t &= (\cup_i\, dealersells_{i,t}),\ \ for\ i \in \{1..n\} \\
mo\_buys_t &= (\cup_i\, dealerbuys_{i,t}),\ \ for\ i \in \{1..n\} \\
lo\_bids_t &= (\cup_i\, dealerbids_{i,t}),\ \ for\ i \in \{1..n\} \\
lo\_asks_t &= (\cup_i\, dealerasks_{i,t}),\ \ for\ i \in \{1..n\}
\end{aligned}
$$

FIGURE 20.9
Recurrence relation model (exchange).

$$dealerbids_{i,t} = \{(bidprice_{i,t}, bidsize_{i,t})\}, \ if\ (lthresh_{i,t} < inv_{i,t} < uthresh_{i,t})\ and\ even(t)$$
$$= \{\}, \ otherwise$$

$$dealerasks_{i,t} = \{(askprice_{i,t}, asksize_{i,t})\}, \ if\ (lthresh_{i,t} < inv_{i,t} < uthresh_{i,t})\ and\ even(t)$$
$$= \{\}, \ otherwise$$

$$dealersells_{i,t} = \{(inv_{i,t} - uthresh_{i,t})\}, \ if\ (inv_{i,t} > uthresh_{i,t})\ and\ even(t)$$
$$= \{\}, \ otherwise$$

$$dealerbuys_{i,t} = \{(lthresh_{i,t} - inv_{i,t})\}, \ if\ (inv_{i,t} < lthresh_{i,t})\ and\ even(t)$$
$$= \{\}, \ otherwise$$

$$inv_{i,t} = inv_{i,t-1} + xbuysizes_{i,t-1} - xsellsizes_{i,t-1} + xbidsizes_{i,t-1} - xasksizes_{i,t-1}, \ if\ even(t)$$
$$= inv_{i,t-1}, \ otherwise$$

$$xbuysizes_{i,t} = \psi_i(xbuys_t)$$

$$xsellsizes_{i,t} = \psi_i(xsells_t)$$

$$xbidsizes_{i,t} = \psi_i(xbids_t)$$

$$xasksizes_{i,t} = \psi_i(xasks_t)$$

FIGURE 20.10
Recurrence relation model (dealer). The function ψi() is described in Section 20.5.2.2.

20.5.2.2 The Dealer Subsystems

To complete this very simple model, Figure 20.10 provides definitions for the individual limit and market orders issued by the dealers (where *lthresh* is the lower threshold and *uthresh* is the upper threshold).

Orders are only issued on even time steps because it takes two time steps for inventory to be updated as a result of the execution of a previous order. Consequently:

$$inv_{i,t} = inv_{i,t-1}, \ if\ odd(t) \tag{20.1}$$

The above definitions assume that a dealer maintains inventory between two thresholds (upper and lower), as described in (Huang et al. 2012). We assume that the upper and lower threshold will vary dynamically, but we do not define them further. The quantities *dealersells* and *dealerbuys* are similarly taken from (Huang et al. 2012). The quantities *bidprice, bidsize, askprice,* and *asksize* will vary according to inventory and perhaps also according to other factors (e.g., to avoid adverse selection risk), but for the moment are not defined further. The function $\psi_i(\ )$ takes a set of executed orders, extracts the subset that was originally issued by dealer *i*, and then extracts and sums the executed sizes within that subset (i.e., it calculates the inventory impact—the number of assets bought or sold).

From inspection of dependencies in this model, we can determine the following:

- A feedback dependency involving a single dealer: a dealer's inventory at time *t* depends on its inventory at a previous time (since the latter determines the orders previously sent to the exchange, from which trades are executed, potentially causing inventory to change). Of course, dealers would not issue orders if they could not influence future inventory!

- A feedback dependency involving two dealers: for example in a two-dealer system $inv_{1,t}$ depends on $xsellsizes_{1,t-1}$, which depends on $xsells_{t-1}$, which depends on ob_bids_{t-2}, which depends on lo_bids_{t-2}, which depends on $dealerbids_{t-2}$, which depends on $inv_{2,t-2}$. In the same fashion $inv_{2,t-2}$ depends on $inv_{1,t-4}$.

These feedback dependencies (and other dependencies) may or may not be easy to see in the recurrence relations, but are amenable to static analysis (see below), and although no value depends on the same value at the same time, these dependencies may have the potential to become destabilizing.

To gain some initial insight into whether the system is actually stable or unstable, we perform a static analysis of the recurrence relations.

Analyzing the values of the recurrence relations requires us to be more specific about certain values, for example the definitions of *bidsize* and *asksize*, and the starting inventories for the dealers. Arbitrarily, for the purposes of example, we will use the variables $UT_{i,t}$ and $LT_{i,t}$ to denote the upper and lower thresholds, and $BP_{i,t}$ and $AP_{i,t}$ will denote the bid and ask prices for dealer *i*—we assume that $max(BP_{i,t}) < min(AP_{i,t})$ throughout so that limit orders never execute against each other (moreover, we require that market orders never execute against each other). The bid and ask sizes can be any value that would not, if executed in full, cause the inventory to exceed $UT_{i,t}$ or $LT_{i,t}$. Thus the upperbound for *bidsize* is given by:

$$bidsize_{i,t} < UT_{i,t} - inv_{i,t} \tag{20.2}$$

and the upperbound for *asksize* is given by

$$asksize_{i,t} < inv_{i,t} - LT_{i,t} \tag{20.3}$$

The lowerbound for both *bidsize* and *asksize* is zero.

20.5.2.2.1 Dealers Issuing Limit Orders

Let there be one or more dealers having inventories between the two thresholds (therefore issuing bid and ask limit orders). These dealers will not initially issue any buy or sell orders. Thus, we can ask, "What is the maximum possible inventory for such a dealer?" The inventory equation can be expanded as illustrated in Figure 20.11.

Thus, those dealers that start with an inventory below the upper threshold will never equal or exceed the upper threshold $UT_{i,t-2}$. By a similar expansion it is possible to show neither will they ever equal or exceed the lower threshold.

20.5.2.2.2 Dealers Issuing Market Orders

Let there be one or more dealers having initial inventories exceeding a threshold. These dealers will not initially issue any bid or ask orders. Thus, we can ask, "What is the maximum possible inventory for a dealer whose initial inventory is below the lower threshold?" The inventory equation can be utilized and expanded as illustrated above, with the result that any dealer whose inventory starts below the lower threshold can raise its inventory at most to the value of the lower threshold (at which point it issues no orders at all). A similar finding holds (mirrored) for those dealers whose inventories start above the upper threshold. Furthermore, any dealer whose inventory starts below the lower threshold only issues

$$
\begin{aligned}
&MAX(inv_{i,t}) \\
&= inv_{i,t-1} + MAX(xbidsizes_{i,t-1} - xasksizes_{i,t-1}),\ \textit{if even(t)} \\
&= inv_{i,t-1} + MAX(xbidsizes_{i,t-1}) && \textit{(maximised if no asks are executed)} \\
&= inv_{i,t-1} + MAX(\psi_i(xbids_{t-1})) \\
&= inv_{i,t-1} + MAX(\psi_i(ob\_bids_{t-1} \oplus mo\_sells_{t-2})) \\
&= inv_{i,t-1} + MAX(\psi_i(((ob\_empty \oplus lo\_bids_{t-2}) \varnothing\ mo\_sells_{t-2}) \oplus mo\_sells_{t-2})) \\
&= inv_{i,t-1} + MAX(\psi_i(lo\_bids_{t-2})) && \textit{(maximised if all bids are executed)} \\
&= inv_{i,t-1} + MAX(\psi_i((\cup_j\ dealerbids_{j,t-2}),\ \textit{for } j \in \{1..n\})) \\
&= inv_{i,t-1} + MAX(\psi_i(dealerbids_{i,t-2})) && (\Psi_i \textit{ only considers bids for this dealer}) \\
&= inv_{i,t-1} + MAX(bidsize_{i,t-2}) && (\Psi_I \textit{ returns the total size of the bids}) \\
&= inv_{i,t-1} + (UT_{i,t-2} - inv_{i,t-2} - 1) && \textit{(from (20.2))} \\
&= UT_{i,t-2} - 1 + (inv_{i,t-1} - inv_{i,t-2}) \\
&= UT_{i,t-2} - 1 && (\textit{even(t) implies } inv_{i,t-1} = inv_{i,t-2})
\end{aligned}
$$

FIGURE 20.11
Static analysis—expanding the inventory equation.

buy orders and can never decrease inventory further (unless the lower threshold itself changes): similarly, those with starting inventories above the upper threshold can never increase their inventory (unless the upper threshold changes).

20.5.3 Dynamic Analysis Using InterDyne

Static analysis of a recurrence relation model of the case study indicates that despite the existence of feedback dependencies the market is stable: if all dealers start with inventories between the thresholds, they will never exceed either threshold and therefore Hot Potato trading will not occur. Furthermore, if there were some way for a dealer's inventory to exceed a threshold, its inventory would monotonically move towards the nearest threshold as fast as possible, so Hot Potato Instability will not occur.

We will now illustrate the importance of dynamic analysis by exploring ways in which an apparently stable (and self-stabilizing) model can nevertheless be induced to exhibit Hot Potato Instability. Further details can be found in (Clack 2013, Clack et al. 2014).

It is straightforward to recast the previous recurrence relation model as an InterDyne simulation:

- One InterDyne agent will represent the exchange subsystem.
- *n* InterDyne agents will represent the *n* dealer subsystems.

To explore the progression of this System of Systems as it moves from stable behavior into unstable behavior (we will explain below how this is possible), we can also add one or more trading agents to provide market supply and demand. Starting inventories can be passed to the dealer subsystems as runtime arguments.

Figure 20.12 provides example code for a System of Systems that comprises thirteen subsystems: an exchange, five dealers, a noise trader (that randomly issues buys and sells), three seller traders (issuing only sell orders), and three buyer traders (issuing only

```
newExperiment
= do
  sim 1000 args agents
  where
  args   = [ (Arg (Str "inventory", 0)) ]
  agents = ( replicate 1 (exchangeWrapper,     [])) ++
           ( replicate 5 (dealerWrapper,       [])) ++
           ( replicate 1 (noiseTraderWrapper,  [])) ++
           ( replicate 3 (sellerWrapper,       [])) ++
           ( replicate 3 (buyerWrapper,        []))
```

FIGURE 20.12
A financial market System of Systems comprising an exchange, five dealers, a noise trader, three sellers, and three buyers.

buy orders). The runtime argument "inventory" provides the same starting inventory for all the dealers (in this example, 0). No subsystem subscribes to a broadcast channel and therefore any broadcast messages will not be received by any subsystem.

The exchange subsystem will maintain *ob_bids* and *ob_asks* as internal state items, receive *lo_bids, lo_asks, mo_buys,* and *mo_sells* as inbound messages, and output *xbids, xasks, xbuys,* and *xsells* as messages. Each dealer subsystem will maintain $inv_{i,t}$, *lthresh,* and *uthresh* internal state items, receive *xbids, xasks, xbuys,* and *xsells* as inbound messages from the exchange, and output *lo_bids, lo_asks, mo_buys,* and *mo_sells* messages to the exchange. The function $\psi_i(\ )$ is partially implemented by the exchange because for example it only sends *xbids* and *xsells* messages to the two dealers involved in each trade.

In our previous static analysis the complexity of the dealer code was hidden—for example, the analysis simply referred to the calculated upper and lower constraints for limit order sizes. In the dynamic analysis we have the opportunity be more specific, for example to provide an accurate implementation of the Huang et al. algorithm previously discussed. However, initially we are interested only in whether it is possible to induce a stable algorithm to become unstable, and so we implement code that always issues the maximum sized orders according to the constraints used in the static analysis (this may be considered a "stress test"). Furthermore, to reduce complexity, we initially set static inventory thresholds. Figure 20.13 is a plot of the five dealer inventories in a simulation with inventory thresholds 3,000 and –3,000, and illustrates (as predicted by the static analysis) that the dealer inventories never exceed either threshold. The inventories tend to stay close to the thresholds due to the extreme "stress test" size calculations for limit orders.

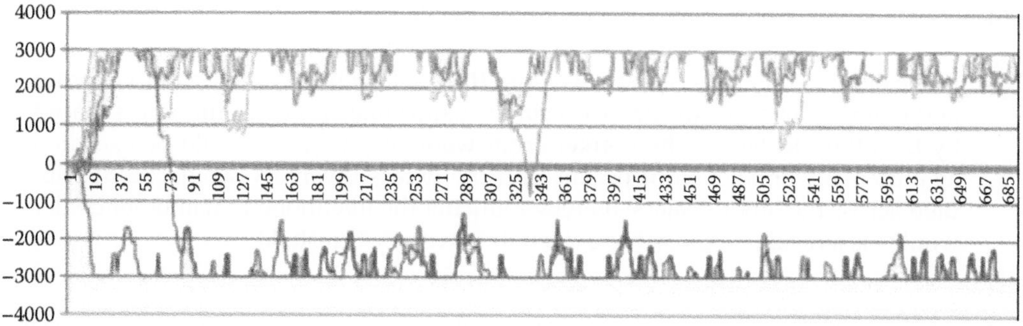

FIGURE 20.13
Dealer inventories.

To explore how such a system might be induced to become unstable, we systematically change elements of the experiment, one at a time, and visualize time-varying data such as inventories, order sizes, message sources, and destinations, and sequences of messages. There are two aspects to this investigation: (i) how to cause a dealer to exceed its inventory threshold, and (ii) once that has been accomplished, how to cause two or more dealers to be recruited into a behavior of repeated inter-dealer trading.

20.5.3.1 Exceeding the Inventory Thresholds

For an inventory threshold to be exceeded, either the inventory must be forced to cross the threshold, or the threshold must be forced to cross the inventory. The former corresponds to loss of inventory control, and the latter corresponds to a dynamically recalculated threshold.

Huang et al. (Huang et al. 2012) discuss inventory control in the context of a currency exchange market and derive an optimal inventory-control strategy using thresholds. This strategy assumes thresholds will be temporarily exceeded because a currency dealer has little control over market supply and demand and must execute inbound market orders. Dealers in an order-book market have more control (as previously discussed), but if a dealer prioritizes a "market-maker rebate" (see Section 20.5) over inventory control, then a bid and an ask will always be submitted to the exchange even when execution of such limit orders would cause a threshold to be exceeded. We note that Huang et al.'s thresholds are varied dynamically and either threshold might be reduced at any time, following which the inventory might exceed the threshold even though the inventory did not change.*

The dealer code for the InterDyne simulation can be modified so that dealers always issue bids and asks of given size regardless of inventory. Figure 20.14 shows that in this scenario all dealer inventories will exceed the upper threshold.

Further exploration shows that if there is both supply and demand, even if net supply is zero, then some dealers can exceed the upper threshold while others exceed the lower threshold. It is straightforward to vary the thresholds during a simulation and explore the circumstances under which this can cause inventory to exceed a threshold, though we do not present the details here.

20.5.3.2 Inducing Hot Potato Instability

Although we have identified two ways in which a dealer's inventory might exceed an inventory threshold, understanding Hot Potato Instability is more problematic (Clack 2013):

- If one dealer were to exceed an inventory threshold, it would issue a market order and that market order would be executed against one or more limit orders placed by the other dealers. If the market order were very large, it would be executed against many smaller limit orders, and the excess inventory would be distributed across the other dealers, thereby reducing the likelihood that another dealer would, as a result, be caused to exceed its inventory threshold.

* Note that if dealers issue limit orders that might cause the inventory threshold to be exceeded, then this departs from the System of Systems specification used in the static analysis and we should not expect stability. Similarly, although that specification does accommodate varying thresholds, if we are unable to predict the threshold in the next time step, this would be another source of potential instability.

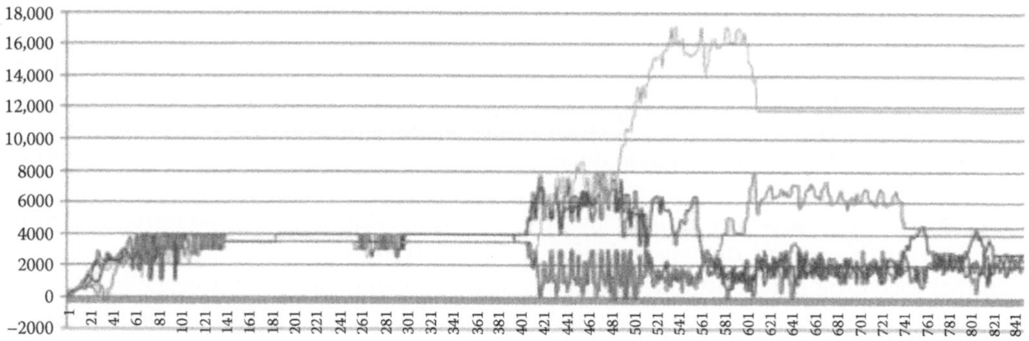

FIGURE 20.14
Inventories for five dealers with inventory thresholds of ± 3,000 and issuing bid and ask orders with minimum sizes of 1,000, together with a trader that issues sell orders of size 500 for each of the first 500 time steps.

- If the first dealer's excess inventory were offloaded, and if no other dealer were caused to exceed a threshold, then there would be no further market orders and no Hot Potato trading.
- Alternatively, if *all* dealers were somehow induced to exceed a threshold, they would all issue market orders. In the absence of any market participants issuing limit orders, there would be no trades executed and no Hot Potato trading. However, if the dealers were issuing fixed-size limit orders even when inventory exceeds the thresholds, then Hot Potato behavior could occur.
- To achieve Hot Potato trading it is necessary to match a market order from one dealer with a limit order from another dealer, i.e., it is necessary for a group of dealers either (i) to continue issuing limit orders when inventory exceeds a threshold (so that trading continues even when *all* dealers exceed a threshold), or (ii) to oscillate between issuing limit orders and issuing market orders, and to do so in synchronized anti-correlation (so that some dealers always issue limit orders at the times that other dealers issue market orders).

Closer inspection of Figure 20.14 provides evidence of both Hot Potato trading and Hot Potato Instability. In the first 50 time steps the inventories of all five dealers increase (non-monotonically), after which point the inventories start to oscillate above and below the upper threshold (set to 3,000). From approximately time steps 140 to 260 the inventories are stationary: all exceed the upper threshold and either no trading is done or in every time step the sizes bought equal the sizes sold (amounts bought will be due to the fixed-size bids continuing to be issued). Following a brief session of trading and oscillating inventories (roughly time steps 260 to 300), the inventories are once again stationary until about time step 400, at which point the inventories start to oscillate with large amplitude (two dealers' inventories drop to below the threshold, while three continue to have excessive inventory). At time step 500 the trader stops issuing sell orders, and from time step 500 to the end of the simulation the dealers are only trading with themselves.

This corresponds to our definition of Hot Potato Instability, and dynamic analysis reveals a complex pattern of behavior that could be further explored with a micro-analysis of the message history.

20.5.3.3 Hot Potato Instability Triggered by Information Delay

An important element in the 2010 Flash Crash was the presence of substantial information delay, due to the technology infrastructure becoming overwhelmed with data (CFTC and SEC 2010, Kirilenko et al. 2014, Nanex 2010a, Nanex 2010b). As part of our hypothesis formulation, we explore information delay to see whether it might play a role in triggering Hot Potato trading.

In real life such communication delays are typically unknown for some initial period of time. During this period, dealers and other traders will continue to operate on the assumption that incoming data is timely (i.e., that it has not been delayed). In InterDyne we simply revert to the first experiment (Section 20.5.3) and set an extra delay of one time step (Section 20.4.4) for all messages from the exchange to the dealers.

Figure 20.15 illustrates how such Hot Potato Instability (induced by a communication delay of just one time step) can persist. The figure plots inventories for five homogeneous dealers over 300 time steps. The instability is triggered by a seller trader that issues sell orders of size 100 for each of the first 40 time steps.

Figure 20.16 shows the dynamic inventories of five homogeneous dealers when a market exhibits an information delay of one additional time step from exchange to dealer. To determine whether this is due to message selection bias, the facility for randomizing message arrival times has been activated (the effect persists regardless of order arrival time). For this example, two of the dealers have an initial inventory that exceeds the upper threshold, two start with inventories below the lower threshold, and the last starts with an initial inventory of zero. Inventories for the first 100 time steps are shown.

In roughly the first 25 steps all dealers trade among themselves, causing periodic jumps to excessive inventory and back (due to information delay). These jumps are undesired because the market orders typically incur financial loss (a dealer loses the spread on each market order trade); the dealers therefore try to avoid those jumps by restraining their limit orders when their inventories approach the upper and lower thresholds (UL and LL).

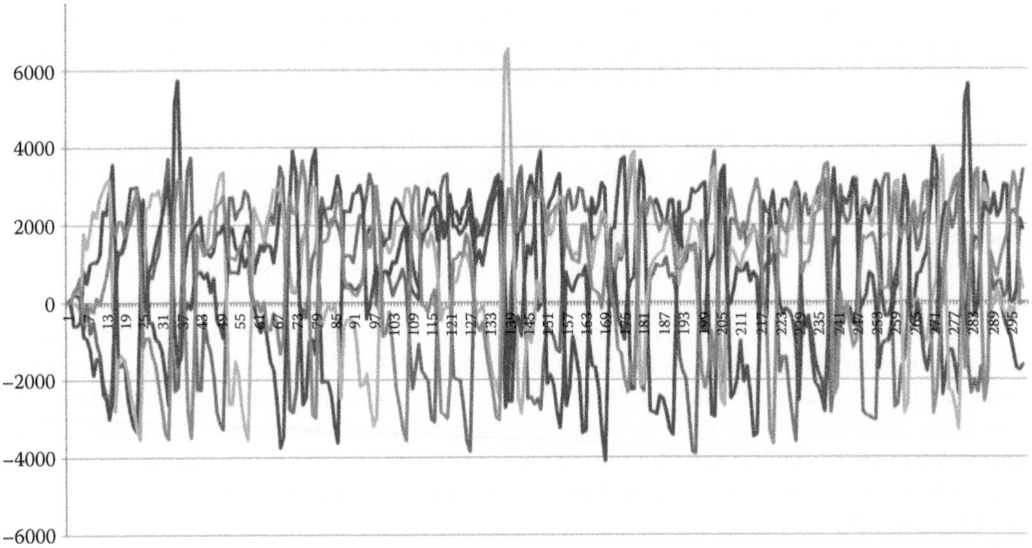

FIGURE 20.15
Hot Potato Instability.

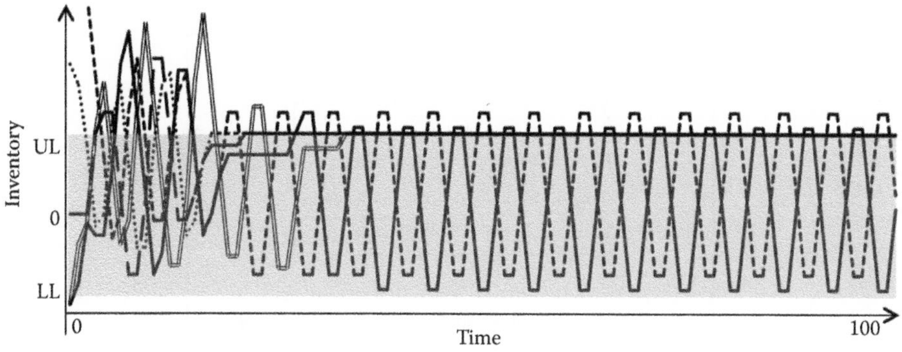

FIGURE 20.16
Inventory changes for a market with five homogeneous dealers. The shaded zone is a stable zone within the inventory thresholds.

In the remaining 75 steps three out of five dealers manage to stabilize their inventories near the upper threshold. At an inventory of exactly UL-1 they do not issue any bids, and if there are no delayed executions in the pipeline, they cannot thereafter exceed the upper threshold. However, the other two dealers remain coupled in a feedback loop and continue to trade with each other. This leads to an infinite oscillation, where the two dealers repeatedly exchange the same inventory. This could create a false impression of continuous market liquidity.

20.5.3.4 Understanding Hot Potato Instability

Figures 20.14, 20.15, and 20.16 illustrate the complexity of Hot Potato Instability, with a broad range of dynamic behavior. At first sight the behavior may appear chaotic, yet exploration of different initial conditions (such as initial inventories) and other factors such as the length of delays and details of the order size equations has helped us to identify several patterns of behavior.

We aim to understand this emergent behavior sufficiently to suggest proposals for reengineering the financial markets infrastructure: to avoid cases of undesirable oscillations where that is possible, and to detect other cases so that remedial intervention can be applied.

Dynamic analysis of the interaction dynamics is essential, and an essential component of that is the ability to observe message-passing histories. With InterDyne it is possible to trace full information about (i) all messages sent by an agent; (ii) all messages received by an agent; and (iii) all messages held in a delay queue awaiting delivery to an agent. The latter is possible because of the way that InterDyne implements delays—the simulator harness receives all messages output by agents, determines their destinations, and constructs queues of messages awaiting delivery to each agent. Although the user-defined delay function (Section 20.4.4) provides delay information for every possible combination of sender and receiver of messages, the simulator harness only needs to maintain n delay queues for n agents, as illustrated in Figure 20.17.

As an example of how observation of messages can be useful, consider a simple two-dealer system: we might wish to investigate the effects of changes in the size of sell orders that a dealer issues when inventory exceeds the upper threshold. One option discussed previously is to sell sufficient inventory to revert inventory (in the best case) to the upper

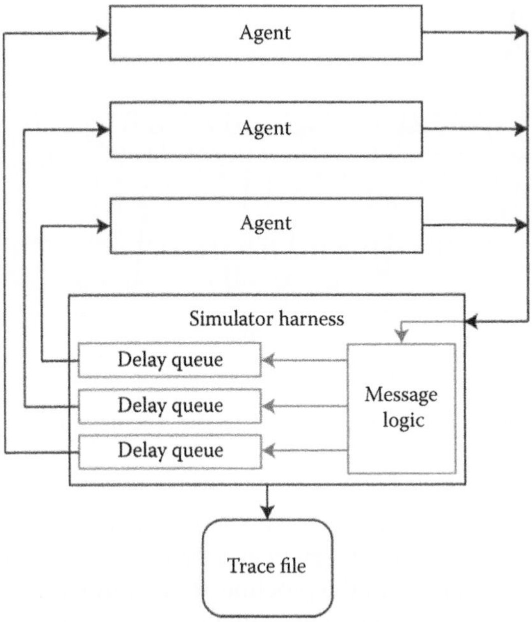

FIGURE 20.17
Message delay queues.

threshold. Another option is to always sell an amount of UT, where UT is the value of the upper threshold: if the inventory were only just above the threshold, this would in the best case revert inventory to be very close to zero (a neutral position if the thresholds are symmetrically arranged above and below zero). By investigating the time history of message-passing given in the InterDyne trace file, we can see which policy might be more stable in the presence of communication delays.

As a concrete example, assume a delay of one time step in trade confirmations reaching the dealers, a static upper threshold of 1,000 for all dealers, and a two-dealer system with starting inventories –998 and 1998. Assume that bids are issued at the "stress test" size of (999 – inv_t) and compare the effect of a sell size of (i) inv_t – 1,000; versus (ii) 1,000. With a sell size of (inv_t – 1,000) the traced message histories are illustrated in Table 20.1: at each time step the inventories of the two dealers are recorded (following receipt of trade

TABLE 20.1
Tracing Message Histories for Sell Size = (inv_t – 1,000)

Time	Inv_1	Inv_2	Message $Order_1$	Message $Order_2$	Executed	Messages Delayed	Messages Received
0	–998	1998	1997	–998			
1	–998	1998			998		
2	–998	1998	1997	–998		998	
3	0	1000			998		998
4	0	1000	999	0		998	
5	998	2					998
6	998	2					

TABLE 20.2

Tracing Message Histories for Sell Size = *1,000*

Time	Inv_1	Inv_2	Message $Order_1$	Message $Order_2$	Executed	Messages Delayed	Messages Received
0	-998	1998	1997	-1000			
1	-998	1998			1000		
2	-998	1998	1997	-1000		1000	
3	2	998			1000		1000
4	2	998				1000	
5	1002	-2					1000
6	1002	-2	-1000	1001			
7	1002	-2			1000		
8	1002	-2	-1000	1001		1000	
9	2	998			1000		1000
10	2	998	997	1		1000	
11	-998	1998					1000
12	-998	1998	1997	-1000			

confirmations in that time step); the next two columns give the sizes of any orders issued in that time step (here we only consider bids and sells: a positive size is a bid, a negative size is a sell); the fifth column indicates the size of a sent trade confirmation message, the sixth column shows a message in a delay queue (in fact there are two confirmation messages—one for each dealer); and the final column indicates that a trade confirmation has been received (by both dealers) in that time step. This example leads to stable inventories (i.e., not exceeding the upper threshold) at time step 5.

By contrast consider Table 20.2 where sell sizes are always 1,000. In this case the two-dealer System of Systems enters a loop where the system state repeats every 12 time steps. If a system state is repeated when InterDyne is operating in deterministic mode, then such a loop will continue forever (i.e., until the end of the simulation).

Tables 20.1 and 20.2, and Figures 20.12 through to 20.15, illustrate our approach to visualizing both beneficial and harmful emergent behavior. We analyze the InterDyne trace file and import data into a spreadsheet, from which we generate tables and graphs, which we then process to focus our search, and regress: thus our workflow is a form of feedback loop (simulate, process, simulate, process, and so on). It is currently a rudimentary methodology and work is underway to provide support for more systematic (and hopefully more automated) methods.

20.5.4 Detection, Classification, Prediction, and Control

Our case study is an example of emergent behavior in a man-made System of Systems, and our "two views" approach provides both a mathematical representation of the system and a visualization of the behavior as it emerges. InterDyne supports the interrogation and exploration of the antecedents of this behavior to facilitate understanding of the phenomenon:

- Emergent behavior is *detected* by tracking the time-varying value of one or more system parameters (for example, dealer inventories or traded prices). Where these values display unexpected (and especially undesirable) behavior, we suspect emergent behavior: if the behavior is not predictable from the encoded functionality of any of the components of the system, then we define it to be an emergent behavior.

- We proceed to *classification* of emergent behavior. As outlined in Chapter 1, Page (Page, 2009) has suggested four high-level classifications (Simple, Weak, Strong, and Spooky) that provide a starting point for classification. In our case study the emergent behavior is unexpected and therefore can be classified as "Strong" (other taxonomies exist; see for example (Fromm 2005)). However, further behavioral classification can aid understanding; for example, in our case study we classify the emergent behavior according to the temporal properties of the magnitude of the behavior. We identify "Decaying, Finite" behavior whose magnitude decreases and disappears after a small number of time steps; "Growing, Finite" behavior whose magnitude increases for a fixed number of time steps and then plateaus; "Growing, Infinite" behavior whose magnitude increases without limit until system failure; "Infinite" behavior whose magnitude holds a constant value from the point of onset of the emergent behavior until the end of the simulation; and "Periodic" behavior whose magnitude may drop to zero for a brief time and then restart (it may do this many times).
- If classification leads to understanding, then we test that understanding by attempting to trigger the emergent behavior. If successful, this permits *prediction* of the circumstances under which an emergent behavior may be manifested. For example, with our case study we are able to predict how the introduction of a communication delay between the exchange and the dealers will lead to Hot Potato Instability. Where message histories are observed and shown to lead to repeated system states (as illustrated in Table 20.2) we can predict emergent behavior that is potentially "Infinite" in character, and future states at any given time step can be calculated. In other examples, such as Figure 20.14, which has a "Periodic" behavior, it is more difficult to predict the values of future states. This is an area of current research.
- InterDyne's ability to visualize emergent behavior and explore its antecedents, albeit currently with a rudimentary implementation, is a step towards the ability to *control* emergence because it provides improved understanding of the constituents of such emergent behavior and how the behavior is triggered. In relation to the financial markets, our long-term aim is to identify how the markets could be re-engineered to minimize undesirable emergent behavior.

20.6 Chapter Summary

In this chapter we have introduced the InterDyne simulator and our simulation method for exploring Systems of Systems and identifying the antecedents of undesirable emergent behavior that derives from interaction dynamics such as feedback loops. InterDyne is an agent-based simulator that differs from other agent-based technology in (i) its focus on supporting experimental observation of message-passing interaction, including broadcast messages, a broad range of message types, and a very flexible treatment of agent-to-agent communication delays; and (ii) our "two-view" approach where the agent-based model is paired with a recurrence-relation model (the latter supports static analysis, whereas the former supports dynamic analysis). A current research topic is to establish a formal equivalence of the semantics of these two views.

InterDyne supports investigation of time-varying interactions between subsystems, to explore complex Systems of Systems that may not be amenable to static analysis, and to encompass a broad range of real-life behavior. InterDyne also supports deterministic and non-deterministic experiments, with randomized message arrival at each agent (to demonstrate that emergent behavior is not an artefact of hard-wired bias in arrival times) and heterogeneous agent strategies (to demonstrate that emergent behavior is not simply due to a resonance between homogeneous algorithms). With InterDyne (as with other simulators for Systems of Systems), we can explore questions such as "how is a given behavior triggered?", "under what circumstances might a repeated behavior continue forever?", and "under what circumstances might it naturally resolve?" Although the antecedents of emergent behavior need not be unique, and may not be applicable to other configurations of a given System of Systems, they often serve to falsify a negative hypothesis in stress testing (e.g., "it is not the case that this can never happen"). Moreover, we view dynamic analysis as an important component of hypothesis formulation, to assist in clarifying hypotheses and their consequences for dynamic behavior, and to assist in communication between domain experts.

Many of the project tools, including the InterDyne simulator and visualization tools, are part of ongoing research. The InterDyne Project team at University College London has included staff and many students, and we gratefully acknowledge their contributions. Members of the project team who have worked on the core technology have included Vikram Bakshi (Bakshi 2015), Aman Chopra, Elias Court (Court 2013), Richard Everett, Yifei Liu (Liu 2014), and Dmitrijs Zaparanuks (Clack et al. 2014).

References

Bakshi, V. 2015. Simulating Interaction in Financial Markets. MSc dissertation. Department of Computer Science, University College London.

Barkhausen, H. 1935. *Lehrbuch der Elektronen-Röhren und ihrer technischen Anwendungen: Rückkopplung*, Hirzel.

Box, G.E.P. 1976. Science and Statistics. *Journal of the American Statistical Association* 71:791–799.

CFTC and SEC. 2010. Findings regarding the market events of May 6, 2010.

Chan, W.K.V. and Son, Y.-J. and Macal, C.M. 2010. Agent-Based Simulation Tutorial—Simulation of Emergent Behavior and Differences Between Agent-Based Simulation and Discrete-Event Simulation. In *Proceedings of the 2010 Winter Simulation Conference*, 135–150.

Clack, C.D. and Peyton Jones, S.L. 1985a. Generating Parallelism from Strictness Analysis. In Proceedings *Workshop on Implementation of Functional Languages,Programming Methodology Group*. Report 17, 92–131. Chalmers University, Sweden.

Clack, C.D. and Peyton Jones, S.L. 1985b. Strictness analysis—A practical approach, *Lecture Notes in Computer Science* 201:35–49.

Clack, C.D. 2013. The Complexity Challenge of Competing Algorithms. Presentation at the HiFREQ TRADE 2013 conference. http://www0.cs.ucl.ac.uk/staff/C.Clack/Clack-HiFreqTrade2013.pdf (accessed July 11, 2017).

Clack, C.D. and Court, E. and Zaparanuks, D. 2014. Dynamic Coupling and Market Instability. Working Paper. Department of Computer Science, University College London. http://www0.cs.ucl.ac.uk/staff/C.Clack/ClackCourtZaparanuks2014.pdf (accessed June 12, 2017).

Clack, C.D. 2016. The InterDyne User Manual. Working Paper. Department of Computer Science, University College London. http://www0.cs.ucl.ac.uk/staff/C.Clack/InterDyne-User-Manual.pdf (accessed June 12, 2017).

Court, E. 2013. The Instability of Market-Making Algorithms. MEng dissertation. Department of Computer Science, University College London.

Day, R.H. and Huang, W. 1990. Bulls, bears and market sheep. *Journal of Economic Behavior & Organization* 14(3):299–329.

Easley, D. and López de Prado, M.M. and O'Hara, M. 2011. The microstructure of the "flash crash": flow toxicity, liquidity crashes, and the probability of informed trading. *The Journal of Portfolio Management* 37.2:118–128.

Faustini, A.A. 1982. An operational semantics for pure dataflow. *Lecture Notes in Computer Science* 140.

Fromm, J. 2005. Types and forms of emergence. arXiv preprint nlin/0506028.

Gould, M. and Porter, M. and Williams, S. 2013. Limit order books. *Quantitative Finance* 13(11):1709–1742.

Horowitz, I. 1959. Fundamental theory of automatic linear feedback control systems. *IRE Transactions on Automatic Control* 4.3:5–19.

Huang, K. and Simchi-Levi, D. and Song, M. 2012. Optimal Market-Making with Risk Aversion. *Operations Research* 60(3):541–565.

Hudak, P. and Peyton Jones, S.L. and Wadler, P. et al. 1992. Report on the programming language Haskell: A non-strict, purely functional language version 1.2. *SIGPLAN Not.* 27(5):1–164.

Kirilenko, A. and Kyle, A.S. and Samadi, M. and Tuzun, T. 2014. The Flash Crash: The Impact of High Frequency Trading on an Electronic Market, Working Paper, SSRN. http://ssrn.com/abstract=1686004 (accessed April 30, 2017).

Leombruni, R. and Richiardi, M. 2005. Why are economists sceptical about agent-based simulations? *Physica A: Statistical Mechanics and Its Applications* 355(1):103–109.

Liu, Y. 2014. Re-engineering agent based simulator with functional language. MSc dissertation. Department of Computer Science, University College London.

Lyons, R.K. 1997. A simultaneous trade model of the foreign exchange hot potato. *Journal of International Economics* 42(3–4):275–298.

Milner, R. 1999. *Communicating and Mobile Systems: The Pi Calculus.* Cambridge University Press.

Nanex 2010a. Nanex Flash Crash Analysis Final Conclusion. Working paper. Nanex LLC. http://www.nanex.net/FlashCrashFinal/FlashCrashAnalysis_Theory.html (accessed April 30, 2017).

Nanex 2010b. Nanex Flash Crash Summary Report. Working paper, Nanex LLC. http://www.nanex.net/FlashCrashFinal/FlashCrashSummary.html and http://www.nanex.net/FlashCrashFinal/FlashCrashSummary_DJIADelay.html (accessed April 30, 2017).

Newton, I. 1687. *Philosophiæ Naturalis Principia Mathematica.*

Page. 2009. *Understanding Complexity.* The Teaching Company, VA, USA.

Rosu, I. 2009. A Dynamic Model of the Limit Order Book. *Review of Financial Studies* 22(11):4601–4641.

Rötzer, G. 2012. The Stochastic Pi Machine and Finance. MSc thesis, Department of Computer Science, University College London.

Skogestad, S. and Postlethwaite, I. 2007. *Multivariable feedback control: Analysis and design. Vol. 2.* Wiley.

Sundman, K. 1912. Memoire sur le probleme des trois corps. *Acta Mathematica* 36:105–179.

Turing, A. 1937. On computable numbers, with an application to the Entscheidungsproblem. *Proceedings of the London Mathematical Society, Series 2* 42:230–265.

Turner, D.A. 1985. Functional programs as executable specifications. In *Proceedings of a discussion meeting of the Royal Society of London on Mathematical logic and programming languages,* ed. C.A.R. Hoare and J.C. Shepherdson. Prentice-Hall, Inc.

Wilberforce, L.R. 1896. On the vibrations of a loaded spiral spring. *Philosophical Magazine* 38:386–392.

Zames, G. 1966. On the input-output stability of time-varying nonlinear feedback systems part one: Conditions derived using concepts of loop gain, conicity, and positivity. *IEEE transactions on automatic control* 11.2:228–238.

21

Emergence in the Context of System of Systems

Charles B. Keating and Polinpapilinho F. Katina

CONTENTS

21.1 Introduction: Background and Problem Space

The design, operation, and maintenance of multiple complex systems (System of Systems) is a hallmark of modern society. We would be hard pressed to identify a system that is truly independent of all other systems, events, and interactions. For example, transportation, energy, healthcare, education, defense, security, and finance all exist in relationship to one another, creating interdependencies. How would a hospital function independent of power from the energy grid or transport of essential supplies and patients? In essence, the notion of an independent system is analogous to a "closed" system, operating in complete isolation. A closed system operates without exchange across boundaries of energy, information, or matter (i.e., with other systems or the environment). In thermodynamic terms, for a closed system, entropy (disorder) approaches maximum and the system evolves to the lowest energy state (maximum disorder). Thus, the system is at rest, in a static equilibrium state where no activity proceeds—nothing in and nothing out. Closed systems are easily achieved in a laboratory beaker; however, this is not the case for complex systems that propel society and must operate in the "real world" (Hammond, 2002; Laszlo, 1996; Martin, 2006). *There are no complex systems presently serving larger societal needs that would fit into the "closed system" realm.* Instead, our systems fall into the realm of complex open systems—existing in interactions with other systems, exchanging (resources, energy, information, matter) with the environment, and continually modifying in response to changing conditions in the environment and their internal operation. In effect, we are in a "system of systems" (SoS) world (Rainey and Tolk, 2014; Keating, 2014; Keating and Katina, 2011) that engages multiple systems as a "construct of convenience," bounded in ways that permit us to proceed with design, execution, and evolution necessary to advance the state and performance prospects of these systems of interest. In this world, our systems do not exist in isolation from one another. Instead, they exist as multiple interrelated open systems—with delineated boundaries and dependent upon other systems as well as having other systems dependent on them.

A hallmark of SoS is the property of *emergence*. At the most basic level, emergence holds that as a complex SoS operates, there are properties (behavior, structure, performance) that only come about and are recognized as the SoS operates. These "emergent" properties cannot be attributed to individual constituent systems comprising the SoS, but rather only come forth (emerge) as the individual constituent systems interact with one another and the environment. It is the interaction that produces an emergent property, not the characteristics of the constituent systems, which are also subject to "local level" emergence from the interaction of their constituent elements. Take away the interaction and the emergent property is lost. Thus, for complex SoS, exploration of emergence is not directed to the constituent systems, but rather to the interaction and interrelationships among those systems. Therefore, emergence recognizes that unpredictable properties, behaviors, structure, and ultimately performance in SoS arise as the constituent systems operate within their local context. Thus, our challenge is to pursue a deeper understanding and appreciation of emergence as a central phenomenon of SoS design, execution, and evolution.

In this chapter, we seek to provide a detailed exploration of emergence within the context of SoS. There are five major elements of this exploration. First, we elaborate the nature of the SoS problem domain. This is essential to establish the context within which modern SoS must operate. All SoS design, execution, and evolution is achieved within, and in response to, this problem domain. Second, we capture the essence of the nature, characteristics, and function of a SoS. Every collection of systems is not a SoS. However, a more precise accounting of criteria for classification as a SoS will be an important backdrop for understanding emergence in relationship to SoS. Third, we survey the phenomenon of emergence from a systems perspective. By scrutinizing a range of definitions and literature, we seek to establish an "emergence perspective" from a systems viewpoint. Fourth, we examine the implications of emergence for design, execution, and development of SoS. The focus of this examination is targeted to developing a set of practitioner guidance through introduction of several emergence strategies. Fifth, we explore three different application scenarios to suggest the relevance of enacting emergence strategies in the context of SoS. The chapter closes with a set of insights for practitioners interested in a deeper sophistication in dealing with emergence in SoS.

21.2 Problem Domain for SoS

Although SoS represent a special case of integrating multiple complex systems, they are not insulated from the domain that defines the landscape for complex systems. This problem domain has been previously identified (Jaradat et al., 2014; Keating, 2014; Keating and Katina, 2011). Based on that identification, the following is a summary of central characteristics of that domain applicable to our current exploration:

- *Exponential Rise in Complexity*—At a most basic level, complexity is evidenced by the sheer number of system variables, richness of interconnections within the system and between the system/environment, unpredictable behavior/consequences, and dynamic shifts in system structure, behavior, context, or performance (Sussman, 2005). These conditions are only escalating for future complex systems and are becoming more recognizable as the norm rather than exception. A cursory examination of the increasing availability, magnitude, and accessibility of information exemplifies the rise in complexity that is challenging our capacity to mount effective responses.

- *Dominance of Emergence*—The appearance of structures, behaviors, performance, or consequences that cannot be known in advance, and only come about as a complex system operates, define the degree of "emergence." For complex systems, emergence is increasingly expected as the norm, rendering approaches based in assumptions of stability ever more doubtful.
- *Ambiguity in Understanding*—Ability to achieve or maintain clarity necessary to fully comprehend system behavior or action-consequences relationships is increasingly doubtful. Continuity of interpretations over time are unstable and subject to drastic shifts. For complex systems, absolute understanding is an illusion.
- *Uncertainty as a Norm*—Lack of knowledge, as well as diminishing confidence in knowledge, is characteristic of complex systems. Knowledge is always incomplete and fallible. This increases doubt concerning how to proceed to produce desired performance. Uncertainty can never be fully removed for a complex system. This stems from shifts over time, as our knowledge and understanding of a system, its behavior, and environment also change over time.
- *Holistic Satisficing Solutions*—Complex system problems are not resolved by "optimal" solutions that can be verified as the singular "best" solution among all possible solutions. Rather they are addressed by "satisficing" solutions that represent a response deemed to be appropriate to the situation. The entire holistic spectrum of technology/technical, organizational/managerial, human/social, and political/policy are in play across spacial, temporal, and social dimensions for complex systems.
- *Contextual Dominance*—All complex systems are embedded in a unique set of circumstances, factors, patterns, and conditions that cast varying degrees of influence on their execution and performance. This "system context" serves to both enable and constrain decision, action, and interpretations taken with respect to the system. For SoS, contextual factors (e.g., politics) become increasingly important as the degree of complexity exponentially increases.

There have been several earlier works that attempt to capture a concise nature of the domain confronted by complex systems and their practitioners (Keating and Katina, 2011; 2012; Katina, 2016) as well as the notion of Ackoff's (1971) "messes" and Rittel and Webber's (1973) "wicked problem." Following earlier work by Keating et al. (2014), we succinctly attempt to capture this state of what we call the complex system problem domain and corresponding implications related to dealing with that domain in Table 21.1.

While these characteristics are not insurmountable for dealing with complex SoS, they delineate a very different problem domain. We suggest four primary implications posed by this problem domain for SoS with relevance to emergence. First, the nature of this problem domain does not lend itself to assumptions of traditional approaches rooted in stability of circumstances, objectivity in classification/interpretation, or assurance of repeatability of past successes. The traditional scientific approach, while it "has been successful in advancing our knowledge in the natural sciences … [it is not equipped] to investigate social phenomena" (Flood and Carson, 1993, p. 248). Instead, doubt is cast on these assumptions as being appropriate for the problem domain for complex SoS. Second, responses to this problem domain which are focused on pursuit of "complication" (increasing regulation, constraint, and controls) as a strategy to confront the domain are misplaced.

TABLE 21.1

Complex System Problem Domain Implications for Emergence in SoS

Domain Characteristic	Explanation	Implications for Emergence in SoS
Rapid expansion of information	The proliferation of information and access to information continues to expand and accelerate exponentially, creating difficulties related to quantity, quality, veracity, and processing capacity	The magnitude, veracity, and quality of information make filtering difficult, increase the likelihood of misinformation, and increase the probability of poorly informed decisions, inconsistent support for actions, and misinterpretations. Information and derivative knowledge produce critical enabling and constraining factors for dealing with emergence in SoS.
Conflicting perspectives and divergence in views across stakeholders	Given the magnitude of information sources, coupled with different interests among interested parties, there is the distinct possibility for conflict in decisions, corresponding actions, and determination of "appropriate" paths forward related to complex systems and their issues.	Disparities in underlying worldviews among stakeholders are inevitably going to produce differences with the potential for conflict. Without doubt, there will be unavoidable difficulties related to the identification, assessment, response, and interpretation of emergent issues in SoS.
Scarce and dynamically shifting resources	Scarcity of resources is not new. However, increasing pressures related to time demands and rapidity of shifts in resources are becoming more pronounced. Taking a "short view" and demands for immediate response to emergent issues can foster a climate of instability and discontinuities concerning resource availability.	Increasing levels of emergence and compressing the time horizon between system disruptions can exacerbate the magnitude and timing of resource instabilities. Lacking resource stability, routine system planning, design, and execution platforms cannot provide traditional stability required to experience their value.
Unintended consequences	All systems can generate unexpected behaviors. However, higher degrees of uncertainty, incomplete knowledge, and shortening of decision time horizons continue to be the norm. This creates conditions capable of accelerating the volume and magnitude of behaviors and patterns that were neither intended nor anticipated.	Under the best of conditions, there has always been emergence in systems. However, as stability in systems continues to diminish, dealing with the corresponding increase in emergence becomes more critical. Failure to incorporate emergence into design, execution, and development for SoS is shortsighted.
Ambiguous boundaries	Boundaries are essential to determine what is included and excluded in a complex system. As the complexity of systems and their domain increases, the degree and rapidity of change in boundary conditions can also be subject to dramatic shifts.	The larger the degree of ambiguity in complex system boundaries, the greater the potential for emergence in boundary conditions. Thus, if boundaries for a SoS become excessively emergent, so too do the instabilities experienced in the SoS.
Politically influenced decisions, actions, and interpretations	Politically charged environments for complex systems are marked by attempts to pursue strategies to increase influence (power). Excessively informal power structures can be detrimental to coherent decisions, actions, and interpretations.	Politics as the pursuit of strategies to influence are a fact of life in complex systems. However, as the problem domain of SoS continues to experience increasing turbulence, the role of informal power structures can extend beyond the rational boundaries and interests of the SoS. Accounting for influences in power structures in a SoS becomes paramount.

(Continued)

TABLE 21.1 (CONTINUED)

Complex System Problem Domain Implications for Emergence in SoS

Domain Characteristic	Explanation	Implications for Emergence in SoS
Solution urgency	The theme of urgency in making decisions and taking decisive action in response to problematic situations is a mainstay for complex systems. However, given the current conditions and operating scenarios of complex systems, the drive for urgency can dominate more stable forms of calculated thinking and analysis. The pursuit of more instant gratification can sacrifice more thoughtful decisions based in more deliberate forms of analysis.	Increasing emergence in SoS can generate a sense of urgency for decisions and corresponding actions. The result might be the premature tradeoffs of time for other essential aspects of complex system problem understanding and resolution. Premature actions in response to increasing emergence are likely to result in incomplete solutions and quite possibly exacerbate the original emergent conditions through "fast but faulty" responses.
Unclear entry point or approach to address issues	The pace, lack of understanding, and compelling need for response creates conditions where the appropriate approach is in question. Assumptions of repeatability of successes achieved by historically productive approaches are suspect at best.	In the case of escalating emergence, the approach and entry to deal with SoS issues is left in distress. SoS activities must appreciate the need for rapid reconfiguration of approaches based on new knowledge, understanding, and effects. Improvisation supplants detailed upfront approaches.

"Complication-based strategies" aimed at confronting complexity by increasing "control" through increased regulation (constraint) are unlikely to achieve the intended outcomes. Excessive constraint is not an effective strategy to deal with complexity, as it only increases complexity while simultaneously limiting autonomy and flexibility to respond to new and novel emergent conditions confronting a system. Third, this problem domain is "holistic" and as such requires a "holistic" treatment. Responding with limited approaches, such as "technology" only strategies, will lack comprehensiveness necessary to understand and address the complex problem. This comprehensiveness must address the holistic spectrum (e.g., social, human, organizational, managerial, policy, political) of dimensions that define the domain. Fourth, there is a tendency to address this domain from an "output" orientation. This orientation is focused on creation of verifiably concrete, observable, tangible, and objective products (outputs). Complex system problem solutions lie beyond "outputs," and require a focus on "outcomes." Outcomes are more focused on meeting expectations for resolution of a problem/need, irrespective of whatever "outputs" might be produced. For example, meeting every requirement specified for a complex system does not ensure that expectations for problem resolution/need fulfillment will be met for stakeholders with the problem/need. Escaping the confusion in distinguishing "outputs" from "outcomes" is vital to dealing with complex system problem domains and their inherent emergence.

Complementing these four primary implications posed by SoS problem domain is a cautionary note: "Our technologies—from websites and trading systems to urban infrastructures, scientific models, and even the supply chains and logistics that power businesses—have become hopelessly interconnected and overcomplicated, such that in many cases even those who build and maintain then on a daily basis can't fully understand them any longer" (Arbesman, 2016, p. 2). Clearly, there is a need for a different path forward. Having set the nature of the problem domain and emergence implications, we now shift our focus to a more detailed examination of SoS.

21.3 System of Systems

System of Systems (SoS) has received considerable attention in the literature (Keating et al., 2003; Jamshidi, 2008) and has been described in a variety of ways (Keating et al., 2014; Keating, 2005; Keating et al., 2003; Keating and Katina, 2011). While there are different formulations of SoS and its counterpart System of Systems Engineering (SoSE), there is considerable agreement as to the essential characteristics for "something" to be considered a SoS. These characteristics generally follow from the original work of Maier (1999) and Sage and Cuppan (2001) presented in Table 21.2.

In meeting these attributes, there are at least 4 types of SoS identified (Guide, U.S. DoD, 2008; Maier, 1999; Dahmann, 2008).

1. *Virtual*—lacking centralized management and integrated purpose and whose maintenance is largely unstructured with large scale emergent behavior;
2. *Collaborative*—voluntary interaction of constituent systems with a common purpose and establishment of expectations without direct power to enforce standards residing in the SoS;
3. *Acknowledged*—common recognized SoS objectives and designation of integration management responsibility and resource allocation, although constituent systems retain a level of independence with changes being achieved through joint collaboration, and
4. *Directed*—a purpose built SoS intended to fulfill a specific intent, with central management where there is a subordination to the higher-level SoS.

This list is intended (1) to recognize different types of SoS, and (2) to acknowledge that there is not a "one size fits all" specification of SoS. Moreover, and at the very essence of the literature and perspectives of SoS, we suggest five important threads with their

TABLE 21.2
SoS Characteristics

SoS Characteristic	Description
Operational independence of constituent systems	Each constituent system has the ability and is explicitly designed to operate independently, supporting a mission unique to the system.
Managerial independence of constituent systems	The constituent systems comprising a SoS can be separately acquired and are independently managed.
Evolutionary development	SoS evolve over time. Constituent system capabilities may be added, removed, or modified as needs change and experience is gained.
Emergent behavior	SoS have emergent capabilities and properties that do not reside in the constituent systems.
Geographical distribution of systems	SoS are comprised of constituent complex systems geographically distributed with the ability to readily exchange information.
Self-organization	SoS produce structural and behavioral patterns that did not exist, and were not conceived, by design. These patterns only emerge and are recognized as a SoS operates within its unique environment.
Adaptation	SoS continually evolve, either by proactive redesign or self-organizing processes, to maintain performance levels in response to external environmental shifts or internal shifts.

corresponding relationship to emergence. First, SoS and its engineering are concerned with integration of multiple diverse constituent systems into a higher-level system. This higher-level system (of systems) pursues a purpose (mission) requiring capabilities that lie beyond the singular capabilities of any of the constituent systems. While there are capabilities sought in SoS design, emergence would also hold the possibility of unpredicted consequences that will appear only after the SoS is in operation. Second, the very nature of integration of multiple constituent systems into a "unity" increases the complexity of the situation exponentially. This renders more traditional approaches (e.g., Systems Engineering, Flood and Carson, 1993) based in well-defined problem domains as suspect concerning their direct applicability to the complexities inherent to the SoS problem domain. A high degree of emergence in increasingly complex SoS suggests that engagement based in traditional approaches to systems (e.g., Systems Engineering) becomes increasingly suspect to confront SoS and their problems. The hope for direct interpolation of traditional Systems Engineering methods, tools, procedures, technologies, or techniques to SoS becomes problematic and increasingly doubtful. Table 21.3 presents the distinction from the more traditional problem perspective for systems versus the SoS problem perspective. As demonstrated in the table, the nature of the SoS perspective is significantly different than that of a traditional systems perspective. The table suggests a sharp contrast between the two perspectives and might be best described as a set of "bookends" for the spectrum of problem spaces which might be appropriate to the perspectives. In reality, a complex system, or SoS, problem may lie at different points along the spectrum of characteristics defined by the "bookend" endpoints. However, we must conclude that the gestalt differences between traditional and SoS problems are significant.

Third, consistent with Table 21.3, more traditional "systems" problems have been viewed as capable of being addressed (Keating, 2014; Katina, 2016) as well bounded, objectively formulated, and amenable to rigorous *systematic* resolution. In contrast, SoS problems are ambiguously bounded, subjectively formulated, and more compatible with *systemic* based approaches. Systemic based approaches are more compatible with the perspective of emergence, where complete knowledge is not possible, is fallible, and will continue to evolve as a SoS operates within its environment. Fourth, with respect to SoS, it is acknowledged that there are different formulations of SoS types. This is an important consideration in understanding that there is not a singular configuration of SoS, but rather a variety of potential types of SoS. Fifth, there has been a propensity of traditional systems approaches (e.g., Systems Engineering) to focus heavily on the "technical" aspects of complex system problems. In contrast, the SoS problem domain extends across the wide-ranging spectrum of not only technology, but also organizational, managerial, human, social, policy, and political dimensions. This calls for a much more "holistic" treatment of SoS problems. The result of overemphasis on technology driven "solutions" to SoS is quite possibly pushing technology to the foreground when in fact the "nontechnical" aspects of the SoS problem may need to be in the foreground.

The heart of SoS approaches lies in appreciating the need for more holistic considerations. As a summary of SoS and what the function of a SoS is with respect to treatment of multiple complex systems, we offer the following perspective based on earlier work of Keating et al. (2003, p. 23), *"The design, deployment, operation, and transformation of higher level metasystems that must function as an integrated complex system to produce desirable results."* In a more recent evolution of SoS engineering, the notion of SoS being integrated by a "metasystem" has matured to a current state of Complex System Governance (Keating and Bradley, 2015; Keating and Katina, 2016). The tenets of CSG are grounded in *Systems Theory* (laws governing the behavior and performance of complex systems) focused on

TABLE 21.3

Delineating the Spectrum of Traditional Systems Versus SoS Perspectives

Attribute	Traditional Systems Perspective	SoS Perspective
Foundational paradigm for understanding	***Reductionism***—Focused on understanding through breaking apart (analysis). Performance of a whole can be understood as an aggregation of the properties of the parts.	***Holism***—A system is only understood at the (irreducible) whole system level. The behavior of the whole cannot be ascertained simply from understanding the parts. Instead, understanding of the system must include the interactions among parts that produce properties beyond those held by the parts.
Objective	***Optimization***—There is one solution which is best (optimal) for system performance. This is the solution or configuration which is sought.	***Learning***—The primary function of system exploration is to learn about the system and be capable of mounting appropriate response(s) based on that learning to improve a situation.
Methodology	***Systematic***—Approach is defined by prescribed processes that can be replicated independent of context—prescriptive.	***Systemic***—Approach is a high-level guide that provides a general set of malleable directions—non-prescriptive and tailorable to local circumstances and conditions.
Goal/objectives	***Clearly defined and agreed upon (unitary)***—The system goals and objectives are assumed to be clear, unambiguous, defined, and stable.	***Ambiguous and shifting (pluralistic)***—Clarity is not assured, and goals/objectives are subject to multiple interpretations and can be unstable.
Preferred representation	***Symbolic/Mathematical***—Preference is always to the mathematical model focused on precision and validity. Known relationships and predictive (mathematically) behavior dominate.	***Interpretative/Non-mathematical***—Preference is to representations providing approximate constructions that offer utility versus absolute certainty focused on precision and validity. Forms of representation are mostly non-quantitative in nature. Behavior is not precise, predictable, or repeatable.
Understanding	***Part/Reducible***—Understanding is capable through successive reduction to the level of parts/elements of the system.	***Whole/Irreducible***—System understanding is not capable through successive reduction and breaking down to increasingly finite elements. Understanding is held in relationships, not parts.
Solution form	***Verifiable Optimal Solution***—Assumes that a singular "best" solution can be selected among competing alternatives.	***Satisficing Response***—Assumes that a response can be developed that is contextually compatible, feasible, and presents an agreeable path forward for improvement in the system/situation.
System knowledge	***Absolute/Defined***—System understanding exists such that prediction and explanation can be made with confidence.	***Emergent/Fallible***—System understanding emerges with new and reformulated knowledge over time. Understanding, interpretations, and knowledge rest on assumptions/logic and are fallible.
Perspectives	***Unitary***—Assumes that there is alignment of perspectives for the problem domain and system objectives.	***Pluralist***—Assumes there exist multiple, potentially divergent, perspectives on the problem domain as well as objectives pursued.
Quantification	***Comprehensive***—System is capable of mathematical reduction with a high level of confidence. Preference for quantitative emphasis.	***Limited***—System is not capable of retaining understanding by reduction to precision required for mathematical representation with a high level of confidence. Preference for qualitative emphasis.

(*Continued*)

TABLE 21.3 (CONTINUED)

Delineating the Spectrum of Traditional Systems Versus SoS Perspectives

Attribute	Traditional Systems Perspective	SoS Perspective
Relationships	***Defined/Apolitical***—Relationships of actors in the system are well defined and understood. Power relationships are not an essential aspect of the system.	***Variable/Political***—Relationships of actors are in flux and power influences can dominate. Politics (influence strategies to increase power/influence) may shift and are not arbitrarily excluded at any point.
Dynamic nature	***Minimal/Controlled***—The problem/system and environment are sufficiently static/stable such that control can be maintained.	***Assured/Emergent***—The problem/system and environment are turbulent with corresponding emergence such that maintenance of absolute control is not possible.
Interpretation	***Objective***—The nature of the system and problems are capable of being known in concrete terms.	***Subjective***—The nature of the system and problems are subjective where multiple and potentially conflicting interpretations can simultaneously exist.
Contextual influences	***Low***—Contextual influences are assumed to be "minimized" by tight bounding of the problem.	***High***—Contextual influences are seen as integral. Systems/problems are not easily separable from context, with boundaries remaining loose, negotiable, and flexible.
Environment	***Stable***—Disturbances in the environment are minimal and rate/depth of changes not considered overbearing on system solutions.	***Turbulent***—Environmental disturbances are potentially extensive, rapidly evolving, and influential in ability to develop system solutions.
System and problem definition	***Simple/Defined***—Low number of variables, interactions well understood, behavior somewhat static/deterministic, environment stable.	***Complex/Emergent***—High number of variables, rich interactions not well understood, dynamic and uncertain (emergent) patterns/behaviors, environment unstable.
Boundaries	***Clearly delineated***—Boundaries are definitive, stable, and understood.	***Unclear and shifting***—Boundaries are evolving, unstable, ambiguous, and interpretative (negotiable).
Worldview	***Aligned***—Divergence in worldviews not made explicit or considered central to understanding.	***Potentially divergent***—Divergence considered highly probable and shifting, with understanding of divergence sources critical to understanding and interpretation.
Defining metaphor emphasis	***Mechanistic/Technical***—Clear understanding of predictable interrelationships.	***Contextual/Sociotechnical***—Lack of clarity in understanding the nature of interrelationships and external influences.
Behavior	***Predictable***—System behavior is deducible from understanding historical patterns/trends and system interactions. It is linear and deterministic.	***Emergent***—Patterns of system behavior/performance cannot be known in advance, emerging only through operation of the system in its unique context and environment. Behavior/performance is uncertain and non-linear.
Structural emphasis	***Elements***—Primary determinant of structure is held in the individual elements comprising the system.	***Relationships***—Primary determinant of structure is held in the relationships between individual elements comprising the system.

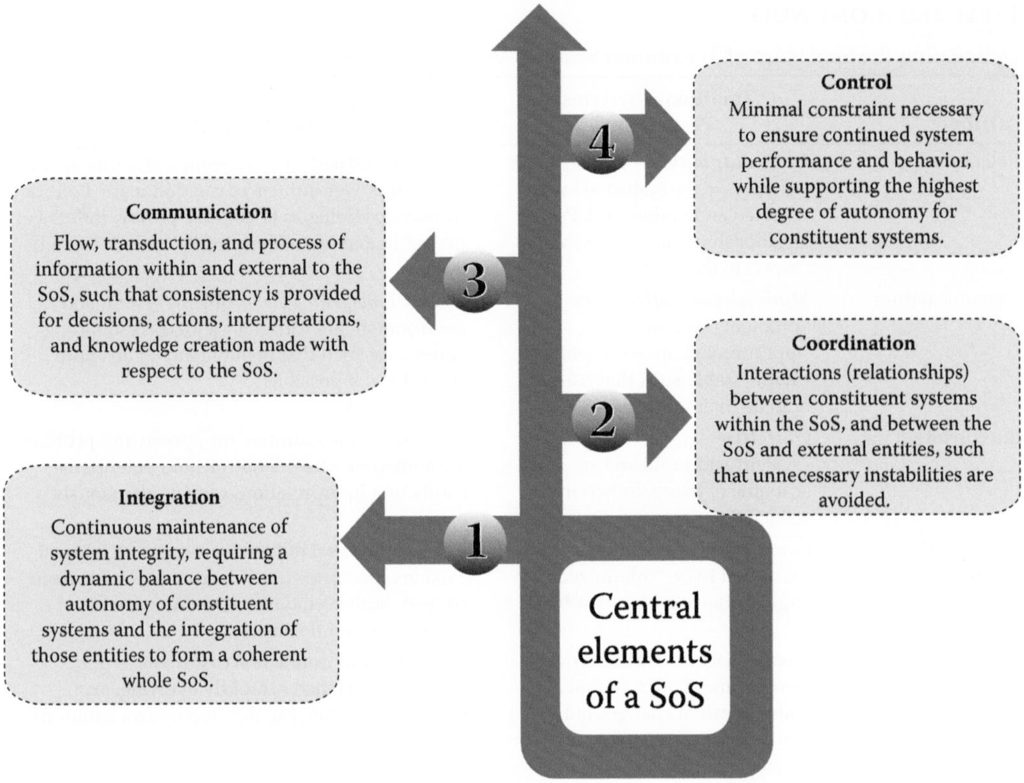

FIGURE 21.1
Four central elements that must be achieved for SoS.

integration and coordination of multiple systems; *Management Cybernetics* (the science of effective system organization) focused on provision of communication and control; and *System Governance* (essence of system direction, oversight, and accountability) focused on higher-level integrative design and evolution of systems. Following the CSG formulation as the extension of SoS (engineering), Figure 21.1 captures four central issues related to what must be achieved by a SoS: Integration, Coordination, Communication, and Control.

- *Integration.* Quite possibly the greatest challenge that a SoS must overcome is achievement and maintenance of integration of constituent systems. SoS constituent systems have not been conceived, deployed, or evolved in deference to the higher-level SoS mission/purpose to which they are to become an integral part. The challenge is to "integrate" the constituent systems into a SoS to perform a purpose/mission which constituents cannot achieve independently. Inevitably, this integration of managerially and operationally independent systems involves surrender of some degree of autonomy by constituents. Autonomy is the degree of freedom and independence of decision, action, and interpretation experienced by a system and its actors. Integration into a SoS requires some level of autonomy to be "surrendered." This is essential to permit the SoS to act in concert as a unity—with a singular purpose/mission and identity that exists beyond the missions and identity of any of the constituent systems.

- *Coordination.* This involves providing for interactions (relationships) between constituent systems within the SoS, and between the SoS and external entities, such that unnecessary instabilities are avoided. SoS have significantly increased complexity in contrast to a single constituent system. For SoS, the interactions (coordination) extend beyond the limited grasp of technology-based exchanges. Instead, the ambiguous and shifting SoS boundaries range across the entire spectrum of organizational, managerial, human, social, policy, and political dimensions that must be coordinated for a SoS to function as a unity. Thus, more traditional approaches (Keating, 2014; Katina, 2016) (e.g., Systems Engineering) based in well-defined problem domains, objective understanding, and *systematic* (reductionist) formulations are rendered as suspect concerning their capacity to effectively deal with the coordination complexities inherent to the SoS problem domain. Inevitably, the SoS problem domain is marked by ambiguity in boundary conditions, subjective formulation, and need for *systemic* (holistic) formulation. Further, the hope for direct interpolation of traditional SE methods, tools, procedures, technologies, or techniques to the SoS problem domain is doubtful at best and potentially catastrophic at worst. This elevates coordination to a central role and challenge for SoS.
- *Communication.* A SoS must provide for effective communication—the flow, transduction, and processing of information within and external to the system. This is necessary to ensure consistency is maintained with respect to decisions, actions, interpretations, and knowledge creation for the SoS. While there has been a propensity of thinking about communication in SoS as purely technical in nature (e.g., interoperability), we maintain that this is shortsighted. For SoS, communications extend beyond a narrow technology formulation which is simplified to *transmission-receiver interactions.* In contrast, communication in SoS must be extended to include a wider formulation, considering such issues as interpretative schema differentials, worldview projections, and conceptual redundancy. Thus, communication in the SoS world exist beyond more simplistic "0's" and "1's" perspectives dominated by technology considerations (e.g., bandwidth capacity) and "noise reduction" strategies. Instead, while technical formulation of communication in SoS is necessary, that formulation alone is not sufficient for SoS development.
- *Control.* The primary function of control in any system is to invoke constraints (regulatory capacity) such that the system purpose/goal can be achieved. Unfortunately, too often control is viewed pejoratively, sometimes with good reason, as a burden to system performance. In contrast, the control challenge for SoS is more appropriately captured in viewing control as "preservation of the greatest possible level of autonomy for constituent systems." In this sense, autonomy is the degree of freedom and independence of decision, action, and interpretation of actors in constituent systems. In system parlance, this is capture in the principle of *Minimal Critical Specification* (Whitney et al., 2016), or for SoS, only specifying those constraints absolutely necessary to ensure that the SoS continues to achieve its intended purpose (mission). Excess constraint wastes resources and diminishes local decision/action efficiencies achievable by constituent systems. Insufficient constraint jeopardizes the SoS capacity to continue to achieve its intended higher-level purpose. The level and nature of constraint (control) in a SoS is key to achieving a balance between constituent system autonomy and SoS level integration necessary to assure sustained performance.

In this section, we have: (1) presented the attributes allowing for classification as a SoS and four types of SoS, (2) identified the distinctions in addressing problems/systems from a SoS perspective versus a traditional systems perspective, and (3) identified four primary functions that must be performed for SoS, including: integration, coordination, communication, and control. With this basis of understanding established for SoS, we now shift direction to development of a perspective for emergence and its implications for SoS.

21.4 Emergence: Introduction and Central Tenets

Emergence is not a new concept. Emergence is a principle that exists in classical systems theory, suggesting that system properties (patterns, capabilities, structure, behaviors) develop from interaction between system elements and between system elements and the environment (Aristotle, 2002; Hitchins, 2008; Sousa-Poza et al., 2008). In a SoS context, properties which emerge are not traceable to any of the particular constituent systems and cannot be understood/predicted from the properties of those constituent systems. This general concept has existed for some time, finding roots back to Aristotle, who suggested that the whole is more than the sum of its parts (Checkland, 1999). With respect to the roots of the concept of emergence, Holman (2010) traces emergence to original work by Lewes (1875, p. 412) who suggested that "there is a co-operation of things of unlike kinds. The emergent is unlike its components . . . and it cannot be reduced to their sum or their difference." Further, Goldstein (1999, p. 49), in introducing the inaugural issue of the journal *Emergence*, captured emergence as "... the arising of novel and coherent structures, patterns, and properties during the process of self-organization in complex systems. Emergent phenomena are conceptualized as occurring on the macro level, in contrast to the micro-level components and processes out of which they arise." Lucas and Milov (1997) suggest that emergence requires development of new categories, not present in existing elements (parts), and requires a new vocabulary to describe properties that did not previously exist at the part level. With respect to emergence in relationship to design, Chalmers (1990) suggests that emergence occurs when a system is designed in accordance with certain principles and expectations, yet "interesting" properties arise that were not intended or conceived in the original design.

While there are many different perspectives on emergence, there does appear to be some level of consensus of central characteristics for emergence. Following the works of (Adams et al., 2014; Kim, 1997; Keating, 2009; Holland, 2000; Goldstein, 1999; El-Hani and Pereira, 1999; Corning, 2002; Arshinov and Fuchs, 2003; De Wolf and Holvoet, 2005), Table 21.4 is provided to consolidate commonly held aspects for the concept of "emergence" in the literature.

Another aspect of emergence is found in the concept of "strength of emergence." Chalmers (2006) explains that *strong emergence* occurs when higher-level phenomena emerge from a lower-level domain but are not deducible from that domain—focused on novelty experienced as irreducibility (incapability of understanding by reduction to phenomena held at a lower level) and downward causation (novel properties which arise at the macro level and generate causation impacting behavior at the micro level). Thus, novel properties arise in a higher-level system which are not attributable to lower-level constituent systems yet can impact behavior of the constituents. In contrast, *weak emergence* exists when a novel property unpredictably/unexpectedly arises in a higher-level

TABLE 21.4

A Variety of Characteristics and Perspectives for Emergence

Emergence Characteristic	Description
Novelty	Properties or features that did not previously exist. The nature of these properties is the source for unpredictability of emergence.
Coherence or Correlation	Emergent characteristics maintain a sense of identity over time. A higher-level unity is maintained from correlation of lower-level elements.
Global Level	Properties exist at the higher (global/macro) level as opposed to the lower level (micro) from which the properties emerge.
Dynamic	Properties evolve over time and are not capable of being predetermined or predicted in advance of their manifestation.
Ostensive	Recognition of emergent properties becoming known when they become exposed—never the same as previously exposed emergent properties/ phenomena.
Interacting Parts	Production of emergent properties requires that elements interact.
Decentralized Control	There is not central control. Macro level behavior is produced by local mechanisms that interact to produce global level behaviors.
Downward Causation	Micro to macro level produces emergent behavior/structure. Macro to micro emergent structure influences the micro level generating causation in a downward direction.
Robustness	Single entities cannot be a single point of failure, as emergents are insensitive to perturbations. This generates robustness.
Flexibility	Degradation due to perturbation/error is gradual, meaning that flexibility permits emergent structure to remain.

system. However, although the phenomenon might exhibit unpredictable results, it may nevertheless be deducible given sufficient knowledge (initial state of the system and governing rules). Notwithstanding this strong-weak emergence distinction, for our purposes we focus on three items: (1) emergent properties not being held at the level of constituents, (2) the nature of the emergent properties being novel, and (3) the inability to "reduce" understanding of emergent properties to constituent entities from the level where they were produced. We offer a common perspective of emergence in SoS as *a novel higher-level SoS property arising from interacting lower-level constituent systems, being incapable of prediction or reduction to properties of lower-level constituent systems.*

Emergence is also related to other system concepts, including resilience and self-governance. According to The Lloyd's Register Foundation (2015) report, "emergence as a property of a system, can be used to describe relevant attributes that allow systems to withstand, respond, and or adapt to vast range of destructive events" (Lloyd, 2015, p. 7). If one takes the definition of resilience as "a process linking a set of adaptive capacities to a positive trajectory of functioning and adaptation after a disturbance" (Norris et al. 2008), one can then look at emergence as being one of those capabilities that enable systems (i.e., individuals, communities, institutions, businesses) "to survive, adapt, and grow no matter what kinds of chronic stresses and acute shocks they experience" (Martin-Breen and Anderies, 2011). Emergence's relation to self-organization can be explained in terms of the foundational concepts of self-organization in complex systems. Literature indicates that complex systems organize themselves (Ashby, 1962; Skyttner, 2005). In this case, "organizing themselves" is an inherent capability that somehow "emerges" to influence structural and behavioral patterns in a complex system, primarily resulting from the interaction among parts of a complex system. Emergent properties are also recognized in traditional systems engineering. INCOSE's handbook (2011, p. 8) notes that "[t]he SE [Systems Engineering] process has an iterative nature that supports learning and continuous improvement. As the processes unfolds, system engineers uncover the real requirements and the emergent properties of the system." Moreover, Sauser et al. (2009) suggest that emergence can be intended or unintended. For intended emergence the broad precepts of desirable (emergent) results from modifications in the system. However, unintended emergence is the experiencing of properties that are neither foreseen nor desirable within a range of anticipated outcomes. This is commonly referred to as "unintended consequences."

Given these perspectives of emergence, we now shift our development to implications, applicability, and utility that emergence offers for dealing with design, execution, and development of SoS.

21.5 Emergence: Implications for SoS

Emergence is a powerful concept with far reaching implications for the design, execution, and development of SoS. To bridge from concept to implications we have focused this section on development of three primary themes for emergence: systems propositions, SoS landscape, and implications. First, we will suggest a set of systems propositions (laws, principles, and concepts) that can provide important insights with respect to implications for emergence in SoS. Second, we suggest a perspective for integrating emergence into the landscape of SoS. This will amplify considerations and implications for emergence in design, execution, and development for SoS. Third, we explore a set of implications for enhancing SoS practices considering emergence. The emphasis of this exploration is on guidance implications for practitioners challenged to deal with emergence in SoS.

General Systems Theory (GST) offers the set of propositions (principles, laws, concepts) that have been continually developed and applied over the past eight decades (von Bertalanffy, 1968; Skyttner, 1996; Adams et al., 2014; Whitney et al., 2015). The propositions have withstood the test of time and application seeking to explain the structure, behavior, and performance of systems. The contributions and implications that system propositions hold for SoS have been previously elaborated (Keating et al., 2016). These contributions can be summarized in three major points: (1) GST provides a rigorous theoretical grounding, which provides a consistent reference point based in an established set of propositions, (2) the language and propositions of GST can inform alternative thinking and utility for application of emergence in SoS, and (3) GST brings a worldview that is much more compatible with emergence and the problem domain being faced for SoS. In application for emergence in SoS, we have identified several applicable Systems Theory propositions and their implications for emergence in SoS (Table 21.5).

With respect to emergence in SoS, we can be certain of three driving factors: (1) emergence is going to happen, (2) neither the precise nature of emergence nor the impacts can be known or predicted in advance of system deployment, and (3) strategies can be invoked in attempts to deal with emergent conditions/events for SoS. It is instructive to consider emergence and its impact across three stages of SoS (Figure 21.2), including *design* (*pre-emergence* that exists prior to deployment of the SoS), *execution* (*intra-emergence* occurring once the SoS is deployed and enters into an operational mode), and *development* (*post-emergence* occurring as the SoS evolves over time to maintain performance). Thinking about emergence in each

TABLE 21.5

Systems Theory Propositions with Applicability for Emergence in SoS (Propositions and Descriptions Based on Whitney et al., 2016; Keating et al., 2016)

Systems Theory Proposition	Description	Implications for Emergence in SoS
Circular causality (Korzybski and Rouben Mamoulian Collection (Library of Congress), 1933)	An effect becomes a causative factor for future "effects," influencing them in a manner particularly subtle, variable, flexible and of an endless number of possibilities.	Considerations must include not only emergence from interaction of constituent systems, but also the impacts that accrue from downward causation stemming from emergent properties at the higher-level system.
Complementarity (Bohr, 1928)	Two different perspectives or models about a system will reveal truths regarding the system that are neither entirely independent nor entirely compatible.	Emergence is subject to interpretation of sources, impacts, and appropriate response. Care must be taken to include consideration of multiple perspectives concerning the nature of emergence, formulation of responses, and interpretation of results.
Control (Checkland, 1993)	The process by means of which a whole entity retains its identity and/or performance under changing circumstances.	Ensuring strength of identity provides for a more robust and resilient system in response to a variety of unknown and unforeseen emergent conditions.
Incompressibility (Cilliers, 1998; Richardson, 2004)	Each element in the system is ignorant of the behavior of the system as a whole and only responds to information that is available to it locally.	At the constituent level, there is not necessarily the information or understanding necessary to make sense of emergent properties held at the higher-level system. Appreciating and acting on the need to bridge this understanding gap is essential.
Dynamic equilibrium (von Bertalanffy, 1968; Miller, 1978)	An entity exists as expressions of a pattern of processes of an ordered system of forces, undergoing fluxes and continuing flows of matter, energy and information in an equilibrium that is not static	As emergence occurs in operation of a system, design should be sufficiently robust across a wide range of potential emergent conditions. Capability for resilience in the face of emergent conditions should be a major emphasis for effective design.
Equifinality (von Bertalanffy, 1950)	If a steady state is reached in an open system, it is independent of the initial conditions and determined by the system parameters, e.g., rates of reaction and transport.	Emergence will occur; however, system design and recognition should realize that maintaining system performance in the face of emergence is possible by following many different paths in response to emergence.
Holism (Smuts, 1926)	A system must be considered as a unique whole, rather than a sum of its parts.	It is shortsighted to proceeding with design, execution, or development of a system without considerations for emergent behavior stemming from interactions.
Homeorhesis (Waddington 1957, 1968)	The concept encompassing dynamical systems that return to an acceptable trajectory through adjustments in dynamic equilibrium controlled by interrelated regulation mechanisms.	Designing for response to emergence should account for understanding the impact of emergent properties/conditions on the trajectory of a system. This should be from both returning to a particular trajectory or revising the trajectory based on emergence.

(Continued)

TABLE 21.5 (CONTINUED)

Systems Theory Propositions with Applicability for Emergence in SoS (Propositions and Descriptions Based on Whitney et al., 2016; Keating et al., 2016)

Systems Theory Proposition	Description	Implications for Emergence in SoS
Homeostasis (Cannon, 1929)	The property of an open system to regulate its internal environment to maintain a stable condition, by means of multiple dynamic equilibrium adjustments controlled by interrelated regulation mechanisms.	Design for emergence should focus on building resilience capacity across a range of emergence categories. Achieving rapid identification, efficient assessment, appropriate response development, deployment, and evaluation of impacts is essential to dynamically maintain system stability in response to emergence.
Minimal critical specification (Cherns, 1976, 1987)	This principle has two aspects, negative and positive. The negative simply states that no more constraint should be specified than is absolutely essential; the positive requires that we identify what is an essential level of constraint to maintain system performance.	In both design for emergence as well as response to emergence care must be taken not to overregulate the system. Emphasis should be on preservation of autonomy for system entities that are situated in the best position to mount a response to emergence. Overregulation wastes resources and can degrade system performance.
Multifinality (Buckley, 1967)	Radically different end states are possible from the same initial conditions.	Emergence can impact the pathway and end state for a system. Identification and understanding of the impact of emergence on pathways (strategies) and end state achievement is essential.
Redundancy (Pahl et al., 2011)	Means of increasing both the safety and reliability of systems by providing superfluous or excess resources.	Emergence is likely to generate the need for additional (redundant) resources to cope with the potential impacts.
Consequent production (Keating and Pyne, 2001)	A system can only produce what it produces, nothing more and nothing less. It is perfectly executing to produce the patterns, performance, behavior, and structure that it is producing.	When confronted with emergent performance or behavior that is undesirable, we must first look to the system that is generating the behavior/performance variance. Understanding and redesign of the underlying structure of a "problem system" is necessary to change the behavior or performance that fails to meet expectations.
Relaxation time (Clemson, 1984)	A system in dynamic equilibrium requires a time period (relaxation time) to return to a state of dynamic equilibrium following a perturbation (change) from internal or external forces. Multiple simultaneous changes (perturbations) in a complex system, if additional perturbations occur at a frequency that does not permit resettling into a state of dynamic equilibrium, will put the system into oscillation and quite possibly on a trajectory from which dynamic equilibrium is no longer achievable.	Intervention into a system that invokes multiple simultaneous changes suggests that: (1) the system may be placed into oscillation without the chance to return to a state of dynamic equilibrium, and (2) the "impact" of a single change might not be understood in the face of multiple simultaneous changes. Care must be taken to question the nature, depth, and interrelationship among change initiatives taken in complex systems, as well as the emergence they generate.

(*Continued*)

TABLE 21.5 (CONTINUED)

Systems Theory Propositions with Applicability for Emergence in SoS (Propositions and Descriptions Based on Whitney et al., 2016; Keating et al., 2016)

Systems Theory Proposition	Description	Implications for Emergence in SoS
Basins of stability (Clemson, 1984)	A system will seek a level of stability (lowest energy state) unless acted on by external forces. The system will move to a new basin of stability (past a threshold) only when sufficient energy (resources) are applied to provide "momentum" necessary to shift to the new basin of stability.	In attempting to make enduring change in a system, there must be an understanding that the change will only be temporary if sufficient energy (e.g., resources, processes, constraint) is not provided to carry the system to a new level. The energy must be in correspondence with that required from the new stability sought, with sufficient indicators to recognize achievement and maintenance of the new stability (performance/behavior measurement). Emergence can impact achievement and maintenance of stability.
Viability (Beer, 1979; Keating 2009)	Viability is the ability of a system to maintain existence. There are three primary axes for which a system must have a suitable fit to the environment to maintain viability: *change* (ranging from stability to adaptation), *design* (ranging from total self-organization to purposeful), and *control* (ranging from integration to autonomy).	A system must be questioned with respect to how it is "appropriately" fit to the environment within which it must exist (remain viable). Does the capability for change correspond to the rate of change of the environment? Is the design sufficiently loose/tight to allow responsiveness to environmental flux and emergence? Is the system minimally constrained to permit system level performance, while permitting constituents the greatest latitude in decision and action (autonomy) in response to local emergence?
Unity of system Purpose (Beer, 1979, 1985; Jackson, 1991)	Maximizing system performance requires that the different subsystems structure their performance, objectives, and approach to be consistent with maximizing higher-level system performance, not subsystem performance. The system cannot achieve the notional goal of optimal performance without imposing some level of constraint on the constituent subsystems.	Assumption of unity of system purpose is a mistake—unity of system purpose must be designed, maintained, and evolved over time. Unity in a system is a precursor to ensure that there is consistency in decisions, actions, and interpretations taken in behalf of the system, particularly in conditions marked by high levels of emergence. Lack of unity is likely to result in fragmentation, unilateral behavior, and pursuit of objectives by constituents that are not necessarily consistent with the overall system identity.
Self-organization (Clemson, 1984)	The majority of the structural and behavioral patterns for a complex system only emerge after operation of the system in its environment (context). Unintended consequences can be mitigated through design for robust feedback, feedforward, and redundancy of critical system functions.	Self-organization should be maximized for constituent systems of a complex SoS. This is achieved by maximizing autonomy (freedom and independence of decision, action, and interpretation) within the minimal system level constraints necessary for system integration. Unnecessary limiting of subsystem autonomy can create waste of system resources and inefficiencies. Emergence will impact maintenance of the autonomy-integration balance between the SoS and constituent systems.

(Continued)

TABLE 21.5 (CONTINUED)

Systems Theory Propositions with Applicability for Emergence in SoS (Propositions and Descriptions Based on Whitney et al., 2016; Keating et al., 2016)

Systems Theory Proposition	Description	Implications for Emergence in SoS
Complementarity (Clemson, 1984)	There are multiple perspectives for any given system. Each perspective is both correct and incorrect, dependent upon the particular vantage point from which the system is viewed. Each vantage point has a set of corresponding assumptions and logic that supports the legitimacy of that vantage point. Multiple perspectives are neither entirely independent nor entirely inclusive.	Multiple views and perspectives are essential, particularly in the formative stages for a system effort, to ensure a robust approach and design. Failure to include multiple perspectives can be limiting to the eventual system solution that is generated. Views are subject to shifts in response to, and understanding of, emergence. New perspectives can emerge as new system knowledge is gained and processed.

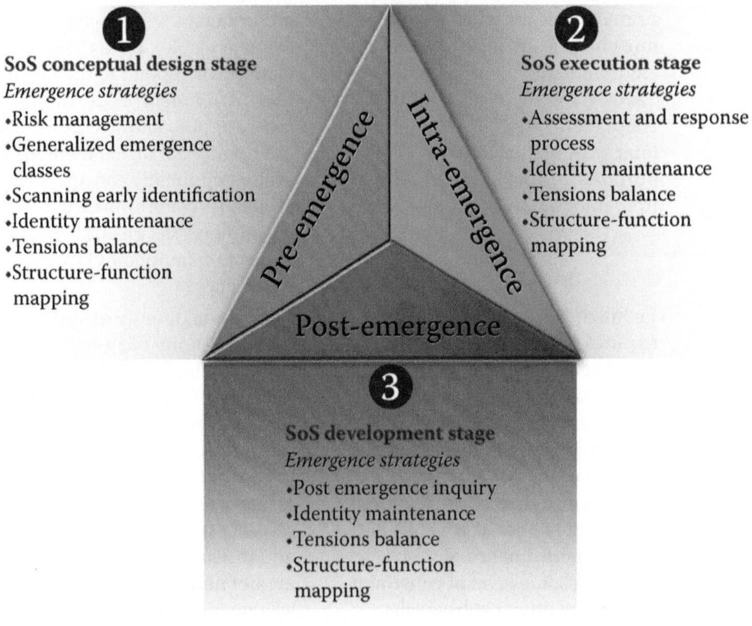

FIGURE 21.2
Emergence and strategies across 3 stages of SoS.

of these different stages of SoS provides insights into thinking and potential responsive strategies to deal more effectively with emergence.

Design is focused on the early stages of bringing a SoS into being. As such it exists first at the conceptual level, prior to commitment to the actual building (even out of existing constituent systems) and deployment of a SoS in response to a problem or need. Responding to emergence (pre-emergence) in this phase of a SoS is difficult, and arguably cannot actually be accomplished because the SoS is not in operation—thus, how could emergence occur since it requires interaction among constituents? However, there are strategies that can be undertaken at this stage to narrow the potential negative consequences that might accrue from future episodes of emergence. Emergence at the execution stage, *intra-emergence*, of a SoS involves dealing with emergent properties/events which occur as the SoS is operating. Decisions and actions at this stage must address actual emergence stemming from an operating SoS. While the emergence itself cannot be stopped, nevertheless it can be dealt with through strategies to lessen negative consequences (i.e., unintended emergence) or amplify positive consequences (i.e., intended emergence). Finally, emergence experienced in development, following occurrence of emergence, or *post-emergence*, also has strategies that might be pursued in response. We now shift focus to suggest several different strategies to assist practitioners in dealing with emergence in SoS. It is important to note that these strategies are not binary (all or nothing) in their application. There are certainly a range of activities and substrategies that might be deployed. In addition, the strategies are not mutually exclusive of one another. There might be occasions where multiple strategies are invoked simultaneously and possibly in conjunction with one another. The strategies include:

- *STRATEGY 1 (pre-emergence): Emphasis on detailed Risk Management at conceptual design can help identify potential "classes" of system emergence and provide design cues for predeployment design modifications.* During conceptual design, the unknowns in a SoS are at the highest point. Risk Management (Antunes and Gonzalez, 2015) is focused on the identification, assessment, and prioritization of risks. Although the precise timing and specific form of emergence is not known in advance, scenario-based analysis and considerations from a risk-based management framework can be effective in identifying "potential" emergent conditions (weak emergence) for which responsive SoS design modifications might be implemented at the design stage.
- *STRATEGY 2 (pre-emergence): Design in response to "generalized classes" of emergence can put into place the structure, processes, and procedures to more effectively manage emergence yet to occur.* Although the specific nature and timing of emergent properties cannot be known in advance, there are design cues that can be installed in a system prior to deployment. The potential "classes" of emergence can be identified, without knowing the specific form or timing of the emergent property/condition. For example, there may be a class of emergent conditions related to a spectrum of "technology" based failures. The potential redesign of the SoS response capability, prior to deployment, might be initiated to increase resilience through installation of redundancies across a spectrum of technology related emergence conditions. Although the specific emergent condition may not be known, it may well fit to a prepared "class" of emergent conditions.
- *STRATEGY 3 (pre-emergence): Install a "scanning system" to provide early identification of emergent conditions that may impact SoS performance.* Mechanisms for identification of emergent conditions for a SoS can range across dimensions of formal/informal, tacit/explicit, and sporadic/routine. The installation of a compatible and appropriately "fit" approach to provide early detection of emergent conditions can be an effective strategy to reduce the time between identification and response, possibly limiting the potential for "SoS damage" incurred from the emergence. In this sense, by early detection of emergent conditions the possible decision space for response might offer more alternatives, including classifications of, for instance, "acceptable" and "unacceptable" emergence.
- *STRATEGY 4 (intra-emergence): Design and install a deliberate process for assessment of impacts, and response development for emergence.* This process can be designed and

installed in the SoS prior to deployment. Therefore, although the precise nature of emergent properties/conditions may not be known, processes/procedures/methods to direct SoS response to occurrences of emergence can be formalized prior to experiencing emergent conditions. It is appropriate to note that not all emergence is negative. There may be instances where emergence produces "good" for the system and these conditions should also be subject to exploration, response, and potential exploitation.

- *STRATEGY 5 (post-emergence): Conduct post-emergence inquiry to make necessary SoS adjustments.* Each case of emergent property/event presents an opportunity to engage in learning that provides potential for: (1) SoS modification by *first order learning* (Keating, 2009), which involves detecting and correcting system errors (deviations from expected results, e.g., emergence) through processes of inquiry that stay within established assumptions, constraints, boundaries, norms, and the present system design; or (2) SoS modification by *second order learning*, which involves detection and correction of system error based in inquiry that calls into question operating assumptions, norms, objectives, and present system design inadequacies which may be driving the aberrant (emergent) condition (Keating, 2009). Of particular significance for emergence is engagement in *deutero learning*, or "learning to learn," where the focus is on getting better at inquiry which applies the correct type of learning to the situation and finding novel responses to familiar circumstances (Keating, 2009). In any case, it is a miscalculation to succumb to the notion that since we cannot predict emergence, then any efforts to try to preclude, or learn from emergent episodes, is futile. On the contrary, a SoS should always be seeking to improve, even in the face of emergence. While unpredictable properties/events will occur, it would be shortsighted not to install SoS improvements that may attenuate future "similar" negative episodes or amplify properties emanating from positive episodes.
- *STRATEGY 6 (pre-intra-post-emergence): Focus on maintenance of a strong identity to invoke stability and guide consistent response to emergent properties/events.* Emergence is inevitable in SoS. In effect, emergence can challenge the stability of a SoS by the introduction of properties/events/conditions with the potential to degrade performance. Identity of a SoS includes the essential aspects that define the uniqueness of the SoS and serve to distinguish it from all other systems. Identity permits the SoS to function and is a grounding force to provide consistency in decisions, actions, and interpretations made on behalf of the SoS. Each SoS has its own unique identity, which may range from strong to weak. A strong and well maintained SoS identity will provide increased consistency and resilience in the face of emergence. Similarly, a weak identity will produce decreased consistency and system vulnerability to emergence. Thus, the instabilities created by emergence can challenge the identity of a SoS. The strength of identity will be a critical factor in the degree to which the SoS is resilient in the face of emergence. It is also possible that emergence will question the continuing "present" identity appropriateness. However, while identity of a SoS can shift over time, engaging in purposeful inquiry in response to emergent conditions can provide a "graceful" examination as opposed to a crisis examination.
- *STRATEGY 7 (pre-intra-post-emergence): Maintaining a balance in tensions (design, control, change) to produce resilience in response to emergence.* (Figure 21.3, following Yuchnovicz, 2016). Emergence can challenge SoS stability. The degree to which

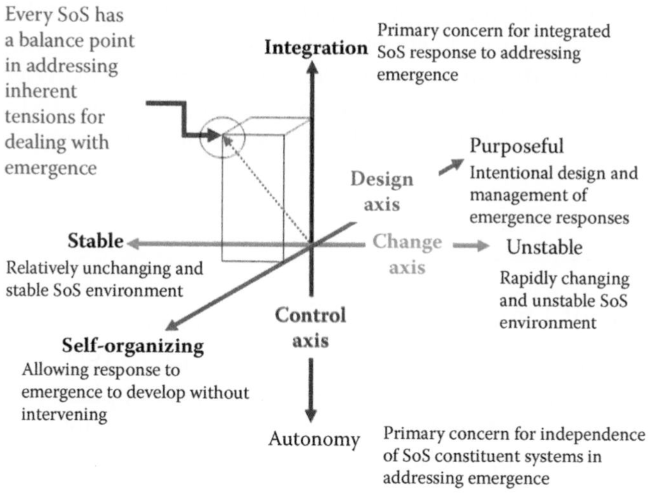

FIGURE 21.3
Tensions and balance in addressing emergence in SoS.

a SoS is "fit" to the particular circumstances in which it operates, the higher the stability. Higher stability can increase resilience capacity for response to perturbations experienced (emergence) as the SoS operates. It is important to note that the balance point may be stressed by emergent properties/events. The degree to which the "balance point" is, and continues to be, appropriately fit to the operating circumstances of the SoS will provide a degree of stability in the face of emergence. In addition, it is entirely possible that the response to emergence may indicate a necessity to shift the balance point. However, it is desirable to make the shift through processes of purposeful inquiry in response to experienced emergence, rather than hastily construed and implemented "crisis mode" responses to emergence.

- *STRATEGY 8 (pre-intra-post-emergence): Explicit mapping of the structure and function of the system, context, and environment.* We are continually astounded to see that something as complex as a SoS is generally designed, operated, and maintained without a clear, comprehensive, or explicit representation that maps the SoS structure and function. This representation should minimally identify the *SoS* (structural system configuration, boundaries, functional operation, processes, e.g., boundary conditions), the *context* (internal circumstances, factors, conditions, patterns that enable/constrain functioning of the SoS, e.g., power relationships), and *environment* (entities/activities/events external to the system that impact and are impacted by the SoS, e.g., suppliers). Without this explicit understanding of "how this thing works," the planning, identification, assessment, response, and evaluation cycle related to emergence across *pre, intra, and post* emergent properties/conditions/events cannot be fully exploited. Simple articulations of the SoS (e.g., organizational charts, process manuals) are incapable of providing the analytic support necessary to either: (1) design out, mitigate, or eliminate undesirable effects of emergence, or (2) exploit potential positive implications of emergence for system development.

In this section, we have offered a systems perspective with applicability to dealing with emergence in SoS. In addition, eight strategies have been suggested to help guide thinking and practices related to addressing emergence in SoS. A SoS can experience emergence in *pre*, *intra*, and *post* phases of SoS, paralleling SoS design, operation, and development. We now shift our focus to examination of application scenarios for emergence in SoS.

21.6 Applications of Emergence in SoS

Emergence appears to be an ephemeral concept in relationship to SoS application. As much as we might appreciate the related emergence concepts, the applications of emergence in SoS context are limited at best, and near non-existent at worst. However, to demonstrate utility of "emergence thinking" for SoS, we have selected three application scenarios from the SoS world to demonstrate the potential utility of emergence for SoS. For each application, the context, SoS in focus, "emergence situation," and the exploration of emergence insights stemming from the eight strategies are presented. Our primary objective is to suggest how an appreciation of "emergence thinking" offers utility for dealing better with the inherent emergence that is a mainstay for the SoS world, present and future.

EMERGENCE APPLICATION SCENARIO 1: *Advanced Manufacturing SoS Introduction of New Warehousing Technology*. This scenario occurred in an advanced manufacturing facility. The product was complex guidance and control systems for an advanced missile program for the Department of Defense. These systems utilized state of the art technologies and were extremely complex "subsystems" of a larger system (of systems) that included all aspects of design, manufacturing, testing, integration, and delivery. At the initial prototype stages, a small group of technicians and engineers worked together in close proximity to produce the initial production units. Production was increasing at a rapid rate, with increased deliveries set to move from an initial level of 5 to a top level of 300 units per month (a 6000% increase) within 3 years. The staff, support systems, and infrastructure (including warehousing) would also have to expand to meet the acceleration in production. While automation was being introduced, each production unit was still guided through production from initial build to final certification testing prior to shipment. The *emergence* focused on for this exploration stemmed from the deployment of a "Just-in-Time" (JIT) inventory system intended to provide more precise tracking, accountability, responsiveness, and automated support related to providing production materials. The JIT inventory system was designed and built by a support group of engineers, the deployment date was announced for the switch to the JIT system, and the system was made operational on the announced deployment date. Within 24 hours the JIT system began having difficulties. Errors were widespread, modifications made "on the fly" further delayed production, and the production rate went to a near standstill. Production workers, engineers, and technicians were idled. The senior program executive announced, "I want that system overridden and production to move forward in the next 2 hours.... If I have to go unplug that system myself, it will be disabled." The result was a disabling of the system and an indefinite delay for installation of a more automated warehousing system. From the perspective of emergence, we offer the Table 21.6 synopsis,

TABLE 21.6

Emergence Perspective for the JIT Inventory System

Strategy for Emergence	Phase (pre, intra, post)	Assessment Red—None Yellow—Some Green—Exist	Scenario Reflection
Strategy 1: Risk management at design	Pre	Yellow	No formalized risk management process was conducted with respect to the entire SoS, although technical risk assessments were conducted for the product.
Strategy 2: Design for generalized classes of Emergence	Pre	Red	No classes of potential emergence identified.
Strategy 3: Scanning system for early identification	Pre	Yellow	SoS had informal scanning mechanisms in place. However, these were directed to the external environment and not to the SoS itself.
Strategy 4: Process for assessment and response	Intra	Green	The process for assessment and response for emergent issues was in place and exercised in this instance. However, the situation escalated such that the routine Engineering Issue Assessment process had to be bypassed due to the urgency and severity of the issue.
Strategy 5: Conduct post-emergence inquiry	Post	Red	There was no visible post-emergence inquiry conducted. Responsible departments were tasked with their own assessments which were neither formal nor accessible to the wider SoS.
Strategy 6: Maintenance of strong identity	Pre-intra-post	Green/yellow	In this SoS, the closeness of the staff provided a strong identity and was helpful in working through the emergent issue. This identity was not by active design or purposeful maintenance—thus, the capability of identity to support effective responses to future episodes of emergence was questionable.
Strategy 7: Balance in tensions	Pre-intra-post	Red	While it might have been arguable that a balance existed between design, control, and change, this was not an active, explicit, or purposeful pursuit for the SoS.
Strategy 8: Mapping of structure and function	Pre-intra-post	Red	There was no explicit mapping of the SoS capable of providing insight to the structural configuration and performance of system functions.

organized by the implications of the strategies suggested (Section 21.5) and their applicability/implications related to this situation. While retrospective considerations of emergence offer insights, we also suggest that had strategies been deployed, the resulting "catastrophic" effects would possibly have been avoided.

EMERGENCE APPLICATION SCENARIO 2: SoS Unit for Integration of Independent Information Systems to Support a Larger SoS Design for Maintenance Reporting. This scenario is built from analysis conducted concerning the operation of a SoS established with the purpose of integrating multiple constituent information systems (total of 23 different constituent systems) to provide a higher-level maintenance reporting capability for equipment readiness. The SoS was being continually "surprised" by emergent requirements and requests with respect to developing and maintaining the integrated SoS capabilities for supplying essential information to support maintenance decisions. In addition, the SoS was frequently directed to provide capabilities and development that were beyond the capacity to develop or deliver. A particular area of focus with respect to the examination of emergence was the continually shifting strategy for long term evolution of the SoS. The results of this particular emergent condition on the system included: (1) a degradation of SoS capability stemming from continual shifts and realignment of system development priorities; (2) personnel turbulence, frustrations, and angst emanating from an unstable planning environment; and (3) instabilities in resources allocated to system development based on constantly shifting priorities. Table 21.7 summarizes the strategies for emergence.

TABLE 21.7

Emergence Perspective for the Maintenance Reporting SoS

Strategy for Emergence	Phase (pre, intra, post)	Assessment Red—None Yellow—Some Green—Exist	Scenario Reflection
Strategy 1: Risk management at design	Pre	Red	No formalized risk assessment was conducted prior to formulation of the SoS.
Strategy 2: Design for generalized classes of emergence	Pre	Red	No classes of potential emergence identified.
Strategy 3: Scanning system for early identification	Pre	Yellow	SoS had several formal scanning mechanisms. However, these were at a frequency that was inconsistent with the rate of change and emergence being experienced from the environment.
Strategy 4: Process for assessment and response	Intra	Red	The process for assessment and response for emergent issues was a "seat of the pants" approach. Not capable of sufficiently processing and developing a coherent response to identified issues and emergence.
Strategy 5: Conduct post-emergence inquiry	Post	Red	There was no visible post-emergence inquiry of emergence episodes conducted. The only examination was from external entities and not directed at emergence in the SoS.
Strategy 6: Maintenance of strong identity	Pre-intra-post	Red	In this SoS, the identity appeared weak and unable to impact degradation from the emergence experienced. Several examinations identified inconsistencies in interpretation of actions taken on behalf of the SoS in response to emergence.
Strategy 7: Balance in tensions	Pre-intra-post	Red	The SoS was in constant oscillation which caused instabilities. The impact on emergence was exacerbation of negative impacts of emergence in the case of shifting strategy. The SoS was out of sync and operated continually in an unbalanced (crisis) mode.
Strategy 8: Mapping of structure and function	Pre-intra-post	Red	There was no explicit mapping of the SoS capable of providing insight into the structural configuration and performance of functions. However, there were mappings of the technological configuration and processes to guide technology integration.

EMERGENCE APPLICATION SCENARIO 3: Hospital SoS Budget Shortfall. This scenario is based on a hospital SoS that experienced a budget shortfall in excess of $5M at the midpoint of the fiscal year. This shortfall was viewed as an unanticipated (unpredicted) condition emergent from operation of the SoS. The shortfall amounted to approximately 5.4% beyond the already allocated (and 50% expended) annual budget. Prior SoS level issues of such emergent "crises" had reverted to responses that were characterized as a "slash and burn" approach to correction. The result of responses to previous crises had lasting consequences on the system, including degrading the SoS function, causing dissension within the staff, and a lack of trust in stability for budget planning. In response, the SoS developed a "new" and "novel" budget process to get through the situation. The new process was quickly designed by the SoS leadership team and instituted. The shortfall was addressed with minimal disruption to SoS services and became a source of pride for the leadership in demonstrating how the response had demonstrated a "coming together" to confront adversity in a way that did not create turmoil. Table 21.8 provides an overview of emergence strategies.

The three application scenarios demonstrate the difficulties as well as opportunities to better deal with emergence in complex SoS. Although no SoS can claim to have mastered emergence, there are strategies at the pre, intra, and post phases of emergence that may serve to better equip SoS and their practitioners to confront emergence.

21.7 Conclusions and Implications

Dealing with emergence is critical to effectiveness in the SOS problem domain. As a high-level concept, the essence of emergence is not difficult to grasp. In effect, we have established several central tenets of emergence, including: (1) emergence is the development of properties, conditions, patterns, structure, or events that come about as a SoS operates; (2) we cannot know the precise form or timing of the arrival of emergence; (3) emergence exists beyond any of the constituent systems of a SoS and cannot be understood from the properties of those constituents; (4) emergence is unpredictable and not capable of being known in advance of manifestation; and (5) emergence can create a sense of hopelessness, or of existing at its mercy. Given this set of circumstances for emergence, it would be easy to suggest that we can do nothing and only respond when emergence is experienced. And, as is the case with unpredictable phenomena, most of application experience is in hindsight, looking backward at emergence. While hindsight is instructive, anticipation of emergence can save many headaches associated with its occurrence. Additionally, we suggested that emergence can be thought of in three phases of preparation and occurrence—*pre* (prior to experiencing the specific case of emergence), *intra* (experience within emergence as it is unfolding), and *post* (review of emergence after it has passed). It is important that two of the three phases are not based in response to emergence as it is unfolding. Present research introduced eight different strategies across those three phases that permits a more active approach to emergence. The present examination was to offer an alternative to succumbing to what we believe are unsupported assumptions concerning emergence that are born out of ignorance, unquestioned acceptance, or surrender. In closing, three critical

TABLE 21.8

Emergence Perspective for the Hospital Budget Shortfall

Strategy for Emergence	Phase (pre, intra, post)	Assessment Red—None Yellow—Some Green—Exist	Scenario Reflection
Strategy 1: Risk management at design	Pre	Yellow	Formalized risk assessment processes were in place for the SoS. However, the experience of such emergence as budget shortfalls were beyond the scope to the assessment process. Risk were more directed toward legal liabilities, patient care, and facilities.
Strategy 2: Design for generalized classes of emergence	Pre	Yellow-Green	No classes of potential emergence specifically identified. However, the SoS was experienced across a broad range of perturbations that might be expected in a 24/7/365 operation that by its very nature deals with emergent conditions (e.g., emergency room).
Strategy 3: Scanning system for early identification	Pre	Yellow	SoS had informal scanning mechanisms in place. However, these were not sufficient to provide adequate early warning of emergent conditions (e.g., budget shortfall) in a timely manner. The result was less time and decision alternatives available for response generation.
Strategy 4: Process for assessment and response	Intra	Green	The process for assessment and response to emergent issues was in place and exercised in this instance. The process was effective in directing an effective and timely response to the situation (emergence).
Strategy 5: Conduct post-emergence inquiry	Post	Yellow	There was not an orchestrated post-emergence inquiry conducted. The learning opportunity to capitalize on successes and implications for emergence episodes were largely lost. However, there were discussion forums related to the specific issue and response.
Strategy 6: Maintenance of strong identity	Pre-intra-post	Green	In this SoS, the closeness of the staff provided a strong identity and was central to developing the approach and response to the issue. For this SoS, identity maintenance was by active design with several formal and informal mechanisms in place to continually and purposefully question and maintain a strong identity.
Strategy 7: Balance in tensions	Pre-intra-post	Red-Yellow	This SoS did in fact appear to achieve a balance between design, control, and change. However, this was not an active, explicit, or purposeful pursuit for the SoS.
Strategy 8: Mapping of structure and function	Pre-intra-post	Red	There was no explicit mapping of the SoS that provided insight to the structural configuration and performance of functions.

emergence insights and their implications are suggested for SoS practitioners who must deal with emergence and its effects.

Insight 1: Emergence is a fact of life for SoS and should be engaged as any other potential impediment to fulfilling the purpose of the SoS. Given the complexity of SoS, it would be naïve to think that we could design, operate, and maintain a SoS free from emergence. It would be foolish to engage SoS without the expectation that there will be significant emergence to be addressed. Engaging emergence can be subject to the rigorous treatment that other system design attributes receive (e.g., reliability). The primary distinction is that for SoS the entire holistic spectrum of technology, human, social, managerial, organizational, policy, and political dimensions are in play and subject to emergence. Therefore, while it is a critical aspect of system design, "emergence design" does not receive the level of attention accorded to other design attributes of SoS (e.g., interoperability). It suffices to say that, without explanation as to why this is the case, we feel it is shortsighted and wrong for SoS to shortchange purposeful design for emergence.

Insight 2: Strategies can be deployed to be proactive in pre, intra, and post activities to more effectively address emergence for SoS. Accepting that emergence is a given for SoS, it is advisable that effective engineering for SoS be vigilant in developing and deploying strategies to prevent, mitigate, and design out negative instances of emergence. It must be accepted that emergence will occur and that all emergence is not negative. Consequences stemming from emergence may introduce positive forces into SoS. These positive forces must also be considered for evolving the SoS. Deployment of emergence strategies for SoS does not assume that emergent patterns/properties are known, or can be predicted, in advance of their occurrence. However, this should not preclude the development of designs to deal with emergent conditions, prior to experiencing the specific nature, form, and timing of the emergence event. For SOS, this proactive stance for emergence is a necessity for generating more robust solutions and increasing SoS resilience. Capabilities to engage the sequence of identify, assess, and respond to emergent conditions are to a large degree a function of effective SoS design. This sequence can be achieved in advance of particular emergent conditions through pursuit of several of the strategies introduced. Although the strategies provided offer a starting point, it would be disingenuous to suggest that they are a complete set. However, failure to think in terms of designing for emergence in SoS is an invitation to engage solely at the reactive level. Reaction alone is a shortsighted strategy for dealing with SoS emergence. Unfortunately, this appears to be the default strategy far more often that it should be for operating SoS. Hence, although the specifics of an emergence episode in a SOS will never be routine (i.e., it is unpredictable), the robust SoS design to support early identification, rigorous assessment, comprehensive response, and critical systems inquiry following, suggests a proactive stance to more effectively engage emergence for SoS.

Insight 3: Each experience of an emergence episode in a SoS should be treated as unique—variable in timing, breadth, impact, response reaction, and evolution. It is shortsighted to think that an emergence episode will be the same as something previously experienced. Each incidence of emergence is unique and requires a unique response tailored specifically to the episode. It is a mistake to assume that an emergence episode is "the same as a previous episode" and will respond in a similar manner to the same response treatment. An emergence episode can vary in *breadth* (expanse of the reach of the emergence episode across the SoS), *impact* (the degree to which the episode might disable or degrade SoS performance), *reaction to treatment* (how the episode responds to actions), and *evolution* (how the episode changes over time). However, this does not suggest that we should not develop generalized frameworks that can be "transportable" across emergence episodes. In effect, a generalizable approach can retain sufficient malleability such that it can be tailored to specific circumstances unique to the emergence episode at hand. For example, an emergence framework might have general "classes" of emergence which can be instructive in understanding the nature and form of a specific emergence episode. While appreciating the uniqueness of an emergence episode, this also allows a level of "generalization" to accelerate considerations to facilitate rapid understanding and response development.

This present exploration of emergence in the context of SoS is intended to provoke more rigorous thinking. Emergence in SoS is a major part of the landscape. The one consistent theme for emergence in SoS is that we must expect to be surprised, not ever knowing the precise nature, form, or timing of emergence episodes. A general take away is that emergence in SoS will occur. However, the effectiveness in dealing with emergence is not a given without purposeful design, execution, and system development. This is the responsibility of practitioners of SoS. An appreciation and commitment to more rigorous treatment of emergence in SoS is a critical first step.

References

Ackoff, R.L. (1971). Toward a system of systems concepts, *Management Science, 17*(11), pp. 661–671.

Adams, K.M., Hester, P.T., Bradley, J.M., Meyers, T.J., & Keating, C.B. (2014). Systems theory as the foundation for understanding systems. *Systems Engineering, 17*(1), pp. 112–123.

Antunes, R., & Gonzalez, V. (2015). A Production model for construction: A theoretical framework. *Buildings. 5*(1), pp. 209–228.

Arbesman, S. (2016). *Overcomplicated: Technology at the limits of comprehension.* New York: Current.

Aristotle. (2002). *Metaphysics: Book H—Form and being at work.* (J. Sachs, Trans.) (2nd ed.). Santa Fe, CA: Green Lion Press.

Arshinov, V., & Fuchs, C. (Eds.) (2003). Causality, emergence, self-organisation. Moscow: NIA-Priroda.

Ashby, W.R. (1962). Principles of the self-organizing system. In H. von Foerster & G. Zopf (Eds.), *Principles of Self-Organization* (pp. 255–278). New York: Pergamon Press.

Beer, S. (1979). *The heart of the enterprise.* New York: John Wiley & Sons.

Beer, S. (1985). *Diagnosing the system for organizations.* New York: John Wiley & Sons Inc.

Bohr, N. (1928). *The quantum postulate and the recent development of atomic theory, Nature, 121*(3050), pp. 580–590.

Buckley, W. (1967). *Sociology and modern systems theory.* Englewood Cliffs, NJ: Prentice-Hall.

Cannon, W.B. (1929). Organization for physiological homeostasis, *Physiological Reviews, 9*(3), pp. 399–431.

Chalmers, D.J. (1990, October 6) Thoughts on emergence. Retrieved from http://consc.net/notes/emergence.html.

Checkland, P. (1999). Systems Thinking. In W. Currie and R. Galliers (Eds.), *Rethinking management information systems* (pp. 45–56). Oxford, UK: Oxford University Press.

Checkland, P.B. (1993). *Systems Thinking, Systems Practice.* New York: John Wiley & Sons.
Cherns, A. (1976). The principles of sociotechnical design, *Human Relations, 29*(8), pp. 783–792.
Cherns, A. (1987). The principles of sociotechnical design revisited, *Human Relations, 40*(3), pp. 153–161.
Cilliers, P. (1998). *Complexity and Postmodernism: Understand Complex Systems.* New York: Routledge.
Clemson, B. (1984). *Cybernetics, A New Management Tool.* Tunbridge Wells: Abacus.
Corning, P.A. (2002). The re-emergence of "emergence": A venerable concept in search of a theory. *Complexity, 7*(6), pp. 18–30.
Dahmann, J., & Baldwin, K. (2008). Understanding the Current State of US Defense Systems of Systems and the Implications for Systems Engineering, Montreal, Canada: IEEE Systems Conference, 7–10 April.
De Wolf, T., & Holvoet, T. (2004). Emergence versus self-organisation: Different concepts but promising when combined. In *International Workshop on Engineering Self-Organising Applications* (pp. 1–15). Heidelberg: Springer Berlin.
El-Hani, C.N., & Pihlström, S. (2002). Emergence theories and pragmatic realism. *Essays in Philosophy, 3*(2), p. 120.
Flood, R.L., & Carson, E.R. (1993). *Dealing with complexity: An introduction to the theory and application of systems science.* New York: Plenum Press.
Goldstein, J. (1999). Emergence as a construct: History and issues. *Emergence, 1*(1), pp. 49–72.
Guide, U.S. DoD (2008). *Systems engineering guide for systems of systems.* Version, 1, 20301-3090.
Hammond, D. (2002). Exploring the genealogy of systems thinking, *Systems Research and Behavioral Science, 19*(5), pp. 429–439.
Hitchins, D.K. (2008). *Systems engineering: A 21st century systems methodology.* New York: John Wiley & Sons.
Holman, P. (2010). *Engaging emergence: Turning upheaval into opportunity.* Berrett-Koehler Publishers.
Holland, J. H. (2000). *Emergence: From chaos to order.* Oxford, UK: Oxford University Press.
INCOSE. (2011). *Systems engineering handbook: A guide for system life cycle processes and activities.* (H. Cecilia, Ed.) (3.2 ed.). San Diego, CA: INCOSE.
Jackson, M.C. (1991). The origins and nature of critical systems thinking. *Systemic Practice and Action Research, 4*(2), pp. 131–149.
Jamshidi, M. (Ed.). (2008). *Systems of systems engineering: Principles and applications.* Boca Raton, FL: CRC Press.
Jaradat, R.M., Keating, C.B., & Bradley, J.M. (2014). A histogram analysis for system of systems. *International Journal of System of Systems Engineering, 5*(3), pp. 193–227.
Katina, P.F. (2016). Systems Theory as a Foundation for Discovery of Pathologies for Complex System Problem Formulation. In *Applications of Systems Thinking and Soft Operations Research in Managing Complexity*, pp. 227–267. Springer International Publishing.
Keating, C., & Katina, P. (2011). Systems of systems engineering:Prospects and challenges for the emerging field, *International Journal of System of Systems Engineering, 2*(2), pp. 234–256.
Keating, C.B. (2009). Emergence in system of systems. In M. Jamshidi (Ed.), *System of Systems Engineering* (pp. 169–190). Hoboken, NJ: John Wiley & Sons, Inc.
Keating, C.B., & Bradley, J.M. (2015). Complex system governance reference model. *International Journal of System of Systems Engineering, 6*(1–2), pp. 33–52.
Keating, C.B., & Katina, P.F. (2012). Prevalence of pathologies in systems of systems. *International Journal of System of Systems Engineering, 3*(3–4), pp. 243–267.
Keating, C.B., & Katina, P.F. (2016). Complex system governance development: A first generation methodology. *International Journal of System of Systems Engineering, 7*(1–3), pp. 43–74.
Keating, C., Rogers, R., Unal, R., Dryer, D., Sousa-Poza, A., Safford, R., Peterson, W., & Rabadi, G. (2003). System of Systems Engineering, *Engineering Management Journal, 15*(3), pp. 36–45.
Keating, C.B., & Pyne, J.C. "Advancing Sociotechnical Systems Theory." Proceedings of the American Society for Engineering Management, 2001, October 11–13, 2001, Huntsville, AL, pp. 336–341.

Keating, C.B. (2005). Research Foundations for System of Systems Engineering, *IEEE International Conference on Systems, Man, and Cybernetics*, Waikoloa, Hawaii, pp. 2720–2725, October 2005.

Keating, C.B. (2014). Governance implications for meeting challenges in the system of systems engineering field. In *IEEE System of Systems Engineering (SOSE), 2014 9th International Conference*, pp. 154–159.

Keating, C.B., Katina, P.F., & Bradley, J.M. (2014). Complex system governance: concept, challenges, and emerging research. *International Journal of System of Systems Engineering, 5*(3), pp. 263–288.

Keating, C.B., Katina, P.F., Bradley, J.M., & Pyne, J.C. (2016). Systems Theory as a Conceptual Foundation for System of Systems Engineering, *INSIGHT, 19*(3), pp. 47–50.

Kim, J. (1997). Supervenience, emergence, and realization in the philosophy of mind. *Mindscapes: Philosophy, science, and the mind, 5*, p. 271.

Korzybski, A., & Rouben Mamoulian Collection (Library of Congress) (1933). *Science and Sanity: An introduction to non-aristotelian systems and general semantics* (1st ed.), New York, Lancaster, Pa.: International Non-Aristotelian Library Pub. Co.; Science Press Printing Co., Distributors.

Laszlo, E. (1996). *The systems view of the world: A holistic vision for our time.* Cresskill, NJ: Hampton Press.

Lewes, G.H. (1875). On actors and the art of acting (Vol. 1533). London Smith, Elder 1875.

Lloyd. (2015). Foresight review of resilience engineering: Designing for the expected and unexpected (No. Report Series: No. 2015.2). London: The Lloyd's Register Foundation.

Lucas, C., & Milov, Y. (1997). Conflicts as Emergent phenomena of complexity. Calresco Group, http://www.calresco.org/group/conflict. htm.

Maier, M. (1999). Architecting principles for systems-of-systems, *Systems Engineering, 1*(4), pp. 267–284.

Martin, J. (2006). *The meaning of the 21st century: A vital blueprint for ensuring our future.* New York: Riverhead Books.

Martin-Breen, P., & Anderies, J.M. (2011). Resilience: A literature review (p. 64). New York: The Rockefeller Foundation. Retrieved from http://www.rockefellerfoundation.org/blog/resilience-literature-review.

Miller, J. (1978). *Living systems*, New York: McGraw-Hill.

Norris, F.H., Stevens, S.P., Pfefferbaum, B., Wyche, K.F., & Pfefferbaum, R.L. (2008). Community resilience as a metaphor, theory, set of capacities, and strategy for disaster readiness. *American Journal of Community Psychology, 41*(1–2), pp. 127–150.

Pahl, G., Beitz, W., Feldhusen, J., & Grote, K.-H. (2011). *Engineering Design: A Systematic Approach*, 3rd ed., Wallace, K. and Blessing, L.T.M. (Trans.). Springer, Darmstadt.

Rainey, L.B., & Tolk, A. (2014). *Modeling and simulation support for system of systems engineering applications* (1st ed.). Wiley.

Richardson, K.A. (2004). Systems theory and complexity: Part 1, *Emergence: Complexity and Organization, 6*(3), pp. 75–79.

Rittel, H.W., & Webber, M.M. (1973). Dilemmas in a general theory of planning. *Policy Sciences, 4*(2), pp. 155–169.

Sage, A., & Cuppan, C. (2001). On the Systems Engineering and Management of Systems of Systems and Federations of Systems. *Information, Knowledge, Systems Management, 2*(4), pp. 325–345.

Sauser, B., Boardman, J., & Gorod, A. (2009). System of systems management. In M. Jamshidi (Ed.), *System of Systems Engineering* (pp. 191–217). Hoboken, NJ: John Wiley & Sons, Inc.

Skyttner, L. (2005). *General Systems Theory*, Danvers, MA: World Scientific Publishing Co. Pte. Ltd.

Skyttner, L. (1996). *General Systems Theory: An Introduction*, Philadelphia, PA: Trans-Atlantic.

Smuts, J. (1926). Holism and Evolution, New York: Greenwood Press.

Snyder, L. (1997) Discoverer's induction, *Philosophy of Science, 64*(4), pp. 580–604.

Sousa-Poza, A.A., Kovacic, S., & Keating, C.B. (2008). System of systems engineering: An emerging multidiscipline. *International Journal of System of Systems Engineering, 1*(1/2), pp. 1–17.

Sussman, J.M. (2005). *Perspectives on intelligent transportation systems.* New York: Springer.

von Bertalanffy, L. (1950). An outline of general systems theory, *The British Journal for the Philosophy of Science, 1*(2), pp. 134–165.

von Bertalanffy, L. (1968). *General System Theory: Foundations, Development, Applications*, rev. ed., New York: George Braziller.

Waddington, C.H. (1957). *The Strategy of Genes: A Discussion of Some Aspects of Theoretical Biology*, London: George Allen and Unwin.

Waddington, C.H. (1968). Towards a theoretical biology, *Nature, 218*(5141), pp. 525–527.

Whitney, K., Bradley, J.M., Baugh, D.E., & Chesterman Jr., C.W. (2015). Systems theory as a foundation for governance of complex systems. *International Journal of System of Systems Engineering*, *6*(1–2), pp. 15–32.

Yuchnovicz, D.E. (2016). Understanding system structural tensions to support complex system governance development methodology, *International Journal of System of Systems Engineering*, pp. 109–129.

Section IV

Summary

22

Lessons Learned and the Proposed Way-Ahead

Larry B. Rainey and Charles B. Keating

CONTENTS

22.1 Lessons Learned and Associated Rationale

This section documents those key observations that were concluded from the perspective of each author and what their reasoning was for that observation. These observations and associated rationale constitute the foundation of thought for this text.

22.1.1 Chapter 3

Lesson Learned 1:

With the anticipated growth of interest in emergence research, the research community would greatly benefit from a common medium to express the plethora of constructs likely to be explored. The common lingua franca would be more productive to the extent that it supports systems-based, formal, and computational formulation of concepts in a uniform framework so that hypotheses can be tested with definiteness. We claim that the behavioral and structural features of Discrete Event Systems Specification (DEVS) make it the right formalism and computational basis to support both the abstraction and elaboration necessary to deal with emergence in complex systems.

Rationale 1:

With the anticipated growth of interest in Emergence research, the research community would greatly benefit from a common medium to express the plethora of constructs likely to be explored. The common lingua franca would be more productive to the extent that it supports systems-based, formal, and computational formulation of concepts in a uniform framework so that hypotheses can be tested with definiteness.

Lesson Learned 2:

We hypothesize that a three-layered architectural framework offers a solid basis to formulate models of emergence in various domains. With such an architecture, it would be possible for each research proposal to test whether the proposed claims for emergence understanding, prediction, or control can be reduced to the prior existence of the conditions manifested in three layers: (1) System of Systems Ecology: the systems that will become component systems of the System of Systems (SoS) already exist as viable autonomous entities in an ecology; however, left unperturbed they would not emerge into the system of systems under consideration. (2) Network Supporting Pragmatic Level of Communication: the ability to communicate among putative component systems of the SoS, not only at the technical level, but a level that supports the coordination needed at the next level. (3) Coordination Economics: a compound term that refers to (a) the coordination required to enable the components to interact in a manner that allows emergence of the SoS with its own purposes, and (b) the economic conditions that enable emergence—the collective benefit that the SoS affords versus the cost to individuals in their own sustainability to contribute to the SoS objectives.

Associated Rationale 2:

A commonly accepted generic architecture would make it possible for research proposals from the emergence community to test their proposed claims for emergence understanding, modeling, prediction, or control against a common reference framework. Such attempts will either confirm this architecture or result in improved frameworks that can be adopted universally.

22.1.2 Chapter 4

Lesson Learned 1:

Biological models of evolution in a natural ecosystem can instruct evolution of man-made systems across a system of systems (SoS).

Associated Rationale 1:

The work by Bak and Sneppen and others to describe complex natural phenomenon with simple models should be considered and further investigated for man-made systems. In the natural world, less fit species may have a lower barrier for change/mutation. Species that are fit are more stable with less likelihood (or less pressure) to evolve. Similar constructs can describe man-made systems of systems. Technological advances make engineered systems less useful—less valued. Likewise, programmatic issues such as life-cycle costs can also make systems less affordable and less valued. Systems must evolve and be modernized, if they are able to, or become extinct, and therefore be replaced. The analogy is clear.

Lesson Learned 2:

Systems within a SoS can be described by a fitness and a fitness barrier to change. Pragmatic rules for evolution and the structure of programmatic interdependencies will drive dynamic changes to these attributes.

Associated Rationale 2:

Evolutionary forces will drive dynamic fitness levels across the SoS. The fitness landscape can be influenced by the replacement rules. However, it can also be influenced by the interdependency graph structure. Chapter 4 examined the effects of Bak-Sneppen dynamics on 2 regular, 4 regular, scale free, and small-world graphs, as well as graphs with explicit weighted interdependency/interoperability measurements. The existence of hubs is hypothesized for greater amounts and frequency of avalanches. These randomly weaken the entire SoS. The addition of explicit interdependency measurement greatly affected the fitness distribution.

Lesson Learned 3:

A SoS will demonstrate complex emergent behaviors; managers and engineers must be mindful of these behaviors even within steady state.

Associated Rationale 3:

System of systems have direct and indirect relationships between systems. When these interconnected systems are driven by pragmatic rules for evolution, the SoS will demonstrate complex behaviors, including: self-organized criticality (SOC), punctuated equilibrium (avalanches), and a steady state fitness landscape, including a critical fitness barrier. In the end, the fitness landscape may be low due to the lessons of Lesson Learned 2 above. Lower fitness barriers allow greater chance for mutation or replacement.

22.1.3 Chapter 5

Lesson Learned 1:

Leveraging complexity is not an accepted practice, and not even considered as an option.

Associated Rationale 1:

Complexity is generally avoided, and current methods for managing complex systems either avoid complexity or attempt to mitigate it by over-simplifying the system. The over-simplification of a system may limit complex behavior and can eliminate the richness of associated emergent behaviors. For a complex adaptive system, this would significantly curtail learning and adaptation to dynamic environments. In summary, the complex system can be over-simplified to produce linear and deterministic models, but at the expense of the full spectrum of desired adaptive behaviors.

Lesson Learned 2:

Discovering mathematical models for unknown equations is hard, but there may be a method to the discovery.

Associated Rationale 2:

Over time, individuals have discovered new mathematical models (equations) either independently by accident, or more recently by using experimental (computational) mathematics. It would seem that a computer could have better results by exhaustively searching the many permutations of current formulae to find useful perturbations or permutations of these. The new formulae that result from this experimental mathematics search could provide new methods for solving currently intractable problems. The primary obstacle to the success of this method is the extreme size of the search space. One possible means to significantly reducing this search space is to apply heuristics and a variety of machine learning methods similar to those applied to data mining.

22.1.4 Chapter 6

Lesson Learned 1:

When carrying out fundamental research involving systems of systems, any predisposition to a particular outcome is perilous. A forethought that is believed to be unbiased, however strongly stated, believed, and adamantly based on phenomenology, undermines the stalwart principles and is completely antithetical to the scientific process.

Associated Rationale 1:

The method of discovery that leads to the most extraordinary science is not the path of deductive logic. Rather the greatness of discovery is cast within recursive and non-recursive logic, eventually leading to reflexive insight. When viewed pluralistically, the conception of science is not discoverable through methodological rules. The logic of greatness follows not in consensus, nor the logical next idea, but rather from the genius insight that is foraged from ideas that others have overlooked or have not yet seen.

Lesson Learned 2:

The first to challenge the author's work is the author; the second to challenge the author's work is someone the author trusts to be straightforward and dogmatic. Others need not provide the author with proof, only inexactitudes which the author needs to ponder in depth.

Associated Rationale 2:

Outstanding research requires critical assessment and evaluation by others, not just by the author. Even critical integrative thinking by one person is often myopic, whereas others who see from alternative perspectives may find items that did not seem important—but are. A true test of a good and sound idea is to capture the opposing ideas in the formulation of logic that should be posed with much vigor and support.

22.1.5 Chapter 7

Lesson Learned 1:

Research is carried out with a general acceptance of phenomenology as the essence of observation and discovery. Unfortunately, observations without principles, theory, approach, method, and an interpretive framework are plausible but false.

Associated Rationale 1:
Building up from a general theory of integration that is premised on solid principles that are applicable acoss all disciplines and fields of work, a lifecycle systems approach illuminates the appropriate methods to construct an interpretive framework from which to view observations. That interpretive framework builds on the Leśniewski model of objects and processes to show emergence in its true demeanor.

Lesson Learned 2:
The ways and manners of systems of systems are steeped in their constituent systems, without which no systems of systems can exist.

Associated Rationale 2:
Removing the notion of system from a system of systems is counterintuitive to the basic logic of systems on which to build systems and systems of systems. A system of systems can be no more than the emergence ascribed to its constituent systems. It is so fundamental to requirements for a system of systems that the quest for a sound, valid definition of systems is rekindled.

22.1.6 Chapter 8

Lesson Learned 1:
Variation among concepts and theories of emergence is a matter of scale (i.e., detail) at which the concepts and theories are considered.

Associated Rationale 1:
When we consider theories and concepts of emergence at a micro level there is great variety and even apparent conflicts: (1) properties, patterns, or behaviors; (2) new or reoccurring/persistent; (3) originates from the bottom up or the top down; (4) provisional or permanent; (5) observer dependent or independent; (6) diachronic or synchronic. However, when we examine the concepts and theories at a macro scale, we find common themes and a level of congruency that helps explain the nature of emergence in terms of eight characteristics (i.e., sub-classes): type of effect; logical relationship between parts and systems; observer's perspective; indicators of occurrence; temporality of occurrence; system structure; knowledge constraints; and application domains. The eight subclasses help to resolve conflicts and define emergence in terms of its essential elements which may contribute to efforts to model emergence.

Lesson Learned 2:
Resolving the variation among concepts and theories of emergence contributes to developing specifications of the concept and contributes to efforts to model emergence.

Associated Rationale 2:
The ontology for emergence provides an unambiguous definition and specifications of the concept; it can be used to inform and validate the design of the simulation models of emergence in engineered systems. As an unambiguous definition of the concept, the ontology can be used to objectively determine if a phenomenon formally corresponds with the

concept of emergence. As a specification, the ontology can be used to identify attributes of a phenomenon that are related to emergence, and to validate that key aspects of emergence are represented in models of the phenomenon.

22.1.7 Chapter 9

Lesson Learned 1:
Complex systems are often the subject of models.

Associated Rationale 1:
Systems with sufficient size and scope to be important are often complex. Their complexity means that they may not be amenable to closed-form mathematical analysis, leaving modeling as the best or only way to study them, and their importance justifies the time and expense to do so.

Lesson Learned 2:
Both model development and model validation are more challenging when the system being modeled is complex.

Associated Rationale 2:
The defining characteristics of complex systems include sensitivity to initial conditions, emergent behavior, and composition of components. Each of these characteristics has problematic implications for model development and model validation. In modeling, for example, sensitivity to initial conditions exacerbates the well-known problem of numerical precision limits in physics-based models. In validation, the fact that models, even if separately valid, cannot be assumed to be valid when composed means that models of complex systems composed of component models must be validated at both the component and composite levels.

Lesson Learned 3:
Model development and model validation will require additional time and effort when the system being modeled is complex.

Associated Rationale 3:
A set of specific modeling and validation challenges are associated with each of the defining characteristics of complexity. Mitigation methods are available for each challenge. For example, the unreliability of face validation for models of complex systems that is caused by emergent behavior can be addressed using structured face validation involving multiple subject matter experts executing carefully pre-planned scenarios designed to elicit the emergent behavior. However, the mitigation methods are in general not guaranteed to completely eliminate the problems, and as with the example, should be expected to add time and effort beyond that required to develop and validate a model of a non-complex system.

22.1.8 Chapter 10

Lesson Learned 1:
Understanding the interchange of variety and constraint is fundamental to understanding emergence in systems.

Associated Rationale 1:

Ashby's Law of Requisite Variety relates the complexity of a system to the complexity of the entities comprising the system. When we model systems, we can explore the relationship between variety and constraint to gain insight to the mechanisms that regulate the influence of constituent entities on the overall system.

Lesson Learned 2:

We can measure emergence.

Associated Rationale 2:

Emergence can be measured if we are careful to define what we mean by emergence. In this chapter, we distinguished between five types of emergence, where each type can be described in terms of the information flows between the system and the entities comprising the system, the complexity of the entities in the system, and the information uncertainty, i.e., entropy.

Lesson Learned 3:

Whereas historically emergence has been considered a phenomenon to be avoided in systems, nature appears to capitalize and even depend on emergence.

Associated Rationale 3:

Simulations of three systems considered in this chapter are based on natural systems where emergence is generally accepted as present. The types of emergence and the types of constraints identified distinguish the three systems. Although differing in complexity of constituent entities and types of constraint, each system is recognized by its exhibition of emergence quantifiable by the measures presented. Although the systems are simulated, the emergence measured in the simulation suggests similar phenomena in the natural systems. If we can design systems to exploit emergence, engineered systems might derive the same sophisticated behaviors and robust qualities replete in natural systems.

22.1.9 Chapter 11

Lesson Learned 1:

Understanding systems of systems emergent behavior is made more tractable by using multiple viewpoints.

Associated Rationale 1:

Myriad aspects of complexity in a system of systems give rise to emergent behaviors that are socio-technical in nature and challenging to characterize, as they are multi-faceted. Examining emergence from multiple viewpoints enriches understanding while retaining a holistic approach.

Lesson Learned 2:

Emergent behaviors triggered by architectural change occur throughout the operational life of system of systems.

Associated Rationale 2:

Over the lifespan of a traditional system, positive emergent behavior becomes increasingly understood and managed, given that its architecture and connectivity is fairly stable.

The system of systems differs from a traditional system in that there is periodic re-architecting at various points throughout its operational life, leading to emergent behaviors—positive and negative—triggered by architectural changes.

Lesson Learned 3:
Research and development of time-based analysis approaches will have substantial impact on the ability to architect systems of systems for desired emergent behaviors.

Associated Rationale 3:
System properties "ilities" (e.g., resilience, survivability) are desired emergent behaviors that necessitate a temporal perspective and time-based analysis. In the case of systems of systems, time-based analysis is urgently needed if these properties are to be purposefully designed, but such analysis is cognitively difficult and computationally intensive. Accordingly, this is a significant area of research for the modeling and simulation community.

22.1.10 Chapter 12

Lesson Learned 1:
Acquisition and development lifecycles focus on effective security within a single system, but there is limited consideration of the security of the system of systems, and attackers are increasingly taking advantage of discrepancies among systems that share information but do not share security features and capabilities.

Associated Rationale 1:
Security policies and practices as well as acquisition and development policies and practices are still focused on individual systems.

Lesson Learned 2:
Engineering security for the system of systems will require a governance capability that can review information about assumptions, trust relationships, and risks made and implemented within each participating system to ensure the choices by one system owner do not create unacceptable risks for every other participating system.

Associated Rationale 2:
Decisions to defer fixing vulnerabilities or to use software containing malware made by a system owner represent risks to all others trusting the information provided by that system, since current analytical capabilities will not provide the ability to determine if a trusted source has been compromised.

22.1.11 Chapter 13

Lesson Learned 1:
Current incentives to produce and field software quickly and cheaply do not provide sufficient cybersecurity.

Associated Rationale 1:
Without specific effort to remove design and coding weaknesses, software is increasingly subject to successful attacks. Removing these weaknesses requires specialized

knowledge, time, and tools not widely used by developers, and all of these add cost to the final product.

Lesson Learned 2:
Security is only as strong as the weakest system in the system of systems because each system typically trusts data received from others in the system of systems, which provides opportunities for attackers to compromise a single system and, using the trust relationships, compromise all other participating systems.

Associated Rationale 2:
Specific effort is required to structure a non-trusting relationship, so by default information is widely shared and used without consideration of its trustworthiness.

22.1.12 Chapter 14

Lesson Learned 1:
Understanding of emergence which is helpful in engineering demands non-traditional research methods.

Associated Rationale 1:
The nature of engineering, as the activity of developing and sustaining technical products and interventions in the world, is inextricably linked with emergence. The practical nature of the engineering task, using knowledge to perform action in the world that achieves intended effects, requires a formulation of knowledge which is focused on usability of the knowledge rather than the purity of the knowledge. The engineering task applies, and relies on, the emergent effects of what happens when the elements of a solution are brought together. The concern, among engineers, with emergence is the cases where unexpected emergent effects of a disadvantageous or disastrous kind are experienced. Predicted emergent effects, what a system is designed to produce, are usually developed as the application of knowledge about things and phenomena.

Lesson Learned 2:
Existing research methods frequently used in engineering produce knowledge of either specific things and phenomena or of historic cases, but this knowledge cannot prevent problematic instances of emergence.

Associated Rationale 2:
Traditional engineering knowledge is developed through case studies and scientific experiments. Case studies, of various kinds, may lead to identification of specific emergent effects, but are limited in not providing more than the observation of confluence of factors and are also limited, without the addition of further types of investigation, in producing understanding of causality. Observational studies advance on case studies by including a significant plurality of cases, although under uncontrolled conditions, and therefore provide knowledge of repetition and expectation of observed outcomes in association with input conditions. Experimental and theoretical research methods provide the opportunity both to explore the phenomenon of emergence and to characterize particular interactions that lead to emergent effects. These research

methods produce knowledge about the phenomenon of emergence in general or specific instances of emergent effects.

Lesson Learned 3:

Engineering, as a field concerned with producing effects in the world, demands investigation of emergent effects that enables recognition of when and where an emergent effect is likely to be manifested when there is no prior knowledge or experience that points to the specific kind of emergent effect.

Associated Rationale 3:

The engineer first needs knowledge which will enable the prediction that an interaction of elements in their design is likely to lead to an emergent effect. Second, the engineer needs knowledge that enables the prediction of what emergent outcome is to be expected when a particular design is realized. The objective of enriching the engineer's capacity to predict the fact of emergent effects demands generation of a different kind of knowledge than is commonly sought in engineering research. Further, the difference in the kind of knowledge sought leads to the reasonable expectation that different research methods will be needed to find helpful knowledge of emergence for engineers.

22.1.13 Chapter 15

Lesson Learned 1:

Emergence must be relevant to the Subject Matter Expert (SME). We cannot simply claim that a system is emergent if what we consider emergent is not an actual concern.

Associated Rationale 1:

The use of a "tie" as the emergent behavior was helpful in devising a method to discover the occurrence of the behavior and to investigate its root causes. However, a "tie" is an outcome that is not of particular interest to the SME. As we illustrated, finding a relevant emergent condition is very challenging.

Lesson Learned 2:

Even if one can find a relevant emergent condition, it must be explainable in a plausible way. If the emergent condition is very rare and only occurs under a highly unlikely combination of inputs, it will be dismissed by SMEs, which then brings into question the utility of the model.

Associated Rationale 2:

The discovery that the model could not find a way to guarantee a win sparked interesting debates on whether this is due to the model being under-specified or being over-constrained, or due to the actual system of systems not being capable of providing such a guarantee. A method to help create a clear explanation of the likely causes of emergent conditions within a model of a system of systems is key for getting SMEs to understand and believe in the results of the model. Our use of statistical debugging is a step towards improving the ability to explain the root causes of emergent behaviors.

Lesson Learned 3:

The large number of elements, parameters, and environmental configurations (both the distribution of elements as well as their geographic layout) within our model of the BMDS greatly muddles the ability to identify emergent behaviors. Additionally, the increased combinations of parameters and geographic placements of elements also increases difficulties in identifying the likely causes of emergent behaviors.

Associated Rationale 3:
We believe that we created a more realistic view of the BMDS by using a combined ABM and DES model because this allowed us to minimize the number of simplifications and constraints imposed on the modeled system. However, this simultaneously increased the number of outputs for us to track during experimentation. Only by applying the combination of Latin Hypercube sampling along with statistical debugging were we able to analyze the outputs on a reasonable timeframe.

22.1.14 Chapter 16

Lesson Learned 1:
We need if not standardized, then at least shared architectures for communication and collaboration in engineering of emergence.

Associated Rationale 1:
Emergent systems research and engineering is on one hand under the influence of agent-based systems concepts, and on the other hand required to find engineering solutions for the upcoming technical systems, which are nowadays named after Internet of Things and Cyber Physical Systems. The research and engineering community around the emergent systems requires shared conceptualizations in the form of architectures to enable communication and collaboration. The Observer-Controller architecture from the Organic Computing project of the German Research Foundation is important in this sense.

Lesson Learned 2:
Upcoming requirements for modeling and simulation of emergent systems are neither well understood nor implemented by the state of the art modeling and simulation tools for technical systems.

Associated Rationale 2:
Modeling and simulation tools are the key means that will enable engineering of emergent systems. However, the upcoming requirements of modeling and simulation of emergence are not yet supported by modeling and simulation tools that are available in the market for technical systems. Further, there is no consensus yet about what the requirements are or which features are required. Before such a challenge is addressed, it will be hard to industrialize the emergent systems. Variable structure modeling can be pronounced as one of these challenges.

22.1.15 Chapter 17

Lesson Learned 1:
It is essential to develop methodologies to correctly describe macroscopic and microscopic features of engineering applications and the relations between them. This results in having a systematic approach that can help designers to discover emergent behaviors of engineering applications in order to avoid failures and unintended situations in the system.

Associated Rationale 1:

Emergence in engineering applications can be observed and described at the macroscopic level. However, complex systems and SoS are designed with many components where each component is defined by its microscopic features. The macroscopic features are the result of the components' interaction with each other which can produce emergence. Hence, investigating and studying macroscopic features are essential to design fault-free systems.

Lesson Learned 2:

Large scale multi-agent simulations are the correct way to study complex behaviors in engineering systems. Researchers are urged to design and develop methods that are based on multi-agent simulations to accurately model and simulate individual components within an acceptable time. DEVS formalism provides a rich modeling and simulation environment that can be used by researchers at a large scale.

Associated Rationale 2:

It is nearly impossible to formulate a closed-form equation that describes complex engineering systems and SoS because of the large number of components and the complex interaction rules. Multi-agent simulations provide a framework that naturally fits to describe each component as an agent. Since there is a large number of components, multi-agent simulation at a large scale is needed in which researchers and engineers can experiment with large number of parameters to discover emergence. We argue that DEVS formalism can be used to model different engineering applications and simulate the behavior and interaction of the different components.

22.1.16 Chapter 18

Lesson Learned 1:

To engineer emergence, we must think in terms of relaxing or restricting control over component or system interactions. In design, positive emergence may be considered as the acceptable behaviors and interactions that remain after negative emergence has been thoroughly exposed and pruned. The next steps should be to elaborate on the impact of controlling emergence in this way on the requirements analysis process.

Associated Rationale 1:

Each system has its own behavior that will eventually express, given enough time, unless explicitly constrained. To coax these behaviors out in simulation, fewer constraints are initially imposed within and among comprehensive system models. The scope-complete scenario generation capability of Monterey Phoenix (MP) provides an approach and tool that regards the independent nature of interacting systems or components, such that all combinations of behaviors may be inspected, and a constraint (requirement) discovery process may employed to shape the overall behavior. Constraints may be imposed or lifted in MP to observe the effects on the overall design, suggesting that this is an appropriate activity to conduct to discover requirements that may not have otherwise been found until much later.

Lesson Learned 2:

Early system design is about building layers of models, and verifying, validating, testing, and debugging them. To encourage expression of emergent behaviors in simulation, the following modeling concepts should be employed: separate behaviors and interactions, model system behaviors and environment behaviors, formalize models for automatic execution, properly allocate each task to a human or to a machine, and use abstraction and refinement to manage large models. The next steps should be to undertake the automation of functions currently being performed manually (such as behavior pattern detection).

Associated Rationale 2:

Separation of behaviors and interactions is what enables the interactions to be layered as constraints over the system behaviors (see Lesson Learned 1). Modeling system behaviors together with the fullest possible description of independent environment behaviors enables more combinations of SoS behaviors to be detected or predicted. Formalization of models for execution makes use of automated tools to generate/detect many (and in MP all, up to the specified scope) scenarios. The thoughtful allocation of each task to a human or to a machine promotes the efficient use of artificial and biological resources on the tasks of design. Without automation of emergent behavior detection, the burden is otherwise on humans to detect all possible behaviors, which is time consuming and subject to errors of omission. The use abstraction and refinement to manage large models breaks up the design work into small and manageable yet integrated pieces so that consequences of the design are more visible for understanding and classification.

Lesson Learned 3:

Certain emergent behaviors of a SoS can be detected, classified, predicted, and controlled through early modeling and simulation with MP. In making this discovery, a need to refine and formalize the current classification taxonomy was also found. Next steps should include developing clear criteria for classifying different types of emergent behavior, as well as addressing fundamentally why we need a classification taxonomy in the first place (e.g., to aid in the conduct of risk analysis, to develop metrics for emergent behaviors in designs).

Associated Rationale 3:

Three examples were provided in Chapter 18 that demonstrated analysis of emergent behavior across different systems in different domains. Enabled by the concepts listed in Lesson Learned 2, MP automatically detected all possible combinations of system behaviors and interactions permitted by the model. During inspection, the emergent behaviors were classified as favorable or unfavorable, simple, weak, strong, and positive or negative. The human inspector used each generated scenario as a canvas for predicting future states of emergence, which influenced classification. The negative emergent behaviors were controlled through modification of the individual behavior models or relaxation/restriction of interaction constraints (see Lesson Learned 1). Now that a collection of models exists, the classification taxonomy can more easily be tested and refined to provide a standard way to assign and substantiate the type of emergent behavior.

22.1.17 Chapter 19

Lesson Learned 1:

Develop the behavior model at a high level of abstraction.

Associated Rationale 1:

Execution of the behavior model at a high level of abstraction enables the developer to focus on the system behaviors and their interactions. The behavior model needs to implement enough detail to expose critical behaviors, without including minutia that has no impact on the behaviors.

Lesson Learned 2:
Control the model through constraints.

Associated Rationale 2:
Logical, simplification, and design constraints enable the developer to establish a complex model, while limiting the potential outcomes.

Lesson Learned 3:
Verify the existence of desirable behaviors and limit undesirable behaviors.

Associated Rationale 3:
The Monterey Phoenix (MP) environment supports positive emergence verification by the powerful means to define model behaviors and interactions, constraints, and small scope hypothesis. Negative emergence can be detected by assertion checking and templates (describing typical mistakes and omissions in models).

22.1.18 Chapter 20

Lesson Learned 1:
Undesirable and complex emergent behavior in systems of systems can arise from the time-varying dynamics of interaction between the component systems.

Associated Rationale 1:
The deleterious effect of feedback loops is a prime example of the importance of interaction dynamics. Although this statement may appear obvious in hindsight, it is noticeable that industry practitioners often overlook the importance of feedback loops in the analysis of real-world systems of systems such as the financial markets (our case study).

Lesson Learned 2:
Interaction between components in system of system models can be effectively investigated using an agent-based model that focuses on the dynamics of a broad range of communication behavior between component systems.

Associated Rationale 2:
The InterDyne Simulator is an existence proof of this statement. Our case study shows how interaction dynamics can be modeled, simulated, explored, and visualized to provide detection, classification, prediction, and a first step towards control.

Lesson Learned 3:
Although the antecedents of emergent behavior need not be unique, and may not be applicable to other configurations of a given system of systems, they often serve to falsify a negative hypothesis in stress testing (e.g., given a hypothesis during stress testing that a certain bad behavior "can never occur," with InterDyne we have a way to show that under certain circumstances it certainly can occur—thus, "it is not the case that this can never happen").

Associated Rationale 3:
This is demonstrated by the case study in Chapter 20.

22.1.19 Chapter 21

Lesson Learned 1:
Proactive design for emergence. We can design for broad classes of emergence (e.g., technology, political maneuvering, resource disruptions, etc.). Similar to a risk-based perspective, we can identify classes of emergence and develop corresponding "processes" to identify, assess, respond, and evaluate prior to experiencing emergence.

Associated Rationale 1:
Although the specific detail of an emergent "event" and subsequent consequences cannot be known in advance, there are things that we absolutely can know in advance about emergence. What we can know is that emergence is going to happen and it will have consequences. Thus, although the precise nature and timing are not known, we can be proactive prior to experiencing emergence.

Lesson Learned 2:
All emergence is not bad. Generally, we think of emergence as having bad consequences for complex systems. However, emergence can also be positive, generating responses in systems, that although unexpected, nevertheless can make a system better.

Associated Rationale 2:
Emergence, resulting in negative or positive consequences, can be a source of innovation in complex systems. In response to emergent conditions, we should look for second order opportunities to generate innovative responses to emergence. Additionally, we should also look for opportunities to make system modifications based on emergence.

Lesson Learned 3:
Emergence is less about what will happen than how the system will respond. Given that emergence will happen, there are three critical aspects: (1) design should be resilient to major classes of emergence, (2) the time between the emergent event and discovery should be minimized to allow greater range of response, and (3) the processing of the event should also question system design which allowed for the emergent condition.

Associated Rationale 3:
Response to emergence should consider two forms. One is to immediately address the emergent event to limit negative consequences or amplify positive consequences. The second form, and the potentially more important one, is to examine the event for system redesign implications suggested from evaluation of analysis and response.

22.2 Summary

This text has essentially been a basic research effort to use various modeling and simulation techniques (i.e., agent-based, MATLAB/Simulink, Monterey Phoenix Behavior Model) to detect, classify, predict, control, and visualize emergent behavior in system of systems.

22.3 Proposed Way-Ahead

Given the foundation that has been laid for this text, as addressed in the above section, this section documents the superstructure for this text. Specifically, it addresses what is the proposed methodology to move this subject forward to facilitate greater understanding, development, and application of emergent behavior in system of systems engineering applications.

On September 12, 2017, at the National Centers for System of Systems Engineering at Old Dominion University in Norfolk, Virginia, the Engineering Emergence Workshop was held to develop a recommended set of research objectives for emergent behavior in system of systems engineering applications. Representatives from each major staff and branch of the Department of Defense (DoD), two Federally Funded Research and Development Center organizations, and one DoD major contractor were present. Each organization provided a briefing on its perspective of emergent behavior in the context of system of systems engineering applications.

Listed below is the summary of the research objectives that were concluded from the closing discussion with workshop participants:

1. ***Classification schema for emergence in systems of systems.***

   Discussion: There is not currently a tailored or accepted approach to classify emergence for systems of systems. This would also need to address "positive" as well as "negative" impacts of emergence. Classification would need to be capable of generalization across different types of emergence and include the entire spectrum of systems of systems where emergence might emanate from (e.g., software). Additionally, emergence classification would need to include virtual aspects as well as more traditional (physics-based) emergence.

2. ***Development of more effective training/education to address current deficiencies in addressing emergence in systems of systems.***

   Discussion: The current state of training/education is not providing development necessary to effectively deal with emergence in systems of systems. Additionally, there is very limited education/training provided in software to prepare professionals to better deal with software related issues (including emergence). Also, exploration of generational differences is an area of concern in developing and maintaining the future professional workforce.

3. ***Inadequacies of the present paradigm for dealing with systems of systems, including emergence, should be explored.***

   Discussion: There is much to be discovered concerning the congruence of the present paradigm for design, operation, maintenance, and development of systems of systems. Questions related to the ability of our current paradigms (legacies of past contexts, e.g., acquisition) to remain appropriate for the complexity of the digitally connected landscape are ripe for exploration. Included in a shifting paradigm is discovery and understanding of internal/external triggers of emergence in systems of systems—as different than those characteristics of past approaches to system acquisition, operation, and maintenance.

4. ***Development of a new generation of methods and tools capable of addressing emergence in systems of systems.***

   Discussion: The environment within which systems of systems are designed, executed, and evolved are fraught with heightened states of uncertainty, complexity, interconnectedness, and emergence. Present tools and methods are more accustomed to stability in understanding achieved through detailed analysis. Present and future demands for systems of systems are driven by much more precipitous and uncertain conditions subject to a different level of thinking (e.g., risk, resilience, fragility, and vulnerability). There is need to evolve capabilities necessary to engage emergence experienced in a systems of systems as well as the enterprises that drive the system of systems.

5. ***The conditions and circumstances that act as "triggers" to evoke emergence in systems of systems would benefit from more rigorous examination.***

   Discussion: Present approaches to emergence are rather limited in a "response" based in reaction to emergent events/conditions. In contrast, better understanding of "antecedent" conditions/circumstances to emergence in systems would provide capability for "practionary" as opposed to "reactionary" modes. This development would need to move beyond "risk" based analysis of potential failure modes in systems of systems. A more "holistic" set of considerations of triggers would also need to address not just technological triggers, but also triggers related to such areas as social aspects of systems of systems.

6. ***Exploration of uncertain future based scenarios to suggest implications for development of system of systems subject to conditions of extreme uncertainty.***

   Discussion: Given increasing levels of uncertainty in the design, development, deployment, operation, and maintenance of systems of systems, examination of future based scenarios might be influential in suggesting breakthroughs in thinking. Exploration of these scenarios can provide foundations to better understand implications and strategies to more effectively deal with emergence in systems of systems. Results might suggest gaps that need to be narrowed to enhance capabilities and practitioner effectiveness.

7. ***Detailed examination of the acquisition process with respect to effectiveness in dealing with increasing emergence in systems of systems.***

   Discussion: The present acquisition system, and constituent processes, is the subject of continual reform. It is noteworthy that the present acquisition system was not designed with either systems of systems or emergence in mind. Therefore, a deeper examination of the congruence of the acquisition system and processes with the present demands of designing, fielding, operating, and maintaining systems of systems is needed. Additionally, the nature of emergence across the lifecycles of constituent systems in systems of systems has broad ramifications for the acquisition system. The more stable, certain, and independent systems that have been the major focus of the acquisition system are now in the minority instead of the majority. Investigation of the impacts and implications for this major shift is necessary.

8. ***Exploration of the value of investment to address emergence in systems of systems.***

   Discussion: While there might be consensus agreement that emergence is something that should be considered in systems of systems, there is not clarity concerning either what the value of investment in emergence might be or how that value might be ascertained. Without movement beyond the conceptual notion of general goodness in appreciation of emergence, there is little substance to provide trade-off determinations when competing for the investment of scarce resources in system development. Without rigorous determination of value of investment in emergence for systems of systems, in relationship to other possible investments, it is not likely that the trade-offs will lean in the favor of emergence investment. Thus, exploration of value and measurement of value in support of trade-off decisions is ripe for investigation.

As of the final writing and publication of this text, various alternatives are being pursued to develop a funded research project to pursue the execution of the above stated research objectives for emergent behavior in system of systems engineering applications.

Index

T

Printed and bound by CPI Group (UK) Ltd, Croydon, CR0 4YY
24/10/2024
01778298-0010